HAZARDOUS MATERIALS REFERENCE BOOK CROSS-INDEX

HAZARDOUS MATERIALS REFERENCE BOOK CROSS-INDEX

Daniel J. Davis   Julie A. Davis

JOHN WILEY & SONS, INC.
New York • Chichester • Weinheim • Brisbane • Singapore • Toronto

Published by John Wiley & Sons, Inc., Hoboken, New Jersey.
Published simultaneously in Canada.

ISBN-13 978-0-471-28681-3
ISBN-10 0-471-28681-8

10 9 8 7 6 5 4 3 2

Contents

Introduction

The *Hazardous Materials Reference Book Cross-Index,* by Daniel J. Davis and Julie A. Davis, was designed as a tool—to be used to assist in the research portion of a hazardous materials incident. Studies have shown that using the information provided herein can save up to 50 percent of the time that might be spent in locating emergency information needed in a hazardous materials incident.

The index is ordered by chemical name, alphabetically, with UN/NA numbers given to confirm that the proper material is being researched. Fourteen of the most useful reference books have been included to provide necessary information for the material involved. You can tell at a glance which books contain information on the hazardous material in question and which page number(s) the information can be retrieved from.

The source references (and other information) are arranged on the page in tabular form. Each table consists of a grid—horizontal and vertical lines. Information is provided in columns. The columns are, in order of their appearance on the page:

1. Chemical or material name (up to the first 56 letters) in alphabetical order, and any tradename or synonym, listed right under the chemical or material name. ["Synonym" on the grid is used to represent the synonym or the tradename.] A material may not be specifically listed under its chemical name in a reference book; it may instead be listed under a tradename or synonym, which you would need to know to locate the information. In some cases you may find some information under the chemical name and some information under the tradename or synonym.

2. UN/NA number from the Department of Transportation's *Emergency Response Guide Book* (1990 Edition).

3. Guide number to be used in conjunction with the *Emergency Response Guide Book* (1994 Edition).

4. The *Firefighter's Hazardous Materials Reference Book,* by Daniel J. Davis, Julie A. Davis, and Grant T. Christianson (2nd Edition, 1992), with 913 entries.

5. The *First Aid Manual for Chemical Accidents,* by Marc J. Lefevre, revised by Shirley A. Conibear (2nd Edition, 1989).

6. *Sax's Dangerous Properties of Industrial Materials,* by Richard J. Lewis, Sr. (8th Edition, 1992).

7. *Hazardous Chemicals Desk Reference,* by Richard J. Lewis, Sr. (2nd Edition, 1991).

8. *Rapid Guide to Hazardous Chemicals in the Workplace,* by Richard J. Lewis, Sr. (3rd Edition, 1993), with over 185 entries.

9. *Fire Protection Guide on Hazardous Materials,* by the National Fire Protection Association (10th Edition, 1991). This particular book is broken down into two sections: (1) NFPA 49: *Hazardous Chemicals Data,* pages 49-13 through 49-185, and (2) NFPA 325M: *Properties of Flammable Liquids, Gases, and Volatile Solids,* pages 325M-11 through 325M-94. Some materials are listed in both sections; an example of a page listing is "173-86." The number before the dash, 173, would indicate that the information is found on page 173 in the NFPA 49: *Hazardous Chemicals Data* section of the NFPA book. The number after the dash, 86, would indicate that the information is found on page 86 in the NFPA 325M: *Properties of Flammable Liquids, Gases, and Volatile Solids* section of the NFPA book.

10. The *Firefighter's Handbook of Hazardous Materials,* by Charles Baker (5th Edition, 1990), with over 5,600 entries.

11. The U.S. Department of Health and Human Services' *Pocket Guide to Chemical Hazards* (NIOSH 1990 Edition), with over 397 entries.

12. The *Chemical Hazard Response Information System,* from the Coast Guard (1993 Edition), with over 1,200 entries. Because of the CHRIS manual's index codes, the authors strongly suggest using their indexing system with the CHRIS manual's. This will make the information retrievable much faster from this reference. In the front of the CHRIS manual you will find an Index of Codes on page 9-1. If you number the codes consecutively from AAC (Acetic Acid) to ZSL (Zinc Silicofluoride), you will have a more efficient CHRIS manual. To assist you in staying in order, the following have been numbered: AAC 1 (Acetic Acid); AMN 56 (Ammonium Nitrate); BAM 118 (n-Butylamine); BUA 183 (tert-Butylamine); CHO 247 (Chloroacetaldehyde); CSN 312 (Copper Sulfate, ammoniated); DDD 373 (DDD); DMO 436 (2,2 Dimethyl-octanoic acid); DZP 498 (Di-(p-chlorobenzoyl peroxide)); ETB 561 (Ethylbenzene); HBR 621 (Hydrogen bromide); JPT 686 (Jet fuels: JP-3); MEA 749 (Monoethanolamine); MTT 813 (Methyl acetate); OET 878 (Octyl epoxy tallate); PDE 943 (1,3-Pentadiene); SAB 1007 (Sodium alkylbenzenesulfonates); SVO 1070 (Silver oxide); TOD 1133 (p-Toluidine); and ZBR 1195 (Zinc bromide).

13. *Emergency Handling of Hazardous Materials in Surface Transportation,* by the American Association of Railroads (1992 Edition), with over 4,500 entries.

14. The *Emergency Action Guides for the American Association of Railroads* (1993 Edition), with over 185 entries.

15. *Chemical Data Notebook: A User's Manual,* by Frank L. Fire (1993 Edition). Because there is no numerical index, the authors suggest using the indexing they have designed for the *Chemical Data Notebook.* Starting with Acetaldehyde, give the chemicals alphanumeric designations: "A1" through "A14" for the "A's," then "B1" through "B4" for the "B's," and so on, until you reach "V1" for Vinyl Chloride.

16. The *Condensed Chemical Dictionary,* by Richard J. Lewis, Sr. (12th Edition).

As you familiarize yourself with the index, you will be able to retrieve information faster on chemicals that are involved in a hazardous materials incident.

There are many other books that could give you the same information you will find here, and this is not the only source of information to be used at a hazardous materials incident. But, because of its content and presentation, we believe that it is a vital step in the right direction for initial assessment, identification, containment, neutralization, and safety in dealing with these chemicals.

Acknowledgments

We wish to thank all persons who helped and gave permission to use their material for reference:

Mr. Jerry L. Hagan, Chem-Tox, 7121 Brentwood Blvd., Brentwood, TN 37027.

Mr. Richard J. Lewis, Sr., Lewis Information Systems, Inc., 2490 Royalview Ct., Cincinnati, OH 45244.

Van Nostrand Reinhold, for the use of *Dangerous Properties of Industrial Materials, Eighth Edition*, by Richard J. Lewis, Sr., copyright 1992.

The National Fire Protection Association, for the use of *Fire Hazard Properties of Flammable Liquids, Gases, and Volatile Solids*, NFPA 325M-1984. Copyright 1991, the National Fire Protection Association, Quincy, MA 02269-9110. This reprinted material is not the official position of the national Fire Protection Association, which is represented only by the standard in its entirety.

The Association of American Railroads, for the use of *Emergency Handling of Hazardous Materials in Surface Transportation*. Copyright 1989, the Association of American Railroads, Hazardous Materials Systems (BOE), 50 F Street NW, Washington, DC 20001.

The Department of Transportation, for the use of the *Emergency Response Guide Book*. Copyright 1990, Office of Hazardous Materials Transportation (DHM-51), Research and Special Programs Administration, U.S. Department of Transportation, Washington, DC 20590.

The National Institute of Occupational Safety and Health, for the use of the *Pocket Guide to Chemical Hazards*. Copyright 1990, the Division of Standards and Technology Transfer, NIOSH, 4676 Columbia Parkway, Cincinnati, OH 45226-1998.

The United States Department of Transportation—U.S. Coast Guard, for the use of the *Chemical Hazard Response Information System* (CHRIS). Copyright 1984, Commandant (GWER-2), the United States Coast Guard, Washington, DC 20593.

The INDEX

CHEMICAL OR MATERIAL NAME	UN/NA Number	Guide Number (DOT)	Firefighter's Hazardous Materials Reference Book	First Aid Manual for Chemical Accidents	Sax's Dangerous Properties of Industrial Materials	Hazardous Chemicals Desk Reference	Rapid Guide to Hazardous Chemicals in the Workplace	Fire Protection Guide on Hazardous Materials (NFPA)	Firefighter's Handbook of Hazardous Materials	Pocket Guide to Chemical Hazards (NIOSH)	Chemical Hazard Response Information System (CHRIS)	Emergency Handling of Hazardous Materials in Surface Transportation (AAR)	Emergency Action Guides (EAG)	Chemical Data Notebook: A User's Manual	Condensed Chemical Dictionary
1,1,1-DIFLUOROCHLOROETHANE	2517	22							112						
1,1,1-TRICHLORO-2,2-BIS(p-CHLOROPHENYL)-ETHANE **Synonym: DDT**	2761	55	255		1023		62		90	80	DDT 379				1169
1,1,1-TRICHLOROETHANE **Synonym: TRICHLOROETHANE**	2831/3082	74		25	3350			173–88	273		TCE1082	919	T4.2		1170
1,1,2,2-TETRABROMOETHANE	2504	58						165–85	261						
1,1,2,2-TETRACHLORO-1,2-DIFLUOROETHANE	1078	12			3207	1096	208			206					
1,1,2,2-TETRACHLOROETHANE **Synonym: TETRACHLOROETHANE**	1702	55	807	25	3208	1097	209		261	208	TEC1099	887			1127
1,1,2-TRICHLOROETHANE	2831/1993	74			3350	1136	218	173–	273	216	TCM1086	918			1170
1,1,2-TRICHLOROETHYLENE **Synonym: TRICHLOROETHYLENE**	1710	74	837	25	3352	1137	218	174–88	273	216	TCL1085	919	T5.0	T 6	1170
1,1,3,3-TETRAMETHOXYPROPANE								–86							1133
1,1,3,3-TETRAMETHYLBUTYL HYDROPEROXIDE	2160	48							264						
1,1,3,3-TETRAMETHYLBUTYLPEROXY-2-ETHYL HEXANOATE	2161/3115	52							264			898			
1,1,3,3-TETRAMETHYLUREA					3249	1106			265						
1,1,3-TRIETHOXYHEXANE					3367			–88							1174
1,1-BIS(p-CHLOROPHENYL)-2,2,2-TRICHLOROETHANOL **Synonym: 4,4-DICHLORO-alpha-TRICHLOROMETHYLBENZHY-DROL**	2761	55			469	158			286		DTM 489				
1,1-DCE **Synonym: VINYLIDENE CHLORIDE**	1303	26	878	26	3498		229	181–93			VCI1173	952	V1.2		1217
1,1-DI(tert-BUTYLPEROXY)-3,3,5-TRIMETHYLCYCLOHEXANE	2145/3101	48							92			296			
1,1-DI(tert-BUTYLPEROXY)-3,3,5-TRIMETHYLCYCLOHEXANE	2147/3106	48							92			295			
1,1-DI(tert-BUTYLPEROXY)-3,3,5-TRIMETHYLCYCLOHEXANE	2146/3107	48							92			295			

HAZARDOUS MATERIALS REFERENCE BOOKS INDEX

CHEMICAL OR MATERIAL NAME	UN/NA Number	Guide Number (DOT)	Firefighter's Hazardous Materials Reference Book	First Aid Manual for Chemical Accidents	Sax's Dangerous Properties of Industrial Materials	Hazardous Chemicals Desk Reference	Rapid Guide to Hazardous Chemicals in the Workplace	Fire Protection Guide on Hazardous Materials (NFPA)	Firefighter's Handbook of Hazardous Materials	Pocket Guide to Chemical Hazards (NIOSH)	Chemical Hazard Response Information System (CHRIS)	Emergency Handling of Hazardous Materials in Surface Transportation (AAR)	Emergency Action Guides (EAG)	Chemical Data Notebook: A User's Manual	Condensed Chemical Dictionary
1,1-DI(tert-BUTYLPEROXY) CYCLOHEXANE	2180/3107	49							92						
1,1-DI(tert-BUTYLPEROXY) CYCLOHEXANE	2897/3105	48							92			297			
1,1-DI(tert-BUTYLPEROXY) CYCLOHEXANE	2179/3101	49							92			297			
1,1-DI(tert-BUTYLPEROXY) CYCLOHEXANE	2885/3106	48							92			297			
1,1-DICHLORO-1-NITRO ETHANE	2650/2810	57			1154	388		−35	100	88		301			382
1,1-DICHLORO-1-NITRO PROPANE								−35	100						
1,1-DICHLORO-1-NITROETHANE	2650/2810	57			1154	388	70	−35	100	88		301			382
1,1-DICHLORO-2,2-BIS(p-CHLOROPHENYL) ETHANE	2761	55	254		3207				100		DDD 373	880			378
Synonym: DDD															
1,1,1,2-TETRACHLORO-2,2-DIFLUOROETHANE	1710	74	837	25	3352		208	174–88	273	216	TCL1085				1170
Synonym: TRICHLOROETHYLENE															
1,1-DICHLORO-2-PROPANONE	2649	55			1023				100						
1,1-DICHLOROACETONE	2649	55			3350				101						
1,1-DICHLOROETHANE	2362/1993	27	1		1141	382	69		103	86	DCH 361	301			380
1,1-DICHLOROETHENE	1303	26	878	26	3498	1175	229	181–93	286		VCI1173	952	V1.2		1217
Synonym: VINYLIDENE CHLORIDE															
1,1-DICHLOROETHENE	1150	29			3207										
1,1-DICHLOROETHYLENE	1303	26	878	26	3498	1175	229	181–93	103		VCI1173	952	V1.2		1217
Synonym: VINYLIDENE CHLORIDE															
1,1-DICHLOROETHYLENE	1150	29			3352				103						
1,1-DICHLOROPROPANE	1279	27			1163						DPB 459				
1,1-DIFLUOROETHANE	1030	22	2						112		DFE 397				399
1,1-DIFLUOROETHYLENE	1959/1954	22							112			303			399
1,1-DIMETHOXYETHANE	2377/1993	27							115			305			
1,1-DIMETHYLHYDRAZINE	1163	28			1381	459	79	−43	122	96	DMH 432				417

CHEMICAL OR MATERIAL NAME	UN/NA Number	Guide Number (DOT)	Firefighter's Hazardous Materials Reference Book	First Aid Manual for Chemical Accidents	Sax's Dangerous Properties of Industrial Materials	Hazardous Chemicals Desk Reference	Rapid Guide to Hazardous Chemicals in the Workplace	Fire Protection Guide on Hazardous Materials (NFPA)	Firefighter's Handbook of Hazardous Materials	Pocket Guide to Chemical Hazards (NIOSH)	Chemical Hazard Response Information System (CHRIS)	Emergency Handling of Hazardous Materials in Surface Transportation (AAR)	Emergency Action Guides (EAG)	Chemical Data Notebook: A User's Manual	Condensed Chemical Dictionary
1,1-ETHYLIDENE DICHLORIDE	2362	27						-53	145						
1,1-IMINODI-2-PROPANOL Synonym: DIISOPROPANOLAMINE					1302			-41			DIP 419				406
1,1-OXYBISETHANE Synonym: ETHYL ETHER	1155	26	354	16	1614	546	96	87 -52	137	112	EET 524	346	E2.8	E 3	492
1,11-DIAMINO-3,6,9-TRIAZAUNDECANE Synonym: TETRAETHYLENEPENTAMINE	2320/1719	60	816			1100		-85	262		TTP1157	892			1129
1,2,2-TRICHLOROETHYLENE Synonym: TRICHLOROETHYLENE	1710	74	837	25	3352	1137	218	174-88	273	216	TCL1085	919	T5.0	T 6	1170
1,2,3,4,5,6-HEXACHLOROCYCLOHEXANE Synonym: BENZENE HEXACHLORIDE	2761	55	116		360	117					BHC 144				595
1,2,3,4,5,6-HEXANNEHEXOL Synonym: SORBITOL			789		3124	1068					SBT1018				1077
1,2,3,4-TETRAHYDROBENZENE	2256	29							263						1131
1,2,3,4-TETRAHYDRONAPHTHALENE Synonym: TETRAHYDRONAPHTHALENE	1993	27	819		3230			-86	263		THN1118	895			1132
1,2,3,4-TETRAMETHYLBENZENE 95%					3240			-86							
1,2,3,5-TETRAMETHYLBENZENE 85.5%					3240			-86			TTB1151				
1,2,3,6-TETRAHYDROBENZALDEHYDE	2498	29						-85	263			894			1131
1,2,3-BENZENETRIOL Synonym: PYROGALLIC ACID			735						245		PGA 951				
1,2,3-PROPANETRIOL Synonym: GLYCERINE	1760	60	417		1794	606	105	-56			GCR 606	474			
1,2,3-TRICHLOROBENZENE	2321	54							272		TBZ1079				1169
1,2,3-TRICHLOROPROPANE					3360	1140	219	-88	274	218	TCN1087				1172
1,2,3-TRICHLOROPROPENE	2810	55			3360				274			921			

HAZARDOUS MATERIALS REFERENCE BOOKS INDEX

CHEMICAL OR MATERIAL NAME	UN/NA Number	Guide Number (DOT)	Firefighter's Hazardous Materials Reference Book	First Aid Manual for Chemical Accidents	Sax's Dangerous Properties of Industrial Materials	Hazardous Chemicals Desk Reference	Rapid Guide to Hazardous Chemicals in the Workplace	Fire Protection Guide on Hazardous Materials (NFPA)	Firefighter's Handbook of Hazardous Materials	Pocket Guide to Chemical Hazards (NIOSH)	Chemical Hazard Response Information System (CHRIS)	Emergency Handling of Hazardous Materials in Surface Transportation (AAR)	Emergency Action Guides (EAG)	Chemical Data Notebook: A User's Manual	Condensed Chemical Dictionary
1,2,3-TRIHYDROXYBENZENE Synonym: PYROGALLIC ACID			735						277		PGA 951				1178
1,2,3-TRIHYDROXYPROPANE Synonym: GLYCERINE	1760	60	417		1794	606	105	-55			GCR 606	474			
1,2,3-TRIMETHYLBENZENE	2325	26			3404			-90	278						1181
1,2,4,5-TETRAMETHYLBENZENE 95%								-86							
1,2,4-TRICHLOROBENZENE	2321	54			3349	1136	218	-88	272		TCB1081				1169
1,2,4-TRIMETHYLBENZENE	2325	26			3404	1147		-90	278		TME1128				1181
1,2,5,6-TETRAHYDROPYRIDINE	2410	26							263						1132
1,2,6-HEXANETRIOL					1862			-58							599
1,2-BENZENEDICARBOXYLIC ACID ANHYDRIDE Synonym: PHTHALIC ANHYDRIDE	2214/1759	60	694	22	2798	954	182	141–80	234	184	PAN 927	761	P4.4		913
1,2-BENZENEDICARBOXYLIC ACID DIETHYL ESTER Synonym: DIETHYL PHTHALATE	9188/3082	31	288		1236	415	74	-39	108		DPH 465	348			396
1,2-BENZENEDIOL Synonym: CATECHOL					718	252	44		67		CTC 318				232
1,2-BUTANEDIOL					575	189		-20	54						
1,2-BUTYLENE OXIDE Synonym: BUTYLENE OXIDE	3020/3022	55						-23			BTO 180	171			185
1,2-DI(DIMETHYLAMINO) ETHANE	2372	26							92			307			
1,2-DI(tert-BUTYLPEROXY) CYCLOHEXANE	2181	48										298			
1,2-DIAMINOETHANE Synonym: ETHYLENEDIAMINE	1604	29	378					84 -50	95	108	EDA 517	437			362
1,2-DIBROMO-2,2-DICHLOROETHYL DIMETHYL PHOSPHATE Synonym: NALED	2783/3082	55	600		2461	835	152		207		NLD 838	673			804
1,2-DIBROMO-3-CHLOROPROPANE	2872	58			1101	369	65		98	82					369

CHEMICAL OR MATERIAL NAME	UN/NA Number	Guide Number (DOT)	Firefighter's Hazardous Materials Reference Book	First Aid Manual for Chemical Accidents	Sax's Dangerous Properties of Industrial Materials	Hazardous Chemicals Desk Reference	Rapid Guide to Hazardous Chemicals in the Workplace	Fire Protection Guide on Hazardous Materials (NFPA)	Firefighter's Handbook of Hazardous Materials	Pocket Guide to Chemical Hazards (NIOSH)	Chemical Hazard Response Information System (CHRIS)	Emergency Handling of Hazardous Materials in Surface Transportation (AAR)	Emergency Action Guides (EAG)	Chemical Data Notebook: A User's Manual	Condensed Chemical Dictionary
1,2-DIBROMOETHANE Synonym: ETHYLENE DIBROMIDE	1605	55	371					85 –	98	110	EDB 518	430	E2.4.1		370
1,2-DICHLORO-1,1,2,2-TETRAFLUOROETHANE	1958	12													
1,2-DICHLOROBENZENE Synonym: DICHLOROBENZENE (-ortho)	1591	58	272	15	1127	376	66	64 –34	101	84	DBO 350	329	D1.0.1		377
1,2-DICHLOROBUTANE								–34	102						
1,2-DICHLOROETHANE Synonym: ETHYLENE DICHLORIDE	1184	26	372	17	1602	540	94	86 –50	103	110	EDC 519	430	E2.4.5	E 5	380
1,2-DICHLOROETHYLENE	1 50/3082	29	3		1144	384	69	–35	103	88	DEL 388	301			
1,2-DICHLOROETHYLENE-cis Synonym: 1,2-DICHLOROETHYLENE	1150/3082	29	3		1144	384	69	–35	103	88	DEL 388	301			
1,2-DICHLOROETHYLENE-trans Synonym: 1,2-DICHLOROETHYLENE	1150/3082	29	3		1144	384	69	–35	103	88	DEL 388	301			
1,2-DICHLOROPROPANE Synonym: PROPYLENE DICHLORIDE	1279	27	4		2919	1001	189	150–82	104	188	DPP 470	798	P8.0.3		383
1,2-DICHLOROPROPENE	2047	29			1164										
1,2-DIETHOXYETHANE Synonym: ETHYLENE GLYCOL DIETHYL ETHER	1153	26	374		1606	542		–51	107		EEE 522	431			488
1,2-DIETHYL HYDRAZINE	2382	57			1226	411									
1,2-DIHYDRO-3,6-PYRIDAZINEDIONE Synonym: MALEIC HYDRAZIDE			529		1278						MLH 770				402
1,2-DIHYDROXYBENZENE Synonym: CATECHOL					718	252	44		67		CTC 318				232
1,2-DIHYDROXYETHANE Synonym: ETHYLENE GLYCOL		27	373	17	1604	541	94	–50	143		EGL 531	434	E2.4.7		487

CHEMICAL OR MATERIAL NAME	UN/NA Number	Guide Number (DOT)	Firefighter's Hazardous Materials Reference Book	First Aid Manual for Chemical Accidents	Sax's Dangerous Properties of Industrial Materials	Hazardous Chemicals Desk Reference	Rapid Guide to Hazardous Chemicals in the Workplace	Fire Protection Guide on Hazardous Materials (NFPA)	Firefighter's Handbook of Hazardous Materials	Pocket Guide to Chemical Hazards (NIOSH)	Chemical Hazard Response Information System (CHRIS)	Emergency Handling of Hazardous Materials in Surface Transportation (AAR)	Emergency Action Guides (EAG)	Chemical Data Notebook: A User's Manual	Condensed Chemical Dictionary
1,2-DIHYDROXYPROPANE Synonym: PROPYLENE GLYCOL			728	23				–82	244		PPG 975				404
1,2-DIMETHOXYETHANE Synonym: ETHYLENE GLYCOL DIMETHYL ETHER	2252/1993	27			1316	349		–51	115		EGD 528	306			410
1,2-DIMETHYLBENZENE Synonym: XYLENE	1307	27	881	26	3520	1182	230	183–	121	226		959	X1.0		413
1,2-DIMETHYLBENZENE Synonym: XYLENE (-ortho)	1307	27	881	26	3521	1183	231	183–94	121	226	XLO1188	959	X1.0		413
1,2-DIMETHYLHYDRAZINE	2382	57			1382	460	80		122		DML 433				
1,2-DINITRO BENZOL	1597	56						–44							
1,2-DINITROBENZENE Synonym: DINITROBENZENE (-ortho)	1597	56	311	16	1428	476	81	76 –	124	98	DNO 448	375			422
1,2-EPOXY-3-CHLOROPROPANE Synonym: EPICHLOROHYDRIN	2023	30	336	16	1525	509	87	79 –46	132	102	EPC 552	401	E1.0	E 1	467
1,2-EPOXY-3-ETHOXYPROPANE	2752	26													
1,2-EPOXY-3-ETHYLOXYPROPANE	2752	26													
1,2-EPOXYETHANE Synonym: ETHYLENE OXIDE	1040	69	377	17	1611	544	96	87 –52	132	112	EOX 550	434	E2.7	E 6	491
1,2-EPOXYPROPANE Synonym: PROPYLENE OXIDE	1280	26	729	23	2922	1003	190	150–82	133	190	POX 971	798	P8.1		971
1,2-ETHANEDIAMINE Synonym: ETHYLENEDIAMINE	1604	29	378		1554	517	88	84 –50	134	108	EDA 517	437			486
1,2-ETHANEDIOL Synonym: CROTONALDEHYDE	1143	28	233	14	964	325	56	57 –29	134	76	CTA 317	271	C7.1		325
1,2-ETHANEDIOL Synonym: ETHYLENE GLYCOL	1993	27	373	17	1604	541	94	–50	134		EGL 531	434	E2.4.7		487

CHEMICAL OR MATERIAL NAME	UN/NA Number	Guide Number (DOT)	Firefighter's Hazardous Materials Reference Book	First Aid Manual for Chemical Accidents	Sax's Dangerous Properties of Industrial Materials	Hazardous Chemicals Desk Reference	Rapid Guide to Hazardous Chemicals in the Workplace	Fire Protection Guide on Hazardous Materials (NFPA)	Firefighter's Handbook of Hazardous Materials	Pocket Guide to Chemical Hazards (NIOSH)	Chemical Hazard Response Information System (CHRIS)	Emergency Handling of Hazardous Materials in Surface Transportation (AAR)	Emergency Action Guides (EAG)	Chemical Data Notebook: A User's Manual	Condensed Chemical Dictionary
1,2-ETHANEDIOL DIFORMATE								−47	134						
1,2-ETHYLENE DIBROMIDE	1605	55			1601	540	93								
1,2-ETHYLENEDICARBOXYLIC ACID-cis Synonym: MALEIC ACID	2215	60	527		2154	736			183		MLI 771				724
1,2-ETHYLENEDICARBOXYLIC ACID-trans Synonym: FUMARIC ACID	9126/9188	31	407		1760	595			152		FUM 599	465			543
1,2-ETHYLIDENE DICHLORIDE								−53	145						
1,2-PROPANEDIOL Synonym: PROPYLENE GLYCOL			728	23	2900	994		−82	244		PPG 975				966
1,2-PROPANEDIOL 1-METHACRYLATE Synonym: HYDROXYPROPYL METHACRYLATE			459								HPM 642				625
1,2-PROPANEDIOL-1-ACRYLATE Synonym: HYDROXYPROPYL ACRYLATE	1760	60	458								HPA 640	535			
1,2-PROPYLENE OXIDE Synonym: PROPYLENE OXIDE	1280	26	729	23	2922	1003	190	150–82	244	190	POX 971	798	P8.1		971
1,2-PROPYLENEIMINE Synonym: PROPYLENEIMINE	1921	30	731		2921	1002	190			188	PII 959	799	P8.0.5		971
1,2-PROPYLENIMINE Synonym: PROPYLENEIMINE	1921	30	731		2921	1002	190		244	188	PII 959	799	P8.0.5		971
1,3,5-TRICHLOROBENZENE	2321	54							272						
1,3,5-TRIETHYLBENZENE Synonym: TRIETHYLBENZENE			847		3369				275		TEB1098				
1,3,5-TRIMETHYLBENZENE	2325/1993	26			3405	1148		−90	278			929			1181
1,3,5-TRINITROBENZENE	0214				3422	1152			279						1185
1,3,6-TRINITROPYRENE					3425	1153	222								

HAZARDOUS MATERIALS REFERENCE BOOKS INDEX

CHEMICAL OR MATERIAL NAME	UN/NA Number	Guide Number (DOT)	Firefighter's Hazardous Materials Reference Book	First Aid Manual for Chemical Accidents	Sax's Dangerous Properties of Industrial Materials	Hazardous Chemicals Desk Reference	Rapid Guide to Hazardous Chemicals in the Workplace	Fire Protection Guide on Hazardous Materials (NFPA)	Firefighter's Handbook of Hazardous Materials	Pocket Guide to Chemical Hazards (NIOSH)	Chemical Hazard Response Information System (CHRIS)	Emergency Handling of Hazardous Materials in Surface Transportation (AAR)	Emergency Action Guides (EAG)	Chemical Data Notebook: A User's Manual	Condensed Chemical Dictionary
1,3-BENZENEDIOL	2876	55	738	23	2984	1019	192	-83	41		RSC1006	813			1003
Synonym: RESORCINOL															
1,3-BUTADIENE	1010	17	139	12	572	187	28	40 -20	53	48	BDI 134	148	B4.0	B 4	177
Synonym: BUTADIENE															
1,3-BUTANEDIAMINE					574	189		-20	54						
1,3-BUTANEDIOL					575	189			54						178
1,3-BUTYLENE GLYCOL									60						185
1,3-CYCLOPENTADIENE					1002	338	60		89						340
1,3-DI(2-tert-BUTYLPEROXYISOPROPYL) BENZENE	2112/3106	48										308	B1.0		
1,3-DIAMINOPROPANE									95						363
1,3-DICHLORO-2,4-HEXADIENE								-35	100						
1,3-DICHLORO-2-BUTENE	2924	29						-34							
1,3-DICHLORO-2-PROPANOL	2750/2810	55			1164	393		-35	100			301			383
1,3-DICHLORO-2-PROPANONE	2649	55							100						
1,3-DICHLORO-5,5-DIMETHYL HYDANTOIN					1139	381	68		101	86					380
1,3-DICHLOROACETONE	2649/2810	55							101			300			
1,3-DICHLOROBENZENE	1592	58			1127										
1,3-DICHLOROBUTENE-2	2924	29						-34							379
1,3-DICHLOROPROPANE	1279	27			1163	392			104		DPC 460				
1,3-DICHLOROPROPANOL-2	2750/2810	55			1164				105			301			
1,3-DICHLOROPROPENE	2047	29			1164	393	70	-35	105		DPU 472				383
1,3-DICHLOROPROPENE-cis					1165	394			105						
1,3-DICHLOROPROPYLENE	2047	29							105						
1,3-DIETHYL-1,3-DIPHENYL UREA					1218	408		-37							

CHEMICAL OR MATERIAL NAME	UN/NA Number	Guide Number (DOT)	Firefighter's Hazardous Materials Reference Book	First Aid Manual for Chemical Accidents	Sax's Dangerous Properties of Industrial Materials	Hazardous Chemicals Desk Reference	Rapid Guide to Hazardous Chemicals in the Workplace	Fire Protection Guide on Hazardous Materials (NFPA)	Firefighter's Handbook of Hazardous Materials	Pocket Guide to Chemical Hazards (NIOSH)	Chemical Hazard Response Information System (CHRIS)	Emergency Handling of Hazardous Materials in Surface Transportation (AAR)	Emergency Action Guides (EAG)	Chemical Data Notebook: A User's Manual	Condensed Chemical Dictionary
1,3-DIISOCYANATOMETHYLBENZENE Synonym: TOLUENE DIISOCYANATE	2078	57	828		3312		215		269		TDI1095	910	T4.1	T 5	1157
1,3-DIMETHYLBENZENE Synonym: XYLENE	1307	27	880	26	3520	1182	230	183–94	121	226	XLM1187	959	X1.0		413
1,3-DIMETHYLBENZENE Synonym: XYLENE (-meta)	1307	27	880	26	3521	1183	231	183–94	121	226	XLM1187	959	X1.0		413
1,3-DIMETHYLBUTYL ACETATE								–42	121						
1,3-DIMETHYLBUTYLAMINE	2379/1993	27			1358	452		–42	116			307			
1,3-DIMETHYLCYCLOHEXANE	2263	27			1365	455		–42	122						
1,3-DINITROBENZENE Synonym: DINITROBENZENE (-meta)	1597	56		16	1428	475	81		124	98	DNB 443	375			422
1,3-DINITROBENZOL Synonym: DINITROBENZENE (-meta)	1597	56		16	1428	475	81		124	98	DNB 443	375			422
1,3-DINITROPYRENE					1437	479									
1,3-DIOXOPHTHALAN Synonym: PHTHALIC ANHYDRIDE	2214/1759	60	694	22	2798	954	182	141–80	234	184	PAN 927	761	P4.4		913
1,3-PENTADIENE					2681			–77	223		PDE 943				880
1,3-PROPANEDIAMINE					2900				241						966
1,4,5,6,7,8,8a-HEPTACHLORODICYCLOPENTADIENE Synonym: HEPTACHLOR	2761/3077	55	424		1826	615	106		156	120	HTC 646	505			591
1,4-BENZENE DIOL Synonym: HYDROQUINONE	2662	53	453	18	1906	647	116	–59	165	128	HDQ 631	534			618
1,4-BENZENEDIOL Synonym: HYDROQUINONE	2662	53	453	18	1906	647	116	–59	165	128	HDQ 631	534			618
1,4-BUTANEDICARBOXYLIC ACID Synonym: ADIPIC ACID	9077/3077	31	35		79	28		–12	20		ADA 28	15	A6.0	A 8	178

HAZARDOUS MATERIALS REFERENCE BOOKS INDEX

CHEMICAL OR MATERIAL NAME	UN/NA Number	Guide Number (DOT)	Firefighter's Hazardous Materials Reference Book	First Aid Manual for Chemical Accidents	Sax's Dangerous Properties of Industrial Materials	Hazardous Chemicals Desk Reference	Rapid Guide to Hazardous Chemicals in the Workplace	Fire Protection Guide on Hazardous Materials (NFPA)	Firefighter's Handbook of Hazardous Materials	Pocket Guide to Chemical Hazards (NIOSH)	Chemical Hazard Response Information System (CHRIS)	Emergency Handling of Hazardous Materials in Surface Transportation (AAR)	Emergency Action Guides (EAG)	Chemical Data Notebook: A User's Manual	Condensed Chemical Dictionary
1,4-BUTANEDIOL	1987	26	5		575	189		−20	54		BDO 136				178
1,4-BUTENEDIOL	1987	26	6								BUD 186				
1,4-BUTYNEDIOL	2716	55			641						BTD 175	172			
1,4-DI(2-tert-BUTYLPEROXYISOPROPYL) BENZENE	2112/3106	48										308	B1.0		
1,4-DICHLORO-2-BUTENE	2924	29			1133	379	68		100						
1,4-DICHLORO-2-BUTENE	1760/2920	60			1133	179	68		100		DCB 358	330			
Synonym: DICHLOROBUTENE															
1,4-DICHLORO-2-BUTENE-cis	1760/2920	60									DCB 358	330			
Synonym: DICHLOROBUTENE															
1,4-DICHLORO-2-BUTENE-trans	1760/2920	60									DCB 358	330			
Synonym: DICHLOROBUTENE															
1,4-DICHLORO-2-BUTYLENE	1760/2920	60									DCB 358	330			
Synonym: DICHLOROBUTENE															
1,4-DICHLOROBENZENE	1592	58	273		1128	377	67				DBP 351	330			378
Synonym: DICHLOROBENZENE-para															
1,4-DICHLOROBUTANE								64 −34	102						378
1,4-DICYANOBUTANE	2205	55	36	11	82	28	7	20 −12	106		ADN 29				385
Synonym: ADIPONITRILE															
1,4-DIHYDROXY-2-BUTENE	1987	26	6								BUD 186				
Synonym: 1,4-BUTENEDIOL															
1,4-DIHYDROXY-2-BUTYNE											BTD 175				
Synonym: 1,4-BUTYNEDIOL															
1,4-DIHYDROXYBUTANE	1987	26	5		575	189		−20			BDO 136				178
Synonym: 1,4-BUTANEDIOL															
1,4-DIMETHYLBENZENE	1307	27	882	26		1182	230	183–94	287	226	XLP1189	959	X1.0		413
Synonym: XYLENE															

CHEMICAL OR MATERIAL NAME	UN/NA Number	Guide Number (DOT)	Firefighter's Hazardous Materials Reference Book	First Aid Manual for Chemical Accidents	Sax's Dangerous Properties of Industrial Materials	Hazardous Chemicals Desk Reference	Rapid Guide to Hazardous Chemicals in the Workplace	Fire Protection Guide on Hazardous Materials (NFPA)	Firefighter's Handbook of Hazardous Materials	Pocket Guide to Chemical Hazards (NIOSH)	Chemical Hazard Response Information System (CHRIS)	Emergency Handling of Hazardous Materials in Surface Transportation (AAR)	Emergency Action Guides (EAG)	Chemical Data Notebook: A User's Manual	Condensed Chemical Dictionary
1,4-DIMETHYLBENZENE Synonym: XYLENE (-para)	1307	27	882	26		1184	231	183–94	115	226	XLP1189	959	X1.0		413
1,4-DIMETHYLCYCLOHEXANE	2263	27			1365	455		-42	121			370			
1,4-DIMETHYLCYCLOHEXANE-cis	2263	27						-42							
1,4-DIMETHYLCYCLOHEXANE-trans	2263	27			1365			-42	121						
1,4-DINITROBENZENE Synonym: DINITROBENZENE (-para)	1597	56		16	1429	476	81		124	98	DNZ 452	375			422
1,4-DIOXANE Synonym: DIOXANE	1165	26	317		1449	482	83	78 –	126		DOX 457	380			426
1,4-HEXADIENE	2458	29			1846			-58	158						596
1,4-THIOXANE								-87	268						
1,5-CYCLOOCTADIENE	2520	27						-30	89						
1,5-DICHLOROPENTANE	1152	27			1156			-35	104						
1,5-DINITRONAPHTHALENE					1434	478			125						
1,5-HEXADIYNE					1848				159						597
1,5-NAPHTHALENE DIISOCYANATE					2465	836	153								805
1,5-PENTANEDIAL Synonym: GLUTARALDEHYDE			416	18	1793	606	105		154		GTA 614				565
1,5-PENTANEDIAMINE					2687	914			224						
1,5-PENTANEDIOL					2688			-77							881
1,6-DIAMINOHEXANE Synonym: HEXAMETHYLENE DIAMINE SOLUTION	1783	60					103		160		HMD 637	515	H1.1		362
1,6-DIAMINOHEXANE Synonym: HEXAMETHYLENEDIAMINE	1783	60					103		160		HMD 637	515	H1.1		598
1,6-DIISOCYANATOHEXANE	2281	55			1301	434	76		114						

HAZARDOUS MATERIALS REFERENCE BOOKS INDEX

CHEMICAL OR MATERIAL NAME	UN/NA Number	Guide Number (DOT)	Firefighter's Hazardous Materials Reference Book	First Aid Manual for Chemical Accidents	Sax's Dangerous Properties of Industrial Materials	Hazardous Chemicals Desk Reference	Rapid Guide to Hazardous Chemicals in the Workplace	Fire Protection Guide on Hazardous Materials (NFPA)	Firefighter's Handbook of Hazardous Materials	Pocket Guide to Chemical Hazards (NIOSH)	Chemical Hazard Response Information System (CHRIS)	Emergency Handling of Hazardous Materials in Surface Transportation (AAR)	Emergency Action Guides (EAG)	Chemical Data Notebook: A User's Manual	Condensed Chemical Dictionary
1,6-HEXANEDIAMINE	2280	60			1859	627			160						
1,6-HEXANEDIAMINE Synonym: HEXAMETHYLENE DIAMINE SOLUTION	1783	60			1859	627	109		160		HMD 637	515	H1.1		
1,6-HEXANEDIAMINE Synonym: HEXAMETHYLENEDIAMINE	1783	60			1859	627	109		160		HMD 637	515	H1.1		598
1,6-HEXANEDIOC ACID Synonym: ADIPIC ACID	9077/3077	31	35		79			−12	20		ADA 28	15	A6.0	A 8	24
1,8-TERPODIENE-delta Synonym: DIPENTENE	2052/1993	27	318					−45	126		DPN 468	381			427
1-(2-AMINOETHYL) PIPERAZINE Synonym: AMINOETHYL PIPERAZINE-n	2815	60			175	51		−14	26		AEP 32	51			57
1-ACETOXYETHYLENE Synonym: VINYL ACETATE	1301	26	871	26	3492	1172	227	179–92	285		VAM1172	948	V1.0		1215
1-ACETOXYPROPANE Synonym: PROPYL ACETATE (n)	1276	26	723	23	2914	999	189	−81	25		PAT 929	794	P8.0		969
1-AMINO-2-METHYLPROPANE Synonym: ISOBUTYLAMINE	1214	68	465		2017	676	122	−61			IAM 659	547			652
1-AMINO-2-NITROBENZENE Synonym: NITROANILINE (-ortho)	1661	55		21	2514	859					NTA 851	687	N3.0		824
1-AMINO-2-NITROBENZENE Synonym: 2-NITROANILINE	1661	55							211		NTA 851				824
1-AMINO-2-PROPANOL Synonym: MONOISOPROPANOLAMINE	1760	60	595					−14	25		MPA 782	658			60
1-AMINO-4-CHLOROBENZENE Synonym: CHLOROANILINE-para	2018/2811	53	202		767				73		CAP 210	218			262
1-AMINO-4-NITROBENZENE Synonym: 4-NITROANILINE	1661	55			2514	272	158		25		NAL 818				

CHEMICAL OR MATERIAL NAME	UN/NA Number	Guide Number (DOT)	Firefighter's Hazardous Materials Reference Book	First Aid Manual for Chemical Accidents	Sax's Dangerous Properties of Industrial Materials	Hazardous Chemicals Desk Reference	Rapid Guide to Hazardous Chemicals in the Workplace	Fire Protection Guide on Hazardous Materials (NFPA)	Firefighter's Handbook of Hazardous Materials	Pocket Guide to Chemical Hazards (NIOSH)	Chemical Hazard Response Information System (CHRIS)	Emergency Handling of Hazardous Materials in Surface Transportation (AAR)	Emergency Action Guides (EAG)	Chemical Data Notebook: A User's Manual	Condensed Chemical Dictionary
1-AMINOBUTANE **Synonym: BUTYLAMINE (-n)**	1125	68		13	594	197	32	41 –	25	52	BAM 118	169	B7.3		55
1-AMINONAPHTHALENE **Synonym: 1-NAPHTHYLAMINE**	2077	55			2471	838	154	–73	208		NAO 820				
1-AMINOPENTANE	1106	68							26						59
1-AMINOPROPANE **Synonym: PROPYLAMINE (-n)**	1277	68	726	23	2915	1000		149–	26		PRA 984	796			969
1-AMYL ALCOHOL **Synonym: AMYL ALCOHOL (-n)**	1105/1987	26	78		242	69			31		AAN 4	74			73
1-AZANAPHTHALENE **Synonym: QUINOLINE**	2656	29	736		2968	1014		–83	246		QNL1005	805			987
1-BENZAZINE **Synonym: QUINOLINE**	2656	29	736			1014		–83	246		QNL1005	805			987
1-BROMO-2-PROPANONE **Synonym: BROMOACETONE**	1569	55			562				51		BRE 166	141	B3.1		170
1-BROMOBUTANE	1126	29			544	180			52		BBU 124				171
1-BROMOPROPANE	2344	29			561	185			52		BPR 162				
1-BUTANAMINE-n-BUTYL **Synonym: DI-n-BUTYLAMINE**	2248/1993	68			1108	370					DBA 342	312	D3.2		371
1-BUTANETHIOL **Synonym: BUTYL MERCAPTAN (-n)**	2347	27			623	208	33	–20	54	54	BTM 178	155			178
1-BUTANOL **Synonym: BUTYL ALCOHOL (-n)**	1120	26			593	196	31	–21	54	52	BAN 119	150	B7.1		178
1-BUTENE **Synonym: *BUTYLENE***	1012	22	150		578	191			55		BTN 179				185

CHEMICAL OR MATERIAL NAME	UN/NA Number	Guide Number (DOT)	Firefighter's Hazardous Materials Reference Book	First Aid Manual for Chemical Accidents	Sax's Dangerous Properties of Industrial Materials	Hazardous Chemicals Desk Reference	Rapid Guide to Hazardous Chemicals in the Workplace	Fire Protection Guide on Hazardous Materials (NFPA)	Firefighter's Handbook of Hazardous Materials	Pocket Guide to Chemical Hazards (NIOSH)	Chemical Hazard Response Information System (CHRIS)	Emergency Handling of Hazardous Materials in Surface Transportation (AAR)	Emergency Action Guides (EAG)	Chemical Data Notebook: A User's Manual	Condensed Chemical Dictionary
1-BUTENE Synonym: BUTENE	1012	22	141		578	191		−20	55		BTN 179		B5.0		
1-BUTOXY BUTANE Synonym: DI-n-BUTYL ETHER	1149	26							55		DBE 344				373
1-BUTOXYBUTANE Synonym: DI-n-BUTYL ETHER	1149	26							55		DBE 344				373
1-BUTOXYETHOXY-2-PROPANOL Synonym: AMINOETHYLPIPERAZINE-n	2815	60			587	194		−21			AEP 32	51			179
1-BUTYNE	2452	17							61						192
1-CHLOR-2,3-EPOXYPROPANE Synonym: EPICHLOROHYDRIN	2023	30	336	16	1525	509	87	79 −46	132	102	EPC 552	401	E1.0	E 1	467
1-CHLORO-1,1-DIFLUORETHANE	2517	22			791	277			70						
1-CHLORO-1-NITROPROPANE					843	289	50	−27	70	68					270
1-CHLORO-2,2-DICHLOROETHYLENE Synonym: TRICHLOROETHYLENE	1710	74	837	25	3352	1137	218	174–88	273	216	TCL1085	919	T5.0	T 6	1170
1-CHLORO-2,3-EPOXYPROPANE Synonym: EPICHLOROHYDRIN	2023	30	336	16	1525	509	87	79 −46	71	102	EPC 552	401	E1.0	E 1	266
1-CHLORO-2,4-DINITROBENZENE	1577	56			799	279			71						265
1-CHLORO-2-METHYL PROPANE					837				71						
1-CHLORO-2-NITROBENZENE Synonym: CHLORONITROBENZENE (-ortho)	1578/2811	55	208		841				76		CNO 273	225	N3.3		269
1-CHLORO-2-PROPANOL	2611	57						−27	71						272
1-CHLORO-2-PROPANONE	1695	59							71						272
1-CHLORO-2-PROPENE Synonym: ALLYL CHLORIDE	1100	57	40	11	106	35	8	22 −13	22	36	ALC 40	36	A7.2		39
1-CHLORO-3-METHYL PROPANE									71						

CHEMICAL OR MATERIAL NAME	UN/NA Number	Guide Number (DOT)	Firefighter's Hazardous Materials Reference Book	First Aid Manual for Chemical Accidents	Sax's Dangerous Properties of Industrial Materials	Hazardous Chemicals Desk Reference	Rapid Guide to Hazardous Chemicals in the Workplace	Fire Protection Guide on Hazardous Materials (NFPA)	Firefighter's Handbook of Hazardous Materials	Pocket Guide to Chemical Hazards (NIOSH)	Chemical Hazard Response Information System (CHRIS)	Emergency Handling of Hazardous Materials in Surface Transportation (AAR)	Emergency Action Guides (EAG)	Chemical Data Notebook: A User's Manual	Condensed Chemical Dictionary
1-CHLORO-3-NITROBENZENE Synonym: NITROCHLOROBENZENE-meta	1578	55	632		840	289			212			690	N3.2		
1-CHLORO-4-METHYLBENZENE Synonym: CHLOROTOLUENE-para	2238	27	213								CRN 304				274
1-CHLOROBUTANE	1127	27			775				74						264
1-CHLOROETHYLENE Synonym: VINYL CHLORIDE	1086	17	872	26	3495	1174	227	180–93	285	224	VCM1174	949	V1.1	V 1	1215
1-CHLOROHEXANE								–27	76						
1-CHLORONAPHTHALENE					839			–27	76						
1-CHLOROPENTANE	1107	26							77						270
1-CHLOROPROPANE	2356	26				297									272
1-CHLOROPROPRENE-2 Synonym: ALLYL CHLORIDE	1100	57	40	11	106	35	8	22 –13	22	36	ALC 40	36	A7.2		39
1-CHLOROPROPYLENE								–27	78						
1-CHLORPENTANE Synonym: AMYL CHLORIDE (-n)	1107	26	79						31		AMY 61	75			74
1-CROTYL BROMIDE								–29	85						
1-CROTYL CHLORIDE								–29	85						
1-DECANOL Synonym: DECYL ALCOHOL (-n)	1987	26	258		1032	347			90		DAN 338	291			349
1-DECENE								–31	91		DCE 359				350
1-DODECANETHIOL Synonym: LAURYL MERCAPTAN	2124	48	498		2092	706	127		130		LRM 707				685
1-DODECENE											DDC 372				
1-ETHOXYPROPANE	2615	26							135						

HAZARDOUS MATERIALS REFERENCE BOOKS INDEX

CHEMICAL OR MATERIAL NAME	UN/NA Number	Guide Number (DOT)	Firefighter's Hazardous Materials Reference Book	First Aid Manual for Chemical Accidents	Sax's Dangerous Properties of Industrial Materials	Hazardous Chemicals Desk Reference	Rapid Guide to Hazardous Chemicals in the Workplace	Fire Protection Guide on Hazardous Materials (NFPA)	Firefighter's Handbook of Hazardous Materials	Pocket Guide to Chemical Hazards (NIOSH)	Chemical Hazard Response Information System (CHRIS)	Emergency Handling of Hazardous Materials in Surface Transportation (AAR)	Emergency Action Guides (EAG)	Chemical Data Notebook: A User's Manual	Condensed Chemical Dictionary
1-FLUOROHEXANE					1732										
1-HENDECANOL Synonym: UNDECANOL			857								UND1165				
1-HEPTADECANECARBOXYLIC ACID Synonym: STEARIC ACID			791		3131	1070		-84			SRA1055				1088
1-HEPTANETHIOL					1831	616									
1-HEPTANOL Synonym: HEPTANOL			426	18					157		HTN 648				592
1-HEPTENE	2278	27							157		HTE 647				592
1-HEPTYLENE	2278	27							157						
1-HEXANETHIOL					1861	628	110								
1-HEXANOL Synonym: HEXANOL (-n)	2282/1993	26	436	18					161		HXN 652	519			599
1-HEXENE	2370	27			1865	629		-58	161		HXE 650				600
1-HYDROXY-1'-HYDROPEROXY DICYCLOHEXYL PEROXIDE	2119	51							165						
1-HYDROXY-1'-HYDROPEROXY DICYCLOHEXYL PEROXIDE	2118	51							165						
1-HYDROXY-1'-HYDROPEROXY DICYCLOHEXYL PEROXIDE	2117	49							165						
1-HYDROXY-2,4-DINITRO BENZENE Synonym: 2,4-DINITROPHENOL	0076				1435	479			125		DNP 449				
1-HYDROXY-2-CYANOETHANE Synonym: ETHYLENE CYANOHYDRIN			370					84 -50	142		ETC 562				486
1-HYDROXY-2-METHYLBENZENE Synonym: CRESOL	2076	55	232		959	324	55	57 -	85	74	CRS 307	270	C7.0		322
1-HYDROXY-3-METHYLBENZENE Synonym: CRESOL	2076	55	232		959	324	55	57 -	85	74	CRS 307	270	C7.0		322

CHEMICAL OR MATERIAL NAME	UN/NA Number	Guide Number (DOT)	Firefighter's Hazardous Materials Reference Book	First Aid Manual for Chemical Accidents	Sax's Dangerous Properties of Industrial Materials	Hazardous Chemicals Desk Reference	Rapid Guide to Hazardous Chemicals in the Workplace	Fire Protection Guide on Hazardous Materials (NFPA)	Firefighter's Handbook of Hazardous Materials	Pocket Guide to Chemical Hazards (NIOSH)	Chemical Hazard Response Information System (CHRIS)	Emergency Handling of Hazardous Materials in Surface Transportation (AAR)	Emergency Action Guides (EAG)	Chemical Data Notebook: A User's Manual	Condensed Chemical Dictionary
1-HYDROXY-6-METHYLBENZENE Synonym: CRESOL	2076	55	232		959	324	55	57 –	85	74	CRS 307	270	C7.0		322
1-HYDROXYBUTANE Synonym: BUTYL ALCOHOL (-n)	1120	26			593	196	31	-21	56	52	BAN 119	150	B7.1		181
1-HYDROXYHEPTANE Synonym: HEPTANOL			426	18							HTN 648				
1-HYDROXYHEXANE Synonym: HEXANOL (-n)	2282	26	436	18					161		HXN 652	519			
1-METHALLYL ALCOHOL	2614	26													
1-METHOXY-2-PROPANOL Synonym: PROPYLENE GLYCOL METHYL ETHER	3092	26			2921			-82	190		PME 966	616			
1-METHYL PIPERAZINE Synonym: METHYLPIPERAZINE-n					2383			-71	196						778
1-METHYL-1-PHENYLETHYLENE Synonym: METHYL STYRENE-alpha	2303	27	580		2398	818	149		197	154	MSR 802				780
1-METHYL-1-PROPYLETHYLENE Synonym: 2-METHYL-1-PENTENE	2288	27			2372			-71	198		MPE 785				777
1-METHYL-2,4-DINITROBENZENE Synonym: 2,4-DINITROTOLUENE	1600	56			1440	480		-44			DTT 493	380			
1-METHYL-2-(3-PYRIDYL) PYRROLIDINE Synonym: NICOTINE	1654	55	622		2500	853	157	-74	209	160	NIC 831	682			820
1-METHYL-2-PYRROLIDINONE Synonym: 1-METHYLPYRROLIDONE					2394						MPY 791				
1-METHYL-3,5-DIETHYLBENZENE								-68							
1-METHYL-4-ISOPROPYLBENZENE Synonym: CYMENE-para	2046/1993	27	252		1011			-31			CMP 269	288			

HAZARDOUS MATERIALS REFERENCE BOOKS INDEX

CHEMICAL OR MATERIAL NAME	UN/NA Number	Guide Number (DOT)	Firefighter's Hazardous Materials Reference Book	First Aid Manual for Chemical Accidents	Sax's Dangerous Properties of Industrial Materials	Hazardous Chemicals Desk Reference	Rapid Guide to Hazardous Chemicals in the Workplace	Fire Protection Guide on Hazardous Materials (NFPA)	Firefighter's Handbook of Hazardous Materials	Pocket Guide to Chemical Hazards (NIOSH)	Chemical Hazard Response Information System (CHRIS)	Emergency Handling of Hazardous Materials in Surface Transportation (AAR)	Emergency Action Guides (EAG)	Chemical Data Notebook: A User's Manual	Condensed Chemical Dictionary
1-METHYLETHYL ALCOHOL Synonym: ISOPROPANOL	1219	26	480						174			555	14.0		659
1-METHYLHYDRAZINE Synonym: METHYL HYDRAZINE	1244	57	586		2327	797	144	114–	195	152	MHZ 762	648	M3.4.6		770
1-METHYLNAPHTHALENE	1993	27			2350			−70	203		MNA 776				
1-METHYLPYRROLIDONE Synonym: METHYLPYRROLIDONE			587		2394						MPY 791				779
1-NAPHTHYL-n-METHYL CARBAMATE	2757	55													808
1-NAPHTHYLAMINE	2077	55			2471	838	154	−73	208		NAO 820				
1-NAPHTYL n-METHYLCARBAMATE Synonym: CARBARYL	2757/3077	55	182		688	241	40		208	58	CBY 224	187			215
1-NITRONAPHTHALENE	2538	32			2544	870	162	−74	214						
1-NITROPROPANE	2608	26			2553	872	163	−74	215	166	NPN 847				831
1-NONANOL Synonym: NONANOL			646								NNN 841				835
1-OCTANOL Synonym: OCTANOL	1987	26	650		2620				218		OTA 909	706	O1.0		847
1-OCTENE	1993	27						−75	218		OTE 913				847
1-OCTYL ALCOHOL Synonym: OCTANOL	1987	26	650		2620						OTA 909	706	O1.0		847
1-PENTADECANOL Synonym: PENTADECANOL			666								PDC 942				
1-PENTAETHIOL Synonym: AMYL MERCAPTAN (-n)	1111	27	80					28 −15	31		AMM 55	75			75
1-PENTANETHIOL Synonym: AMYL MERCAPTAN (-n)	1111	27	80		2689	914	173	28 −15	224		AMM 55	75			75

CHEMICAL OR MATERIAL NAME	UN/NA Number	Guide Number (DOT)	Firefighter's Hazardous Materials Reference Book	First Aid Manual for Chemical Accidents	Sax's Dangerous Properties of Industrial Materials	Hazardous Chemicals Desk Reference	Rapid Guide to Hazardous Chemicals in the Workplace	Fire Protection Guide on Hazardous Materials (NFPA)	Firefighter's Handbook of Hazardous Materials	Pocket Guide to Chemical Hazards (NIOSH)	Chemical Hazard Response Information System (CHRIS)	Emergency Handling of Hazardous Materials in Surface Transportation (AAR)	Emergency Action Guides (EAG)	Chemical Data Notebook: A User's Manual	Condensed Chemical Dictionary
1-PENTANOL **Synonym: AMYL ALCOHOL (-n)**	1105/1987	26	78		242	69		-77	224		AAN 4	74			881
1-PENTANOL ACETATE	1104	26							224						
1-PENTENE	1108	26			2691			-77	224		PTE 994				882
1-PENTYL CHLORIDE **Synonym: AMYL CHLORIDE (-n)**	1107	26	79						31		AMY 61	75			74
1-PENTYNE					2697			-77							
1-PHENYLBUTANE	2709	27							229						897
1-PHENYLDECANE **Synonym: DECYLBENZENE-n**											DBZ 356				
1-PHENYLDODECANE **Synonym: DODECYLBENZENE**					2749						DDB 371				441
1-PHENYLTETRADECANE **Synonym: TETRADECYLBENZENE**											TDB1093				
1-PHENYLUNDECANE **Synonym: UNDECYLBENZENE (-n)**			859								UDB1163				
1-PROPANETHIOL **Synonym: PROPYL MERCAPTAN-n**	2704										PMN 967				972
1-PROPANOL **Synonym: PROPYL ALCOHOL (-n)**	1274	26	724	23	2915	1000	189		242	188	PAL 925	794			966
1-PROPENOL-3 **Synonym: ALLYL ALCOHOL**	1098	57	38	11	102	33	7	21 -13	22	34	ALA 39	35	A7.0		38
1-PROPYL ACETATE **Synonym: PROPYL ACETATE (-n)**	1276	26	723	23	2914	999	189	-81	243		PAT 929	794	P8.0		969
1-PROPYLAMINE **Synonym: PROPYLAMINE (-n)**	1277	68	726	23	2915	1000		149-	244		PRA 984	796			969

HAZARDOUS MATERIALS REFERENCE BOOKS INDEX

CHEMICAL OR MATERIAL NAME	UN/NA Number	Guide Number (DOT)	Firefighter's Hazardous Materials Reference Book	First Aid Manual for Chemical Accidents	Sax's Dangerous Properties of Industrial Materials	Hazardous Chemicals Desk Reference	Rapid Guide to Hazardous Chemicals in the Workplace	Fire Protection Guide on Hazardous Materials (NFPA)	Firefighter's Handbook of Hazardous Materials	Pocket Guide to Chemical Hazards (NIOSH)	Chemical Hazard Response Information System (CHRIS)	Emergency Handling of Hazardous Materials in Surface Transportation (AAR)	Emergency Action Guides (EAG)	Chemical Data Notebook: A User's Manual	Condensed Chemical Dictionary
1-TARTARIC ACID AMMONIUM SALT Synonym: **AMMONIUM TARTRATE**	9091/3077	31	72						30		ATR 106	69			70
1-TETRADECANOL Synonym: **TETRADECANOL**			809		3215	1100		-85			TTN1156				1128
1-TETRADECENE								-85			TTD1152				1128
1-TRIDECANOL Synonym: **TRIDECANOL**			842		3365			-88	274		TDN1096				1173
1-TRIDECENE											TDC1094				
1-UNDECANOL Synonym: **UNDECANOL**			857								UND1164				1199
1-UNDECENE											UDC1164				
10,11-DIMETHOXYSTRYCHNINE Synonym: **BRUCINE**	1570	53	138		566	186			53		BRU 169	147			174
10-AZAANTHRACENE Synonym: **ACRIDINE**	2713/1325	32	30		59	25	5		19		ACD 16	11			17
2,1,4-CHLORODINITROBENZENE	1577	56													
2,2'-AMINODIETHANOL Synonym: **DIETHANOLAMINE**			286		1184	399	73	-36			DEA 381		D1.1.2		388
2,2'-AZODI(2,4-DIMETHYL-4-METHOXYVALERONITRILE)	2955	70							37			103			
2,2'-AZODI(2,4-DIMETHYLVALERONITRILE)	2953	70							37						
2,2'-AZODI(2-METHYLBUTYRONITRILE)	3030	70													
2,2'-DIHYDROXYDIETHYL AMINE Synonym: **DIETHANOLAMINE**			286		1184	399	73	-36			DEA 381		D1.1.2		388
2,2'-DIHYDROXYDIETHYLAMINE Synonym: **DIETHANOLAMINE**			286		1184	399	73	-36			DEA 381		D1.1.2		388

CHEMICAL OR MATERIAL NAME	UN/NA Number	Guide Number (DOT)	Firefighter's Hazardous Materials Reference Book	First Aid Manual for Chemical Accidents	Sax's Dangerous Properties of Industrial Materials	Hazardous Chemicals Desk Reference	Rapid Guide to Hazardous Chemicals in the Workplace	Fire Protection Guide on Hazardous Materials (NFPA)	Firefighter's Handbook of Hazardous Materials	Pocket Guide to Chemical Hazards (NIOSH)	Chemical Hazard Response Information System (CHRIS)	Emergency Handling of Hazardous Materials in Surface Transportation (AAR)	Emergency Action Guides (EAG)	Chemical Data Notebook: A User's Manual	Condensed Chemical Dictionary
2,2'-IMINOBISETHANOL Synonym: DIETHANOLAMINE			286		1184	399	73	−36			DEA 381		D1.1.2		388
2,2'-IMINODIETHANOL Synonym: DIETHANOLAMINE			286		1184	399	73	−36			DEA 381		D1.1.2		388
2,2'-METHYLENE-BIS(3,4,6-TRICHLOROPHENOL)	2875	53							201						
2,2'-OXYBISPROPANE Synonym: DIISOPROPYL ETHER	1159	26						72 -	114			357	D2.1		407
2,2,3,3-TETRAMETHYL PENTANE								−86	264						
2,2,3,4-TETRAMETHYLPENTANE								−86	264						
2,2,3-TRIMETHYLBUTANE								−90							1181
2,2,3-TRIMETHYLPENTANE								−91							
2,2,4-TRIMETHYL-1,3-PENTANEDIOL					3414			−91							1183
2,2,4-TRIMETHYL-1,3-PENTANEDIOL ISOBUTYRATE								−91							
2,2,4-TRIMETHYLPENTANE	1262	27				1150	221	−91	279						1183
2,2,5-TRIMETHYLHEXANE								−90	279						1182
2,2-AMINOETHOXY ETHANOL	1760	60			170	51			26						
2,2-BIS(4-HYDROXYPHENYL) PROPANE Synonym: BISPHENOL A					511	169			48		BPA 154				154
2,2-BIS(HYDROXYMETHYL)-1,3-PROPANEDIOL Synonym: PENTAERYTHRITOL			667		2682	912	172				PET 949				880
2,2-BIS(p-HYDROXYPHENYL) PROPANE DIGLYCIDYL ETHER Synonym: BISPHENOL A DIGLYCIDYL ETHER					511	169					BDE 133				
2,2-BIS(p-METHOXYPHENYL)-1,1,1-TRICHLOROETHANE Synonym: METHOXYCHLOR	2761/3077	55	550		2224	763	133		191	140	MOC 780	615			753
2,2-DI(4,4-DI-tert-BUTYLPEROXYCYCLOHEXYL) PROPANE	2168/3106	48							92			311			

CHEMICAL OR MATERIAL NAME	UN/NA Number	Guide Number (DOT)	Firefighter's Hazardous Materials Reference Book	First Aid Manual for Chemical Accidents	Sax's Dangerous Properties of Industrial Materials	Hazardous Chemicals Desk Reference	Rapid Guide to Hazardous Chemicals in the Workplace	Fire Protection Guide on Hazardous Materials (NFPA)	Firefighter's Handbook of Hazardous Materials	Pocket Guide to Chemical Hazards (NIOSH)	Chemical Hazard Response Information System (CHRIS)	Emergency Handling of Hazardous Materials in Surface Transportation (AAR)	Emergency Action Guides (EAG)	Chemical Data Notebook: A User's Manual	Condensed Chemical Dictionary
2,2-DI(tert-BUTYLPEROXY) BUTANE	2111/3103	48			118				92			296			
2,2-DI(tert-BUTYLPEROXY) PROPANE	2884/3106	48							92			299			
2,2-DI(tert-BUTYLPEROXY) PROPANE	2883/3105	48							93			300			
2,2-DIAMINODIETHYLAMINE	2079/1760	29	293		1222	409	74	69 −38			DET 394	352			393
Synonym: DIETHYLENETRIAMINE															
2,2-DICHLORO ISOPROPYL ETHER	2490	59						−35	103		DCI 362				
2,2-DICHLOROACETYL CHLORIDE	1765	60			1125	376			101			328			377
Synonym: DICHLOROACETYL CHLORIDE															
2,2-DICHLOROBENZIDINE					1129				101						
2,2-DICHLOROETHYL ETHER	1916/2810	55	276		1145	384	69	66 −35	103		DEE 385	334			380
Synonym: DICHLOROETHYL ETHER															
2,2-DICHLOROETHYLAMINE					1143				103						
2,2-DICHLOROPROPANE					1163				104						
2,2-DICHLOROPROPANOIC ACID	1760	60	253				71				DLP 426				345
Synonym: DALAPON															
2,2-DICHLOROPROPIONIC ACID	1760	60	253		1166		71		105		DLP 426	338			383
Synonym: DALAPON															
2,2-DIHYDROPEROXY PROPANE	2178/3102	49										354			
2,2-DIHYDROXYDIPROPYLAMINE					1302			−41			DIP 419				406
Synonym: DIISOPROPANOLAMINE															
2,2-DIMETHYL-3-METHYLENENORBORANE	9011/3077	58	178		675	236			64		CPH 289	185			211
Synonym: CAMPHENE															
2,2-DIMETHYLBUTANE	1208	27	610		1357	451	78				NHX 830	676			414
Synonym: NEOHEXANE															
2,2-DIMETHYLBUTANE	2457	27				452		−42	116						414
2,2-DIMETHYLPROPANE	2044/1954	22						−44	117			307			420

CHEMICAL OR MATERIAL NAME	UN/NA Number	Guide Number (DOT)	Firefighter's Hazardous Materials Reference Book	First Aid Manual for Chemical Accidents	Sax's Dangerous Properties of Industrial Materials	Hazardous Chemicals Desk Reference	Rapid Guide to Hazardous Chemicals In the Workplace	Fire Protection Guide on Hazardous Materials (NFPA)	Firefighter's Handbook of Hazardous Materials	Pocket Guide to Chemical Hazards (NIOSH)	Chemical Hazard Response Information System (CHRIS)	Emergency Handling of Hazardous Materials In Surface Transportation (AAR)	Emergency Action Guides (EAG)	Chemical Data Notebook: A User's Manual	Condensed Chemical Dictionary
2,2-ETHYLENEDIOXYDIETHANOL Synonym: TRIETHYLENE GLYCOL			848	25	3370	1142		-89			TEG1101				1175
2,2-OXYBISETHANOL Synonym: DIETHYLENE GLYCOL			292	15	1219	409		-37	110		DEG 386				391
2,3,3-TRIMETHYL-1-BUTENE								-90	278						
2,3,3-TRIMETHYLPENTANE								-91	279						
2,3,4-TRIMETHYL-1-PENTENE								-91	278						
2,3,5-TRICHLOROPHENOL	2020	53							274						
2,3-BUTANEDIOL						190		-20	54						178
2,3-BUTANEDIONE	2346	26				190		-20	54						
2,3-DICHLORO-1,4-NAPHTHOQUINONE	2761	55			1152	387			100						382
2,3-DICHLORO-1-PROPANE Synonym: 2,3-DICHLOROPROPENE	2047	29			1165	394		-35	105		DPF 463				
2,3-DICHLOROBUTADIENE-1,3					1132			-34							
2,3-DICHLOROBUTANE								-34	102						
2,3-DICHLOROPROPENE	2047	29			1165	394		-35	105		DPF 463				
2,3-DICHLOROPROPIONALDEHYDE					1165	394			105						
2,3-DICHLOROPROPIONIC ACID	1760	60							105						
2,3-DICHLOROPROPYLENE Synonym: 2,3-DICHLOROPROPENE	2047	29			1165	394		-35	105		DPF 463				
2,3-DIHYDROPYRAN	2376	26							113						402
2,3-DIMETHYL-1-BUTENE								-42							
2,3-DIMETHYL-2-BUTENE								-42							
2,3-DIMETHYLBUTANE	2457	27			1357	452	78	-42	116			369			414
2,3-DIMETHYLCYCLOHEXYL AMINE	2264	60													

HAZARDOUS MATERIALS REFERENCE BOOKS INDEX

CHEMICAL OR MATERIAL NAME	UN/NA Number	Guide Number (DOT)	Firefighter's Hazardous Materials Reference Book	First Aid Manual for Chemical Accidents	Sax's Dangerous Properties of Industrial Materials	Hazardous Chemicals Desk Reference	Rapid Guide to Hazardous Chemicals in the Workplace	Fire Protection Guide on Hazardous Materials (NFPA)	Firefighter's Handbook of Hazardous Materials	Pocket Guide to Chemical Hazards (NIOSH)	Chemical Hazard Response Information System (CHRIS)	Emergency Handling of Hazardous Materials in Surface Transportation (AAR)	Emergency Action Guides (EAG)	Chemical Data Notebook: A User's Manual	Condensed Chemical Dictionary
2,3-DIMETHYLHEXANE					1380			-43	122						
2,3-DIMETHYLOCTANE								-43	123						
2,3-DIMETHYLPENTALDEHYDE								-43							419
2,3-DIMETHYLPENTANE					1404			-44	123						
2,3-DINITROPHENOL	0076				1435	479			125						
2,3-DINITROTOLUENE					1440				125						
2,3-DITHIABUTANE	2381	27													
2,3-EPOXY-1-PROPANOL									132						468
2,3-EPOXYPROPYL ACRYLATE					1535	511			133						
2,3-EPOXYPROPYL CHLORIDE	2023	30	336	16	1525	509	87	79 -46	132	102	EPC 552	401	E1.0	E 1	467
Synonym: EPICHLOROHYDRIN															
2,4,4-TRIMETHYL-1-PENTENE	2050/1993	26	298					-40	114		DBL 347	356	D2.0		405
Synonym: DIISOBUTYLENE															
2,4,4-TRIMETHYLPENTENE	2050/1993	26	298		3415			-40	114		DBL 347	356	D2.0		405
Synonym: DIISOBUTYLENE															
2,4,4-TRIMETHYLPENTENE-1	2050/1993	26	298					-40	114		DBL 347	356	D2.0		1184
Synonym: DIISOBUTYLENE															
2,4,4-TRIMETHYLPENTENE-2	2050/1993	26	298					-40	279		DBL 347	356	D2.0		1184
Synonym: DIISOBUTYLENE															
2,4,4-TRIMETHYLPENTYL-2-PEROXYPHENOXYACETATE	2961	52							279						
2,4,5-T	2765	55	8		3176	1088	206		259	202	TCA1080				1113
Synonym: 2,4,5-TRICHLOROPHENOXYACETIC ACID															
2,4,5-T AMINE	2765	55							259						
2,4,5-T ESTERS	2765	55							259		TES1105				
2,4,5-T SALT	2765	55							259						

CHEMICAL OR MATERIAL NAME	UN/NA Number	Guide Number (DOT)	Firefighter's Hazardous Materials Reference Book	First Aid Manual for Chemical Accidents	Sax's Dangerous Properties of Industrial Materials	Hazardous Chemicals Desk Reference	Rapid Guide to Hazardous Chemicals in the Workplace	Fire Protection Guide on Hazardous Materials (NFPA)	Firefighter's Handbook of Hazardous Materials	Pocket Guide to Chemical Hazards (NIOSH)	Chemical Hazard Response Information System (CHRIS)	Emergency Handling of Hazardous Materials in Surface Transportation (AAR)	Emergency Action Guides (EAG)	Chemical Data Notebook: A User's Manual	Condensed Chemical Dictionary
2,4,5-TP	2765	55			3358										
2,4,5-TP ESTER	2765/3077	55										921			
2,4,5-TRICHLOROPHENOL Synonym: TRICHLOROPHENOL	2020	53	839		3357	1139			274		TPH1139				1171
2,4,5-TRICHLOROPHENOXYACETIC ACID	2765	55	8						274		TCA1080				1172
2,4,5-TRICHLOROPHENOXYACETIC ACID AMINE	9188	31	7												
2,4,5-TRICHLOROPHENOXYPROPIONIC ACID	2765	55	9		3358				274						
2,4,5-TRIMETHYLANILINE					3401	1146	221								1181
2,4,6,2',4',6'-HEXANITRODIPHENYLAMINE									161						
2,4,6-TRICHLOROPHENOL	2020	53			3357	1140									1172
2,4,6-TRIMETHYL-1,3,5-TRIOXANE Synonym: PARALDEHYDE	1264	26	662		2665	906		132–76	278		PDH 944	726			1184
2,4,6-TRINITRO-m-CRESOL	0216				3423	1152			279						1185
2,4,6-TRINITROBENZOIC ACID	0215								280						1185
2,4,6-TRINITROTOLUENE	0209	33			3425	1153	223								1185
2,4,6-TRINITROTOLUENE	1356	33				1153	223			220					1185
2,4,8-TRIMETHYL-6-NONANOL								–90	278						
2,4-D Synonym: 2,4-DICHLOROPHENOXYACETIC ACID	2765/2810	55			1017	342	61		90	80	DCA 357	336			382
2,4-D ESTERS	2765/3077	55	10						90		DES 393	336			
2,4-DIAMINOANISOLE					1067	357	63								
2,4-DIAMYLPHENOL								–33							
2,4-DICHLOROBENZOYL PEROXIDE	2137	48							102						
2,4-DICHLOROBENZOYL PEROXIDE	2138	48	11						102						

CHEMICAL OR MATERIAL NAME	UN/NA Number	Guide Number (DOT)	Firefighter's Hazardous Materials Reference Book	First Aid Manual for Chemical Accidents	Sax's Dangerous Properties of Industrial Materials	Hazardous Chemicals Desk Reference	Rapid Guide to Hazardous Chemicals in the Workplace	Fire Protection Guide on Hazardous Materials (NFPA)	Firefighter's Handbook of Hazardous Materials	Pocket Guide to Chemical Hazards (NIOSH)	Chemical Hazard Response Information System (CHRIS)	Emergency Handling of Hazardous Materials in Surface Transportation (AAR)	Emergency Action Guides (EAG)	Chemical Data Notebook: A User's Manual	Condensed Chemical Dictionary
2,4-DICHLOROBENZOYL PEROXIDE	2139	48							102						
2,4-DICHLOROPHENOL	2020	53	12		2256	390		-35	104		DCP 366				382
2,4-DICHLOROPHENOXYACETIC ACID	2765/2810	55									DCA 357	336			382
2,4-DICHLOROPHENOXYACETIC ACID BUTOXYETHYL ESTER	2765/3077	55	10						90		DES 393	336			
Synonym: 2,4-D ESTERS															
2,4-DICHLOROPHENOXYACETIC ACID ESTER	9188	31													
2,4-DIMETHYL-3-ETHYLPENTANE								-43	118						
2,4-DIMETHYL-3-PENTANOL								-44	118						
2,4-DIMETHYLHEXANE					1381			-43	122						
2,4-DIMETHYLPENTANE					1404			-44	123						419
2,4-DINITRANILINE	1596	56			1427	475		76 -44	124		DNT 450				422
Synonym: 2,4-DINITROANILINE															
2,4-DINITRO-6-CYCLOHEXYL PHENOL	9026	53							124						
2,4-DINITROANILINE	1596	56			1427	475		76 -44	124		DNT 450				422
2,4-DINITROBENZENE SULFENYL CHLORIDE									124						
2,4-DINITROCHLOROBENZENE	1577	56							125						
2,4-DINITROPHENOL	0076				1435	479			125		DNP 449				
2,4-DINITROTOLUENE	1600	56			1440	480		-44			DTT 493				
2,4-DINITROTOLUENE	2038	56			1440	480			125		DTT 493				
2,4-DINITROTOLUOL	1600	56			1440	480		-44			DTT 493				
Synonym: 2,4-DINITROTOLUENE															
2,4-HEXADIENAL								-58	158						
2,4-PENTANEDIONE	2310/1993	26	28		37	17			224		ATA 97	730			881
Synonym: ACETYLACETONE															

CHEMICAL OR MATERIAL NAME	UN/NA Number	Guide Number (DOT)	Firefighter's Hazardous Materials Reference Book	First Aid Manual for Chemical Accidents	Sax's Dangerous Properties of Industrial Materials	Hazardous Chemicals Desk Reference	Rapid Guide to Hazardous Chemicals in the Workplace	Fire Protection Guide on Hazardous Materials (NFPA)	Firefighter's Handbook of Hazardous Materials	Pocket Guide to Chemical Hazards (NIOSH)	Chemical Hazard Response Information System (CHRIS)	Emergency Handling of Hazardous Materials in Surface Transportation (AAR)	Emergency Action Guides (EAG)	Chemical Data Notebook: A User's Manual	Condensed Chemical Dictionary
2,4-TOLUENE DIISOCYANATE Synonym: TOLUENE DIISOCYANATE	2078	57	828		3313		215		269		TDI1095	910	T4.1	T 5	1157
2,4-TOLYLENE DIISOCYANATE Synonym: TOLUENE-2,4-DIISOCYANATE	2078	57	828	25	3313	1128	215	170–87	269	214	TDI1095	910	T4.1		1157
2,4-XYLIDINE	1711	55				1186	233								
2,5,5-TRIMETHYLHEPTANE								−90	279						
2,5-DICHLOROANILINE					1126				101						377
2,5-DIETHOXY-4-MORPHOLINEBENZENEDIAZONIUM ZINC CHLORIDE	3036	72							107			302			
2,5-DIHYDROPEROXY-2,5-DIMETHYLHEXANE Synonym: DIMETHYLHEXANE DIHYDROPEROXIDE	2174	49	307		1381				122		DDW 380				
2,5-DIMETHYL-2,5-DI(2-ETHYLHEXANOYLPEROXY) HEXANE	2157/3115	52							118			366	H2.0		
2,5-DIMETHYL-2,5-DI(BENZOYLPEROXY) HEXANE	2959	49							118				H2.0		
2,5-DIMETHYL-2,5-DI(BENZOYLPEROXY) HEXANE	2172/3102	49							118			364	H2.0		
2,5-DIMETHYL-2,5-DI(BENZOYLPEROXY) HEXANE	2173/3106	49							118			364	H2.0		
2,5-DIMETHYL-2,5-DI(tert-BUTYLPEROXY) HEXANE	2156/3106	48			1366				118			365	H2.0		
2,5-DIMETHYL-2,5-DI(tert-BUTYLPEROXY) HEXANE	2155/3105	48			1366				118			365	H2.0		
2,5-DIMETHYL-2,5-DI(tert-BUTYLPEROXY) HEXYNE-3	2159	48													
2,5-DIMETHYL-2,5-DI(tert-BUTYLPEROXY) HEXYNE-3	2159	49													
2,5-DIMETHYL-2,5-DIHYDROPEROXY HEXANE	2174/3104	49										366			
2,5-DIMETHYLFURAN								−43	116						
2,5-DIMETHYLHEPTANE					1380				122						
2,5-DIMETHYLHEXANE-2,5-DIHYDROPEROXIDE Synonym: DIMETHYLHEXANE DIHYDROPEROXIDE	2174	49	307		1381				122		DDW 380				
2,5-DIMETHYLPIPERAZINE					1411				123						

CHEMICAL OR MATERIAL NAME	UN/NA Number	Guide Number (DOT)	Firefighter's Hazardous Materials Reference Book	First Aid Manual for Chemical Accidents	Sax's Dangerous Properties of Industrial Materials	Hazardous Chemicals Desk Reference	Rapid Guide to Hazardous Chemicals in the Workplace	Fire Protection Guide on Hazardous Materials (NFPA)	Firefighter's Handbook of Hazardous Materials	Pocket Guide to Chemical Hazards (NIOSH)	Chemical Hazard Response Information System (CHRIS)	Emergency Handling of Hazardous Materials in Surface Transportation (AAR)	Emergency Action Guides (EAG)	Chemical Data Notebook: A User's Manual	Condensed Chemical Dictionary
2,5-DIMETHYLPYRAZINE					1413	471		-44	123						
2,5-DINITROPHENOL	1599	57			1435						DNE 445				
2,5-DINITROTOLUENE					1441				125						
2,5-DNP Synonym: 2,5-DINITROPHENOL	1599	57			1435						DNE 445				
2,5-FURANDIONE Synonym: MALEIC ANHYDRIDE	2215	60	528	19	2155	736	130	106–64	184	138	MLA 769	592	M1.0		545
2,5-FURANEDIONE Synonym: MALEIC ANHYDRIDE	2215	60	528	19	2155	736	130	106–64	184	138	MLA 769	592	M1.0		724
2,5-HEXANEDIOL					1860			-58							
2,5-HEXANEDIONE					1861				160						
2,6,8-TRIMETHYL-4-NONANOL					3412			-90	278						
2,6-DI-tert-BUTYLPHENOL Synonym: DIBUTYLPHENOL											DBT 354				373
2,6-DICHLOROBENZONITRILE	2769	55			1130				101						378
2,6-DIETHYLANILINE	2432	57			1209						DMN 435				
2,6-DIMETHYL-4-HEPTANE Synonym: DIISOBUTYL KETONE	1157	26	296		1300	433	75	-40		92	DIK 414	355			405
2,6-DIMETHYL-4-HEPTANOL Synonym: DIISOBUTYLCARBINOL	1993/1987	27	297	15	1299	433		-40	113		DBC 343	355			405
2,6-DIMETHYL-4-HEPTANONE	1157	26							118						
2,6-DIMETHYLMORPHOLINE					1394			-43							418
2,6-DINITRO-n,n-DIPROPYL-4-TRIFLUOROMETHYLANILINE Synonym: TRIFLURALIN	1609								277		TFR1110				1178
2,6-DINITRO-o-CRESOL Synonym: DINITROCRESOL	1598	53	313	16	1430						DNC 444				

CHEMICAL OR MATERIAL NAME	UN/NA Number	Guide Number (DOT)	Firefighter's Hazardous Materials Reference Book	First Aid Manual for Chemical Accidents	Sax's Dangerous Properties of Industrial Materials	Hazardous Chemicals Desk Reference	Rapid Guide to Hazardous Chemicals in the Workplace	Fire Protection Guide on Hazardous Materials (NFPA)	Firefighter's Handbook of Hazardous Materials	Pocket Guide to Chemical Hazards (NIOSH)	Chemical Hazard Response Information System (CHRIS)	Emergency Handling of Hazardous Materials in Surface Transportation (AAR)	Emergency Action Guides (EAG)	Chemical Data Notebook: A User's Manual	Condensed Chemical Dictionary
2,6-DINITROPHENOL	1599	57			1435	479					DNH 446				
2,6-DINITROTOLUENE	2038	56			1441				125		DNL 447				
2,6-DNT Synonym: 2,6-DINITROTOLUENE	2038	56			1441				125		DNL 447				
2,6-XYLENOL Synonym: XYLENOL	2261	55	883		3523	1184			287		XYL1190	962			
2-(2-AMINOETHOXY) ETHANOL	1760	60	13		170	51			26			50			56
2-(2-BUTOXYETHOXY) ETHANOL Synonym: DIETHYLENE GLYCOL MONOBUTYL ETHER					1221			−38	111		DME 430				392
2-(2-BUTOXYETHOXY) ETHANOL ACETATE Synonym: DIETHYLENE GLYCOL MONOBUTYL ETHER ACE-TATE								−38			DEM 389				392
2-(2-ETHOXYETHOXY) ETHANOL Synonym: DIETHYLENE GLYCOL MONOETHYL ETHER	1993	27						−38	111		DGE 401	351			392
2-(2-METHOXYETHOXY) ETHANOL Synonym: DIETHYLENE GLYCOL MONOMETHYL ETHER					1221			−38	111		DGM 402				393
2-(DIETHYLAMINO) ETHYL ACRYLATE					1196			−36	110						
2-ACETYLAMINOFLUORENE										32					
2-AMINO-1-BUTANOL Synonym: 2-AMINOBUTAN-1-OL					155	49		−13	25						55
2-AMINO-1-METHYLBENZENE Synonym: TOLUIDINE (-ortho)	1708	55	830	25	3317	1129	216	171–87		216	TLI1124				1159
2-AMINO-2-METHYL-1-PROPANOL	1993	27						−14	25		APR 79	51			58
2-AMINO-4-METHYLPENTANE	2379	27							25						
2-AMINO-5-AZOTOLUENE					144	46									

HAZARDOUS MATERIALS REFERENCE BOOKS INDEX

CHEMICAL OR MATERIAL NAME	UN/NA Number	Guide Number (DOT)	Firefighter's Hazardous Materials Reference Book	First Aid Manual for Chemical Accidents	Sax's Dangerous Properties of Industrial Materials	Hazardous Chemicals Desk Reference	Rapid Guide to Hazardous Chemicals in the Workplace	Fire Protection Guide on Hazardous Materials (NFPA)	Firefighter's Handbook of Hazardous Materials	Pocket Guide to Chemical Hazards (NIOSH)	Chemical Hazard Response Information System (CHRIS)	Emergency Handling of Hazardous Materials in Surface Transportation (AAR)	Emergency Action Guides (EAG)	Chemical Data Notebook: A User's Manual	Condensed Chemical Dictionary
2-AMINO-5-CHLOROTOLUENE	2239	55			879	299	51				CTD 319				274
Synonym: 4-CHLORO-o-TOLUIDINE															
2-AMINO-5-DIETHYLAMINOPENTANE	2946	31										50			
2-AMINOBENZENETHIOL					146	47			25						
2-AMINOETHANOL	2491	60	594					118–	26		MEA 749		M5.0		56
Synonym: MONOETHANOLAMINE															
2-AMINOETHOXYETHANOL	1760	60	13		170	51			26			50			56
Synonym: 2-(2-AMINOETHOXY) ETHANOL															
2-AMINOETHYL ETHANOLAMINE	1760	60						–14							
2-AMINOPROPANE	1221	68	487	18	2014	688	124	102–63	26	134	IPP 678	563			60
Synonym: ISOPROPYLAMINE															
2-AMINOPYRIDINE	2671	55			209	56	10		27	36					61
2-AMINOTHIAZOLE					216				27						61
2-AMINOTOLUENE	1708/2810	55	830	25	3317	1129	216	171–87	27	216	TLI1124	911			1159
Synonym: TOLUIDINE (-ortho)															
2-ANILINOETHANOL					256	74		–16	32						
2-beta-BUTOXYETHOXYETHYL CHLORIDE								–21							
2-BIPHENYLAMINE					433										
2-BROMOBUTANE	2339/1993	27			545	181			52		BBT 123	143			171
2-BUTANETHIOL								–20	54						178
2-BUTANOL	1120	26			593	197	31	–21	54	52	BAS 120	150	B7.1		181
Synonym: BUTYL ALCOHOL (-sec)															
2-BUTANONE	1193	26	566	20	2319	793	143	–69	54		MEK 751	627	M3.2	M 5	178
Synonym: METHYL ETHYL KETONE															
2-BUTANONE	1232	26								48					178

CHEMICAL OR MATERIAL NAME	UN/NA Number	Guide Number (DOT)	Firefighter's Hazardous Materials Reference Book	First Aid Manual for Chemical Accidents	Sax's Dangerous Properties of Industrial Materials	Hazardous Chemicals Desk Reference	Rapid Guide to Hazardous Chemicals in the Workplace	Fire Protection Guide on Hazardous Materials (NFPA)	Firefighter's Handbook of Hazardous Materials	Pocket Guide to Chemical Hazards (NIOSH)	Chemical Hazard Response Information System (CHRIS)	Emergency Handling of Hazardous Materials in Surface Transportation (AAR)	Emergency Action Guides (EAG)	Chemical Data Notebook: A User's Manual	Condensed Chemical Dictionary
2-BUTANONE PEROXIDE	2550	51									BNP 150				
2-BUTEN-1-OL	2614	26			580	192			54						
2-BUTENAL Synonym: CROTONALDEHYDE	1143	28	233	14	964	325	56	57 −29	55	76	CTA 317	271	C7.1		178
2-BUTENAL-trans Synonym: CROTONALDEHYDE	1143	28	233	14	964	325	56	57 −29	55	76	CTA 317	271	C7.1		325
2-BUTENE-1,4-DIOL Synonym: 1,4-BUTENEDIOL	1987	26	6								BUD 186				
2-BUTENE-1,4-DIOL-cis Synonym: 1,4-BUTENEDIOL	1987	26	6								BUD 186				
2-BUTENE-cis					579	191		−20	55						178
2-BUTENE-trans	1012	22			579	192		−20	55						179
2-BUTOXYETHANOL Synonym: ETHYLENE GLYCOL MONOBUTYL ETHER	2369	26			585	193	29	−51	55	50	EGM 532	432			179
2-BUTOXYETHANOL ACETATE Synonym: ETHYLENE GLYCOL MONOBUTYL ETHER ACE-TATE	1172	26						−51	143		EMA 542	432			489
2-BUTOXYETHYL ACETATE Synonym: ETHYLENE GLYCOL MONOBUTYL ETHER ACE-TATE	1172	26			587	194	29	−51	143		EMA 542	432			489
2-BUTYLAMINOETHANOL					597	198			60						
2-BUTYLENE DICHLORIDE Synonym: DICHLOROBUTENE	760/2920	60			615				60		DCB 358	330			
2-BUTYNE-1,4-DIOL Synonym: 1,4-BUTYNEDIOL	2716	55			641	216			61		BTD 175	172			

CHEMICAL OR MATERIAL NAME	UN/NA Number	Guide Number (DOT)	Firefighter's Hazardous Materials Reference Book	First Aid Manual for Chemical Accidents	Sax's Dangerous Properties of Industrial Materials	Hazardous Chemicals Desk Reference	Rapid Guide to Hazardous Chemicals In the Workplace	Fire Protection Guide on Hazardous Materials (NFPA)	Firefighter's Handbook of Hazardous Materials	Pocket Guide to Chemical Hazards (NIOSH)	Chemical Hazard Response Information System (CHRIS)	Emergency Handling of Hazardous Materials In Surface Transportation (AAR)	Emergency Action Guides (EAG)	Chemical Data Notebook: A User's Manual	Condensed Chemical Dictionary
2-CHLORETHANOL Synonym: ETHYLENE CHLOROHYDRIN	1135	55	369	16	1599	538	93	83 –	142	108	ECH 513	429			486
2-CHLORO-1,3-BUTADIENE Synonym: CHLOROPRENE (-beta)	1991	30	211					–26	70	70	CRP 306	230	C4.4		272
2-CHLORO-1-NITROBENZENE Synonym: NITROCHLOROBENZENE-ortho	1578	55	633		841				212			690	N3.3		
2-CHLORO-1-PROPANOL	2611	57			869	297		–27	71						
2-CHLORO-2-NITROPROPANE					844	290		–27	71						
2-CHLORO-4-(HYDROXYMERCURI) PHENOL					819				72						267
2-CHLORO-5-NITROBENZOTRIFLUORIDE								–27	72						
2-CHLORO-6-(TRICHLOROMETHYL) PYRIDINE					884	301	52		72						274
2-CHLORO-6-NITROTOLUENE	2433	53							72						270
2-CHLOROALLYL CHLORIDE Synonym: 2,3-DICHLOROPROPENE	2047	29			1165	394		–35	105		DPF 463				
2-CHLOROBUTA-1,3-DIENE Synonym: CHLOROPRENE (-beta)	1991	30	211						77	70	CRP 306	230	C4.4		272
2-CHLOROBUTADIENE Synonym: CHLOROPRENE (-beta)	1991	30	211						77	70	CRP 306	230	C4.4		272
2-CHLOROBUTENE-2	1127	27						–26	74						
2-CHLOROETHANOL Synonym: ETHYLENE CHLOROHYDRIN	1135	55	369	16	1599	538	93	83 –26	75	108	ECH 513	429			266
2-CHLOROETHYL ACETATE	1181	27						–26	75						
2-CHLOROETHYL ALCOHOL Synonym: ETHYLENE CHLOROHYDRIN	1135	55	369	16	1599	538	93	83 –	75	108	ECH 513	429			486
2-CHLOROETHYL VINYL ETHER					814	282			75						266
2-CHLOROMETHYLTHIOPHENE					838	289			76						

CHEMICAL OR MATERIAL NAME	UN/NA Number	Guide Number (DOT)	Firefighter's Hazardous Materials Reference Book	First Aid Manual for Chemical Accidents	Sax's Dangerous Properties of Industrial Materials	Hazardous Chemicals Desk Reference	Rapid Guide to Hazardous Chemicals in the Workplace	Fire Protection Guide on Hazardous Materials (NFPA)	Firefighter's Handbook of Hazardous Materials	Pocket Guide to Chemical Hazards (NIOSH)	Chemical Hazard Response Information System (CHRIS)	Emergency Handling of Hazardous Materials in Surface Transportation (AAR)	Emergency Action Guides (EAG)	Chemical Data Notebook: A User's Manual	Condensed Chemical Dictionary
2-CHLOROPROPANE	2356/1993	26			868				77			231			
2-CHLOROPROPENE	2456	27			870				78			231			272
2-CHLOROPROPYLENE OXIDE Synonym: EPICHLOROHYDRIN	2023	30	336	16	1525	509	87	79 −46	78	102	EPC 552	401	E1.0	E 1	467
2-CHLOROPYRIDINE	2822/2810	54			873	298			78			232			272
2-CRESOL Synonym: CRESOL (-ortho)	2076	55	232	14	960	325	55	57 −29	84	74	CRO 305	270	C7.0		322
2-CYANOETHANOL Synonym: ETHYLENE CYANOHYDRIN			370					84 −50	142		ETC 562				486
2-CYCLOHEXYL-4,6-DINITROPHENOL	9026	53			997	336									338
2-CYCLOPENTENE-1-OL					1005				89						
2-DIAZO-1-NAPHTHOL-4-SULFOCHLORIDE	3042	71							96						
2-DIAZO-1-NAPHTHOL-5-SULFOCHLORIDE	3043	71							96						
2-DIETHYLAMINOETHANOL	2686	29			1192	402	73		110	92					
2-DIMETHYLETHANOLAMINE	2051	29							122						416
2-ETHOXY-3,4-DIHYDRO-2-PYRAN								−47	135						
2-ETHOXY-3,4-DIHYDRO-2H-PYRAN Synonym: ETHOXYDIHYDROPYRAN					1562	521			135		EHP 538				476
2-ETHOXYETHANOL Synonym: ETHYLENE GLYCOL MONOETHYL ETHER	1171	26	375		1563	521	89	−51	135	104	EGE 529	432	E2.5		476
2-ETHOXYETHANOL ACETATE Synonym: ETHYLENE GLYCOL MONOETHYL ETHER ACETATE	1172	26	376					−51	143		EGA 526	432	E2.6		489
2-ETHOXYETHYL ACETATE Synonym: ETHYLENE GLYCOL MONOETHYL ETHER ACETATE	1172	26	376		1563	522	89	−47	135	104	EGA 526	432	E2.6		477

HAZARDOUS MATERIALS REFERENCE BOOKS INDEX

CHEMICAL OR MATERIAL NAME	UN/NA Number	Guide Number (DOT)	Firefighter's Hazardous Materials Reference Book	First Aid Manual for Chemical Accidents	Sax's Dangerous Properties of Industrial Materials	Hazardous Chemicals Desk Reference	Rapid Guide to Hazardous Chemicals in the Workplace	Fire Protection Guide on Hazardous Materials (NFPA)	Firefighter's Handbook of Hazardous Materials	Pocket Guide to Chemical Hazards (NIOSH)	Chemical Hazard Response Information System (CHRIS)	Emergency Handling of Hazardous Materials in Surface Transportation (AAR)	Emergency Action Guides (EAG)	Chemical Data Notebook: A User's Manual	Condensed Chemical Dictionary
2-ETHYL BUTYL GLYCOL								–49	136						
2-ETHYL HEXALDEHYDE Synonym: ETHYLHEXALDEHYDE	1191	26	356						144		EHA 535	438			493
2-ETHYL HEXANOL	1993	27			1619				144		EHX 540	416			493
2-ETHYL HEXANOL Synonym: OCTANOL	1987	55	650		1619				144		EHX 540	416	O1.0		493
2-ETHYL HEXYL ALCOHOL Synonym: OCTANOL	1987	26	650		2620				145		OTA 909	706	O1.0		493
2-ETHYL-1-BUTANOL Synonym: ETHYLBUTANOL	2275	26	347		1582	530			140		EBT 509				
2-ETHYL-1-BUTENE					1582	530		–49	140						481
2-ETHYL-1-HEXANOL Synonym: 2-ETHYL HEXANOL	1987	26			1619				140		EHX 540	416			493
2-ETHYL-1-HEXANOL HYDROGEN PHOSPHATE Synonym: DI(2-ETHYLHEXYL) PHOSPHORIC ACID	1902	60	259						92		DEP 391				
2-ETHYL-1-HEXENE					1620	547			140						493
2-ETHYL-2-HEXENAL Synonym: 2-ETHYL-3-PROPYLACROLEIN	1191	26	14					–54	140		EPA 551				499
2-ETHYL-3-PROPYL ACRYLALDEHYDE Synonym: 2-ETHYL-3-PROPYLACROLEIN	1191	26	14					–54	140		EPA 551				499
2-ETHYL-3-PROPYLACROLEIN	1191	26	14					–54			EPA 551				499
2-ETHYL-3-PROPYLACRYLIC ACID								–54							499
2-ETHYLAMINOETHANOL					1575	527			136						
2-ETHYLANILINE	2273	55			1577	528			141			409			
2-ETHYLBUTANOL	2275	26			1582	530			142			411			481
2-ETHYLBUTYL ACETATE	1177	26						–49	142						481

CHEMICAL OR MATERIAL NAME	UN/NA Number	Guide Number (DOT)	Firefighter's Hazardous Materials Reference Book	First Aid Manual for Chemical Accidents	Sax's Dangerous Properties of Industrial Materials	Hazardous Chemicals Desk Reference	Rapid Guide to Hazardous Chemicals in the Workplace	Fire Protection Guide on Hazardous Materials (NFPA)	Firefighter's Handbook of Hazardous Materials	Pocket Guide to Chemical Hazards (NIOSH)	Chemical Hazard Response Information System (CHRIS)	Emergency Handling of Hazardous Materials in Surface Transportation (AAR)	Emergency Action Guides (EAG)	Chemical Data Notebook: A User's Manual	Condensed Chemical Dictionary
2-ETHYLBUTYL ACRYLATE					1583	530		−49	142						
2-ETHYLBUTYL ALCOHOL Synonym: ETHYLBUTANOL	2275/1987	26	347					−49	142		EBT 509	411			481
2-ETHYLBUTYRIC ACID								−49	142						482
2-ETHYLBUTYRIC ALDEHYDE	1178	26							142						
2-ETHYLCAPROALDEHYDE Synonym: ETHYLHEXALDEHYDE	1191	26	356						142		EHA 535	438			
2-ETHYLHEXALDEHYDE	1191	26							144						493
2-ETHYLHEXANAL Synonym: ETHYLHEXALDEHYDE	1191	26	356					−52	144		EHA 535	438			493
2-ETHYLHEXANOIC ACID								−52	144		EHO 537				
2-ETHYLHEXANOL	1987	26			1619			−52	144		EHX 540	416			
2-ETHYLHEXYL 2-PROPENOATE Synonym: 2-ETHYLHEXYL ACRYLATE								88 −52	145		EAI 502				493
2-ETHYLHEXYL ACETATE								−52	145						493
2-ETHYLHEXYL ACRYLATE	1993	27						88 −52	145		EAI 502		E2.7.5		493
2-ETHYLHEXYL CHLORIDE								−52	145						494
2-ETHYLHEXYL CYCLOHEXYLAMINE-n					1621			−52	145						
2-ETHYLHEXYL-1-CHLORIDE					1621				145						
2-ETHYLHEXYLAMINE	2276/1760	29			1620	548		−52	145		EHM 536	417			494
2-ETHYLISOHEXANOL					1629			−53	145						495
2-ETHYNYL-2-BUTANOL					1670				146						
2-FORMYLPHENOL Synonym: SALICYLALDEHYDE								−83	248		SAL1009				1019

HAZARDOUS MATERIALS REFERENCE BOOKS INDEX

CHEMICAL OR MATERIAL NAME	UN/NA Number	Guide Number (DOT)	Firefighter's Hazardous Materials Reference Book	First Aid Manual for Chemical Accidents	Sax's Dangerous Properties of Industrial Materials	Hazardous Chemicals Desk Reference	Rapid Guide to Hazardous Chemicals in the Workplace	Fire Protection Guide on Hazardous Materials (NFPA)	Firefighter's Handbook of Hazardous Materials	Pocket Guide to Chemical Hazards (NIOSH)	Chemical Hazard Response Information System (CHRIS)	Emergency Handling of Hazardous Materials in Surface Transportation (AAR)	Emergency Action Guides (EAG)	Chemical Data Notebook: A User's Manual	Condensed Chemical Dictionary
2-FURALDEHYDE Synonym: FURFURAL	1199	29	408	17	1765	597	103	92 –55	152	118	FFA 583	467			545
2-FURANCARBINOL Synonym: FURFURYL ALCOHOL	2874/1987	55	409		1766	597	104	–55	152	118	FAL 575	467	F4.0		546
2-FURANCARBONAL	1199	29							152						
2-FURANMETHANOL Synonym: FURFURYL ALCOHOL	2874/1987	55	409		1766	597	104	–55	152	118	FAL 575	467	F4.0		546
2-FURANMETHYLAMINE	2526	28							152						
2-FURYLCARBINOL Synonym: FURFURYL ALCOHOL	2874/1987	55	409		1766	597	104	–55	153	118	FAL 575	467	F4.0		546
2-HEPTANOL					1831	616		–57	157						592
2-HEPTANONE Synonym: AMYL METHYL KETONE (-n)	1110	26	81						157		AMK 53	76			592
2-HEPTENE	2278	27			1832				157						592
2-HEXANONE Synonym: METHYL n-BUTYL KETONE	1993/1224	27	576		1863	628	110		161	122	MBK 729	622			599
2-HYDRAZINOETHANOL					1889	637			163						
2-HYDROXY-1,2,3-PROPANETRICARBOXYLIC ACID Synonym: CITRIC ACID					916	313					CIT 254				286
2-HYDROXY-2-METHYLPROPANENITRILE Synonym: ACETONE CYANOHYDRIN	1541	55	21	11				15 –11	17		ACY 27	4	A3.1		9
2-HYDROXY-2-METHYLPROPIONITRILE Synonym: ACETONE CYANOHYDRIN	1541	55	21	11				15 –11	165		ACY 27	4	A3.1		9
2-HYDROXY-m-XYLENE Synonym: XYLENOL	2261	55	883		3523	1184			288		XYL1190	962			

CHEMICAL OR MATERIAL NAME	UN/NA Number	Guide Number (DOT)	Firefighter's Hazardous Materials Reference Book	First Aid Manual for Chemical Accidents	Sax's Dangerous Properties of Industrial Materials	Hazardous Chemicals Desk Reference	Rapid Guide to Hazardous Chemicals in the Workplace	Fire Protection Guide on Hazardous Materials (NFPA)	Firefighter's Handbook of Hazardous Materials	Pocket Guide to Chemical Hazards (NIOSH)	Chemical Hazard Response Information System (CHRIS)	Emergency Handling of Hazardous Materials in Surface Transportation (AAR)	Emergency Action Guides (EAG)	Chemical Data Notebook: A User's Manual	Condensed Chemical Dictionary
2-HYDROXYBUTANE **Synonym: BUTYL ALCOHOL (-sec)**	1120	26			593	197	31	−21	56	52	BAS 120		B7.1		181
2-HYDROXYETHYL 2-PROPENOATE **Synonym: 2-HYDROXYETHYL ACRYLATE**								−60	166		HAI 617				621
2-HYDROXYETHYL ACRYLATE								−60			HAI 617				621
2-HYDROXYETHYL CYCLOHEXYLAMINE-n								−60							
2-HYDROXYETHYLAMINE **Synonym: MONOETHANOLAMINE**	2491	60	594					118−	166		MEA 749		M5.0		621
2-HYDROXYMETHYL FURAN **Synonym: FURFURYL ALCOHOL**	2874	55	409		1766	597	104	−55	152	118	FAL 575		F4.0		546
2-HYDROXYNITROBENZENE **Synonym: 2-NITROPHENOL**	1663	55			2548				215		NTP 859				
2-HYDROXYPROPANOIC ACID **Synonym: LACTIC ACID**	1760	60	496		2083	703					LTA 711	569			679
2-HYDROXYPROPYLAMINE **Synonym: MONOISOPROPANOLAMINE**	1760	60	595						167		MPA 782	658			625
2-HYDROXYTRIETHYLAMINE **Synonym: DIETHYLETHANOLAMINE**	2686	29							111		DAE 334				393
2-IMIDAZOLIDINETHIONE					1972	661	117								
2-ISOPROPOXY PROPANE **Synonym: ISOPROPYL ETHER**	1159	26	483		2050	691	125	−63	174	134	IPE 673		D2.1		661
2-ISOPROPOXYPROPANE **Synonym: DIISOPROPYL ETHER**	1159	26			2050		125	72 −	174			357	D2.1		407
2-KETOHEPTANE **Synonym: AMYL METHYL KETONE (-n)**	1110	26	81						31		AMK 53	76			

HAZARDOUS MATERIALS REFERENCE BOOKS INDEX

CHEMICAL OR MATERIAL NAME	UN/NA Number	Guide Number (DOT)	Firefighter's Hazardous Materials Reference Book	First Aid Manual for Chemical Accidents	Sax's Dangerous Properties of Industrial Materials	Hazardous Chemicals Desk Reference	Rapid Guide to Hazardous Chemicals in the Workplace	Fire Protection Guide on Hazardous Materials (NFPA)	Firefighter's Handbook of Hazardous Materials	Pocket Guide to Chemical Hazards (NIOSH)	Chemical Hazard Response Information System (CHRIS)	Emergency Handling of Hazardous Materials in Surface Transportation (AAR)	Emergency Action Guides (EAG)	Chemical Data Notebook: A User's Manual	Condensed Chemical Dictionary
2-KETOHEXAMETHYLENIMINE				13	680	238	39				CLS 259		C1.5		
Synonym: CAPROLACTAM															
2-MERCAPTOETHANOIC ACID	1940	60							185						
2-MERCAPTOETHANOL	2966	53			2184	746		-64	186						739
2-METHOXY-4-METHYL PHENOL	2022	55							190						753
2-METHOXYETHANOL	1188	26		17				-51	191		EME 544	433			489
Synonym: ETHYLENE GLYCOL MONOMETHYL ETHER															
2-METHYHL PENTENE-1	2288	27			2372			-71	198		MPE 785				777
Synonym: 2-METHYL-1-PENTENE															
2-METHYL AMINOETHANOL					2255	771			191						
2-METHYL BUTADIENE-1,3	1218	27							192						
2-METHYL CYCLOHEXANOL	2617	26							193						
2-METHYL NITROBENZENE	1664	55		21	2591	882	166	130-75	216		NIE 832	699			832
Synonym: NITROTOLUENE (-ortho)															
2-METHYL OCTANE					2367										
2-METHYL PROPANOIC ACID	2529	29							196						
2-METHYL PROPENAL	2045/1993	26	467	18	2022	679		-61	172		BAD 115	548	11.1		653
Synonym: ISOBUTYRALDEHYDE															
2-METHYL PYRIDINE	2313	27			2391	815		-72			MPR 789				
2-METHYL-1,3-BUTADIENE	1218	27	479		2037	685		102-62	174		IPR 679	554	13.0	I I	759
Synonym: ISOPRENE															
2-METHYL-1,3-PENTADIENE	2461	26			2370			-70	198						
2-METHYL-1,3-PENTANEDIOL								-71	198						776
2-METHYL-1-BUTANOL	1105	26						-67	198						760
2-METHYL-1-BUTENE	2459	26			2276	777		-67	198			622			

CHEMICAL OR MATERIAL NAME	UN/NA Number	Guide Number (DOT)	Firefighter's Hazardous Materials Reference Book	First Aid Manual for Chemical Accidents	Sax's Dangerous Properties of Industrial Materials	Hazardous Chemicals Desk Reference	Rapid Guide to Hazardous Chemicals in the Workplace	Fire Protection Guide on Hazardous Materials (NFPA)	Firefighter's Handbook of Hazardous Materials	Pocket Guide to Chemical Hazards (NIOSH)	Chemical Hazard Response Information System (CHRIS)	Emergency Handling of Hazardous Materials in Surface Transportation (AAR)	Emergency Action Guides (EAG)	Chemical Data Notebook: A User's Manual	Condensed Chemical Dictionary
2-METHYL-1-BUTENE-3-ONE Synonym: **METHYL ISOPROPENYL KETONE**	1246	26	573		2336	799		−70	195		MPK 787	631			772
2-METHYL-1-PENTENE	2288	27						−71	198		MPE 785				777
2-METHYL-1-PROPANAL Synonym: **ISOBUTYRALDEHYDE**	2045/1993	26	467	18	2022	679		−61	172		BAD 115	548	11.1		653
2-METHYL-1-PROPANOL Synonym: **ISOBUTYL ALCOHOL**	1212/1120	26			2016	676	121	−61	198	132	IAL 658	545			652
2-METHYL-1-PROPYL ACETATE Synonym: **ISOBUTYL ACETATE**	1213	26	463	18	2015	675	121	−61	171	130	IBA 661	544	11.0		652
2-METHYL-2,4-PENTANEDIOL Synonym: **HEXYLENE GLYCOL**	2030	59	438		1873		111	−59	198		HXG 651				777
2-METHYL-2-BUTANOL	1105	26						−67	198						760
2-METHYL-2-BUTENE	2460	26						−67							760
2-METHYL-2-ETHYL-1,3-DIOXOLANE								−68	198						768
2-METHYL-2-PROPANETHIOL	2347	27			2388	814		−72	199						
2-METHYL-2-PROPANOL Synonym: **BUTYL ALCOHOL (-tert)**	1120	26			594	197	31	−21	199	52	BAT 121	150	B7.1		779
2-METHYL-3-ETHYLPENTANE								−69	199						
2-METHYL-4,6-DINITROPHENOL	1598	53							199						
2-METHYL-4-ETHYLHEXANE								−69	199						
2-METHYL-4-PENTANONE Synonym: **METHYL ISOBUTYL KETONE**	1245/1993	26	571	20				−70	195		MIK 765	630	M3.3		771
2-METHYL-5-ETHYLPIPERIDINE								−69	199						
2-METHYL-5-ETHYLPYRIDINE	2300	60						−69	199			628			768
2-METHYL-5-PYRIDINE	2300	60						116–							

HAZARDOUS MATERIALS REFERENCE BOOKS INDEX

CHEMICAL OR MATERIAL NAME	UN/NA Number	Guide Number (DOT)	Firefighter's Hazardous Materials Reference Book	First Aid Manual for Chemical Accidents	Sax's Dangerous Properties of Industrial Materials	Hazardous Chemicals Desk Reference	Rapid Guide to Hazardous Chemicals in the Workplace	Fire Protection Guide on Hazardous Materials (NFPA)	Firefighter's Handbook of Hazardous Materials	Pocket Guide to Chemical Hazards (NIOSH)	Chemical Hazard Response Information System (CHRIS)	Emergency Handling of Hazardous Materials in Surface Transportation (AAR)	Emergency Action Guides (EAG)	Chemical Data Notebook: A User's Manual	Condensed Chemical Dictionary
2-METHYLACETONITRILE Synonym: ACETONE CYANOHYDRIN	1541	55	21	11				15 –11	17		ACY 27	4	A3.1		9
2-METHYLANILINE Synonym: TOLUIDINE (-ortho)	1708	55	830	25	3317	1129	216	171–87	201	216	TLI1124				1159
2-METHYLAZACYCLOPROPANE Synonym: PROPYLENEIMINE	1921	30	731		2921	1002	190		244	188	PII 959	799	P8.0.5		971
2-METHYLAZIRIDINE Synonym: PROPYLENEIMINE	1921	30	731		2921	1002	190		192	188	PII 959	799	P8.0.5		971
2-METHYLBUTADIENE Synonym: ISOPRENE	1218	27	479		2037	685		102–62			IPR 679	554	13.0	11	658
2-METHYLBUTANE Synonym: ISOPENTANE	1265	27	476					–62	201		IPT 680	552	12.0		759
2-METHYLBUTYRALDEHYDE								–67	201						761
2-METHYLCYCLOHEXANONE	2297	26			2295	786	140		193						
2-METHYLDECANE								–68	193						
2-METHYLETHYLENEIMINE Synonym: PROPYLENEIMINE	1921	30	731		2921	1002	190		244	188	PII 959	799	P8.0.5		971
2-METHYLFURAN	2301	26			2324	796		–69	194						769
2-METHYLHEPTANE	1262	27							202						770
2-METHYLHEXANE	2287	27			2326			–69	202						770
2-METHYLLACTONITRILE	1541	55				800	146		203						
2-METHYLNAPHTHALENE					2350				203						
2-METHYLPENTANE Synonym: ISOHEXANE	1208	27	473		2028	681	122	–62			IHA 667				776
2-METHYLPENTANE	2462	26						–71	203						776

CHEMICAL OR MATERIAL NAME	UN/NA Number	Guide Number (DOT)	Firefighter's Hazardous Materials Reference Book	First Aid Manual for Chemical Accidents	Sax's Dangerous Properties of Industrial Materials	Hazardous Chemicals Desk Reference	Rapid Guide to Hazardous Chemicals in the Workplace	Fire Protection Guide on Hazardous Materials (NFPA)	Firefighter's Handbook of Hazardous Materials	Pocket Guide to Chemical Hazards (NIOSH)	Chemical Hazard Response Information System (CHRIS)	Emergency Handling of Hazardous Materials in Surface Transportation (AAR)	Emergency Action Guides (EAG)	Chemical Data Notebook: A User's Manual	Condensed Chemical Dictionary
2-METHYLPHENOL **Synonym: CRESOL (-ortho)**	2076	55	232		960	325	55	57 -29	84	74	CRO 305	270	C7.0		322
2-METHYLPROPANAL	2045	26							204						
2-METHYLPROPANE **Synonym: ISOBUTANE**	1369/1075	22	462		2388	814		-61	204		IBT 665	543	10.5		779
2-METHYLPROPANENITRILE **Synonym: ISOBUTYRONITRILE**	2284/1993	28	469	18	2023	680	122	-61	172		IBN 663	549			779
2-METHYLPROPENAL	2396	28						-72	197						
2-METHYLPROPENE **Synonym: ISOBUTYLENE**	1055/1075	22						-72	197		IBL 662	547	11.0.5		779
2-METHYLPROPENOIC ACID **Synonym: METHYL METHACRYLATE**	1247	26	575		2342	803	147	115-70		154	MMM 775	633	M3.6		773
2-METHYLPROPIONALDEHYDE **Synonym: ISOBUTYRALDEHYDE**	2045/1993	26	467	18	2022	679		-61			BAD 115	548	11.1		653
2-METHYLPROPIONITRILE **Synonym: ISOBUTYRONITRILE**	2284/1993	28	469	18	2023	680	122	-61	172		IBN 663	549			654
2-METHYLPROPYL ACETATE **Synonym: ISOBUTYL ACETATE**	1213	26	463	18	2015	675	121	-61	171	130	IBA 661	544	11.0		652
2-METHYLPYRAZINE					2391	815		-72	204						
2-METHYLPYRIDINE	2313	27			2391	815			204		MPR 789				
2-METHYLVALERALDEHYDE	2367	27						-73	197						
2-n-DIBUTYLAMINOETHANOL	2873	55				371	65		99						
2-n-DIETHYLAMINOETHANOL **Synonym: DIETHYLETHANOLAMINE**	2686	29			1192						DAE 334				393
2-NITRO-2-NONENE					2546				211						

HAZARDOUS MATERIALS REFERENCE BOOKS INDEX

CHEMICAL OR MATERIAL NAME	UN/NA Number	Guide Number (DOT)	Firefighter's Hazardous Materials Reference Book	First Aid Manual for Chemical Accidents	Sax's Dangerous Properties of Industrial Materials	Hazardous Chemicals Desk Reference	Rapid Guide to Hazardous Chemicals in the Workplace	Fire Protection Guide on Hazardous Materials (NFPA)	Firefighter's Handbook of Hazardous Materials	Pocket Guide to Chemical Hazards (NIOSH)	Chemical Hazard Response Information System (CHRIS)	Emergency Handling of Hazardous Materials in Surface Transportation (AAR)	Emergency Action Guides (EAG)	Chemical Data Notebook: A User's Manual	Condensed Chemical Dictionary
2-NITROANILINE Synonym: NITROANILINE (-ortho)	1661	55		21	2514	859			211		NTA 851	687	N3.0		824
2-NITRONAPHTHALENE					2545	871	162		214						
2-NITROPHENOL	1663	55			2548				215		NTP 859				
2-NITROPROPANE	2608	26			2554	872	163	−74	215	166	NPP 848				831
2-NITROTOLUENE Synonym: NITROTOLUENE (-ortho)	1664	55		21	2591	882	166	130–75	216		NIE 832	699			832
2-NP Synonym: 2-NITROPROPANE	2608	26			2554	872	163	−74	215	166	NPP 848				831
2-OXETANONE Synonym: PROPIOLACTONE-beta	1993	27	719		2907	997	187		220	186	PLT 964				
2-OXOHEXAMETHYLENIMINE Synonym: CAPROLACTAM				13	680	238	36				CLS 259		C1.5		861
2-PENTANOL ACETATE	1104	26							224						
2-PENTANONE	1249	26			2690	914	173		224	174					881
2-PENTENE					2691				224						882
2-PENTENE-3-CARBOXYLIC ACID	1862	26							224						
2-PENTYLACETATE Synonym: AMYL ACETATE (-sec)	1104	26		12	241	68	13	−14	30	38	AAS 5	73			73
2-PHENYLBUTANE	2709	27							229						897
2-PHENYLPROPANE	1918	28							231						
2-PICOLINE Synonym: 2-METHYLPYRIDINE	2313	27			2391	815			234		MPR 789				
2-PROPANETHIOL Synonym: ISOPROPYL MERCAPTAN	2402	27				994			175		IPM 677	562			661

CHEMICAL OR MATERIAL NAME	UN/NA Number	Guide Number (DOT)	Firefighter's Hazardous Materials Reference Book	First Aid Manual for Chemical Accidents	Sax's Dangerous Properties of Industrial Materials	Hazardous Chemicals Desk Reference	Rapid Guide to Hazardous Chemicals in the Workplace	Fire Protection Guide on Hazardous Materials (NFPA)	Firefighter's Handbook of Hazardous Materials	Pocket Guide to Chemical Hazards (NIOSH)	Chemical Hazard Response Information System (CHRIS)	Emergency Handling of Hazardous Materials in Surface Transportation (AAR)	Emergency Action Guides (EAG)	Chemical Data Notebook: A User's Manual	Condensed Chemical Dictionary
2-PROPANOL Synonym: ISOPROPANOL	1219	26	480			687			242			555	14.0		966
2-PROPANOL Synonym: ISOPROPYL ALCOHOL	1219/1993	26	482	18	2040		124	−63	242	132	IPA 670	560		I 2	966
2-PROPANOL	1274	26													966
2-PROPANONE Synonym: ACETONE	1090	26	20	11	22	10	2	−11	242	30	ACT 26	4	A3.0	A 3	967
2-PROPEN-1-AMINE Synonym: ALLYLAMINE	2334	28			103	34	8	21 −13	22			40	A7.1		38
2-PROPEN-1-OL Synonym: ALLYL ALCOHOL	1098	57	38	11	102	33	7	21 −13	242	34	ALA 39	35	A7.0		38
2-PROPEN-1-ONE Synonym: ACROLEIN	1092	30	31	11	63	25	5	17 −12	242	32	ARL 85	12	A4.0	A 5	18
2-PROPENAL Synonym: ACROLEIN	1092	30	31	11	63	25	5	17 −12	242	32	ARL 85	12	A4.0	A 5	967
2-PROPENAMINE Synonym: ALLYLAMINE	2334	28			103	34	8	21 −13	22			40	A7.1		38
2-PROPENENITRILE Synonym: ACRYLONITRILE	1093	30	34	11	68	27	6	19 −12	20	34	ACN 23	13	A5.1	A 7	19
2-PROPENOATE Synonym: BUTYL ACRYLATE	2348/1993	26	146		592	196	30	41 −21	56		BTC 174	150	B7.0		181
2-PROPENOIC ACID Synonym: ACRYLIC ACID	2218	29	33		65	26	6	18 −12	242		ACR 25	13	A5.0	A 6	181
2-PROPENOIC ACID BUTYL ESTER Synonym: BUTYL ACRYLATE (-n)	2348/1993	26	146		592	196	30	41 −21	56		BTC 174	150	B7.0		181
2-PROPENOIC ACID ETHYL ESTER Synonym: ETHYL ACRYLATE	1917	26	343	16	1571	524	90	80 −48	136	106	EAC 500	407	E2.1		478

HAZARDOUS MATERIALS REFERENCE BOOKS INDEX

CHEMICAL OR MATERIAL NAME	UN/NA Number	Guide Number (DOT)	Firefighter's Hazardous Materials Reference Book	First Aid Manual for Chemical Accidents	Sax's Dangerous Properties of Industrial Materials	Hazardous Chemicals Desk Reference	Rapid Guide to Hazardous Chemicals in the Workplace	Fire Protection Guide on Hazardous Materials (NFPA)	Firefighter's Handbook of Hazardous Materials	Pocket Guide to Chemical Hazards (NIOSH)	Chemical Hazard Response Information System (CHRIS)	Emergency Handling of Hazardous Materials in Surface Transportation (AAR)	Emergency Action Guides (EAG)	Chemical Data Notebook: A User's Manual	Condensed Chemical Dictionary
2-PROPENYL CHLORIDE Synonym: ALLYL CHLORIDE	1100	57	40	11	106	35	8	22 −13	242	36	ALC 40	36	A7.2		39
2-PROPENYL ETHANOATE	2333	28							242						
2-PROPENYLAMINE Synonym: ALLYLAMINE	2334	28			103	34	8	21 −13	242			40	A7.1		967
2-PROPYL ACETATE Synonym: ISOPROPYL ACETATE	1220	26	481	18	2039	686	124	−63	243	132	IAC 656	560			659
2-PROPYN-1-OL	1986	28							244						
2-sec-BUTYL-4,6-DINITROPHENOL					614	204			60						
2-tert-BUTYL-p-CRESOL					609	203			60						
2-THIAPROPANE Synonym: DIMETHYL SULFIDE	1164	27	303					75 −44	117		DSL 478	363			420
2-[(2-AMINOETHYL) AMINO] ETHANOL Synonym: AMINOETHYLETHANOLAMINE	1760	60	47		172				26		AEA 30	50			57
3',3'-DICHLOROBENZIDINE					1129	378	67		101	86					378
3,3'-DIMETHOXYBENZIDINE DIHYDROCHLORIDE					1314	438	76								
3,3'-IMINO-BIS-PROPYLAMINE	2269/1760	60							168			538			
3,3,6,6,9,9-HEXAMETHYL-1,2,4,5-TETRAOXOCYCLONONANE	2167/3105	48							160			514			
3,3,6,6,9,9-HEXAMETHYL-1,2,4,5-TETRAOXOCYCLONONANE	2166/3106	48							160			514			
3,3,6,6,9,9-HEXAMETHYL-1,2,4,5-TETRAOXOCYCLONONANE	2165/3102	49							160			514			
3,3-DIETHYLPENTANE					1233			−39	111						
3,3-DIMETHOXYBENZIDINE DIHYDROCHLORIDE					1314										
3,3-DIMETHYL-2-METHYLENE NORCAMPHANE Synonym: CAMPHENE	9011/3077	58	178		675	236			64		CPH 289	185			211
3,3-DIMETHYLHEPTANE								−43	122						

CHEMICAL OR MATERIAL NAME	UN/NA Number	Guide Number (DOT)	Firefighter's Hazardous Materials Reference Book	First Aid Manual for Chemical Accidents	Sax's Dangerous Properties of Industrial Materials	Hazardous Chemicals Desk Reference	Rapid Guide to Hazardous Chemicals in the Workplace	Fire Protection Guide on Hazardous Materials (NFPA)	Firefighter's Handbook of Hazardous Materials	Pocket Guide to Chemical Hazards (NIOSH)	Chemical Hazard Response Information System (CHRIS)	Emergency Handling of Hazardous Materials in Surface Transportation (AAR)	Emergency Action Guides (EAG)	Chemical Data Notebook: A User's Manual	Condensed Chemical Dictionary
3,4,4-TRIMETHYL-2-PENTENE								–91							
3,4,5-TRIHYDROXYBENZOIC ACID *Synonym:* **GALLIC ACID**			410		1776	600					GLA 608				1179
3,4-DICHLOROANILINE	1590	55			1126			–34	101						377
3,4-DICHLOROBUTENE-1					1133			–34	102						
3,4-DIMETHYLOCTANE								–43	123						
3,4-DINITROTOLUENE	2038	56			1441				125		DNU 451				
3,4-DNT *Synonym:* **3,4-DINITROTOLUENE**	2038	56			1441				125		DNU 451				
3,5,5-TRIMETHYL-2-CYCLOHEXANE-1-ONE *Synonym:* **ISOPHORONE**	1993	27	477		2034	684	123	101–62	174	132	IPH 674	553			658
3,5,5-TRIMETHYLHEXANOL								–90	279						
3,5-DIMETHYLHEPTANE					1380				122						
3,5-DINITRO-o-CRESOL *Synonym:* **DINITROCRESOL**	1598	53	313	16	1431	477	82				DNC 444				
3,5-DINITRO-p-CRESOL	1598	53			1431	477									
3,6,9-TRIOXAUNDECAN-1,11-DIOL *Synonym:* **TETRAETHYLENE GLYCOL**	1993	27	815		3218	1100		–85			TTG1155	892			1129
3,7,7-TRIMETHYLBICYCLO-0,1,4-HEPT-3-ENE *Synonym:* **CARENE**											CAR 211				
3-(1-METHYL-2-PYRROLIDYL) PYRIDINE *Synonym:* **NICOTINE**	1654	55	622		2500	853	157	–74	209	160	NIC 831	682			820
3-(2-ETHYLBUTOXY) PROPIONIC ACID					1582			–49							
3-(2-HYDROXYETHOXY)-4-PYRROLIDIN-1-YLBENZENE-DIAZONIUM	3035	70							166			535			
3-DIETHYLAMINO PROPYLAMINE	2684/1760	29			1207			–36	110			303			389

CHEMICAL OR MATERIAL NAME	UN/NA Number	Guide Number (DOT)	Firefighter's Hazardous Materials Reference Book	First Aid Manual for Chemical Accidents	Sax's Dangerous Properties of Industrial Materials	Hazardous Chemicals Desk Reference	Rapid Guide to Hazardous Chemicals in the Workplace	Fire Protection Guide on Hazardous Materials (NFPA)	Firefighter's Handbook of Hazardous Materials	Pocket Guide to Chemical Hazards (NIOSH)	Chemical Hazard Response Information System (CHRIS)	Emergency Handling of Hazardous Materials in Surface Transportation (AAR)	Emergency Action Guides (EAG)	Chemical Data Notebook: A User's Manual	Condensed Chemical Dictionary
3-(DIMETHYLAMINO) PROPIONITRILE					1341			-41	121						
3-AMINO-9-ETHYLCARBAZOLE					171	51									
3-AMINOPROPANOL					204	54		-14	26						
3-AMINOPROPENE Synonym: ALLYLAMINE	2334	28			103	34	8	21 -13	22			40	A7.1		38
3-AMINOPROPIONITRILE					205	55			26						60
3-AMINOPROPYL CYCLOHEXYLAMINE-n								-14	27						
3-AMINOPROPYLENE Synonym: ALLYLAMINE	2334	28			130	34	8	21 -13	22			40	A7.1		38
3-BROMOPROPENE Synonym: ALLYL BROMIDE	1099	57	39		105	34	8	22 -13	53		ABR 10	35			173
3-BROMOPROPYLENE Synonym: ALLYL BROMIDE	1099	57	39		105	34	8	22 -13	53		ABR 10	35			39
3-BUTEN-2-ONE Synonym: METHYL VINYL KETONE	1251	28	582		580	192		117-73	54		MVK 814	639			179
3-BUTENE-2-ONE	1251	28							55						
3-CARENE Synonym: CARENE					711						CAR 211				225
3-CHLORO-1,2-EPOXYPROPANE Synonym: EPICHLOROHYDRIN	2023	30	336	16	1525	509	87	79 -46	132	102	EPC 552	401	E1.0	E 1	467
3-CHLORO-1,2-PROPYLENE OXIDE Synonym: EPICHLOROHYDRIN	2023	30	336	16	1525	509	87	79 -46	132	102	EPC 552	401	E1.0	E 1	467
3-CHLORO-1-PROPENE Synonym: ALLYL CHLORIDE	1100	57	40	11	106	35	8	22 -13	22	36	ALC 40	36	A7.2		39
3-CHLORO-4-DIETHYLAMINOBENZENEDIAZONIUM ZINC CHLORIDE	3033	72							72			221			

CHEMICAL OR MATERIAL NAME	UN/NA Number	Guide Number (DOT)	Firefighter's Hazardous Materials Reference Book	First Aid Manual for Chemical Accidents	Sax's Dangerous Properties of Industrial Materials	Hazardous Chemicals Desk Reference	Rapid Guide to Hazardous Chemicals in the Workplace	Fire Protection Guide on Hazardous Materials (NFPA)	Firefighter's Handbook of Hazardous Materials	Pocket Guide to Chemical Hazards (NIOSH)	Chemical Hazard Response Information System (CHRIS)	Emergency Handling of Hazardous Materials in Surface Transportation (AAR)	Emergency Action Guides (EAG)	Chemical Data Notebook: A User's Manual	Condensed Chemical Dictionary
3-CHLOROPEROXYBENZOIC ACID	2755/3102	49			864				77			226			
3-CHLOROPROPANOL	2849	53							78						
3-CHLOROPROPANOL-1	2849	53							78			231			
3-CHLOROPROPENE	1100	57	40	11	106	35	8	22 −13	78	36	ALC 40	36	A7.2		272
Synonym: ALLYL CHLORIDE															
3-CHLOROPROPENE-1	1100	57	40	11	106	35	8	22 −13	78	36	ALC 40	36	A7.2		39
Synonym: ALLYL CHLORIDE															
3-CHLOROPROPIONITRILE	2511	60			871			−27	78						272
3-CHLOROPROPYLENE	1100	57	40	11	106	35	8	22 −13	22	36	ALC 40	36	A7.2		39
Synonym: ALLYL CHLORIDE															
3-CHLOROPROPYLENE-1,2-OXIDE	2023	30													
3-CRESOL	2076	55	232		960	324	55	57 −29	85	74	CRL 303	270	C7.0		322
Synonym: CRESOL (-meta)															
3-DIMETHYLAMINOPROPIONITRILE					1341				121						
3-DIMETHYLAMINOPROPYLAMINE									121						413
3-ETHOXYPROPANAL								−47	135						
3-ETHOXYPROPIONALDEHYDE								−47	135						
3-ETHOXYPROPIONIC ACID								−47	135						
3-ETHYLOCTANE								−53	145						
3-HEPTANOL					1832			−57	157						592
3-HEPTANONE									157						592
3-HEPTENE (mixed cis and trans)	2278	27			1833			−57							592
3-HEXANONE					1863			−58	161						599
3-HEXENOL-cis								−58							
3-HYDROXYBUTANAL	2839	55							166						

CHEMICAL OR MATERIAL NAME	UN/NA Number	Guide Number (DOT)	Firefighter's Hazardous Materials Reference Book	First Aid Manual for Chemical Accidents	Sax's Dangerous Properties of Industrial Materials	Hazardous Chemicals Desk Reference	Rapid Guide to Hazardous Chemicals in the Workplace	Fire Protection Guide on Hazardous Materials (NFPA)	Firefighter's Handbook of Hazardous Materials	Pocket Guide to Chemical Hazards (NIOSH)	Chemical Hazard Response Information System (CHRIS)	Emergency Handling of Hazardous Materials in Surface Transportation (AAR)	Emergency Action Guides (EAG)	Chemical Data Notebook: A User's Manual	Condensed Chemical Dictionary
3-HYDROXYNITROBENZENE Synonym: 3-NITROPHENOL	1663	55			2548	871			215		NIP 833				
3-HYDROXYPROPANENITRILE Synonym: ETHYLENE CYANOHYDRIN			370					84 -50	142		ETC 562				486
3-HYDROXYTOLUENE Synonym: CRESOL (-meta)	2076	55	232		960	324	55	57 -29	85	74	CRL 303	270	C7.0		322
3-ISOPROPOXYPROPIONITRILE								-62	174						
3-METHOXYBUTANOL								-65	190						
3-METHOXYBUTYRALDEHYDE					2223			-65	191						
3-METHOXYPROPIONITRILE					2242			-65							
3-METHOXYPROPYLAMINE					2242	766		-65	191						754
3-METHYL CHOLANTHRENE					2289	783									
3-METHYL OCTANE					2367				203						
3-METHYL-1,3-BUTADIENE Synonym: ISOPRENE	1218	27	479			685		102-62	174		IPR 679	554	13.0	1 1	658
3-METHYL-1-BUTANOL Synonym: ISOAMYL ALCOHOL	1105	26	461		2011	673	120	-60	198	130	IAA 655				760
3-METHYL-1-BUTENE	2561	26			2276	778		-67	198			622			760
3-METHYL-1-PENTYNOL								-71							
3-METHYL-2-BUTANETHIOL								-67	198						
3-METHYL-4-ETHYLHEXANE								-69	199						
3-METHYLBUTANAL Synonym: ISOVALERALDEHYDE	1989	26	488						201		IVA 683				664
3-METHYLBUTYL NITRITE Synonym: AMYL NITRITE-iso	1113	26	83								ANI 63				75

CHEMICAL OR MATERIAL NAME	UN/NA Number	Guide Number (DOT)	Firefighter's Hazardous Materials Reference Book	First Aid Manual for Chemical Accidents	Sax's Dangerous Properties of Industrial Materials	Hazardous Chemicals Desk Reference	Rapid Guide to Hazardous Chemicals in the Workplace	Fire Protection Guide on Hazardous Materials (NFPA)	Firefighter's Handbook of Hazardous Materials	Pocket Guide to Chemical Hazards (NIOSH)	Chemical Hazard Response Information System (CHRIS)	Emergency Handling of Hazardous Materials in Surface Transportation (AAR)	Emergency Action Guides (EAG)	Chemical Data Notebook: A User's Manual	Condensed Chemical Dictionary
3-METHYLBUTYRALDEHYDE Synonym: ISOVALERALDEHYDE	1989	26	488		2279				176		IVA 683				761
3-METHYLCYCLOHEXANOL	2617	26						-68	193						
3-METHYLHEXANE								-69	202						770
3-METHYLNITROBENZENE Synonym: NITROTOLUENE (-meta)	1664	55	643	21	2591	882	165	130–74	195	168	NTR 860				832
3-METHYLPENTANE	1208	27			2371	810		-71							776
3-METHYLPROPANAL Synonym: ISOBUTYRALDEHYDE	2045/1993	26	467	18	2022	679		-61	172		BAD 115	548	11.1		653
3-NITRO-3-HEXENE					2541	869			211						
3-NITROPHENOL	1663	55			2548	871			215		NIP 833				
3-NITROPYRENE					2555	873	163								
3-NITROTOLUENE Synonym: NITROTOLUENE (-meta)	1664	55	643	21	2591	882	165	130–74	216	168	NTR 860				832
3-NITROTOLUOL Synonym: NITROTOLUENE (-meta)	1664	55	643	21	2591	882	165	130–74	216	168	NTR 860				832
3-OXA-1,5-PENTANEDIOL Synonym: DIETHYLENE GLYCOL			292	15	1219	409		-37	110		DEG 386				391
3-PENTANOL	2706							-77	224						881
3-PENTANONE	1156	26							224						881
3-PENTENENITRILE	1993	27										731			
300 DEGREE OIL Synonym: OILS MISCELLANEOUS MINERAL SEAL	1270	27									OMS 889				
3a,4,7,7a-TETRAHYDRO-4,7-METHANOINDENE Synonym: DICYCLOPENTADIENE	2048/1993	26	283		1178	398	72	-36	106		DPT 471	343			385
4,4′-DICHLORO-alpha-TRICHLOROMETHYLBENZHYDROL											DTM 489				

HAZARDOUS MATERIALS REFERENCE BOOKS INDEX

CHEMICAL OR MATERIAL NAME	UN/NA Number	Guide Number (DOT)	Firefighter's Hazardous Materials Reference Book	First Aid Manual for Chemical Accidents	Sax's Dangerous Properties of Industrial Materials	Hazardous Chemicals Desk Reference	Rapid Guide to Hazardous Chemicals in the Workplace	Fire Protection Guide on Hazardous Materials (NFPA)	Firefighter's Handbook of Hazardous Materials	Pocket Guide to Chemical Hazards (NIOSH)	Chemical Hazard Response Information System (CHRIS)	Emergency Handling of Hazardous Materials in Surface Transportation (AAR)	Emergency Action Guides (EAG)	Chemical Data Notebook: A User's Manual	Condensed Chemical Dictionary
4,4'-METHYLENE BIS(2-METHYLANILINE)					2310	789	141								
4,4'-METHYLENE BIS(N,N'-DIMETHYLANILINE)					2309	789	141								
4,4'-METHYLENEDIANILINE	2651	53			2312	791	143		202						767
4,4'-OXYDIANILINE					2650	901	169								
4,4'-THIODIANILINE					3278	1114	213								
4,4-DIAMINODIPHENYL METHANE	2651	53							95						
4,4-DIMETHYLHEPTANE					1380				122						
4,4-ISOPROPYLIDENDIPHENOL Synonym: BISPHENOL A					511	169			48		BPA 154				154
4,4-ISOPROPYLIDENEDIPHENOLEPICHLOROHYDRIN RESIN Synonym: BISPHENOL A DIGLYCIDYL ETHER					511	169					BDE 133				
4,6-DINITRO-o-CRESOL Synonym: DINITROCRESOL	1598	53	313	16	1430				124		DNC 444	308			
4,6-DINITRO-p-CRESOL	1598	53													
4,7,7-TRIMETHYL-3-NORCARENE Synonym: CARENE											CAR 211				
4-(BENZYL(METHYL)AMINO)-3-ETHOXY-BENZENEDIAMMONIUM ZINC	3038	70													
4-(METHYLSULFONYL)-2,6-DINITRO Synonym: NITRALIN	1609		625								NTL 856				822
4-AMINO-2-NITROANILINE					195	53	10								
4-AMINODIPHENYL					168	50				36					
4-BENZYL(ETHYL)AMINO-3-ETHOXYBENZENDIAZONIUM ZINC CHLORI	3037	70							43			120			
4-CHLORO-1-METHYLBENZENE Synonym: CHLOROTOLUENE-para	2238	27	213						70		CRN 304				274

CHEMICAL OR MATERIAL NAME	UN/NA Number	Guide Number (DOT)	Firefighter's Hazardous Materials Reference Book	First Aid Manual for Chemical Accidents	Sax's Dangerous Properties of Industrial Materials	Hazardous Chemicals Desk Reference	Rapid Guide to Hazardous Chemicals in the Workplace	Fire Protection Guide on Hazardous Materials (NFPA)	Firefighter's Handbook of Hazardous Materials	Pocket Guide to Chemical Hazards (NIOSH)	Chemical Hazard Response Information System (CHRIS)	Emergency Handling of Hazardous Materials in Surface Transportation (AAR)	Emergency Action Guides (EAG)	Chemical Data Notebook: A User's Manual	Condensed Chemical Dictionary
4-CHLORO-1-NITROBENZENE Synonym: NITROCHLOROBENZENE (-para)	1578	55	634	21	2525	863	159	−74	212	164		691	N3.4		826
4-CHLORO-2-METHYLANILINE Synonym: 4-CHLORO-o-TOLUIDINE	2239	55			879	299	51				CTD 319				274
4-CHLORO-3-HYDROXYTOLUENE	2669	55							71						
4-CHLORO-o-TOLUIDINE	2239	55			879	299	51				CTD 319				274
4-CHLORO-o-TOLUIDINE HYDROCHLORIDE	1579	53			880				72			215			274
4-CHLOROANILINE Synonym: CHLOROANILINE-para	2018/2811	53	202		767	272			73		CAP 210	218			262
4-CHLOROBUTYRONITRILE											CBN 219				
4-CHLORONITROBENZENE Synonym: NITROCHLOROBENZENE (-para)	1578	55	634	21	2525	863	159	−74	212	164		691	N3.4		826
4-CHLOROPHENOL Synonym: CHLOROPHENOL-para	2020	53	209		848	291		−27	77		CPN 291	228			270
4-CHLOROPHENYLAMINE Synonym: CHLOROANILINE-para	2018/2811	53	202		767				73		CAP 210	218			262
4-CHLOROTOLUENE Synonym: CHLOROTOLUENE-para	2238/1993	27	213								CRN 304	234			274
4-CRESOL Synonym: CRESOL-para	2076	55			961		55				CSO 313	270			322
4-DIMETHYLAMINOAZOBENZENE					1327	444	77			94					
4-DIPROPYLAMINOBENZENEDIAZONIUM ZINC CHLORIDE	3034	72													
4-ETHYLANILINE	2273	55			1577	528			141						
4-ETHYLMORPHOLINE Synonym: ETHYLMORPHOLINE-n					1642			−53	138						497
4-ETHYLOCTANE								−53	145						

HAZARDOUS MATERIALS REFERENCE BOOKS INDEX

CHEMICAL OR MATERIAL NAME	UN/NA Number	Guide Number (DOT)	Firefighter's Hazardous Materials Reference Book	First Aid Manual for Chemical Accidents	Sax's Dangerous Properties of Industrial Materials	Hazardous Chemicals Desk Reference	Rapid Guide to Hazardous Chemicals in the Workplace	Fire Protection Guide on Hazardous Materials (NFPA)	Firefighter's Handbook of Hazardous Materials	Pocket Guide to Chemical Hazards (NIOSH)	Chemical Hazard Response Information System (CHRIS)	Emergency Handling of Hazardous Materials in Surface Transportation (AAR)	Emergency Action Guides (EAG)	Chemical Data Notebook: A User's Manual	Condensed Chemical Dictionary
4-HEPTANONE	2710	26						-57	157						592
4-HYDROXY-4-METHYL-2-PENTANONE Synonym: DIACETONE ALCOHOL	1148	26	260	15	1059	354	62	-32	165	82	DAA 332	316			623
4-HYDROXYNITROBENZENE Synonym: 4-NITROPHENOL	1663	55			2548	871			215		NPH 846				
4-HYDROXYTOLUENE Synonym: CRESOL (-para)	2076	55	232		961	325	55	57 -29	85	74	CSO 313	270	C7.0		322
4-ISOPROPYLHEPTANE								-63	176						
4-METHOXYPHENOL					2236	765	134		190						753
4-METHYL OCTANE					2367				203						
4-METHYL-1,3-PENTADIENE	2461	26			2370			-70	198						
4-METHYL-1-PENTENE					2372			-71	198		MTN 810				777
4-METHYL-2-OXOPENTANE Synonym: METHYL ISOBUTYL KETONE	1245/1993	26	571	20				-70	195		MIK 765	630	M3.3		771
4-METHYL-2-PENTANOL Synonym: METHYL AMYL ALCOHOL	2053	26	557						191		MAA 718				777
4-METHYL-2-PENTANOL Synonym: METHYL ISOBUTYL CARBINOL	2053/1993	26	570		2334	799	145	-70	195	152	MIC 763	630			777
4-METHYL-2-PENTANOL ACETATE Synonym: METHYL AMYL ACETATE	1233	26	556					-71	199		MAC 719	642			757
4-METHYL-2-PENTANONE Synonym: METHYL ISOBUTYL KETONE	1245/1993	26	571	20				-70	199		MIK 765	630	M3.3		777
4-METHYL-2-PENTENE					2372			-71	199						777
4-METHYL-2-PENTYL ACETATE Synonym: METHYL AMYL ACETATE	1233	26	556								MAC 719	642			757

CHEMICAL OR MATERIAL NAME	UN/NA Number	Guide Number (DOT)	Firefighter's Hazardous Materials Reference Book	First Aid Manual for Chemical Accidents	Sax's Dangerous Properties of Industrial Materials	Hazardous Chemicals Desk Reference	Rapid Guide to Hazardous Chemicals in the Workplace	Fire Protection Guide on Hazardous Materials (NFPA)	Firefighter's Handbook of Hazardous Materials	Pocket Guide to Chemical Hazards (NIOSH)	Chemical Hazard Response Information System (CHRIS)	Emergency Handling of Hazardous Materials in Surface Transportation (AAR)	Emergency Action Guides (EAG)	Chemical Data Notebook: A User's Manual	Condensed Chemical Dictionary
4-METHYL-3-PENTENE-2-ONE **Synonym: MESITYL OXIDE**	1229	26	545		2205	757	132	107–64	188	140	MSO 801	606			746
4-METHYL-4-PENTENE **Synonym: 2-METHYL-1-PENTENE**	2288	27			2372			−71	198		MPE 785				777
4-METHYLCYCLOHEXANOL	2617	26						−68	193						
4-METHYLCYCLOHEXENE					2296			−68							
4-METHYLMORPHOLINE **Synonym: METHYL MORPHOLINE**	2335/2924	29			2349			−70	195			650			773
4-METHYLNITROBENZENE **Synonym: NITROTOLUENE (-para)**	1664	55		21	2591	882	166	130–75	216		NTT 861				833
4-METHYLPENTYL-2-ACETATE	1233	26							203						
4-METHYLPYRIDINE	2313	27			2392	816			204		MPF 786				
4-NITROANILINE	1661	55			2514				211		NAL 818				
4-NITROBIPHENYL					2524	862	159			162					
4-NITROPHENOL	1663	55			2548	871			215		NPH 846				
4-NITROSOMORPHOLINE					2577	879	165		215						
4-NITROTOLUOL **Synonym: NITROTOLUENE (-para)**	1664	55		21	2591	882	166	130–75	216		NTT 861				833
4-tert-BUTYLPHENOL					632	212	35								
5-CHLORO-2-AMINOTOLUENE **Synonym: 4-CHLORO-o-TOLUIDINE**	2239	55			879	299	51				CTD 319				274
5-DIISOPROPYLACETONE **Synonym: DIISOBUTYL KETONE**	1157	26	296		1300	433	75	−40	114	92	DIK 414	355			405
5-ETHYL-2-METHYL PYRIDINE **Synonym: METHYLETHYLPYRIDINE**	2300	60	567						194		MEP 753	648			

HAZARDOUS MATERIALS REFERENCE BOOKS INDEX

CHEMICAL OR MATERIAL NAME	UN/NA Number	Guide Number (DOT)	Firefighter's Hazardous Materials Reference Book	First Aid Manual for Chemical Accidents	Sax's Dangerous Properties of Industrial Materials	Hazardous Chemicals Desk Reference	Rapid Guide to Hazardous Chemicals in the Workplace	Fire Protection Guide on Hazardous Materials (NFPA)	Firefighter's Handbook of Hazardous Materials	Pocket Guide to Chemical Hazards (NIOSH)	Chemical Hazard Response Information System (CHRIS)	Emergency Handling of Hazardous Materials in Surface Transportation (AAR)	Emergency Action Guides (EAG)	Chemical Data Notebook: A User's Manual	Condensed Chemical Dictionary
5-ETHYL-2-PICOLINE Synonym: METHYLETHYLPYRIDINE	2300	60	567						194		MEP 753	648			499
5-ETHYLIDENEBICYCLO (2,2,1) HEPT-2-ENE Synonym: ETHYLIDENE NORBORNENE					1627	549	97				ENB 545				
5-METHYL-2-HEXANONE	2302/1993	26							198			629			770
5-METHYL-3-HEPTANONE	2271	26							199	150					
5-NITROACENAPHTHENE					2511	858	158								
5-NONANONE Synonym: DI-n-BUTYL KETONE	1224	26			2597						DBK 346				
6-HYDROXY-3-(2H)-PYRIDAZINONE Synonym: MALEIC HYDRAZIDE			529								MLH 770				724
6-METHYL-1-HEPTANAL Synonym: ISOOCTALDEHYDE											IOC 669				
6-METHYL-1-HEPTANOL Synonym: ISOOCTYL ALCOHOL	1987	26	475		2032	683	123	–62			IOA 668	552			657
8-HEPTADECYLENECARBOXYLIC ACID-cis Synonym: OLEIC ACID			654		2630	896		–76			OLA 883				852
9-OCTADECENOIC ACID-cis Synonym: OLEIC ACID			654			896		–76			OLA 883				852
9-PENTABORON NONAHYDRIDE Synonym: PENTABORANE	1380	75	664	22		910	171	132–76	222	172	PTB 991	728			879
AATREX HERBICIDE Synonym: ATRAZINE	1609				310	98	18		36		ATZ 110				105
ABATE									16						1
ABIETIC ACID					1				16						1
ABRIN					1	3			16						

CHEMICAL OR MATERIAL NAME	UN/NA Number	Guide Number (DOT)	Firefighter's Hazardous Materials Reference Book	First Aid Manual for Chemical Accidents	Sax's Dangerous Properties of Industrial Materials	Hazardous Chemicals Desk Reference	Rapid Guide to Hazardous Chemicals in the Workplace	Fire Protection Guide on Hazardous Materials (NFPA)	Firefighter's Handbook of Hazardous Materials	Pocket Guide to Chemical Hazards (NIOSH)	Chemical Hazard Response Information System (CHRIS)	Emergency Handling of Hazardous Materials in Surface Transportation (AAR)	Emergency Action Guides (EAG)	Chemical Data Notebook: A User's Manual	Condensed Chemical Dictionary
ABSORBENT OIL Synonym: OILS MISCELLANEOUS ABSORPTION	1270	27									OAS 869				
ACCUMULATORS	1956	12							16			1			
ACCUMULATORS Synonym: AMYL ACETATE (-n)	1956	26	77	12	241	68	12	−14	31		AML 53	1			73
ACETAL	1088	26	15		5	3		−11	16		AEL 31	1			5
ACETALDEHYDE	1089	26	16	11	5	3	1	13 −11	16	30	AAD 2	1	A1.0	A 1	5
ACETALDEHYDE AMMONIA	1841	31			6	4			16			2			5
ACETALDEHYDE OXIME	2332	26			7	4			16			2			
ACETALDEHYDE-meta Synonym: METALDEHYDE	1332	32						−65	16			609			747
ACETALDEHYDE-para Synonym: PARALDEHYDE	1264	26	662	11	2665	873		132–76	16		PDH 944	726			871
ACETALDOL Synonym: ALDOL	2839/1993	55			7	5		−11	16			20			32
ACETAMIDE					7	5	1		16						5
ACETAMIDINE HYDROCHLORIDE					8				16						6
ACETANILIDE Synonym: ANTIFEBRIN					13	5		−11	16						6
ACETENE Synonym: ETHYLENE	1038	22		16	1597			83 −50	142		ETL 568	436	E2.4	E 4	485
ACETENE Synonym: ETHYLENE	1962	22	368	16	1597	536		83 −50	142		ETL 568	436	E2.4	E 4	485
ACETIC ACID Synonym: AMMONIUM ACETATE	9079/3077	31	49	11	221				16		AAT 6	55			7

CHEMICAL OR MATERIAL NAME	UN/NA Number	Guide Number (DOT)	Firefighter's Hazardous Materials Reference Book	First Aid Manual for Chemical Accidents	Sax's Dangerous Properties of Industrial Materials	Hazardous Chemicals Desk Reference	Rapid Guide to Hazardous Chemicals in the Workplace	Fire Protection Guide on Hazardous Materials (NFPA)	Firefighter's Handbook of Hazardous Materials	Pocket Guide to Chemical Hazards (NIOSH)	Chemical Hazard Response Information System (CHRIS)	Emergency Handling of Hazardous Materials in Surface Transportation (AAR)	Emergency Action Guides (EAG)	Chemical Data Notebook: A User's Manual	Condensed Chemical Dictionary
ACETIC ACID 2-ETHOXYETHYL ESTER Synonym: ETHYLENE GLYCOL MONOETHYL ETHER ACETATE	1172	26	376					−51	143		EGA 526	432	E2.6		489
ACETIC ACID 2-METHYLPROPYL ESTER Synonym: ISOBUTYL ACETATE	1213	26	463	18	2015	675	121	−61	171	130	IBA 661	544	11.0		652
ACETIC ACID ANHYDRIDE Synonym: ACETIC ANHYDRIDE	1715	39	19	11	18	8	2	14 −11	16	30	ACA 13	3	A2.1		7
ACETIC ACID BUTYL ESTER Synonym: BUTYL ACETATE (-n)	1123	26	142	13	591	195	29	−21	56	50	BCN 128	149	B6.0		170
ACETIC ACID CHROMIUM SALT Synonym: CHROMIC ACETATE	9101	31	214		898	305			80		CRT 308	237			279
ACETIC ACID CUPRIC SALT Synonym: COPPER ACETATE	9106/3077	31	221		944	319					COP 279	257			310
ACETIC ACID DIMETHYLAMIDE Synonym: DIMETHYLACETAMIDE (n,n)					1321	441	77	−41	16		DAC 333				411
ACETIC ACID ETHENYL ESTER Synonym: VINYL ACETATE	1301	26	871	26	3492	1172	227	179–92	285		VAM1172	948	V1.0		1215
ACETIC ACID ETHYL ESTER Synonym: ETHYL ACETATE	1173	26	340	16	1570	523	89	−47	135	104	ETA 560	406	E2.0		477
ACETIC ACID GLACIAL Synonym: ACETIC ACID	2789	29	18	11	15	6	1	14 −11	16	30	AAC 1	3	A2.0	A 2	7
ACETIC ACID ISOBUTYL ESTER Synonym: ISOBUTYL ACETATE	1213	26	463	18	2015	675	121	−61	171	130	IBA 661	544	11.0		652
ACETIC ACID ISOPROPYL ESTER Synonym: ISOPROPYL ACETATE	1220	26	481	18	2039	686	124	−63	16	132	IAC 656	560			659
ACETIC ACID METHYL ESTER Synonym: METHYL ACETATE	1231	26	551	20	2247	767	134	−65	16	142	MTT 813	617			754

CHEMICAL OR MATERIAL NAME	UN/NA Number	Guide Number (DOT)	Firefighter's Hazardous Materials Reference Book	First Aid Manual for Chemical Accidents	Sax's Dangerous Properties of Industrial Materials	Hazardous Chemicals Desk Reference	Rapid Guide to Hazardous Chemicals in the Workplace	Fire Protection Guide on Hazardous Materials (NFPA)	Firefighter's Handbook of Hazardous Materials	Pocket Guide to Chemical Hazards (NIOSH)	Chemical Hazard Response Information System (CHRIS)	Emergency Handling of Hazardous Materials in Surface Transportation (AAR)	Emergency Action Guides (EAG)	Chemical Data Notebook: A User's Manual	Condensed Chemical Dictionary
ACETIC ACID n-BUTYL ESTER Synonym: BUTYL ACETATE (-n)	1123	26	142	13	591	195	29	–21	56	50	BCN 128	149	B6.0		180
ACETIC ACID n-PROPYL ESTER Synonym: PROPYL ACETATE (-n)	1276	26	723	23	2914	999	189	–81	243	188	PAT 929	794	P8.0		969
ACETIC ACID NICKEL (II) SALT Synonym: NICKEL ACETATE			611		2489						NKA 834				817
ACETIC ACID PROPYL ESTER Synonym: PROPYL ACETATE (-n)	1276	26	723	23	2914	999	189	–81	243		PAT 929	794	P8.0		969
ACETIC ACID sec-BUTYL ESTER Synonym: BUTYL ACETATE (-sec)	1123	26		13	591	195	29	–21	56	50	BTA 172	149	B6.0		180
ACETIC ACID SOLUTION	2790	60	17	11	15	6	1		16	30	AAC 1	2	A2.0	A 2	7
ACETIC ACID SOLUTION	2789	29		11	15	6	1		16	30	AAC 1	3	A2.0	A 2	7
ACETIC ACID tert-BUTYL ESTER Synonym: BUTYL ACETATE (-tert)	1123	26		13	591	195	30		56	50	BYA 189	149	B6.0		180
ACETIC ACID VINYL ESTER Synonym: VINYL ACETATE	1301	26	871	26	3492	8	227	179–92	16		VAM1172	948	V1.0		1215
ACETIC ACID ZINC SALT Synonym: ZINC ACETATE	9153/3077	31	886		3536	1191			288		ZNA1211	964			1242
ACETIC ACID-4,6-DINITRO-o-CRESYL ESTER					16	7	2								
ACETIC ALDEHYDE Synonym: ACETALDEHYDE	1089	26	16	11	5	3	1	13 –11	16	30	AAD 2	1	A1.0	A 1	7
ACETIC ANHYDRIDE	1715	39	19	11	18	8	2	14 –11	17	30	ACA 13	3	A2.1		7
ACETIC ESTER Synonym: ETHYL ACETATE	1173	26	340	16	1570	523	89	–47	17	104	ETA 560	406	E2.0		7
ACETIC OXIDE Synonym: ACETIC ANHYDRIDE	1715	39	19	11	18	8	2	14 –11	17	30	ACA 13	3	A2.1		7

HAZARDOUS MATERIALS REFERENCE BOOKS INDEX

CHEMICAL OR MATERIAL NAME	UN/NA Number	Guide Number (DOT)	Firefighter's Hazardous Materials Reference Book	First Aid Manual for Chemical Accidents	Sax's Dangerous Properties of Industrial Materials	Hazardous Chemicals Desk Reference	Rapid Guide to Hazardous Chemicals in the Workplace	Fire Protection Guide on Hazardous Materials (NFPA)	Firefighter's Handbook of Hazardous Materials	Pocket Guide to Chemical Hazards (NIOSH)	Chemical Hazard Response Information System (CHRIS)	Emergency Handling of Hazardous Materials in Surface Transportation (AAR)	Emergency Action Guides (EAG)	Chemical Data Notebook: A User's Manual	Condensed Chemical Dictionary
ACETOACET-ortho-TOLUIDIDE					20	9		−11							8
ACETOACET-para-CHLORANILIDE					19				17						8
ACETOACETANILIDE					19	9		−11	17						8
ACETOACETIC ACID ETHYL ESTER Synonym: ETHYL ACETOACETATE	1993	27	341					−47	17		EAA 499	406			478
ACETOACETIC ESTER Synonym: ETHYL ACETOACETATE	1993	27	341					−47	17		EAA 499	406			478
ACETOL					21	10									8
ACETONE	1090	26	20	11	22	10	2	−11	17	30	ACT 26	4	A3.0	A 3	9
ACETONE CHLOROFORM					23	11			17						9
ACETONE CYANOHYDRIN	1541	55	21	11				15 −11	17		ACY 27	4	A3.1		9
ACETONE OIL	1091	26				11			17			5			
ACETONE PEROXIDE					23	11			17						
ACETONITRILE	1648	28	22	11	24	12	3	15 −12	17	32	ATN 104	5			9
ACETONYL ACETONE								−12	17						9
ACETONYL BROMIDE Synonym: BROMOACETONE	1569	55							51		BRE 166	141	B3.1		170
ACETOPHENONE	9207	27	23		26	13		−12	17		ACP 24				10
ACETOTOLUIDIDE-ortho					27				18						10
ACETOTOLUIDIDE-para					27	13		−12	18						10
ACETOXYL Synonym: BENZOYL PEROXIDE	2085	49	122		392	130	23		42	44			B2.0	B 2	133
ACETOZONE	2081	48							18						10
ACETPHENARSINE					37	17			18						
ACETYL ACETONE PEROXIDE	2080/3105	48							18			6			

CHEMICAL OR MATERIAL NAME	UN/NA Number	Guide Number (DOT)	Firefighter's Hazardous Materials Reference Book	First Aid Manual for Chemical Accidents	Sax's Dangerous Properties of Industrial Materials	Hazardous Chemicals Desk Reference	Rapid Guide to Hazardous Chemicals in the Workplace	Fire Protection Guide on Hazardous Materials (NFPA)	Firefighter's Handbook of Hazardous Materials	Pocket Guide to Chemical Hazards (NIOSH)	Chemical Hazard Response Information System (CHRIS)	Emergency Handling of Hazardous Materials in Surface Transportation (AAR)	Emergency Action Guides (EAG)	Chemical Data Notebook: A User's Manual	Condensed Chemical Dictionary
ACETYL ANHYDRIDE Synonym: ACETIC ANHYDRIDE	1715	39	19	11	18	8	2	14 –11	18	30	ACA 13	3	A2.1		7
ACETYL BENZOYL PEROXIDE	2081	48			39	18			18						11
ACETYL BROMIDE	1716	60	24		40	18			18		ABM 9	6			11
ACETYL CHLORIDE	1717	29	25		41	19		16 –12	18		ACC 15	7			12
ACETYL CYCLOHEXANE SULPHONYL PEROXIDE	2083/3115	52										7			
ACETYL CYCLOHEXANE SULPHONYL PEROXIDE	2082/3112	52										7			
ACETYL ETHER Synonym: ACETIC ANHYDRIDE	1715	39	19	11	18	8	2	14 –11	18	30	ACA 13	3	A2.1		7
ACETYL HYDROPEROXIDE Synonym: PERACETIC ACID	2131	51	670	22				133–	18		PAA 921	732			883
ACETYL IODIDE	1898	60			47	22			18			8			13
ACETYL KETENE	2521	57							19						13
ACETYL METHYL BROMIDE Synonym: BROMOACETONE	1569	55							51		BRE 166	141	B3.1		170
ACETYL METHYL CARBINOL	2621	26							19			8			13
ACETYL NITRATE					50	23			19						13
ACETYL OXIDE Synonym: ACETIC ANHYDRIDE	1715	39	19	11	18	8	2	14 –11	19	30	ACA 13	3	A2.1		13
ACETYL PEROXIDE	2084	49	26		51	23	4	–12	19		APS 80				13
ACETYLACETONE	2310	26	28		37	17			19		ATA 97				11
ACETYLBENZENE Synonym: ACETOPHENONE	9207	27			26	13		–12	18		ACP 24				11
ACETYLENE	1001	17	29	11	43	20	3	17 –12	19		ACE 17	9		A 4	12
ACETYLENE CHLORIDE					44	20			19						

HAZARDOUS MATERIALS REFERENCE BOOKS INDEX

CHEMICAL OR MATERIAL NAME	UN/NA Number	Guide Number (DOT)	Firefighter's Hazardous Materials Reference Book	First Aid Manual for Chemical Accidents	Sax's Dangerous Properties of Industrial Materials	Hazardous Chemicals Desk Reference	Rapid Guide to Hazardous Chemicals in the Workplace	Fire Protection Guide on Hazardous Materials (NFPA)	Firefighter's Handbook of Hazardous Materials	Pocket Guide to Chemical Hazards (NIOSH)	Chemical Hazard Response Information System (CHRIS)	Emergency Handling of Hazardous Materials in Surface Transportation (AAR)	Emergency Action Guides (EAG)	Chemical Data Notebook: A User's Manual	Condensed Chemical Dictionary
ACETYLENE DICHLORIDE Synonym: 1,2-DICHLOROETHYLENE	1150/3082	29	3		1144	384	69	–35	19	88	DEL 388	301			13
ACETYLENE DICHLORIDE-trans	1150	29			45	21	3		19						
ACETYLENE DISSOLVED Synonym: ACETYLENE	1001	17	29	11	43	20	3	17 –12	19		ACE 17	9		A 4	12
ACETYLENE TETRABROMIDE	2504	58			45	21	4		19	32		8			13
ACETYLENE TETRACHLORIDE Synonym: TETRACHLOROETHANE	1702	55	807	25	3207	1096	208		19		TEC1099	887			13
ACETYLENE TRICHLORIDE Synonym: TRICHLOROETHYLENE	1710	74	837	25	3352	1137	218	174–88	273	216	TCL1085	919	T5.0	T 6	1170
ACETYLENOGEN Synonym: CALCIUM CARBIDE	1402	40	167	13	662	229		45 –	63		CCB 226	179	C1.0	C 1	13
ACETYLIDES					47	22			19						13
ACETYLSALICYLIC ACID					53	24	4								14
ACID	1760	60							19			11			15
ACID AMMONIUM CARBONATE Synonym: AMMONIUM BICARBONATE	9081/3077	31	51		221	58			27		ABC 7	56	·		63
ACID AMMONIUM FLUORIDE Synonym: AMMONIUM BIFLUORIDE	1727	60	52						28		ABF 8				63
ACID BUTYL PHOSPHATE	1718	60			57	25			19						16
ACID ETHYL SULFATE	2571	60							19						16
ACID FISH SCRAP	1374	32													
ACID LIQUID HCN Synonym: HYDROCYANIC ACID	1051	13	442	18	1896	641	114	–59	164				H4.1	H 2	613
ACID MIXTURE	1826	60							19						
ACID MIXTURE	1786	59							19						

CHEMICAL OR MATERIAL NAME	UN/NA Number	Guide Number (DOT)	Firefighter's Hazardous Materials Reference Book	First Aid Manual for Chemical Accidents	Sax's Dangerous Properties of Industrial Materials	Hazardous Chemicals Desk Reference	Rapid Guide to Hazardous Chemicals in the Workplace	Fire Protection Guide on Hazardous Materials (NFPA)	Firefighter's Handbook of Hazardous Materials	Pocket Guide to Chemical Hazards (NIOSH)	Chemical Hazard Response Information System (CHRIS)	Emergency Handling of Hazardous Materials in Surface Transportation (AAR)	Emergency Action Guides (EAG)	Chemical Data Notebook: A User's Manual	Condensed Chemical Dictionary
ACID MIXTURES Synonym: NITRATING ACID	1796	73	626			824			19			683	N1.0		822
ACID NEC DRY AND ORGANIC	1759	60										10			
ACID SLUDGE	1906/1760	60							19			9			
ACIDS INORGANIC	1759	60										10			
ACIDS INORGANIC	1760	60													
ACIDS NEC LIQUID AND ORGANIC	1760	60							19			11			
ACIFLOCTIN Synonym: ADIPIC ACID	9077/3077	31	35		79	28		−12	20		ADA 28	15	A6.0	A 8	24
ACINETTEN Synonym: ADIPIC ACID	9077/3077	31	35		79	28		−12	20		ADA 28	15	A6.0	A 8	24
ACN Synonym: ACETONE CYANOHYDRIN	1541	55	21	11				15 −11	17		ACY 27	4	A3.1		9
ACNEGEL Synonym: BENZOYL PEROXIDE	2085	49	122		392	130	23		42	44			B2.0	B 2	133
ACONITE									19						17
ACONITINE					58				19						17
ACQUINITE Synonym: CHLOROPICRIN	1580	56	210	14	865	296	50	54 −	77	70	CPL 290	229	C4.3		271
ACQUINITE Synonym: ACROLEIN	1092	30	31	11	63	25	5	17 −12	20	32	ARL 85	12	A4.0	A 5	18
ACRALDEHYDE Synonym: ACROLEIN	1092	30	31	11	63	25	5	17 −12	20	32	ARL 85	12	A4.0	A 5	17
ACRIDINE	2713/1325	32	30		59	25	5		19		ACD 16	11			17
ACRITET Synonym: ACRYLONITRILE	1093	30	34	11	68	27	6	19 −12	20	34	ACN 23	13	A5.1	A 7	19

CHEMICAL OR MATERIAL NAME	UN/NA Number	Guide Number (DOT)	Firefighter's Hazardous Materials Reference Book	First Aid Manual for Chemical Accidents	Sax's Dangerous Properties of Industrial Materials	Hazardous Chemicals Desk Reference	Rapid Guide to Hazardous Chemicals in the Workplace	Fire Protection Guide on Hazardous Materials (NFPA)	Firefighter's Handbook of Hazardous Materials	Pocket Guide to Chemical Hazards (NIOSH)	Chemical Hazard Response Information System (CHRIS)	Emergency Handling of Hazardous Materials in Surface Transportation (AAR)	Emergency Action Guides (EAG)	Chemical Data Notebook: A User's Manual	Condensed Chemical Dictionary
ACROLEIC ACID Synonym: ACRYLIC ACID	2218	29	33		65	26	6	18 −12	20		ACR 25	13	A5.0	A 6	18
ACROLEIN	1092	30	31	11	63	25	5	17 −12	20	32	ARL 85	12	A4.0	A 5	18
ACROLEIN DIMER	2607/1993	26			64	26		−12	20			11			18
ACROLEIN-trans Synonym: ACROLEIN	1092	30	31	13	63	25	5	17 −12	20	32	ARL 85	12	A4.0	A 5	18
ACRYLALDEHYDE Synonym: ACROLEIN	1092	30	31	11	63	25	5	17 −12	20	32	ARL 85	12	A4.0	A 5	18
ACRYLAMIDE	2074	55	32	11	64	26	5	18 −	20	34	AAM 3	12			18
ACRYLAMIDE SOLUTION (COMBUSTIBLE LIQUID)	1993	27		11											
ACRYLAMIDE SOLUTION (FLAMMABLE LIQUID)	1993	27		11					20			12			
ACRYLIC ACID	2218	29	33		65	26	6	18 −12	20		ACR 25	13	A5.0	A 6	18
ACRYLIC ACID-2-ETHYLHEXYL ESTER Synonym: 2-ETHYLHEXYL ACRYLATE					66			88 −52	20		EAI 502				493
ACRYLIC ACID AMIDE (50%) Synonym: ACRYLAMIDE	2074	55	32	11	64	26	5	18 −	20	34	AAM 3	12			18
ACRYLIC ACID BUTYL ESTER Synonym: BUTYL ACRYLATE (-n)	2348	26	146		592	196	30	41 −21	20		BTC 174	150	B7.0		181
ACRYLIC ACID ETHYL ESTER Synonym: ETHYL ACRYLATE	1917	26	343	16	1571	524	90	80 −48	20	106	EAC 500	407	E2.1		478
ACRYLIC ACID ISOBUTYL ESTER Synonym: BUTYL ACRYLATE-iso	2527	27	145								BAI 116				
ACRYLIC ACID METHYL ESTER Synonym: METHYL ACRYLATE	1919	26	553	20	2250	769	135	−66	20	142	MAM 723	618			755
ACRYLIC ACID n-BUTYL ESTER Synonym: BUTYL ACRYLATE (-n)	2348	26	146		592	196	30	41 −21	56		BTC 174	150	B7.0		181

CHEMICAL OR MATERIAL NAME	UN/NA Number	Guide Number (DOT)	Firefighter's Hazardous Materials Reference Book	First Aid Manual for Chemical Accidents	Sax's Dangerous Properties of Industrial Materials	Hazardous Chemicals Desk Reference	Rapid Guide to Hazardous Chemicals in the Workplace	Fire Protection Guide on Hazardous Materials (NFPA)	Firefighter's Handbook of Hazardous Materials	Pocket Guide to Chemical Hazards (NIOSH)	Chemical Hazard Response Information System (CHRIS)	Emergency Handling of Hazardous Materials in Surface Transportation (AAR)	Emergency Action Guides (EAG)	Chemical Data Notebook: A User's Manual	Condensed Chemical Dictionary
ACRYLIC ALDEHYDE Synonym: ACROLEIN	1092	30	31	11	63	25	5	17 –12	20	32	ARL 85	12	A4.0	A 5	18
ACRYLIC AMIDE (50%) Synonym: ACRYLAMIDE	2074	55	32	11	64	26	5	18 –	20	34	AAM 3	12			18
ACRYLIC RESIN MONOMER Synonym: METHYL METHACRYLATE	1247	26	575		2342	803	147	115–70	195	154	MMM 775	633	M3.6		19
ACRYLON Synonym: ACRYLONITRILE	1093	30	34	11	68	27	6	19 –12	20	34	ACN 23	13	A5.1	A 7	19
ACRYLONITRILE	1093	30	34	11	68	27	6	19 –12	20	34	ACN 23	13	A5.1	A 7	19
ACRYLONITRILE MONOMER Synonym: ACRYLONITRILE	1093	30	34	11	68	27	6	19 –12	20	34	ACN 23	13	A5.1	A 7	19
ACTIVATED CARBON	1362	32							20						21
ACTIVATED CHARCOAL Synonym: CHARCOAL	1361	32	195		739	258			68		CHC 244	203			248
ACTIVE PRIMARY AMYL ALCOHOL	1987/1105	26										14			
ACTIVE SEC AMYL ALCOHOL	1987	26													
ACY Synonym: ACETONE CYANOHYDRIN	1541	55	21	11				15 –11	17		ACY 27	4	A3.1		9
ADACENE 12 Synonym: 1-DODECENE											DDC 372				
ADAMSITE	1698	55							20						22
ADDITIVES FUEL OIL GASOLINE OR LUBRICATING OIL	1993	27										14			
ADENINE					76										22
ADHESIVE	1133	26							20			14			23
ADILACTETTEN Synonym: ADIPIC ACID	9077/3077	31	35		79	28		–12	20		ADA 28	15	A6.0	A 8	24

HAZARDOUS MATERIALS REFERENCE BOOKS INDEX

CHEMICAL OR MATERIAL NAME	UN/NA Number	Guide Number (DOT)	Firefighter's Hazardous Materials Reference Book	First Aid Manual for Chemical Accidents	Sax's Dangerous Properties of Industrial Materials	Hazardous Chemicals Desk Reference	Rapid Guide to Hazardous Chemicals in the Workplace	Fire Protection Guide on Hazardous Materials (NFPA)	Firefighter's Handbook of Hazardous Materials	Pocket Guide to Chemical Hazards (NIOSH)	Chemical Hazard Response Information System (CHRIS)	Emergency Handling of Hazardous Materials in Surface Transportation (AAR)	Emergency Action Guides (EAG)	Chemical Data Notebook: A User's Manual	Condensed Chemical Dictionary
ADIPIC ACID	9077/3077	31	35		79	28		-12	20		ADA 28	15	A6.0	A 8	24
ADIPIC ACID BIS[2-ETHYLHEXYL] ESTER Synonym: DIOCTYL ADIPATE					80			-44			DOA 453				424
ADIPIC KETONE	2245	26							20						
ADIPINIC ACID Synonym: ADIPIC ACID	9077/3077	31	35		79	28		-12	20		ADA 28	15	A6.0	A 8	24
ADIPOL 2EH Synonym: DIOCTYL ADIPATE								-44			DOA 453				424
ADIPONITRILE	2205	55	36	11	82	28	7	20 -12	20		ADN 29				24
ADIPOYL CHLORIDE								-12	20						
ADIPYL CHLORIDE Synonym: ADIPOYL CHLORIDE								-12	21						
ADIPYLDINITRILE	2205	55						-13	21						
ADRONAL Synonym: CYCLOHEXANOL	1993	27	245	14	991	333	59	-30	88	78	CHN 246	280			337
AEROSOL DISPENSER	1760	60													
AEROSOL DISPENSER	1954	22													
AEROSOL DISPENSER	2810	55													
AEROSOL DISPENSER	1956	12													
AEROSOL DISPENSER	1993	27													
AEROSOL PRODUCT	1956	12													26
AEROSOL PRODUCT	1954	22													26
AEROSOL SURFACTANT Synonym: DIOCTYL SODIUM SULFOSUCCINATE											DSS 481				425

CHEMICAL OR MATERIAL NAME	UN/NA Number	Guide Number (DOT)	Firefighter's Hazardous Materials Reference Book	First Aid Manual for Chemical Accidents	Sax's Dangerous Properties of Industrial Materials	Hazardous Chemicals Desk Reference	Rapid Guide to Hazardous Chemicals in the Workplace	Fire Protection Guide on Hazardous Materials (NFPA)	Firefighter's Handbook of Hazardous Materials	Pocket Guide to Chemical Hazards (NIOSH)	Chemical Hazard Response Information System (CHRIS)	Emergency Handling of Hazardous Materials in Surface Transportation (AAR)	Emergency Action Guides (EAG)	Chemical Data Notebook: A User's Manual	Condensed Chemical Dictionary
AEROTHENE Synonym: TRICHLOROETHANE	2831	74		25							TCE1082		T4.2		26
AETHYLIS Synonym: ETHYL CHLORIDE	1037	27	350		1586	532	92	82 -49	136	108	ECL 514	412	E2.3		483
AETHYLIS CHLORIDUM Synonym: ETHYL CHLORIDE	1037	27	350		1586	532	92	82 -49	136	108	ECL 514	412	E2.3		483
AFTER SHAVE LOTIONS	1993	27										16			
AGENTS ETIOLOGIC	2814	24													
AHF Synonym: HYDROGEN FLUORIDE	1052	15	449					99 -	164	126	HFX 636	530	H4.4	H 4	616
AIR, COMPRESSED	1002	12			89	31			21			17			28
ALBONE Synonym: HYDROGEN PEROXIDE	2014	45		18	1901	644	115	99 -	165	126		531	H4.5	H 5	30
ALBONE Synonym: HYDROGEN PEROXIDE	2015	47	450	18	1901	644	115	100-	165	126	HPO 643	532	H4.5	H 5	30
ALBUS Synonym: MERCURIC AMMONIUM CHLORIDE	1630	53							186		MCC 735	598			
ALCOHOL Synonym: ETHYL ALCOHOL	1170/3065	26	344	16	92	31	90	-48	21		EAL 504	18	E2.2	E 2	30
ALCOHOL	1987	26			92	31			21			18			30
ALCOHOL BEVERAGE	1170/3065	26				31			21			18			31
ALCOHOL C 10 Synonym: DECYL ALCOHOL (-n)	1987	26	258		1032	347			91		DAN 338	291			350
ALCOHOL C 11 Synonym: UNDECANOL			857								UND1165				

CHEMICAL OR MATERIAL NAME	UN/NA Number	Guide Number (DOT)	Firefighter's Hazardous Materials Reference Book	First Aid Manual for Chemical Accidents	Sax's Dangerous Properties of Industrial Materials	Hazardous Chemicals Desk Reference	Rapid Guide to Hazardous Chemicals in the Workplace	Fire Protection Guide on Hazardous Materials (NFPA)	Firefighter's Handbook of Hazardous Materials	Pocket Guide to Chemical Hazards (NIOSH)	Chemical Hazard Response Information System (CHRIS)	Emergency Handling of Hazardous Materials in Surface Transportation (AAR)	Emergency Action Guides (EAG)	Chemical Data Notebook: A User's Manual	Condensed Chemical Dictionary
ALCOHOL C 8 Synonym: OCTANOL	1987	26	650		2620						OTA 909	706	O1.0		847
ALCOHOL DENATURED	1987	26			92	31			21			18			31
ALCOHOL DENATURED (toxic)	1986	28			92	31			21			19			
ALCOHOL DISTILLATES SYNTHETIC (COMBUSTIBLE LIQUID)	1987	26				31									
ALCOHOL DISTILLATES SYNTHETIC (FLAMMABLE LIQUID)	1987	26				31						17			
ALCOHOL ETHYL	1170	26				31			21						
ALCOHOL IN BOND FREE OF INTERNAL REVENUE TAX (COMBUST)	1170	26				31									
ALCOHOL IN BOND FREE OF INTERNAL REVENUE TAX (FLAMMABLE)	1170	26				31						17			
ALCOHOL TOXIC	1986	28				31			21			18			
ALCOHOLIC BEVERAGE	1170/3065	26				31			21			19			
ALDEHYDE	1989	26			93	31			21			20			31
ALDEHYDE AMMONIA	1841	31				31			21						31
ALDEHYDE C 10 Synonym: DECALDEHYDE						31					DAL 336				
ALDEHYDE COLLIDINE Synonym: METHYLETHYLPYRIDINE	2300	60	567						194		MEP 753	648			31
ALDEHYDE TOXIC	1988	28				31			21			20			
ALDEHYDINE Synonym: METHYLETHYLPYRIDINE	2300	60	567						194		MEP 753	648			31
ALDIFEN Synonym: 2,4-DINITROPHENOL	0076				1435	479			125		DNP 449				
ALDOL	2839/1993	55						−13	21			20			32

CHEMICAL OR MATERIAL NAME	UN/NA Number	Guide Number (DOT)	Firefighter's Hazardous Materials Reference Book	First Aid Manual for Chemical Accidents	Sax's Dangerous Properties of Industrial Materials	Hazardous Chemicals Desk Reference	Rapid Guide to Hazardous Chemicals in the Workplace	Fire Protection Guide on Hazardous Materials (NFPA)	Firefighter's Handbook of Hazardous Materials	Pocket Guide to Chemical Hazards (NIOSH)	Chemical Hazard Response Information System (CHRIS)	Emergency Handling of Hazardous Materials in Surface Transportation (AAR)	Emergency Action Guides (EAG)	Chemical Data Notebook: A User's Manual	Condensed Chemical Dictionary
ALDRIN	2761	55	37	11	94	31	7		21	34	ALD 41	21			32
ALFA TOX Synonym: DIAZINON	2783/2810	55	261	15	1083	363	64		96		DZN 497	319			366
ALGEON 22 Synonym: MONOCHLORODIFLUOROMETHANE	1018	12	593						205				M4.0		795
ALGOFRENE TYPE 2 Synonym: DICHLORODIFLUOROMETHANE	1028	12	275	15	1137	380	68		102	86	DCF 360	331	D1.1		379
ALGOFRENE TYPE 6 Synonym: MONOCHLORODIFLUOROMETHANE	1018	12	593						205				M4.0		795
ALGRAIN Synonym: ETHYL ALCOHOL	1170	26	344	16	1572	525	90	-48	136		EAL 504	408	E2.2	E 2	478
ALGYLEN Synonym: TRICHLOROETHYLENE	1710	74	837	25	3352	1137	218	174-88	273	216	TCL1085	919	T5.0	T 6	1170
ALKALI METAL	1421	40			96	32			21						34
ALKALI METAL ALLOY	1421	40			96	32			21			24			34
ALKALI METAL AMIDE	1390	40			96	32			21			25			
ALKALIES					96	32			21						
ALKALINE BOILER TREATMENT COMPOUND	1760	60							21						
ALKALINE CORROSIVE LIQUID	1719	60							21						
ALKALINE EARTH METAL ALLOY	1393	40							21						34
ALKALOID	1544	55			96				21						35
ALKALOID SALT	3140	53			96	32									
ALKALOID SALT	1544	55			96	32			21			26			
ALKANESULFONIC ACID CORROSIVE MATERIAL	2584	60													35

CHEMICAL OR MATERIAL NAME	UN/NA Number	Guide Number (DOT)	Firefighter's Hazardous Materials Reference Book	First Aid Manual for Chemical Accidents	Sax's Dangerous Properties of Industrial Materials	Hazardous Chemicals Desk Reference	Rapid Guide to Hazardous Chemicals in the Workplace	Fire Protection Guide on Hazardous Materials (NFPA)	Firefighter's Handbook of Hazardous Materials	Pocket Guide to Chemical Hazards (NIOSH)	Chemical Hazard Response Information System (CHRIS)	Emergency Handling of Hazardous Materials in Surface Transportation (AAR)	Emergency Action Guides (EAG)	Chemical Data Notebook: A User's Manual	Condensed Chemical Dictionary
ALKAWAY LIQUID ALKALINE DERUSTER Synonym: BOILER COMPOUND	1760	60			527				48		BCP 129				159
ALKRON Synonym: METHYL PARATHION	2783	55	577	20	2369	809	148		196		MPT 790	635			776
ALKYL ALUMINUM HALIDES	2845	40										27			
ALKYL PHENOL	2430	55							21			28			
ALKYL SULFONIC ACID	2585	60							22						
ALKYL SULFONIC ACID	2586	60							22						
ALKYL SULFONIC ACID	2584	60							22			29			
ALKYL SULFONIC ACID	2583	60							22						
ALKYLALUMINUM	2845	40		11				20 –							
ALKYLAMINE	2734	29							22			31			
ALKYLAMINE	2733	29							22			31			
ALKYLAMINE	2735	60							22			31			
ALKYLBENZENESULFONIC ACID											ABS 11				
ALKYLBENZENESULFONIC ACID SODIUM SALT Synonym: SODIUM ALKYLBENZENESULFONATES			753		97						SAB1007				
ALLENE	2200	22			99				22						36
ALLENE METHYLACETYLENE MIXTURE Synonym: METHYL ACETYLENE PROPADIENE MIXTURE	1060	17	552		2248	769	134			142	MAP 725	640			755
ALLETHRIN	2902	55			99	33			22			34			37
ALLICIN					100				22						37
ALLOMALEIC ACID Synonym: FUMARIC ACID	9126/9188	31	407		1760	595			152		FUM 599	465			37
ALLYL ACETATE	2333	28			102	33		–13	22			34			

CHEMICAL OR MATERIAL NAME	UN/NA Number	Guide Number (DOT)	Firefighter's Hazardous Materials Reference Book	First Aid Manual for Chemical Accidents	Sax's Dangerous Properties of Industrial Materials	Hazardous Chemicals Desk Reference	Rapid Guide to Hazardous Chemicals in the Workplace	Fire Protection Guide on Hazardous Materials (NFPA)	Firefighter's Handbook of Hazardous Materials	Pocket Guide to Chemical Hazards (NIOSH)	Chemical Hazard Response Information System (CHRIS)	Emergency Handling of Hazardous Materials in Surface Transportation (AAR)	Emergency Action Guides (EAG)	Chemical Data Notebook: A User's Manual	Condensed Chemical Dictionary
ALLYL ALCOHOL	1098	57	38	11	102	33	7	21 −13	22	34	ALA 39	35	A7.0		38
ALLYL ALDEHYDE Synonym: ACROLEIN	1092	30	31	11	63	25	5	17 −12	22	32	ARL 85	12	A4.0	A 5	18
ALLYL BENZENE					104										
ALLYL BROMIDE	1099	57	39		405	34		22 −13	22		ABR 10	35			39
ALLYL CAPROATE					106	35		−13	22						
ALLYL CHLORIDE	1100	57	40	11	106	35	8	22 −13	22	36	ALC 40	36	A7.2		39
ALLYL CHLOROCARBONATE Synonym: ALLYL CHLOROFORMATE	1722	57	41		107	36		23 −13	22		ACF 18	37			
ALLYL CHLOROFORMATE	1722	57	41					−13	22		ACF 18	37			
ALLYL DIMETHYL ARSINE					110				22						
ALLYL ETHER	2335	28						−13							
ALLYL ETHYL ETHER	2335/1992	28			111				22			38			
ALLYL FLUORIDE					111	36			22						
ALLYL FORMATE	2336/1992	28			111	37			22			38			
ALLYL GLYCIDYL ETHER	2219/1993	29		11	112	37	8		22	36		38			39
ALLYL IODIDE	1723/2924	29			113	37				23		39			
ALLYL ISOSULFOCYANATE	1545	57													
ALLYL ISOTHIOCYANATE INHIBITED	1545/2810	57			114	38			23			39			39
ALLYL ISOTHIOCYANATE STABILIZED	1545/2810	57			114	38			23			39			39
ALLYL MERCAPTAN					114	38			23						39
ALLYL PROPENYL	2458	29							23						
ALLYL PROPYL DISULFIDE				11	120	39	8		23						40
ALLYL SULFIDE					121				23						40

HAZARDOUS MATERIALS REFERENCE BOOKS INDEX

CHEMICAL OR MATERIAL NAME	UN/NA Number	Guide Number (DOT)	Firefighter's Hazardous Materials Reference Book	First Aid Manual for Chemical Accidents	Sax's Dangerous Properties of Industrial Materials	Hazardous Chemicals Desk Reference	Rapid Guide to Hazardous Chemicals in the Workplace	Fire Protection Guide on Hazardous Materials (NFPA)	Firefighter's Handbook of Hazardous Materials	Pocket Guide to Chemical Hazards (NIOSH)	Chemical Hazard Response Information System (CHRIS)	Emergency Handling of Hazardous Materials in Surface Transportation (AAR)	Emergency Action Guides (EAG)	Chemical Data Notebook: A User's Manual	Condensed Chemical Dictionary
ALLYL TRICHLORIDE									23						
ALLYL TRICHLOROSILANE	1724	29	42		122	39		-13	23		ATC 99	39			40
ALLYL VINYL ETHER					123				23						
ALLYLAMINE	2334	28			103			21 -13	22			40	A7.1		38
ALLYLIC ALCOHOL Synonym: ALLYL ALCOHOL	1098	57	38	11	102	33	7	21 -13	22	34	ALA 39	35	A7.0		38
ALLYLIDENE DIACETATE								-13	23						
ALLYLSILICONE TRICHLORIDE Synonym: ALLYL TRICHLOROSILANE	1724	29	42		122	39		-13	23		ATC 99	39			40
ALPEROX C	2124	48													
ALROWET D65 Synonym: DIOCTYL SODIUM SULFOSUCCINATE											DSS 481				425
ALUMINUM	1396/1309	40		11	125	39	9	24 –				48			42
ALUMINUM ALKYL	2003	40		11					24						43
ALUMINUM ALKYL	2845/3051	40		11								42			43
ALUMINUM ALKYL	3051/2845	40		11					23			42			43
ALUMINUM ALKYL CHLORIDE	2221	40		11					23						
ALUMINUM ALKYL HALIDE	3052	40		11					23			42			
ALUMINUM ALKYL HALIDE	2221	40		11					23						
ALUMINUM ALKYL HALIDE SOLUTION	2220	40		11					23						
ALUMINUM BOROHYDRIDE	2870	37			126	40			24						43
ALUMINUM BOROHYDRIDE IN DEVICES	2870	37							24			42			
ALUMINUM BROMIDE	1725	39			127	40			24			43			43
ALUMINUM BROMIDE	2580/1760	60			127	40			24			43			43

CHEMICAL OR MATERIAL NAME	UN/NA Number	Guide Number (DOT)	Firefighter's Hazardous Materials Reference Book	First Aid Manual for Chemical Accidents	Sax's Dangerous Properties of Industrial Materials	Hazardous Chemicals Desk Reference	Rapid Guide to Hazardous Chemicals in the Workplace	Fire Protection Guide on Hazardous Materials (NFPA)	Firefighter's Handbook of Hazardous Materials	Pocket Guide to Chemical Hazards (NIOSH)	Chemical Hazard Response Information System (CHRIS)	Emergency Handling of Hazardous Materials in Surface Transportation (AAR)	Emergency Action Guides (EAG)	Chemical Data Notebook: A User's Manual	Condensed Chemical Dictionary
ALUMINUM CARBIDE	1394	40			127				24			41			44
ALUMINUM CHLORATE					127				24						44
ALUMINUM CHLORIDE	1760	60		11											44
ALUMINUM CHLORIDE	1726	39	43	11	127	41		24 –	24		ACL 21	44			44
ALUMINUM CHLORIDE	2581	60		11	127	41			24			43			44
ALUMINUM DIETHYL MONOCHLORIDE	1101								24						44
ALUMINUM ETHYL DICHLORIDE Synonym: ETHYLALUMINUM DICHLORIDE	1924	40						–48	136		EAD 501				479
ALUMINUM ETHYLATE					128	41			24						45
ALUMINUM FERROSILICON	1395	41							24			44			
ALUMINUM FLUORIDE					129				24		ALF 42				45
ALUMINUM HYDRIDE	2463	40			129	41			24			44			45
ALUMINUM HYDROXIDE				11	129										45
ALUMINUM METALLIC	1396/1325	40										48			42
ALUMINUM METHYL	1103	40			130	42			24						
ALUMINUM NITRATE	1438	35	44		130	42			24		ALN 44	45			46
ALUMINUM NITRATE NONAHYDRATE Synonym: ALUMINUM NITRATE	1438	35	44		131	42			24		ALN 44	45			46
ALUMINUM OXIDE				11	131	42	10								47
ALUMINUM PAINT	1263	26													
ALUMINUM PHOSPHATE SOLUTION	1760	60							24			45			47
ALUMINUM PHOSPHIDE	1397	41			131	42		24 –	24		APH 75	45			47
ALUMINUM PHOSPHIDE PESTICIDE	3048	53										46			
ALUMINUM PICRATE					132	43			24						47

HAZARDOUS MATERIALS REFERENCE BOOKS INDEX

CHEMICAL OR MATERIAL NAME	UN/NA Number	Guide Number (DOT)	Firefighter's Hazardous Materials Reference Book	First Aid Manual for Chemical Accidents	Sax's Dangerous Properties of Industrial Materials	Hazardous Chemicals Desk Reference	Rapid Guide to Hazardous Chemicals in the Workplace	Fire Protection Guide on Hazardous Materials (NFPA)	Firefighter's Handbook of Hazardous Materials	Pocket Guide to Chemical Hazards (NIOSH)	Chemical Hazard Response Information System (CHRIS)	Emergency Handling of Hazardous Materials in Surface Transportation (AAR)	Emergency Action Guides (EAG)	Chemical Data Notebook: A User's Manual	Condensed Chemical Dictionary
ALUMINUM POWDER Synonym: ALUMINUM	1396	40			125		9		24			46			42
ALUMINUM POWDER Synonym: ALUMINUM	1309	32			125		9		24			48			42
ALUMINUM RESINATE	2715	32							24			46			47
ALUMINUM SALT Synonym: ALUMINUM NITRATE	1438	35	44		130	42			24		ALN 44	45			46
ALUMINUM SILICON POWDER	1398	40							24			46			48
ALUMINUM SULFATE	9188/3077	31	46	31								47	A7.1.01		48
ALUMINUM SULFATE	1760	60	45		132	43			24				A7.1.01		48
ALUMINUM SULFATE	9078/3077	31	46			43			24		ALM 43	47	A7.1.01		48
ALUMINUM THALLIUM SULFATE					133	43			24						
ALUMINUM TRIBUTYL	2003	40							24						48
ALUMINUM TRIETHYL Synonym: TRIETHYLALUMINUM	1102			25	3368	1141		175–88			TAL1074				49
ALUMINUM TRIISOBUTYL Synonym: TRIISOBUTYL ALUMINUM	1930/3051	40		26	3391	1145		176–89	277		TIA1120	927			
ALUMINUM TRIMETHYL	1103	40							25						49
ALUMINUM TRIPROPYL	1103	40			133				25						
AM FOL Synonym: ANHYDROUS AMMONIA	1005	15	48	11	220	57	11	25 –	32		AMA 47	78	A9.0	A 9	62
AMATOL					134	44			25			48			49
AMCHLOR Synonym: AMMONIUM CHLORIDE	9085/3077	31	56		223	60	11		28		AMC 49	57			64
AMCHLORIDE Synonym: AMMONIUM CHLORIDE	9085/3077	31	56		223	60	11		28		AMC 49	57			64

CHEMICAL OR MATERIAL NAME	UN/NA Number	Guide Number (DOT)	Firefighter's Hazardous Materials Reference Book	First Aid Manual for Chemical Accidents	Sax's Dangerous Properties of Industrial Materials	Hazardous Chemicals Desk Reference	Rapid Guide to Hazardous Chemicals in the Workplace	Fire Protection Guide on Hazardous Materials (NFPA)	Firefighter's Handbook of Hazardous Materials	Pocket Guide to Chemical Hazards (NIOSH)	Chemical Hazard Response Information System (CHRIS)	Emergency Handling of Hazardous Materials in Surface Transportation (AAR)	Emergency Action Guides (EAG)	Chemical Data Notebook: A User's Manual	Condensed Chemical Dictionary
AMERICAN PALM KERNEL OIL Synonym: OILS EDIBLE TUCUM											OTC 911				
AMERICIUM					135	44			25						52
AMIDOETHANE	1036	68							25						
AMINOBENZENE Synonym: ANILINE	1547	57	85	12	254	72	13	29 –16	25	40	ANL 64	79		A12	54
AMINOBENZENE Synonym: ANILINE OIL	1547	57	85	12	254	72	13	29 –16	32	40	ANL 64	79	A10.0	A12	54
AMINOBENZOIC ACID-meta					147										54
AMINOCAPROIC LACTAM Synonym: CAPROLACTAM				13	680	238	39				CLS 259		C1.5		55
AMINOCHLOROPHENOL	2673	53							26						
AMINOCYCLOHEXANE Synonym: CYCLOHEXYLAMINE	2357	68	249		994	336	60	60 –30	26		CHA 243	285			56
AMINOETHANE Synonym: ETHYLAMINE	1036	68	366	16	1573	526	90	–48	26	106	EAM 505	427	E2.2.1		56
AMINOETHOXYETHANOL	3055	60							26						
AMINOETHOXYETHANOL	1760	60													
AMINOETHYL ALCOHOL Synonym: MONOETHANOLAMINE	2491	60	594					118–	206		MEA 749		M5.0		795
AMINOETHYL ALCOHOL-beta Synonym: MONOETHANOLAMINE	2491	60	594					118–	26		MEA 749		M5.0		795
AMINOETHYL MORPHOLINE-n					174				26						
AMINOETHYLENE Synonym: ETHYLENEIMINE	1185	30	380	17	1610	544	95	86 –51	144	110	ETI 567	437	E2.6.1		490
AMINOETHYLETHANOLAMINE (-n)	1760	60	47		172				26		AEA 30	50			57

HAZARDOUS MATERIALS REFERENCE BOOKS INDEX

CHEMICAL OR MATERIAL NAME	UN/NA Number	Guide Number (DOT)	Firefighter's Hazardous Materials Reference Book	First Aid Manual for Chemical Accidents	Sax's Dangerous Properties of Industrial Materials	Hazardous Chemicals Desk Reference	Rapid Guide to Hazardous Chemicals in the Workplace	Fire Protection Guide on Hazardous Materials (NFPA)	Firefighter's Handbook of Hazardous Materials	Pocket Guide to Chemical Hazards (NIOSH)	Chemical Hazard Response Information System (CHRIS)	Emergency Handling of Hazardous Materials in Surface Transportation (AAR)	Emergency Action Guides (EAG)	Chemical Data Notebook: A User's Manual	Condensed Chemical Dictionary
AMINOETHYLPIPERAZINE (-n)	2815	60			175	51			26		AEP 32	51			57
AMINOFORM Synonym: **HEXAMETHYLENETETRAMINE**	1328	32	434		1856	625			160		HMT 639				57
AMINOMERCURIC CHLORIDE Synonym: **MERCURIC AMMONIUM CHLORIDE**	1630	53							186		MCC 735	598			58
AMINOMETHANE Synonym: **METHYLAMINE**	1235	68		20	2253	771	136	109–66	200	144	MTA 804				58
AMINOMETHANE Synonym: **MONOMETHYLAMINE**	1061	19	596						26				M6.0		796
AMINOMETHANE Synonym: **METHYLAMINE**	1061	68	583	20	2253	771	136	109–66	26	144	MTA 804	641			58
AMINOPHEN Synonym: **ANILINE OIL**	1547	57	85		254	72	13		32		ANL 64	79	A10.0	A12	79
AMINOPHENOL	2512	55							26			52			
AMINOPHENOL-para	2512	55							26						59
AMINOPROPYL MORPHOLINE-n	1760/2735	60			207				27			53			60
AMINOPROPYLDIETHANOLAMINE	1760/2735	60			206	55			27			52			
AMINOPROPYLMORPHOLINE (-n)	1760/2735	60			207				27			53			60
AMINOPROPYLPIPERAZINE	1760	60							27						
AMINOPYRIDINE	2671/2810	55		11					27	36		53			
AMINOTOLUENE-alpha Synonym: **BENZYLAMINE**			126		396				43		BZM 197				134
AMIODOXYL BENZOATE					219										
AMMATE Synonym: **AMMONIUM SULFAMATE**	9089/3077	31	69		234	66	12		30	38	ASM 92	68			62

CHEMICAL OR MATERIAL NAME	UN/NA Number	Guide Number (DOT)	Firefighter's Hazardous Materials Reference Book	First Aid Manual for Chemical Accidents	Sax's Dangerous Properties of Industrial Materials	Hazardous Chemicals Desk Reference	Rapid Guide to Hazardous Chemicals in the Workplace	Fire Protection Guide on Hazardous Materials (NFPA)	Firefighter's Handbook of Hazardous Materials	Pocket Guide to Chemical Hazards (NIOSH)	Chemical Hazard Response Information System (CHRIS)	Emergency Handling of Hazardous Materials in Surface Transportation (AAR)	Emergency Action Guides (EAG)	Chemical Data Notebook: A User's Manual	Condensed Chemical Dictionary
AMMONERIC Synonym: AMMONIUM CHLORIDE	9085/3077	31	56		223	60	11		28		AMC 49	57			64
AMMONIA Synonym: ANHYDROUS AMMONIA	1005	15	48	11	220	57	11	25 –14	27	38	AMA 47	78	A9.0	A 9	62
AMMONIA ANHYDROUS Synonym: AMMONIA	1005	15	48	11	220	57	11	25 –14	27	38	AMA 47	78	A9.0	A 9	62
AMMONIA AQUA Synonym: AMMONIUM HYDROXIDE	2672	60	62	12	228	62	11		27		AMH 52	60	A8.0	A10	66
AMMONIA GAS Synonym: ANHYDROUS AMMONIA	1005	15	48	11	220	57	11	25 –	27		AMA 47	78	A9.0	A 9	62
AMMONIA HYDROXIDE Synonym: AMMONIUM HYDROXIDE	2672	60	62	12	228	62	11		29		AMH 52	60	A8.0	A10	66
AMMONIA SOAP Synonym: AMMONIUM OLEATE											AOL 69				67
AMMONIA SOLUTION Synonym: AMMONIUM HYDROXIDE	2672	60	62	11	220	57	11		27	38	AMH 52	54	A8.0	A10	66
AMMONIA SOLUTION	1005	15	48	11	220	57	11	25 –	27	38	AMA 47			A 9	
AMMONIA SOLUTION	2073	15		11	220	57			27	38		53			
AMMONIA WATER Synonym: AMMONIUM HYDROXIDE	2672	60	62	11	228	57	11		27		AMH 52	60	A8.0	A10	66
AMMONIATED MERCURY Synonym: MERCURIC AMMONIUM CHLORIDE	1630	53							186		MCC 735	598			
AMMONIOFORMALDEHYE Synonym: HEXAMETHYLENETETRAMINE	1328	32	434		1856	625			160		HMT 639				598
AMMONIUM ACAMMONIUM ACETATE	9188														
AMMONIUM ACETATE	9079/3077	31	49		221	58					AAT 6	55			63

HAZARDOUS MATERIALS REFERENCE BOOKS INDEX

CHEMICAL OR MATERIAL NAME	UN/NA Number	Guide Number (DOT)	Firefighter's Hazardous Materials Reference Book	First Aid Manual for Chemical Accidents	Sax's Dangerous Properties of Industrial Materials	Hazardous Chemicals Desk Reference	Rapid Guide to Hazardous Chemicals in the Workplace	Fire Protection Guide on Hazardous Materials (NFPA)	Firefighter's Handbook of Hazardous Materials	Pocket Guide to Chemical Hazards (NIOSH)	Chemical Hazard Response Information System (CHRIS)	Emergency Handling of Hazardous Materials in Surface Transportation (AAR)	Emergency Action Guides (EAG)	Chemical Data Notebook: A User's Manual	Condensed Chemical Dictionary
AMMONIUM ACID FLUORIDE	1727	60	52						28		ABF 8				63
Synonym: AMMONIUM BIFLUORIDE															
AMMONIUM AMIDOSULFONATE	9089/3077	31	69		234	66	12		30	38	ASM 92	68			69
Synonym: AMMONIUM SULFAMATE															
AMMONIUM AMIDOSULPHATE	9089/3077	31	69		234	66	12		30	38	ASM 92	68			69
Synonym: AMMONIUM SULFAMATE															
AMMONIUM AMINOFORMATE	9083/3077	31	54		223	59			28		ACM 22	57			64
Synonym: AMMONIUM CARBAMATE															
AMMONIUM ARSENATE	1546	53							27			55			63
AMMONIUM AZIDE					221	58			27						
AMMONIUM BENZOATE	9080/3077		50						27		ABZ 12	55			63
AMMONIUM BICARBONATE	9081/3077		51		221	58			27		ABC 7	56			63
AMMONIUM BICHROMATE	1439	35	59		222	59		26 –	27		AMD 50	58			63
Synonym: AMMONIUM DICHROMATE															
AMMONIUM BIFLUORIDE	2817	60							28						63
AMMONIUM BIFLUORIDE	1727	60	52						28		ABF 8				63
AMMONIUM BISULFATE	2506	60							28						64
AMMONIUM BISULFIDE	2683	28	70	12					30		ASF 90	68	A8.2		64
Synonym: AMMONIUM SULFIDE SOLUTION															
AMMONIUM BISULFITE SOLID	2693/3077	60	53						28		ASU 96	56			
AMMONIUM BISULFITE SOLUTION	2693	60	53						28		ASU 96				
AMMONIUM BOROFLUORIDE	9088/3077	31	60		226	61			28		AFB 33	59			
Synonym: AMMONIUM FLUOBORATE															
AMMONIUM BROMATE					222	59			28						
AMMONIUM BROMIDE					222	59					ANB 62				64

CHEMICAL OR MATERIAL NAME	UN/NA Number	Guide Number (DOT)	Firefighter's Hazardous Materials Reference Book	First Aid Manual for Chemical Accidents	Sax's Dangerous Properties of Industrial Materials	Hazardous Chemicals Desk Reference	Rapid Guide to Hazardous Chemicals in the Workplace	Fire Protection Guide on Hazardous Materials (NFPA)	Firefighter's Handbook of Hazardous Materials	Pocket Guide to Chemical Hazards (NIOSH)	Chemical Hazard Response Information System (CHRIS)	Emergency Handling of Hazardous Materials in Surface Transportation (AAR)	Emergency Action Guides (EAG)	Chemical Data Notebook: A User's Manual	Condensed Chemical Dictionary
AMMONIUM CARBAMATE	9083/3077	31	54		223	59			28		ACM 22	57			64
AMMONIUM CARBONATE	9084/3077	31	55	11	223	60			28		ACB 14	57			64
AMMONIUM CHLORATE	1445	42		11	223	60			28						64
AMMONIUM CHLORIDE	9085/3077	31	56		223	60	11		28		AMC 49	57			64
AMMONIUM CHLOROPLATINATE					224	60			28						65
AMMONIUM CHROMATE	9086/3077	31	57		224	61			28		ACH 19	58			65
AMMONIUM CITRATE	9087/3077		58		225	61			28		ACI 20	58			65
AMMONIUM CITRATE DIBASIC	9087/3077	31	58		225	61			28		ACI 20	58			65
Synonym: AMMONIUM CITRATE															
AMMONIUM CUPRIC SULFATE	9110										CSN 312				315
Synonym: COPPER SULFATE AMMONIATED															
AMMONIUM CYANIDE					225	61			28						
AMMONIUM DECABORATE OCTAHYDRATE											APB 71				68
Synonym: AMMONIUM PENTABORATE															
AMMONIUM DICHROMATE	1439	35	59					26 –	28		AMD 50	58			65
AMMONIUM DINITRO-o-CRESOLATE	1843	42							28						
AMMONIUM DISULFATONICKELATE (II)	9138/3077	31	612		2490				209		NAS 821	676			818
Synonym: NICKEL AMMONIUM SULFATE															
AMMONIUM FERRIC OXALATE TRIHYDRATE	9119/9188	31	385						147		FAO 578	445			512
Synonym: FERRIC AMMONIUM OXALATE															
AMMONIUM FERROUS SULFATE	9122/3077	31	392						147		FAS 579	449			517
Synonym: FERROUS AMMONIUM SULFATE															
AMMONIUM FLUOBORATE	9088/3077	31	60		226	61			28		AFB 33	59			
AMMONIUM FLUORIDE	2505	54	61		226	61		26 –	28		AFR 35	59			65
AMMONIUM FLUOROBORATE	9088	31							28						

HAZARDOUS MATERIALS REFERENCE BOOKS INDEX

CHEMICAL OR MATERIAL NAME	UN/NA Number	Guide Number (DOT)	Firefighter's Hazardous Materials Reference Book	First Aid Manual for Chemical Accidents	Sax's Dangerous Properties of Industrial Materials	Hazardous Chemicals Desk Reference	Rapid Guide to Hazardous Chemicals in the Workplace	Fire Protection Guide on Hazardous Materials (NFPA)	Firefighter's Handbook of Hazardous Materials	Pocket Guide to Chemical Hazards (NIOSH)	Chemical Hazard Response Information System (CHRIS)	Emergency Handling of Hazardous Materials in Surface Transportation (AAR)	Emergency Action Guides (EAG)	Chemical Data Notebook: A User's Manual	Condensed Chemical Dictionary
AMMONIUM FLUOROSILICATE Synonym: AMMONIUM SILICOFLUORIDE	2854	53	68						28		ASL 91	67			69
AMMONIUM FORMATE					226						AFM 34				66
AMMONIUM GLUCONATE											AGC 36				66
AMMONIUM HEXANITRO COBALTATE					227				28						
AMMONIUM HYDRATE Synonym: AMMONIUM HYDROXIDE	2672	60	62	12	228	62	11		28		AMH 52	60	A8.0	A10	66
AMMONIUM HYDROGEN CARBONATE Synonym: AMMONIUM BICARBONATE	9081/3077	31	51		221	58			27		ABC 7	56			66
AMMONIUM HYDROGEN FLUORIDE Synonym: AMMONIUM BIFLUORIDE	1727	60	52		227	62			28		ABF 8	59			66
AMMONIUM HYDROGEN FLUORIDE	2817	60			227	62						59			66
AMMONIUM HYDROGEN SULFATE	2506	60			227				28			60			66
AMMONIUM HYDROGEN SULFIDE Synonym: AMMONIUM SULFIDE SOLUTION	2683/1993	28	70	12					28		ASF 90	68	A8.2		
AMMONIUM HYDROGEN SULFIDE SOLUTION Synonym: AMMONIUM SULFIDE	2683/1993	28	70	12					28		ASF 90	68	A8.2		
AMMONIUM HYDROGEN SULFITE Synonym: AMMONIUM BISULFITE	2693/3077	60	53						28		ASU 96	56			
AMMONIUM HYDROSULFIDE Synonym: AMMONIUM SULFIDE SOLUTION	2683	28	70	12	227	62			28		ASF 90	68	A8.2		66
AMMONIUM HYDROSULFIDE SOLUTION	2683	28			227	62			28			60			66
AMMONIUM HYDROSULFITE Synonym: AMMONIUM BISULFITE	2693/3077	60	53						28		ASU 96	56			
AMMONIUM HYDROXIDE	2672	60	62	12	228	62	11				AMH 52	60	A8.0	A10	66
AMMONIUM HYPOPHOSPHITE					228				29		AHP 37				66

CHEMICAL OR MATERIAL NAME	UN/NA Number	Guide Number (DOT)	Firefighter's Hazardous Materials Reference Book	First Aid Manual for Chemical Accidents	Sax's Dangerous Properties of Industrial Materials	Hazardous Chemicals Desk Reference	Rapid Guide to Hazardous Chemicals in the Workplace	Fire Protection Guide on Hazardous Materials (NFPA)	Firefighter's Handbook of Hazardous Materials	Pocket Guide to Chemical Hazards (NIOSH)	Chemical Hazard Response Information System (CHRIS)	Emergency Handling of Hazardous Materials in Surface Transportation (AAR)	Emergency Action Guides (EAG)	Chemical Data Notebook: A User's Manual	Condensed Chemical Dictionary
AMMONIUM HYPOSULFITE **Synonym: AMMONIUM THIOSULFATE**	9093/3077	31							30		ATF 100	70			
AMMONIUM IODIDE					228	63			29		AID 38				66
AMMONIUM LACTATE											ALT 46				
AMMONIUM LACTATE SYRUP **Synonym: AMMONIUM LACTATE**											ALT 46				
AMMONIUM LAURYL SULFATE											ALS 45				
AMMONIUM MERCAPTAN **Synonym: AMMONIUM SULFIDE SOLUTION**	2683/1993	28	70	12					30		ASF 90	68	A8.2		
AMMONIUM METAVANADATE	2859/2811	53							29			61			67
AMMONIUM MOLYBDATE					229	63					AMB 48				67
AMMONIUM MONOSULFITE **Synonym: AMMONIUM BISULFITE**	2693/3077	60	53						28		ASU 96	56			
AMMONIUM MURIATE **Synonym: AMMONIUM CHLORIDE**	9085/3077		56		223	60	11		28		AMC 49	57			64
AMMONIUM NICKEL SULFATE **Synonym: NICKEL AMMONIUM SULFATE**	9138/3077		612		2490						NAS 821	676			67
AMMONIUM NITRATE	1942	43	63		229	64		26 –	29		AMN 56	62	A8.1	A11	67
AMMONIUM NITRATE	0222	46			229	64			29			61		A11	67
AMMONIUM NITRATE CARBONATE MIXTURE	2068	43										64	A8.1		67
AMMONIUM NITRATE FERTILIZER	2072	43							29			63		A11	
AMMONIUM NITRATE FERTILIZER	0223	46							29			62	A8.1	A11	
AMMONIUM NITRATE FERTILIZER	2067/1942	43							29			63	A8.1	A11	
AMMONIUM NITRATE FERTILIZER	2071	35							29			63	A8.1	A11	

HAZARDOUS MATERIALS REFERENCE BOOKS INDEX

CHEMICAL OR MATERIAL NAME	UN/NA Number	Guide Number (DOT)	Firefighter's Hazardous Materials Reference Book	First Aid Manual for Chemical Accidents	Sax's Dangerous Properties of Industrial Materials	Hazardous Chemicals Desk Reference	Rapid Guide to Hazardous Chemicals in the Workplace	Fire Protection Guide on Hazardous Materials (NFPA)	Firefighter's Handbook of Hazardous Materials	Pocket Guide to Chemical Hazards (NIOSH)	Chemical Hazard Response Information System (CHRIS)	Emergency Handling of Hazardous Materials in Surface Transportation (AAR)	Emergency Action Guides (EAG)	Chemical Data Notebook: A User's Manual	Condensed Chemical Dictionary
AMMONIUM NITRATE FERTILIZER WITH AMMONIUM SULFATE	2069	43							29		ANS 66		A8.1	A11	
AMMONIUM NITRATE FERTILIZER WITH CALCIUM CARBONATE	2068	43							29				A8.1		
AMMONIUM NITRATE FERTILIZER WITH PHOSPHATE OR POTASH	2070	43							29		ANP 65			A11	
AMMONIUM NITRATE FERTILIZERS TYPE B	1760	60													
AMMONIUM NITRATE FUEL OIL MIXTURES	0331	46							29			62			
AMMONIUM NITRATE HOT CONCENTRATED SOLUTION	2426	35							29				A8.1	A11	
AMMONIUM NITRATE MAGNESIA AND PULP MIXTURE	1479/3139	35										64			
AMMONIUM NITRATE PHOSPHATE MIXTURE	2070	43	64						29		ANP 65	64		A11	
AMMONIUM NITRATE SODIUM NITRATE MIXTURE	1479	35													
AMMONIUM NITRATE SOLUTION	2426	35			229				29			65	A8.1	A11	
AMMONIUM NITRATE SULFATE MIXTURE	2069	43							29		ANS 66		A8.1	A11	
AMMONIUM NITRATE UREA SOLUTION											ANU 68				
AMMONIUM NITRITE					230										
AMMONIUM OLEATE											AOL 69				67
AMMONIUM OXALATE	2449/3077	54	65		230	64			29		AOX 70	65			67
AMMONIUM OXALATE HYDRATE Synonym: AMMONIUM OXALATE	2449/3077	54	65		230	64			29		AOX 70	65			67
AMMONIUM PENTABORATE											APB 71				68
AMMONIUM PENTABORATE TETRAHYDRATE Synonym: AMMONIUM PENTABORATE											APB 71				68
AMMONIUM PENTACHLOROZINCATE Synonym: ZINC AMMONIUM CHLORIDE	9154/3077		887						288		ZAC1191	964			1242

CHEMICAL OR MATERIAL NAME	UN/NA Number	Guide Number (DOT)	Firefighter's Hazardous Materials Reference Book	First Aid Manual for Chemical Accidents	Sax's Dangerous Properties of Industrial Materials	Hazardous Chemicals Desk Reference	Rapid Guide to Hazardous Chemicals in the Workplace	Fire Protection Guide on Hazardous Materials (NFPA)	Firefighter's Handbook of Hazardous Materials	Pocket Guide to Chemical Hazards (NIOSH)	Chemical Hazard Response Information System (CHRIS)	Emergency Handling of Hazardous Materials in Surface Transportation (AAR)	Emergency Action Guides (EAG)	Chemical Data Notebook: A User's Manual	Condensed Chemical Dictionary
AMMONIUM PERCHLORATE	0402	46		12	231	64			29			65			68
AMMONIUM PERCHLORATE	1442	43	66	12	231	64		27 –	29		AMP 57	66			68
AMMONIUM PERFLUOROOCTANOATE					231	65	12		30						68
AMMONIUM PERMANGANATE	9190/1482	43				65		27 –	30			66			68
AMMONIUM PEROXYCHROMATE					232	65			30						
AMMONIUM PEROXYDISULFATE Synonym: AMMONIUM PERSULFATE	1444	35	67		232	65	12		30		APE 73	66			68
AMMONIUM PERSULFATE	1444	35	67		232	65	12		30		APE 73	66			68
AMMONIUM PHOSPHATE					232	66					APP 78				68
AMMONIUM PHOSPHATE DIBASIC Synonym: AMMONIUM PHOSPHATE					232	66					APP 78				68
AMMONIUM PHOSPHIDE					233	66			30						
AMMONIUM PICRATE	1310	33			233	66			30		API 76	66			69
AMMONIUM PICRATE	0004	46			233	66			30			66			69
AMMONIUM POLYSULFIDE	2818	60			233				30			67			69
AMMONIUM POLYVANADATE	2861/2811	55							30			67			
AMMONIUM RHODANIDE Synonym: AMMONIUM THIOCYANATE	9092/3077		73		235	67			30		AMT 60	69			70
AMMONIUM SALT Synonym: AMMONIUM NITRATE	1942	43	63		234			26 –	29		AMN 56	62	A8.1	A11	67
AMMONIUM SALT Synonym: AMMONIUM BENZOATE	9080/3077		50			64			27		ABZ 12	55			63
AMMONIUM SALT Synonym: AMMONIUM ACETATE	9079/3077		49		234	58					AAT 6	55			63
AMMONIUM SILICOFLUORIDE	2854	53	68						30		ASL 91	67			69

CHEMICAL OR MATERIAL NAME	UN/NA Number	Guide Number (DOT)	Firefighter's Hazardous Materials Reference Book	First Aid Manual for Chemical Accidents	Sax's Dangerous Properties of Industrial Materials	Hazardous Chemicals Desk Reference	Rapid Guide to Hazardous Chemicals in the Workplace	Fire Protection Guide on Hazardous Materials (NFPA)	Firefighter's Handbook of Hazardous Materials	Pocket Guide to Chemical Hazards (NIOSH)	Chemical Hazard Response Information System (CHRIS)	Emergency Handling of Hazardous Materials in Surface Transportation (AAR)	Emergency Action Guides (EAG)	Chemical Data Notebook: A User's Manual	Condensed Chemical Dictionary
AMMONIUM STEARATE											AMR 58				69
AMMONIUM STEARATE DISPERSION Synonym: AMMONIUM STEARATE											AMR 58				69
AMMONIUM SULFAMATE	9089/3077	31	69			66	12		30	38	ASM 92	68			69
AMMONIUM SULFATE	2506	60				67					AMS 59		A8.1.5		70
AMMONIUM SULFATE NITRATE	1477/3077	35										68			70
AMMONIUM SULFHYDRATE Synonym: AMMONIUM SULFIDE SOLUTION	2683/1993	28	70	12					30		ASF 90	68	A8.2		
AMMONIUM SULFHYDRATE SOLUTION Synonym: AMMONIUM SULFIDE	2683/1993	28	70						30		ASF 90	68	A8.2		
AMMONIUM SULFIDE SOLUTION	2683/1993	28	70	12					30		ASF 90	68	A8.2		
AMMONIUM SULFITE	9090/3077		71						30		AMF 51	69			70
AMMONIUM SULFOCYANIDE Synonym: AMMONIUM THIOCYANATE	9092/3077		73		235	67			30		AMT 60	69			70
AMMONIUM TARTRATE	9091/3077		72						30		ATR 106	69			70
AMMONIUM TETRAFLUOBORATE Synonym: AMMONIUM FLUOBORATE	9088/3077	31	60		226	61			28		AFB 33	59			
AMMONIUM THIOCYANATE	9092/3077		73		235	67			30		AMT 60	69			70
AMMONIUM THIOCYANATE LIQUOR	9188/3082	31	74									70			
AMMONIUM THIOGLYCOLATE									30						70
AMMONIUM THIOSULFATE	9093/3077	31							30		ATF 100	70			70
AMMONIUM THIOSULFATE	9188	31	75												70
AMMONIUM TRICHLOROACETATE					235	67			30						
AMMONIUM TRIOXALATOFERRATE III TRIHYDRATE Synonym: FERRIC AMMONIUM OXALATE	9119/9188	31	385						147		FAO 578	445			512

CHEMICAL OR MATERIAL NAME	UN/NA Number	Guide Number (DOT)	Firefighter's Hazardous Materials Reference Book	First Aid Manual for Chemical Accidents	Sax's Dangerous Properties of Industrial Materials	Hazardous Chemicals Desk Reference	Rapid Guide to Hazardous Chemicals in the Workplace	Fire Protection Guide on Hazardous Materials (NFPA)	Firefighter's Handbook of Hazardous Materials	Pocket Guide to Chemical Hazards (NIOSH)	Chemical Hazard Response Information System (CHRIS)	Emergency Handling of Hazardous Materials in Surface Transportation (AAR)	Emergency Action Guides (EAG)	Chemical Data Notebook: A User's Manual	Condensed Chemical Dictionary
AMMONIUM ZINC CHLORIDE **Synonym: ZINC AMMONIUM CHLORIDE**	9154/3077		887						288		ZAC1191	964			1242
AMMONIUM-m-VANADATE	2859	53			235	67			30						
AMMUNITION	2016	15							30			73			
AMMUNITION	2017	58							30						
AMORPHOUS PHOSPHORUS **Synonym: PHOSPHORUS RED**	1338	32	692		2791	948	180	138–	232		PPR 980	759			
AMP					237				30						71
AMPHETAMINE					237										72
AMS **Synonym: AMMONIUM SULFAMATE**	9089/3077	31	69		234	66	12		30	38	ASM 92	68			69
AMYL ACETATE MIXED ISOMERS **Synonym: AMYL ACETATE (-n)**	1104	26	77	12	241	69	12	-14	31	38	AML 54	73			73
AMYL ACETATE-Iso				12					31						
AMYL ACETATE-n **Synonym: AMYL ACETATE**	1104	26	77	12	241	68	12	-14	31	38	AML 54	73			73
AMYL ACETATE-sec	1104	26		12	241	68	13	-14	30	38	AAS 5				73
AMYL ACETATE-tert	1104	26		12							AYA 111				
AMYL ACID PHOSPHATE	2819	60							31			74			73
AMYL ALCOHOL	1105/1987	26	78		242	69		-14	31		AAN 4	74			73
AMYL ALCOHOL-n **Synonym: AMYL ALCOHOL**	1105/1987	26	78		242	69		-14	31		AAN 4	74			73
AMYL ALCOHOL-N-tert-REFINED	1105	26							31						
AMYL ALCOHOL-sec	1987	26						-14							74
AMYL ALCOHOL-tert	1987	26													

CHEMICAL OR MATERIAL NAME	UN/NA Number	Guide Number (DOT)	Firefighter's Hazardous Materials Reference Book	First Aid Manual for Chemical Accidents	Sax's Dangerous Properties of Industrial Materials	Hazardous Chemicals Desk Reference	Rapid Guide to Hazardous Chemicals in the Workplace	Fire Protection Guide on Hazardous Materials (NFPA)	Firefighter's Handbook of Hazardous Materials	Pocket Guide to Chemical Hazards (NIOSH)	Chemical Hazard Response Information System (CHRIS)	Emergency Handling of Hazardous Materials in Surface Transportation (AAR)	Emergency Action Guides (EAG)	Chemical Data Notebook: A User's Manual	Condensed Chemical Dictionary
AMYL ALDEHYDE	2058/1993	26	867		3475	1167	226	−92	31		VAL1171	945			74
Synonym: VALERALDEHYDE (-n)															
AMYL BENZENE-tert					242				31						
AMYL BIPHENYL					243				31						
AMYL BROMIDE	2343	27						−15	31						
AMYL BROMIDE-d	2343	27			243				31						
AMYL BUTYRATE-n	2620/1993	27			243			−15	31			74			74
AMYL CHLORIDE	1107	26	79					−15	31		AMY 61	75			74
Synonym: AMYL CHLORIDE-n															
AMYL CHLORIDE-n	1107	26	79					−15	31		AMY 61	75			74
Synonym: AMYL CHLORIDE															
AMYL CHLORIDE-tert	1107	26							31						
AMYL ETHER (-n)								−15	31						75
AMYL FORMATE (-n)	1109	26			245	70		−15	31			75			75
AMYL HYDRATE	1105								31						75
AMYL HYDRIDE	1265	27	668	22	2687	913	172	−77	31		PTA 990	730	P1.0		75
Synonym: PENTANE															
AMYL HYDROSULFIDE	1111	27	80					28 −15	31		AMM 55	75			75
Synonym: AMYL MERCAPTAN (-n)															
AMYL LACTATE					245	70		−15	31						
AMYL LAURATE					245	70		−15							
AMYL MERCAPTAN	1111	27	80					28 −15	31		AMM 55	75			75
AMYL MERCAPTAN-n	1111	27	80					28 −15	31		AMM 55	75			75
Synonym: AMYL MERCAPTAN															
AMYL METHYL ALCOHOL	2053	26			246	71			31						

CHEMICAL OR MATERIAL NAME	UN/NA Number	Guide Number (DOT)	Firefighter's Hazardous Materials Reference Book	First Aid Manual for Chemical Accidents	Sax's Dangerous Properties of Industrial Materials	Hazardous Chemicals Desk Reference	Rapid Guide to Hazardous Chemicals in the Workplace	Fire Protection Guide on Hazardous Materials (NFPA)	Firefighter's Handbook of Hazardous Materials	Pocket Guide to Chemical Hazards (NIOSH)	Chemical Hazard Response Information System (CHRIS)	Emergency Handling of Hazardous Materials in Surface Transportation (AAR)	Emergency Action Guides (EAG)	Chemical Data Notebook: A User's Manual	Condensed Chemical Dictionary
AMYL METHYL KETONE	1110	26	81						31		AMK 53	76			
AMYL METHYL KETONE-n Synonym: AMYL METHYL KETONE	1110	26	81						31		AMK 53	76			
AMYL NITRATE	1112/1993	26	82		246	71		29 –15	31		ANT 67	76			75
AMYL NITRATE-n Synonym: AMYL NITRATE	1112/1993	26	82		246	71		29 –15	31		ANT 67	76			75
AMYL NITRITE (-n) Synonym: AMYL NITRITE-iso	1113	26	83		246	71		–15	31		ANI 63	76			75
AMYL NITRITE-iso	1113	26	83		246	71		–15			ANI 63	76			75
AMYL OLEATE								–15	31						
AMYL PEROXY-2-ETHYLHEXANOATE-tert	2398/3115	52										76			
AMYL PEROXYNEODECANOATE-tert	2891	52													
AMYL PHENOL	1760	60										77			
AMYL PHENOL-ortho								–16	31						
AMYL PHENOL-tert-para								–16	31						76
AMYL PHENYL ETHER								–16	31						
AMYL PROPIONATE					248	72		–16	31						76
AMYL STEARATE								–16	31						
AMYL SULFHYDRATE Synonym: AMYL MERCAPTAN (-n)	1111	27	80					28 –15	31		AMM 55	75			75
AMYL SULFIDES MIXED								–16	32						
AMYL THIOALCOHOL Synonym: AMYL MERCAPTAN	1111	27	80					28 –15	31		AMM 55	75			75
AMYL TOLUENE								–16	32						
AMYL XYLYL ETHER								–16	32						

HAZARDOUS MATERIALS REFERENCE BOOKS INDEX

CHEMICAL OR MATERIAL NAME	UN/NA Number	Guide Number (DOT)	Firefighter's Hazardous Materials Reference Book	First Aid Manual for Chemical Accidents	Sax's Dangerous Properties of Industrial Materials	Hazardous Chemicals Desk Reference	Rapid Guide to Hazardous Chemicals in the Workplace	Fire Protection Guide on Hazardous Materials (NFPA)	Firefighter's Handbook of Hazardous Materials	Pocket Guide to Chemical Hazards (NIOSH)	Chemical Hazard Response Information System (CHRIS)	Emergency Handling of Hazardous Materials in Surface Transportation (AAR)	Emergency Action Guides (EAG)	Chemical Data Notebook: A User's Manual	Condensed Chemical Dictionary
AMYLAMINE (-n)	1106	68						28 −14				77			74
AMYLAMINE-sec								−14	32						74
AMYLBENZENE (-n)								−15	31						74
AMYLCARBINOL Synonym: HEXANOL (-n)	2282/1993	26	436	18					31		HXN 652	519			74
AMYLCARBINOL-n Synonym: HEXANOL (-n)	2282/1993	26	436	18					31		HXN 652	519			74
AMYLENE	1108	26			245	70			32			77			
AMYLENE CHLORIDE	1152	27							32						
AMYLENE-cis-beta								−15	32						
AMYLENE-n-alpha Synonym: 1-PENTENE	1108	26			245				32		PTE 994				75
AMYLENE-trans-beta								−15	32						
AMYLPEROXYBENZOATE-tert	3044	48							32						
AMYLPEROXYPIVALATE-tert	2957	52							32						
AMYLPHENOL-sec-para								−16							
AMYLTRICHLOROSILANE	1728	29	84					−16	32		ATS 107	78			76
AMYLTRICHLOROSILANE (-n)	1728	29	84					−16	32		ATS 107	78			76
ANABASINE					249				32						76
ANESTHESIA ETHER Synonym: ETHYL ETHER	1155	26	354	16	1614	546	96	87 −52	32	112	EET 524	346	E2.8	E 3	492
ANESTHETIC ETHER Synonym: ETHYL ETHER	1155	26	354	16	1614	546	96	87 −52	137	112	EET 524	346	E2.8	E 3	492
ANGLISLITE Synonym: LEAD SULFATE	1794	60			2105	716			179		LSF 709	575			692

CHEMICAL OR MATERIAL NAME	UN/NA Number	Guide Number (DOT)	Firefighter's Hazardous Materials Reference Book	First Aid Manual for Chemical Accidents	Sax's Dangerous Properties of Industrial Materials	Hazardous Chemicals Desk Reference	Rapid Guide to Hazardous Chemicals in the Workplace	Fire Protection Guide on Hazardous Materials (NFPA)	Firefighter's Handbook of Hazardous Materials	Pocket Guide to Chemical Hazards (NIOSH)	Chemical Hazard Response Information System (CHRIS)	Emergency Handling of Hazardous Materials in Surface Transportation (AAR)	Emergency Action Guides (EAG)	Chemical Data Notebook: A User's Manual	Condensed Chemical Dictionary
ANHYDRIDE OF AMMONIUM CARBONATE Synonym: AMMONIUM CARBAMATE	9083/3077	31	54		223	59			28		ACM 22	57			64
ANHYDROL Synonym: ETHYL ALCOHOL	1170	26	344	16	1572	525	90	-48	136		EAL 504	408	E2.2	E 2	478
ANHYDRONE Synonym: MAGNESIUM PERCHLORATE	1475	35	525						183		MPC 783	590			721
ANHYDROUS ALUMINUM CHLORIDE Synonym: ALUMINUM CHLORIDE	1726	39	43	11	127	41		24 –	24		ACL 21	44			44
ANHYDROUS AMMONIA Synonym: AMMONIA ANHYDROUS	1005	15	48	11	220	57	11	25 –14	32	38	AMA 47	78	A9.0	A 9	62
ANHYDROUS ETHANOL Synonym: ETHYL ALCOHOL	1170	26	344	16	1572	525	90	-48	136		EAL 504	408	E2.2	E 2	478
ANHYDROUS HYDRAZINE	2029	28							32						
ANHYDROUS HYDROBROMIC ACID Synonym: HYDROGEN BROMIDE	1048	15	446	18					164	124	HBR 621	528	H4.2.1		616
ANHYDROUS HYDROCHLORIC ACID Synonym: HYDROGEN CHLORIDE	1050	15	447	18	1900	643	115	98 –	164	126	HDC 630	528	H4.3	H 3	616
ANHYDROUS HYDROCYANIC ACID Synonym: HYDROCYANIC ACID	1051	13	442	18	1896	641	114	-59	164				H4.1	H 2	613
ANHYDROUS HYDROFLUORIC ACID Synonym: HYDROGEN FLUORIDE	1052	15	449					99 –	164	126	HFX 636	530	H4.4	H 4	616
ANHYDROUS HYDROGEN BROMIDE Synonym: HYDROGEN BROMIDE	1048	15	446	18					164	124	HBR 621	528	H4.2.1		616
ANILIN Synonym: ANILINE OIL	1547	57	85	12	254	72	13	29 –16	32	40	ANL 64	79	A10.0	A12	79
ANILINE Synonym: ANILINE OIL	1547	57	85	12	254	72	13	29 –16	32	40	ANL 64	79	A10.0	A12	79

CHEMICAL OR MATERIAL NAME	UN/NA Number	Guide Number (DOT)	Firefighter's Hazardous Materials Reference Book	First Aid Manual for Chemical Accidents	Sax's Dangerous Properties of Industrial Materials	Hazardous Chemicals Desk Reference	Rapid Guide to Hazardous Chemicals in the Workplace	Fire Protection Guide on Hazardous Materials (NFPA)	Firefighter's Handbook of Hazardous Materials	Pocket Guide to Chemical Hazards (NIOSH)	Chemical Hazard Response Information System (CHRIS)	Emergency Handling of Hazardous Materials in Surface Transportation (AAR)	Emergency Action Guides (EAG)	Chemical Data Notebook: A User's Manual	Condensed Chemical Dictionary
ANILINE 2-NITRO Synonym: NITROANILINE (-ortho)	1661	55		21	2514	859			211		NTA 851	687	N3.0		824
ANILINE CHLORIDE	1548	53							32						79
ANILINE HYDROCHLORIDE	1548	53						–16	32			79			80
ANILINE OIL Synonym: ANILINE	1547	57	85	12	254	72	13	29 –16	32	40	ANL 64	79	A10.0	A12	79
ANILINOBENZENE Synonym: DIPHENYLAMINE			319	16	1457	485	84	–45	32		DAM 337				428
ANILINOMETHANE Synonym: METHYLANILINE (-n)	2294	57	584		2257	772	137		201		MAN 724				757
ANILITE					257				32						
ANIMAL CARBON Synonym: CHARCOAL	1361	32	195		739	258			68		CHC 244	203			248
ANISIDINE	2431	55								40		80			
ANISIDINE-ortho	2431	55		12	259	75	13	–16	33	40					81
ANISIDINE-para	2431	55		12	259	75	14		32	40					81
ANISOLE	2222	26			260	75		–17	33			81			81
ANISOYL CHLORIDE	1729	60	86		261	76			33		ASC 89	81			81
ANISOYL CHLORIDE-para Synonym: ANISOYL CHLORIDE	1729	60	86		261	76			33		ASC 89	81			81
ANISYL CHLORIDE-para	1729	60							33						
ANNULENE Synonym: BENZENE	1114	27	115	12	356	115	21	35 –17	40	44	BNZ 152	112	B1.0	B 1	128
ANODYNON Synonym: ETHYL CHLORIDE	1037	27	350		1586	532	92	82 –49	136	108	ECL 514	412	E2.3		483

CHEMICAL OR MATERIAL NAME	UN/NA Number	Guide Number (DOT)	Firefighter's Hazardous Materials Reference Book	First Aid Manual for Chemical Accidents	Sax's Dangerous Properties of Industrial Materials	Hazardous Chemicals Desk Reference	Rapid Guide to Hazardous Chemicals in the Workplace	Fire Protection Guide on Hazardous Materials (NFPA)	Firefighter's Handbook of Hazardous Materials	Pocket Guide to Chemical Hazards (NIOSH)	Chemical Hazard Response Information System (CHRIS)	Emergency Handling of Hazardous Materials in Surface Transportation (AAR)	Emergency Action Guides (EAG)	Chemical Data Notebook: A User's Manual	Condensed Chemical Dictionary
ANOL Synonym: CYCLOHEXANOL	1993	27	245	14	991	333	59	-30	33	78	CHN 246	280			337
ANONE Synonym: CYCLOHEXANONE	1915	26	246	14	991	334	59	-30	88	78	CCH 228	281	C8.2		337
ANPROLENE Synonym: ETHYLENE OXIDE	1040	69	377	17	1611	544	96	87 -52	144	112	EOX 550	434	E2.7	E 6	491
ANSAR Synonym: CACODYLIC ACID	1572/2811	53	156						62		CDA 235	175			194
ANSUL ETHER 12 Synonym: ETHYLENE GLYCOL DIMETHYL ETHER	2252	27						-51	143		EGD 528				488
ANSUL ETHER 121 Synonym: ETHYLENE GLYCOL DIMETHYL ETHER	2252	27						-51	143		EGD 528				488
ANTHION	1492	35							33						
ANTHON Synonym: TRICHLORFON	2783/3018	55	835						272		TRC1146	917			1169
ANTHRACENE			87		262	76	14	-17	33		ATH 101				83
ANTHRACENE OIL Synonym: CREOSOTE COAL TAR	1137	27	231						33		CCT 233		C6.0		83
ANTHRACIN Synonym: ANTHRACENE			87		262	76	14	-17	33		ATH 101				83
ANTHRANILIC ACID					263	77									84
ANTHRAQUINONE					265	78		-17							84
ANTI FREEZE	1142	26							33						86
ANTIKNOCK COMPOUND Synonym: MOTOR FUEL ANTI KNOCK COMPOUND	1649	56	598					121-	33		MFA 756	660	M7.0		

HAZARDOUS MATERIALS REFERENCE BOOKS INDEX

CHEMICAL OR MATERIAL NAME	UN/NA Number	Guide Number (DOT)	Firefighter's Hazardous Materials Reference Book	First Aid Manual for Chemical Accidents	Sax's Dangerous Properties of Industrial Materials	Hazardous Chemicals Desk Reference	Rapid Guide to Hazardous Chemicals in the Workplace	Fire Protection Guide on Hazardous Materials (NFPA)	Firefighter's Handbook of Hazardous Materials	Pocket Guide to Chemical Hazards (NIOSH)	Chemical Hazard Response Information System (CHRIS)	Emergency Handling of Hazardous Materials in Surface Transportation (AAR)	Emergency Action Guides (EAG)	Chemical Data Notebook: A User's Manual	Condensed Chemical Dictionary
ANTIMONOUS BROMIDE	1549	60	91	12	277				33		ATB 98	84			89
Synonym: ANTIMONY TRIBROMIDE															
ANTIMONY	2871	53		12	269	78	14		33						88
ANTIMONY (III) CHLORIDE	1733	60	92	12	270	78			33		ATM 103	85			89
Synonym: ANTIMONY TRICHLORIDE															
ANTIMONY (V) CHLORIDE	1730	60	88	12	270	79		30 –	33		APC 72				88
Synonym: ANTIMONY PENTACHLORIDE															
ANTIMONY CHLORIDE	1733	60		12	271	78			33						88
ANTIMONY COMPOUND	3141	53		12	271	79						82			88
ANTIMONY COMPOUND	1549	60		12	271	79			33			82			88
ANTIMONY FLUORIDE	1732	59		12					33						88
ANTIMONY HYDRIDE	2676	18		12					33						88
ANTIMONY LACTATE	1550	53		12	271	79			33			82			88
ANTIMONY OXIDE				12	272	80	15								88
ANTIMONY PENTACHLORIDE	1730	60	88	12		79		30 –	33		APC 72				88
ANTIMONY PENTACHLORIDE SOLUTION	1731	60		12				30 –	33			82			88
ANTIMONY PENTAFLUORIDE	1732	59	89	12	273	80		30 –	33		APF 74	83			89
ANTIMONY PENTASULFIDE				12	273				33						89
ANTIMONY PERCHLORIDE	1730	60	88	12		79		30 –	33		APC 72				88
Synonym: ANTIMONY PENTACHLORIDE															
ANTIMONY PERCHLORIDE	1731	60		12											
ANTIMONY POTASSIUM TARTRATE	1551	53	90	12	274	80			33		APT 81	83			89
ANTIMONY POWDER	2871	53		12			14		33	40		83			88
ANTIMONY SULFIDE	1325/1549	32		12				31 –	33			84			
ANTIMONY TRIBROMIDE	1549	60	91	12	277				33		ATB 98	84			89

CHEMICAL OR MATERIAL NAME	UN/NA Number	Guide Number (DOT)	Firefighter's Hazardous Materials Reference Book	First Aid Manual for Chemical Accidents	Sax's Dangerous Properties of Industrial Materials	Hazardous Chemicals Desk Reference	Rapid Guide to Hazardous Chemicals in the Workplace	Fire Protection Guide on Hazardous Materials (NFPA)	Firefighter's Handbook of Hazardous Materials	Pocket Guide to Chemical Hazards (NIOSH)	Chemical Hazard Response Information System (CHRIS)	Emergency Handling of Hazardous Materials in Surface Transportation (AAR)	Emergency Action Guides (EAG)	Chemical Data Notebook: A User's Manual	Condensed Chemical Dictionary
ANTIMONY TRIBROMIDE SOLUTION	1549/1760	60	91	12	277				33		ATB 98	84			
ANTIMONY TRICHLORIDE	1733	60	92	12					33		ATM 103	85			89
ANTIMONY TRICHLORIDE SOLUTION	1733	60	92	12					34		ATM 103	85			
ANTIMONY TRIETHYL				12	277	81			34						
ANTIMONY TRIFLUORIDE	1549/1760	60	93	12					34		ATT 108	86			89
ANTIMONY TRIFLUORIDE SOLUTION	1549/1760	60	93	12					34		ATT 108	86			
ANTIMONY TRIMETHYL				12	277	81			34						
ANTIMONY TRIOXIDE	1549/3077	60	94	12							ATX 109	87			90
ANTIMONY TRISULFIDE	1325	32		12	277	81			34						90
ANTIMONY TRITELLURIDE				12	278	81			34						
ANTU	1651	53			279	82	15		34	40					91
ANYVIM Synonym: ANILINE OIL	1547	57	85		254	72	13		32		ANL 64	79	A10.0	A12	79
AOUARA OIL Synonym: OILS EDIBLE TUCUM											OTC 911				
APOATROPINE					280										
APOCODEINE					281	82			34						
AQUA AMMONIA Synonym: AMMONIUM HYDROXIDE	2672	60	62	12	228	62	11		34		AMH 52	60	A8.0	A10	92
AQUA FORTIS Synonym: NITRIC ACID FUMING	2032	44		21	2509	856		123–		160		684	N2.0	N 1	823
AQUA REGIA	1798	60							34						93
AQUALIN Synonym: ACROLEIN	1092	30	31	11	63	25	5	17 –12	20	32	ARL 85	12	A4.0	A 5	18

HAZARDOUS MATERIALS REFERENCE BOOKS INDEX

CHEMICAL OR MATERIAL NAME	UN/NA Number	Guide Number (DOT)	Firefighter's Hazardous Materials Reference Book	First Aid Manual for Chemical Accidents	Sax's Dangerous Properties of Industrial Materials	Hazardous Chemicals Desk Reference	Rapid Guide to Hazardous Chemicals in the Workplace	Fire Protection Guide on Hazardous Materials (NFPA)	Firefighter's Handbook of Hazardous Materials	Pocket Guide to Chemical Hazards (NIOSH)	Chemical Hazard Response Information System (CHRIS)	Emergency Handling of Hazardous Materials in Surface Transportation (AAR)	Emergency Action Guides (EAG)	Chemical Data Notebook: A User's Manual	Condensed Chemical Dictionary
AQUEOUS AMMONIA Synonym: AMMONIUM HYDROXIDE	2672	60	62	12	228	62	11		34		AMH 52	60	A8.0	A10	66
AQUEOUS HF Synonym: HYDROFLUORIC ACID SOLUTION	1790	59	443	18	1897	642	114		164		HFA 635	527	H4.1.1		
ARARSENOUS AND MERCURIC IODIDE SOLUTION	2810	55													
ARCTON 6 Synonym: DICHLORODIFLUOROMETHANE	1028	12	275	15	1137	380	68		102	86	DCF 360	331	D1.1		379
ARCTON 9 Synonym: TRICHLOROFLUOROMETHANE	3082/9188	31	838	25	3353	1138	219		273		TCF1083	919			1170
ARECOLINE					286	83			34						
ARGENTOUS FLUORIDE Synonym: SILVER FLUORIDE			747		3048	1039			250		SVF1067				1039
ARGENTOUS OXIDE Synonym: SILVER OXIDE			750		3051				251		SVO1070				1040
ARGON, LIQUID, REFRIGERATED	1951	21			287	84			34			89	A10.1		
ARGON, COMPRESSED	1006	12			287	84			34			89	A10.1	A13	95
ARGON CARBON DIOXIDE GAS MIXTURE	1956	12										87			
ARGON HYDROGEN GAS MIXTURE	1954	22										87			
ARGON METHANE GAS MIXTURE	1954	22										88			
ARGON NITROGEN GAS MIXTURE	1956	12										88			
ARGON OXYGEN GAS MIXTURE	1956	12										88			
ARNICA					289	84			34						
AROCHLOR Synonym: POLYCHLORINATED BIPHENYL	2315/9188	31	698	23	2831	964	185	142–	34		PCB 934	773			931
AROMATIC SPIRITS OF AMMONIA					289	85			34						

CHEMICAL OR MATERIAL NAME	UN/NA Number	Guide Number (DOT)	Firefighter's Hazardous Materials Reference Book	First Aid Manual for Chemical Accidents	Sax's Dangerous Properties of Industrial Materials	Hazardous Chemicals Desk Reference	Rapid Guide to Hazardous Chemicals in the Workplace	Fire Protection Guide on Hazardous Materials (NFPA)	Firefighter's Handbook of Hazardous Materials	Pocket Guide to Chemical Hazards (NIOSH)	Chemical Hazard Response Information System (CHRIS)	Emergency Handling of Hazardous Materials in Surface Transportation (AAR)	Emergency Action Guides (EAG)	Chemical Data Notebook: A User's Manual	Condensed Chemical Dictionary
AROTON 4 Synonym: MONOCHLORODIFLUOROMETHANE	1018	12	593										M4.0		795
ARSANILIC ACID					290	85			34						96
ARSECODILE Synonym: SODIUM CACODYLATE	1688/2811	53	765						252		SCD1020	834			1051
ARSENIC	1561	53		12		85	16		34						96
ARSENIC	2811	53		12			16					94			96
ARSENIC (III) OXIDE Synonym: ARSENIC TRIOXIDE	1561	53	100		299	92	16	33 –	35		ATO 105	93	A11.1	A14	98
ARSENIC (III) TRICHLORIDE Synonym: ARSENIC TRICHLORIDE	1560	55	99	12			16		35		AST 95	92	A11.0		97
ARSENIC ACID Synonym: ARSENIC ACID-ortho	1554	53	96		292	86	16		35		ASA 88	91			97
ARSENIC ACID Synonym: ARSENIC ACID-ortho	1553	55	95		292	86	16		35			91			97
ARSENIC ACID ANHYDRIDE Synonym: ARSENIC PENTAOXIDE	1559	53					16		35		APO 77				97
ARSENIC BISULFIDE	1557	53			295	89	16		35						
ARSENIC BROMIDE	1555	53			295	89	16		35			91			97
ARSENIC CHLORIDE Synonym: ARSENIC TRICHLORIDE	1560	55	99	12	296	89	16		35		AST 95	92	A11.0		97
ARSENIC COMPOUND	1557	53		12	296	90	16		35						96
ARSENIC COMPOUND	1556	55		12	296	90	16		35						96
ARSENIC DIETHYL					297	90	16		35						
ARSENIC DIMETHYL					297	91	16		35						
ARSENIC DISULFIDE	1557	53	97				16		35		ARD 83				97

HAZARDOUS MATERIALS REFERENCE BOOKS INDEX

CHEMICAL OR MATERIAL NAME	UN/NA Number	Guide Number (DOT)	Firefighter's Hazardous Materials Reference Book	First Aid Manual for Chemical Accidents	Sax's Dangerous Properties of Industrial Materials	Hazardous Chemicals Desk Reference	Rapid Guide to Hazardous Chemicals in the Workplace	Fire Protection Guide on Hazardous Materials (NFPA)	Firefighter's Handbook of Hazardous Materials	Pocket Guide to Chemical Hazards (NIOSH)	Chemical Hazard Response Information System (CHRIS)	Emergency Handling of Hazardous Materials in Surface Transportation (AAR)	Emergency Action Guides (EAG)	Chemical Data Notebook: A User's Manual	Condensed Chemical Dictionary
ARSENIC IODIDE	1557	53			297	91	16		35			91			
ARSENIC METAL	1558	53			291		15		35	42	ARX 87	94			96
ARSENIC OXIDE Synonym: ARSENIC PENTAOXIDE	1559	53					16		35		APO 77				97
ARSENIC OXIDE Synonym: ARSENIC TRIOXIDE	1561	53	100		299	92	16	33 –	35		ATO 105	93	A11.1	A14	98
ARSENIC PENTAOXIDE	1559	53			297	91		32 –	35		APO 77				97
ARSENIC PENTOXIDE Synonym: ARSENIC PENTAOXIDE	1559	53	98		297	91	16	32 –	35		APO 77	92			97
ARSENIC PENTOXIDE Synonym: ARSENIC ACID	1554	53	96				16		35		ASA 88	91			97
ARSENIC SESQUIOXIDE Synonym: ARSENIC TRIOXIDE	1561	53	100		299	92	16	33 –	35		ATO 105	93	A11.1	A14	97
ARSENIC SULFIDE	1557	53			298	92	16		35			92			97
ARSENIC TRICHLORIDE	1560	55	99	12			16	31 –	35		AST 95	92	A11.0		97
ARSENIC TRICHLORIDE	1325	32		12			16								97
ARSENIC TRIOXIDE	1561	53	100		299	92	16	33 –	35		ATO 105	93	A11.1	A14	98
ARSENIC TRISULFIDE	1557	53	101				16	33 –	35		ART 86	93			98
ARSENIC WHITE	1561	53					16		34			93			
ARSENIC YELLOW Synonym: ARSENIC TRISULFIDE	1557	53	101				16	33 –	35		ART 86	93			98
ARSENICAL COMPOUND	1557	53		12								94			
ARSENICAL DUST	1562	53		12	295	89			35			95			
ARSENICAL FLUE DUST	1562	53		12	295	89			35						
ARSENICAL ORE CRUDE	2811	53													

CHEMICAL OR MATERIAL NAME	UN/NA Number	Guide Number (DOT)	Firefighter's Hazardous Materials Reference Book	First Aid Manual for Chemical Accidents	Sax's Dangerous Properties of Industrial Materials	Hazardous Chemicals Desk Reference	Rapid Guide to Hazardous Chemicals in the Workplace	Fire Protection Guide on Hazardous Materials (NFPA)	Firefighter's Handbook of Hazardous Materials	Pocket Guide to Chemical Hazards (NIOSH)	Chemical Hazard Response Information System (CHRIS)	Emergency Handling of Hazardous Materials in Surface Transportation (AAR)	Emergency Action Guides (EAG)	Chemical Data Notebook: A User's Manual	Condensed Chemical Dictionary
ARSENICAL PESTICIDE	2759	55							36			96			
ARSENICAL PESTICIDE	2760	28							35			95			
ARSENICAL PESTICIDE	2993	28							35			96			
ARSENICAL PESTICIDE	2994/2759	55							35			95			
ARSENIOUS ACID POTASSIUM SALT Synonym: POTASSIUM ARSENITE	1678	54	704		2857	977			237		POA 969	776			949
ARSENIOUS CHLORIDE Synonym: ARSENIC TRICHLORIDE	1560	55	99	12			16		36		AST 95	92	A11.0		97
ARSENIOUS OXIDE Synonym: ARSENIC TRIOXIDE	1561	53	100		299	92	16	33 –	35		ATO 105	93	A11.1	A14	98
ARSENIOUS TRIOXIDE Synonym: ARSENIC TRIOXIDE	1561	53	100		299	92	16	33 –	35		ATO 105	93	A11.1	A14	98
ARSENITE Synonym: ARSENIC TRIOXIDE	1561	53	100		299	92	16	33 –	35		ATO 105	93	A11.1	A14	98
ARSENOLITE Synonym: ARSENIC TRIOXIDE	1561	53	100		299	92	16	33 –	35		ATO 105	93	A11.1	A14	98
ARSENOUS ACID Synonym: ARSENIC TRIOXIDE	1561	53	100		299	92	16	33 –	35		ATO 105	93	A11.1	A14	98
ARSENOUS ACID ANHYDRIDE Synonym: ARSENIC TRIOXIDE	1561	53	100		299	92	16	33 –	35		ATO 105	93	A11.1	A14	98
ARSENOUS ACID CALCIUM SALT Synonym: CALCIUM ARSENITE	1574	53	166		661	228			63		CAS 212	178			200
ARSENOUS ANHYDRIDE Synonym: ARSENIC TRIOXIDE	1561	53	100		299	92	16	33 –	35		ATO 105	93	A11.1	A14	98
ARSENOUS CHLORIDE Synonym: ARSENIC TRICHLORIDE	1560	55	99	12			16		36		AST 95	92	A11.0		97

CHEMICAL OR MATERIAL NAME	UN/NA Number	Guide Number (DOT)	Firefighter's Hazardous Materials Reference Book	First Aid Manual for Chemical Accidents	Sax's Dangerous Properties of Industrial Materials	Hazardous Chemicals Desk Reference	Rapid Guide to Hazardous Chemicals in the Workplace	Fire Protection Guide on Hazardous Materials (NFPA)	Firefighter's Handbook of Hazardous Materials	Pocket Guide to Chemical Hazards (NIOSH)	Chemical Hazard Response Information System (CHRIS)	Emergency Handling of Hazardous Materials in Surface Transportation (AAR)	Emergency Action Guides (EAG)	Chemical Data Notebook: A User's Manual	Condensed Chemical Dictionary
ARSENOUS OXIDE Synonym: ARSENIC TRIOXIDE	1561	53	100		299	92	16	33 –	35		ATO 105	93	A11.1	A14	98
ARSICODILE Synonym: SODIUM CACODYLATE	1688/2811	53	765						252		SCD1020	834			1051
ARSINE	2188	18	102	12	300	94	16	33 –	36	42		96			98
ARSODENT Synonym: ARSENIC TRIOXIDE	1561	53	100		299	92	16	33 –	35		ATO 105	93	A11.1	A14	98
ARSYCODILE Synonym: SODIUM CACODYLATE	1688/2811	53	765						252		SCD1020	834			1051
ARTHODIBROM Synonym: NALED	2783/3082	55	600		2461	835	152		207		NLD 838	673			804
ARTIC Synonym: METHYL CHLORIDE	1063	18	561	20	2284	781	138	111–67	193	146	MTC 806	623	M3.1	M 4	762
ARTIFICIAL ALMOND OIL	1989	26							36						
ARTIFICIAL CINNABAR Synonym: MERCURIC SULFIDE	2025	53	539						186		MSF 800				98
ARYL SULFONIC ACID	2585	60							36						
ARYL SULFONIC ACID	2586	60							36						
ARYL SULFONIC ACID	2584	60							36						
ARYL SULFONIC ACID	2583	60							36						
ASBESTOS BLUE OR BROWN	2212	31	103	12	303	94	17		36	42		99			98
ASBESTOS WHITE	2590	31	104	12	303	94	17		36	42		99			98
ASCARIDOLE					305	96			36						99
ASPHALT	1999	27	105	12	307	97	17	–17	36		ASP 93	99	A12.0		100
ASPHALT BLENDING STOCK ROOFERS FLUX	1999	27		12	307		17		36		ARF 84	99	A12.0		

CHEMICAL OR MATERIAL NAME	UN/NA Number	Guide Number (DOT)	Firefighter's Hazardous Materials Reference Book	First Aid Manual for Chemical Accidents	Sax's Dangerous Properties of Industrial Materials	Hazardous Chemicals Desk Reference	Rapid Guide to Hazardous Chemicals in the Workplace	Fire Protection Guide on Hazardous Materials (NFPA)	Firefighter's Handbook of Hazardous Materials	Pocket Guide to Chemical Hazards (NIOSH)	Chemical Hazard Response Information System (CHRIS)	Emergency Handling of Hazardous Materials in Surface Transportation (AAR)	Emergency Action Guides (EAG)	Chemical Data Notebook: A User's Manual	Condensed Chemical Dictionary
ASPHALT BLENDING STOCK STRAIGHT RUN RESIDUE	1999	27		12	307		17		36		ASR 94	99	A12.0		
ASPHALT CEMENTS Synonym: ASPHALT	1999	27	105	12	307	97	17	−17	36		ASP 93	99	A12.0		100
ASPHALT PAVEMENT SURFACE SEALER Synonym: ASPHALT	1999	27	105	12	307	97	17	−17	36		ASP 93	100	A12.0		100
ASPHALTIC BITUMEN Synonym: ASPHALT	1999	27	105	12	307	97	17	−17	36		ASP 93	99	A12.0		100
ASPHALTIC ROAD MATERIALS Synonym: ASPHALT	1999	27	105	12	307	97	17	−17	36		ASP 93	99	A12.0		100
ASPHALTUM Synonym: ASPHALT BLENDING STOCK ROOFERS FLUX	1999	27	105	12	307		17		36		ARF 84	99	A12.0		
ASPHALTUM Synonym: ASPHALT	1999	27	105	12	307	97	17	−17	36		ASP 93	99	A12.0		100
ASPHALTUM OIL Synonym: ASPHALT BLENDING STOCK ROOFERS FLUX	1999	27	105	12	307		17		36		ARF 84	99	A12.0		
ASPHALTUM OR COAL TAR PAINT OR VARNISH	1263	26										101			
ATE Synonym: TRIETHYLALUMINUM	1102			25	3368	1141		175–88			TAL1074				103
ATRAZINE	1609				310	98	18		36		ATZ 110				105
ATROPINE					311	99			36						105
AUTOMOTIVE GASOLINE Synonym: GASOLINE	1203	27	412	18	1779	601	104	−56	153		GAT 603	473	G1.0	G 1	554
AVIATION GASOLINE Synonym: GASOLINE	1203	27	412	18	1779	601	104	−56	153		GAV 604	473	G1.0	G 1	107
AZABENZENE Synonym: PYRIDINE	1282	26	734		2946	1008	192	152–83	245	190	PRD 985	800	P9.0		979

HAZARDOUS MATERIALS REFERENCE BOOKS INDEX

CHEMICAL OR MATERIAL NAME	UN/NA Number	Guide Number (DOT)	Firefighter's Hazardous Materials Reference Book	First Aid Manual for Chemical Accidents	Sax's Dangerous Properties of Industrial Materials	Hazardous Chemicals Desk Reference	Rapid Guide to Hazardous Chemicals in the Workplace	Fire Protection Guide on Hazardous Materials (NFPA)	Firefighter's Handbook of Hazardous Materials	Pocket Guide to Chemical Hazards (NIOSH)	Chemical Hazard Response Information System (CHRIS)	Emergency Handling of Hazardous Materials in Surface Transportation (AAR)	Emergency Action Guides (EAG)	Chemical Data Notebook: A User's Manual	Condensed Chemical Dictionary
AZACYCLOHEPTANE Synonym: HEXAMETHYLENIMINE	2493	29	443						160		HMI 638				
AZACYCLOPROPANE Synonym: ETHYLENEIMINE	1185	30	380	17	1610		95	86 -51	144	110	ETI 567	437	E2.6.1		490
AZIDES					321	100			37						108
AZINE Synonym: PYRIDINE	1282	26	734		2946	1008	192	152–83	245	190	PRD 985	800	P9.0		979
AZINPHOS ETHYL	2783	55							37						
AZINPHOS METHYL	2783	55	106		323	100	18		37	42	AZM 112	103			109
AZIRAN Synonym: ETHYLENEIMINE	1185	30	380	17	1610		95	86 -51	144	110	ETI 567	437	E2.6.1		490
AZIRANE Synonym: ETHYLENEIMINE	1185	30	380	17	1610		95	86 -51	144	110	ETI 567	437	E2.6.1		490
AZIRIDINE Synonym: ETHYLENEIMINE	1185	30	380	17	1610		95	86 -51	37	110	ETI 567	437	E2.6.1		109
AZOBENZENE					325	101			37						109
AZOBISISOBUTYRONITRILE Synonym: AZOBISISOBUTYLONITRILE	2952/1325	70			325	102		-17	37			104			109
AZOCHLORAMIDE					326				37						
AZODIISOBUTYRONITRILE Synonym: AZOBISISOBUTYLONITRILE	2952/1325	70			325	102			37			104			
AZODRIN	2783	55			326	102			37						109
AZOIC DIAZO COMPONENT 37 Synonym: 4-NITROANILINE	1661	55					158		211		NAL 818				
AZOIC DIAZO COMPONENT 6 Synonym: 2-NITROANILINE	1661	55									NTA 851	687			

CHEMICAL OR MATERIAL NAME	UN/NA Number	Guide Number (DOT)	Firefighter's Hazardous Materials Reference Book	First Aid Manual for Chemical Accidents	Sax's Dangerous Properties of Industrial Materials	Hazardous Chemicals Desk Reference	Rapid Guide to Hazardous Chemicals in the Workplace	Fire Protection Guide on Hazardous Materials (NFPA)	Firefighter's Handbook of Hazardous Materials	Pocket Guide to Chemical Hazards (NIOSH)	Chemical Hazard Response Information System (CHRIS)	Emergency Handling of Hazardous Materials in Surface Transportation (AAR)	Emergency Action Guides (EAG)	Chemical Data Notebook: A User's Manual	Condensed Chemical Dictionary
AZOTIC ACID Synonym: NITRIC ACID FUMING	2032	44		21	2509	856		123–		160		684	N2.0	N 1	110
AZOXYBENZENE					327	103			37						110
BACILLUS SUBTILIS BPN					329		19								
BACILLUS SUBTILIS CARLSBERG					329		19								
BAGASSE DUST					331										
BAGS SODIUM NITRATE EMPTY AND UNWASHED	1498	35										104			
BAL					331	104									113
BANANA OIL Synonym: AMYL ACETATE (-sec)	1104	26		12	241	68	13	–14	38	38	AAS 5	73			114
BANANA OIL Synonym: ISOAMYL ACETATE	1104	26			2011	673	120	–60	38	130	IAT 660				114
BANOMITE					332										
BARBITURIC ACID					333										115
BARIUM	1400	40		12	333	105	19		38			104			116
BARIUM ACETATE				12	333		20								116
BARIUM ALLOY	1854	37					20		38			104			
BARIUM ALLOY	1399	40					20		38			105			
BARIUM AZIDE	1571	36			334	105	19		38			105			116
BARIUM BINOXIDE Synonym: BARIUM PEROXIDE	1449	42	113		339	109	20		38		BPO 161	108			116
BARIUM BROMATE	2719/1479	42			334	106	20		38			105			116
BARIUM CARBIDE					335	106	20		38						
BARIUM CARBONATE	1564	55	107	12	335		20				BRC 165				117
BARIUM CHLORATE	1445	42	108		335	106	20	34 –	38		BCR 130	106			117

HAZARDOUS MATERIALS REFERENCE BOOKS INDEX

CHEMICAL OR MATERIAL NAME	UN/NA Number	Guide Number (DOT)	Firefighter's Hazardous Materials Reference Book	First Aid Manual for Chemical Accidents	Sax's Dangerous Properties of Industrial Materials	Hazardous Chemicals Desk Reference	Rapid Guide to Hazardous Chemicals in the Workplace	Fire Protection Guide on Hazardous Materials (NFPA)	Firefighter's Handbook of Hazardous Materials	Pocket Guide to Chemical Hazards (NIOSH)	Chemical Hazard Response Information System (CHRIS)	Emergency Handling of Hazardous Materials in Surface Transportation (AAR)	Emergency Action Guides (EAG)	Chemical Data Notebook: A User's Manual	Condensed Chemical Dictionary
BARIUM CHLORATE MONOHYDRATE	1445	42	108		335	106	20	34 –	38		BCR 130	106			117
Synonym: **BARIUM CHLORATE**															
BARIUM CHLORIDE	1564/2811	55		12	336		20					106			117
BARIUM CHROMATE (VI)					336	106	20								117
BARIUM COMPOUND	1564	55		12	336	107	20		38						116
BARIUM COMPOUNDS	1446	42		12	336	107	20			44					116
BARIUM CYANIDE	1565	53	109		337	107	20		38		BCY 132	107			117
BARIUM CYANIDE SOLID	1565	53	109		337	107	20		38		BCY 132	107			117
Synonym: **BARIUM CYANIDE**															
BARIUM DICHROMATE					337	107	20		38						117
BARIUM DIOXIDE	1449	42	113		339	109	20		38		BPO 161	108			117
Synonym: **BARIUM PEROXIDE**															
BARIUM HYDRIDE					338	107	20		38						
BARIUM HYDROXIDE				12	338		20								118
BARIUM HYDROXIDE MONOHYDRATE	1759	60					20					107			118
BARIUM HYPOCHLORITE	2741	45					20		38						
BARIUM METAL	1400	40		12		105	19		38			104			116
BARIUM NITRATE	1446	42	110	12	338	108	20		38	44	BNT 151	107			119
BARIUM NITRIDE					339	108	20		38						
BARIUM OXIDE	1884	53		12	339	108	20		38			108			119
BARIUM PERCHLORATE	1447	42	111				20		39		BPC 155	108			119
BARIUM PERCHLORATE TRIHYDRATE	1447	42	111				20		39		BPC 155	108			119
Synonym: **BARIUM PERCHLORATE**															
BARIUM PERMANGANATE	1448	42	112				20		39		BPM 160	108			119
BARIUM PEROXIDE	1449	42	113		339	109	20		39		BPO 161	108			119

CHEMICAL OR MATERIAL NAME	UN/NA Number	Guide Number (DOT)	Firefighter's Hazardous Materials Reference Book	First Aid Manual for Chemical Accidents	Sax's Dangerous Properties of Industrial Materials	Hazardous Chemicals Desk Reference	Rapid Guide to Hazardous Chemicals in the Workplace	Fire Protection Guide on Hazardous Materials (NFPA)	Firefighter's Handbook of Hazardous Materials	Pocket Guide to Chemical Hazards (NIOSH)	Chemical Hazard Response Information System (CHRIS)	Emergency Handling of Hazardous Materials in Surface Transportation (AAR)	Emergency Action Guides (EAG)	Chemical Data Notebook: A User's Manual	Condensed Chemical Dictionary
BARIUM SELENATE	2630	53					20		39						
BARIUM SELENITE	2630	53					20		39						
BARIUM SULFATE					340	109	20								120
BARIUM SULFIDE				12	340	109	20		39						120
BARIUM SUPEROXIDE Synonym: BARIUM PEROXIDE	1449	42	113		339	109	20		39		BPO 161	108			121
BASIC BISMUTH CHORIDE Synonym: BISMUTH OXYCHLORIDE						167					BOC 153				152
BASIC COPPER ACETATE Synonym: COPPER SUBACETATE							55				CST 315				315
BASIC ZIRCONIUM CHLORIDE Synonym: ZIRCONIUM OXYCHLORIDE			910		3550	1198	235				ZCO1200				1252
BASILIN Synonym: HYDROGEN CHLORIDE	1050	15	447	18	1900	643	115	98 –	164	126	HDC 630	528	H4.3	H 3	616
BATTERY ACID Synonym: SULFURIC ACID	1830	39	802	24	3163	1082	204	164–	39	200	SFA1037	871	S4.2	S 8	124
BATTERY BOX COMPOUNDS	1993	27										109			
BATTERY ELECTRIC STORAGE DRY CONTAINING POTASSIUM HYDROX	1813	60													
BATTERY ELECTRIC STORAGE WET FILED WITH ALKALI	2795	60													
BATTERY ELECTRIC STORAGE WET FILLED WITH ACID	2794	60													
BATTERY FLUID ACID	2796	39							39			109			
BATTERY FLUID ACID WITH ELECTRONIC EQUIPMENT	1693	58													
BATTERY FLUID ALKALI	2797	60							39			110			
BATTERY FLUID ALKALI WITH BATTERY	2797	60													

HAZARDOUS MATERIALS REFERENCE BOOKS INDEX

CHEMICAL OR MATERIAL NAME	UN/NA Number	Guide Number (DOT)	Firefighter's Hazardous Materials Reference Book	First Aid Manual for Chemical Accidents	Sax's Dangerous Properties of Industrial Materials	Hazardous Chemicals Desk Reference	Rapid Guide to Hazardous Chemicals in the Workplace	Fire Protection Guide on Hazardous Materials (NFPA)	Firefighter's Handbook of Hazardous Materials	Pocket Guide to Chemical Hazards (NIOSH)	Chemical Hazard Response Information System (CHRIS)	Emergency Handling of Hazardous Materials in Surface Transportation (AAR)	Emergency Action Guides (EAG)	Chemical Data Notebook: A User's Manual	Condensed Chemical Dictionary
BATTERY FLUID ALKALI WITH ELECTRONIC EQUIPMENT OR ACTUAT	2797	60													
BATTERY WET FILLED WITH ACID	2794	60										111			
BATTERY WET FILLED WITH ALKALI	2795	60										111			
BAY 37344 Synonym: MERCAPTODIMETHUR	2757	55	530						185		MCD 736				
BAY RUM	1993	27													
BAYER 13/59 Synonym: TRICHLORFON	2783/3018	55	835						272		TRC1146	917			1169
BAYGON									39						124
BEARING OIL Synonym: OILS MISCELLANEOUS SPINDLE	1270	27						−83			OSD 904				
BEET SUGAR Synonym: SUCROSE			796		3151	1077	196				SRS1056				125
BELLADONNA					344	110			39						126
BENARES SUNN Synonym: JUTE	1372	32													667
BENOXYL Synonym: BENZOYL PEROXIDE	2085	49	122		392	130	23		42	44			B2.0	B 2	
BENZAL CHLORIDE	1886	55			348	112	21		39		BZL 196				127
BENZALDEHYDE	1989	26	114		348	112		34 −17	39		BZD 193	112			127
BENZEDRINE					355	114		−17	40						128
BENZENAMINE Synonym: ANILINE OIL	1547	57	85		254	72	13		32		ANL 64	79	A10.0	A12	
BENZENE	1114	27	115	12	356	115	21	35 −17	40	44	BNZ 152	112	B1.0	B 1	128
BENZENE CARBONYL CHLORIDE	1736	39							40						

CHEMICAL OR MATERIAL NAME	UN/NA Number	Guide Number (DOT)	Firefighter's Hazardous Materials Reference Book	First Aid Manual for Chemical Accidents	Sax's Dangerous Properties of Industrial Materials	Hazardous Chemicals Desk Reference	Rapid Guide to Hazardous Chemicals in the Workplace	Fire Protection Guide on Hazardous Materials (NFPA)	Firefighter's Handbook of Hazardous Materials	Pocket Guide to Chemical Hazards (NIOSH)	Chemical Hazard Response Information System (CHRIS)	Emergency Handling of Hazardous Materials in Surface Transportation (AAR)	Emergency Action Guides (EAG)	Chemical Data Notebook: A User's Manual	Condensed Chemical Dictionary
BENZENE CHLORIDE Synonym: CHLOROBENZENE	1134	27	203	13	768	273	47	51 −26	40	64	CRB 298	219	C4.1		263
BENZENE DIAZONIUM CHLORIDE					359				40						
BENZENE DIAZONIUM NITRATE					359				40						
BENZENE HEXACHLORIDE	2761	55	116		360	117			40		BHC 144				129
BENZENE HEXACHLORIDE-a-isomer					361	118	21								
BENZENE HEXACHLORIDE-a-trans					362	119	22								
BENZENE HEXACHLORIDE-y-isomer	2761	55			361	118	22								
BENZENE HEXAHYDRIDE Synonym: CYCLOHEXANE	1145	26	244	14	990	333	58	−30	88	76	CHX 251	280	C8.1		336
BENZENE NITRO Synonym: NITROBENZENE	1662	55	631	21	2518	861	158	124–74	212	162	NTB 852		N3.1		825
BENZENE PHOSPHORUS DICHLORIDE	2798	39	117						40		BPD 156				129
BENZENE PHOSPHORUS THIODICHLORIDE	2799	39	118						40		BPT 163				
BENZENE SULFOHYDRAZIDE	2970	72							40			113			
BENZENE SULFONIC ACID					363	119			40						129
BENZENE SULFONYL CHLORIDE	2225/1760	59			363	119			40		BSC 171	113			
BENZENE-1,3-DICARBOXYLIC ACID Synonym: ISOPHTHALIC ACID			478		2035						IPL 676				658
BENZENE-1,3-DISULFOHYDRAZIDE	2971	72							40			114			
BENZENE-1-CHLORO-2-NITRO Synonym: NITROCHLOROBENZENE (-ortho)	1578	55	633						212			690	N3.3		826
BENZENE-1-CHLORO-3-NITRO Synonym: NITROCHLOROBENZENE (-meta)	1578	55	632						212			690	N3.2		826

HAZARDOUS MATERIALS REFERENCE BOOKS INDEX

CHEMICAL OR MATERIAL NAME	UN/NA Number	Guide Number (DOT)	Firefighter's Hazardous Materials Reference Book	First Aid Manual for Chemical Accidents	Sax's Dangerous Properties of Industrial Materials	Hazardous Chemicals Desk Reference	Rapid Guide to Hazardous Chemicals in the Workplace	Fire Protection Guide on Hazardous Materials (NFPA)	Firefighter's Handbook of Hazardous Materials	Pocket Guide to Chemical Hazards (NIOSH)	Chemical Hazard Response Information System (CHRIS)	Emergency Handling of Hazardous Materials in Surface Transportation (AAR)	Emergency Action Guides (EAG)	Chemical Data Notebook: A User's Manual	Condensed Chemical Dictionary
BENZENE-1-CHLORO-4-NITRO	1578	55	634	21	2525	863	159	−74	212	164		691	N3.4		826
Synonym: NITROCHLOROBENZENE (-para)															
BENZENECARBANAL	1989	26							40						
BENZENECARBINOL	1987	26			395	131		−18	40		BAL 117	118			134
Synonym: BENZYL ALCOHOL															
BENZENECARBONAL	1989	26							40						
BENZENECARBONYL CHLORIDE	1736	39	121		390	129		36 −18	40		BZC 192	118			133
Synonym: BENZOYL CHLORIDE															
BENZENECARBOXYLIC ACID	9094/3077		119		375	124		−17	41		BZA 191	115			129
Synonym: BENZOIC ACID															
BENZENETHIOPHOSPHONYL CHLORIDE	2799	39	118						40		BPT 163				
Synonym: BENZENE PHOSPHORUS THIODICHLORIDE															
BENZENOL	1671	55	185						65				C2.0		217
Synonym: CARBOLIC ACID															
BENZIDINE	1885	53		12	367	121	22		41	44	BZI 195	114			130
BENZIDINE BASE	1885	53			367				41		BZI 195				
BENZIDINE SULFATE					368	121	22								130
BENZIL					368										130
BENZIN	1203	27	412	18	1779	601	104	−56	153		GAK 602	473	G1.0	G 1	554
Synonym: GASOLINE															
BENZINE	1115/1993	26							41			114			131
BENZINE	1255	27	676				152		227		PTN 999	739	P2.0		
Synonym: PETROLEUM NAPHTHA															
BENZINOFORM	1846	55	192	13	701	246	43	48 –	41	60	CBT 223	193	C2.3	C 5	221
Synonym: CARBON TETRACHLORIDE															
BENZO (a) PYRENE	3077/9188	31				126	23					114			

CHEMICAL OR MATERIAL NAME	UN/NA Number	Guide Number (DOT)	Firefighter's Hazardous Materials Reference Book	First Aid Manual for Chemical Accidents	Sax's Dangerous Properties of Industrial Materials	Hazardous Chemicals Desk Reference	Rapid Guide to Hazardous Chemicals in the Workplace	Fire Protection Guide on Hazardous Materials (NFPA)	Firefighter's Handbook of Hazardous Materials	Pocket Guide to Chemical Hazards (NIOSH)	Chemical Hazard Response Information System (CHRIS)	Emergency Handling of Hazardous Materials in Surface Transportation (AAR)	Emergency Action Guides (EAG)	Chemical Data Notebook: A User's Manual	Condensed Chemical Dictionary
BENZO (b) PYRIDINE Synonym: QUINOLINE	2656	29	736		2968	1014		-83	246		QNL1005	805			987
BENZO (b) QUINOLINE Synonym: ACRIDINE	2713/1325	32	30		59	25	5		19		ACD 16	11			17
BENZOIC ACID Synonym: AMMONIUM BENZOATE	9080/3077	31	50		375	124			27		ABZ 12	55			63
BENZOIC ACID	9094/3077		119		375	124		-17	41		BZA 191	115			131
BENZOIC ACID NITRILE Synonym: BENZONITRILE	2224	55	120		377	125			41		BZN 198	117			132
BENZOIC ACID PEROXIDE Synonym: BENZOYL PEROXIDE	2085	49	122		392	130	23		41	44			B2.0	B 2	133
BENZOIC ALDEHYDE	1989	26							41						131
BENZOIC DERIVATIVE PESTICIDE	3003	28							41			116			
BENZOIC DERIVATIVE PESTICIDE	2770	28							41						
BENZOIC DERIVATIVE PESTICIDE	2769	55							41			115			
BENZOIN					376	125									131
BENZOL Synonym: BENZENE	1114	27	115	12	356	115	21	35 -17	41	44	BNZ 152	112	B1.0	B 1	132
BENZOL DILUENT								-17	41						
BENZOL FORERUNNINGS CRUDE	1993	27													
BENZOLE Synonym: BENZENE	1114	27	115	12	356	115	21	35 -17	40	44	BNZ 152	112	B1.0	B 1	128
BENZONITRILE	2224	55	120		377	125			41		BZN 198	117			132
BENZOPHENOL Synonym: CARBOLIC ACID	1671	55	185						65				C2.0		132
BENZOPHENONE					379	125					BZP 200				132

CHEMICAL OR MATERIAL NAME	UN/NA Number	Guide Number (DOT)	Firefighter's Hazardous Materials Reference Book	First Aid Manual for Chemical Accidents	Sax's Dangerous Properties of Industrial Materials	Hazardous Chemicals Desk Reference	Rapid Guide to Hazardous Chemicals in the Workplace	Fire Protection Guide on Hazardous Materials (NFPA)	Firefighter's Handbook of Hazardous Materials	Pocket Guide to Chemical Hazards (NIOSH)	Chemical Hazard Response Information System (CHRIS)	Emergency Handling of Hazardous Materials in Surface Transportation (AAR)	Emergency Action Guides (EAG)	Chemical Data Notebook: A User's Manual	Condensed Chemical Dictionary
BENZOQUINONE	2587/2811	55				127			41			117			132
BENZOQUINONE-para	2587	55						–18	41		BZQ 201				
BENZOTRICHLORIDE	2226/1760	60						–18	41			117			133
BENZOTRIFLUORIDE	2338/1993	28			388	129		35 –18	41			118			133
BENZOYL BENZENE Synonym: BENZOPHENONE					379	125					BZP 200				132
BENZOYL CHLORIDE	1736	39	121		390	129		36 –18	42		BZC 192	118			133
BENZOYL PEROXIDE	2089	49			392	130	23		42	44			B2.0	B 2	133
BENZOYL PEROXIDE Synonym: DIBENZOYL PEROXIDE	2085	49	122		392	130	23		42	44	DPO 469		B2.0	B 2	133
BENZOYL PEROXIDE	2087	49			392	130	23		42	44			B2.0	B 2	133
BENZOYL PEROXIDE	2086	49			392	130	23		42	44			B2.0	B 2	133
BENZOYL PEROXIDE	2090	49			392	130	23		42	44			B2.0	B 2	133
BENZOYL PEROXIDE	2088	49			392	130	23		42	44			B2.0	B 2	133
BENZOYL SUPEROXIDE Synonym: BENZOYL PEROXIDE	2085	49	122		392	130	23		42	44			B2.0	B 2	133
BENZOYL SUPEROXIDE Synonym: DIBENZOYL PEROXIDE	2085	49	262			130			97		DPO 469				368
BENZYL ACETATE					395	131		–18	42		BZE 194				134
BENZYL ALCOHOL	1987	26			395	131		–18	42		BAL 117	118			134
BENZYL BROMIDE	1737	59	123		398	132			42		BBR 122	119			135
BENZYL BUTYL PHTHALATE					398	132		–18							
BENZYL CHLORIDE	1738	59	124	12	400	133	23	–18	42	46	BCL 127	119			135
BENZYL CHLOROCARBONATE Synonym: BENZYL CHLOROFORMATE	1739	39	125		400	133			42		BCF 126	120			135

CHEMICAL OR MATERIAL NAME	UN/NA Number	Guide Number (DOT)	Firefighter's Hazardous Materials Reference Book	First Aid Manual for Chemical Accidents	Sax's Dangerous Properties of Industrial Materials	Hazardous Chemicals Desk Reference	Rapid Guide to Hazardous Chemicals in the Workplace	Fire Protection Guide on Hazardous Materials (NFPA)	Firefighter's Handbook of Hazardous Materials	Pocket Guide to Chemical Hazards (NIOSH)	Chemical Hazard Response Information System (CHRIS)	Emergency Handling of Hazardous Materials in Surface Transportation (AAR)	Emergency Action Guides (EAG)	Chemical Data Notebook: A User's Manual	Condensed Chemical Dictionary
BENZYL CHLOROFORMATE	1739	39	125		400	133			42		BCF 126	120			135
BENZYL CYANIDE	2470	55						−18	42						135
BENZYL DICHLORIDE	1886	55							42						135
BENZYL DIETHYLAMINE-n	2619	26						−18	42						
BENZYL DIMETHYLAMINE	2619/1719	26							42		BDM 135	120			
BENZYL ETHER					405	134			42						136
BENZYL IODIDE	2653	53							42			121			136
BENZYL n-BUTYL PHTHALATE Synonym: BUTYL BENZYL PHTHALATE	3082/9188	31									BPH 159	151			183
BENZYL TRICHLORIDE	2226	60			413	136	23		43						
BENZYLAMINE	1993	27	126		396				43		BZM 197				134
BENZYLCARBONYL CHLORIDE Synonym: BENZYL CHLOROFORMATE	1739	39	125		400	133			42		BCF 126	120			135
BENZYLDIMETHYLOCTADECYLAMMONIUM CHLORIDE											BZO 199				
BENZYLDIMETHYLSTEARYLAMMONIUM CHLORIDE Synonym: BENZYLDIMETHYLOCTADECYLAMMONIUM CHLORIDE											BZO 199				
BENZYLIDINE CHLORIDE	886/2810	55							43			121			136
BENZYLTRIMETHYLAMMONIUM CHLORIDE					413						BMA 146				138
BERBERINE AND COMPOUNDS					414	136			43						138
BERYLLIA Synonym: BERYLLIUM OXIDE	1566	53	130		421	143	24			46	BEO 141				139
BERYLLIUM CHLORIDE	1566	53	127		417	139	24		43	46	BEC 137	122			140
BERYLLIUM COMPOUND	1566	53			418	139	24		43	46		122			139
BERYLLIUM FLUORIDE	1566	53	128		419	141	24		43	46	BEF 138	123			140

CHEMICAL OR MATERIAL NAME	UN/NA Number	Guide Number (DOT)	Firefighter's Hazardous Materials Reference Book	First Aid Manual for Chemical Accidents	Sax's Dangerous Properties of Industrial Materials	Hazardous Chemicals Desk Reference	Rapid Guide to Hazardous Chemicals in the Workplace	Fire Protection Guide on Hazardous Materials (NFPA)	Firefighter's Handbook of Hazardous Materials	Pocket Guide to Chemical Hazards (NIOSH)	Chemical Hazard Response Information System (CHRIS)	Emergency Handling of Hazardous Materials in Surface Transportation (AAR)	Emergency Action Guides (EAG)	Chemical Data Notebook: A User's Manual	Condensed Chemical Dictionary
BERYLLIUM HYDRIDE					420	142	24		43						140
BERYLLIUM HYDROXIDE	1566	53			420	142	24			46		123			140
BERYLLIUM NITRATE	2464	42	129		420	143	24		43		BEN 140	123			140
BERYLLIUM OXIDE	1566	53	130		421	143	24			46	BEO 141				141
BERYLLIUM PHOSPHATE	1566	53					24			46		123			
BERYLLIUM POWDER Synonym: BERYLLIUM	1567/2811	32			416	137	24	37 –	43	46	BEM 139	122			139
BERYLLIUM SULFATE	1566	53	131		422	144	24		43	46	BES 142	124			141
BERYLLIUM SULFATE TETRAHYDRATE Synonym: BERYLLIUM SULFATE	1566	53	130		422	144	24		43	46	BES 142	124			141
BETRAPRONE Synonym: PROPIOLACTONE-beta			719		2907	997	187		242	186	PLT 964				968
BFE Synonym: BROMOTRIFLUOROETHYLENE	2419/1954	17							43			146			142
BHC Synonym: BENZENE HEXACHLORIDE	2761	55	116		426	145			43		BHC 144				142
BHT					427	146	24								143
BHUSA	1327	32							43						
BICARBURRETTED HYDROGEN Synonym: ETHYLENE	1038	22		16	1597	536		83 -50	142		ETL 568	436	E2.4	E 4	143
BICHROME Synonym: POTASSIUM DICHROMATE	1479	35	710	23					238		PTD 993	779			952
BICYCLO [4.4.0] DECANE Synonym: DECAHYDRONAPHTHALENE	1147	27	257		1027	345		-31	90		DHN 405	289			349
BICYCLOHEXYL	2565	60							44						143

CHEMICAL OR MATERIAL NAME	UN/NA Number	Guide Number (DOT)	Firefighter's Hazardous Materials Reference Book	First Aid Manual for Chemical Accidents	Sax's Dangerous Properties of Industrial Materials	Hazardous Chemicals Desk Reference	Rapid Guide to Hazardous Chemicals in the Workplace	Fire Protection Guide on Hazardous Materials (NFPA)	Firefighter's Handbook of Hazardous Materials	Pocket Guide to Chemical Hazards (NIOSH)	Chemical Hazard Response Information System (CHRIS)	Emergency Handling of Hazardous Materials in Surface Transportation (AAR)	Emergency Action Guides (EAG)	Chemical Data Notebook: A User's Manual	Condensed Chemical Dictionary
BIEBERITE Synonym: **COBALT SULFATE**					935	318					CBS 222				296
BIETHYLENE Synonym: **BUTADIENE**	1010	17	139	12				40 –	53	48	BDI 134	148	B4.0	B 4	
BIFLUORIDE	1740	60							44						
BIFORMYL Synonym: **GLYOXAL**			421		1802	609			44		GOS 610				570
BIMETHYL	1035	22							44						
BIOCIDE Synonym: **ACROLEIN**	1092	30	31	11	63	25	5	17 –12	44	32	ARL 85	12	A4.0	A 5	145
BIPHENYL	1993	27			433	147	25		44			124			147
BIPHENYL DIPHENYL ETHER Synonym: **DOWTHERM**											DTH 487				444
BIPYRIDILIUM PESTICIDE	3016/2781	55							44			125			
BIPYRIDILIUM PESTICIDE	2782	28							44			124			
BIPYRIDILIUM PESTICIDE	2781	55							44			125			
BIPYRIDILIUM PESTICIDE	3015	28							44			126			
BIS(2-AMINOETHYL) AMINE Synonym: **DIETHYLENETRIAMINE**	2079/1760	29	293		1222	409	74	69 –38	111		DET 394	352			393
BIS(2-AMINOETHYL) ETHYLENEDIAMINE-N,N Synonym: **TRIETHYLENE TETRAMINE**	2259/1760	60			3372	1143			275		TET1106	924			1176
BIS(2-CHLOROETHYL) ETHER Synonym: **DICHLOROETHYL ETHER**	1916/2810	55	276			384	69	66 –18	103		DEE 385	334			380
BIS(2-CHLOROETHYL) FORMAL								–19	45						
BIS(2-CHLRORETHYL) SULFIDE					462	155	25								
BIS(2-ETHYLHEXYL) AMINE								–19	45						

CHEMICAL OR MATERIAL NAME	UN/NA Number	Guide Number (DOT)	Firefighter's Hazardous Materials Reference Book	First Aid Manual for Chemical Accidents	Sax's Dangerous Properties of Industrial Materials	Hazardous Chemicals Desk Reference	Rapid Guide to Hazardous Chemicals in the Workplace	Fire Protection Guide on Hazardous Materials (NFPA)	Firefighter's Handbook of Hazardous Materials	Pocket Guide to Chemical Hazards (NIOSH)	Chemical Hazard Response Information System (CHRIS)	Emergency Handling of Hazardous Materials in Surface Transportation (AAR)	Emergency Action Guides (EAG)	Chemical Data Notebook: A User's Manual	Condensed Chemical Dictionary
BIS(2-ETHYLHEXYL) HYDROGEN PHOSPHATE Synonym: DI(2-ETHYLHEXYL) PHOSPHORIC ACID	1902	60	259						92		DEP 391				395
BIS(2-ETHYLHEXYL) MALEATE					488			-19							
BIS(2-ETHYLHEXYL) PHTHALATE Synonym: DIOCTYL PHTHALATE	3082/9188	31			1445	481		-45			DOP 455	127			425
BIS(2-ETHYLHEXYL) SODIUM SULFOSUCCINATE Synonym: DIOCTYL SODIUM SULFOSUCCINATE											DSS 481				425
BIS(2-HYDROXYETHYL) AMINE Synonym: DIETHANOLAMINE			286		1184	399	73	-36			DEA 381		D1.1.2		388
BIS(2-HYDROXYETHYL) ETHER Synonym: DIETHYLENE GLYCOL			292	15	1219	409		-37	110		DEG 386				391
BIS(2-METHOXYETHYL) ETHER Synonym: DIETHYLENE GLYCOL DIMETHYL ETHER					502			-38	45		DGD 400				392
BIS(2-[2-HYDROXYETHOXY] ETHYL) ETHER Synonym: TETRAETHYLENE GLYCOL	1993	27	815		3218	1100		-85			TTG1155	892			1129
BIS(a-METHYL BENZYL) AMINE					504										
BIS(ACETATE) DIOXOURANIUM Synonym: URANYL ACETATE	9180/2982	62	862						283		URA1167	942			1203
BIS(AMINOPROPYL) AMINE	1760	60							45						
BIS(AMINOPROPYL) PIPERAZINE CORROSIVE MATERIAL	1760	60										52			
BIS(beta-CHLOROETHYL) METHYLAMINE					459		25		47						
BIS(beta-CHLOROISOPROPYL) ETHER	2490	59							47						
BIS(CHLOROMETHYL) ETHER	2249	55		12	465	156	26								148
BIS(DIMETHYLAMINO) ETHANE	2372	26							46						
BIS(DIMETHYLTHIOCARBAMYL) DISULFIDE Synonym: THIRAM	2771	55	822		3285		213		46	212	THR1119	904			1146

CHEMICAL OR MATERIAL NAME	UN/NA Number	Guide Number (DOT)	Firefighter's Hazardous Materials Reference Book	First Aid Manual for Chemical Accidents	Sax's Dangerous Properties of Industrial Materials	Hazardous Chemicals Desk Reference	Rapid Guide to Hazardous Chemicals in the Workplace	Fire Protection Guide on Hazardous Materials (NFPA)	Firefighter's Handbook of Hazardous Materials	Pocket Guide to Chemical Hazards (NIOSH)	Chemical Hazard Response Information System (CHRIS)	Emergency Handling of Hazardous Materials in Surface Transportation (AAR)	Emergency Action Guides (EAG)	Chemical Data Notebook: A User's Manual	Condensed Chemical Dictionary
BIS(GLYCINATO) COPPER Synonym: COPPER GLYCINATE							55				CPG 288				312
BIS(p-CHLOROBENZOYL) PEROXIDE Synonym: DI(p-CHLOROBENZOYL) PEROXIDE	0149										DZP 498				
BIS(TRIFLUOROMETHYL) CHLORO PHOSPHINE					518										
BIS(TRIFLUOROMETHYL) CYANO PHOSPHINE					518				47						
BISMUTH					506	167									151
BISMUTH CHLORIDE OXIDE Synonym: BISMUTH OXYCHLORIDE											BOC 153				152
BISMUTH OXYCHLORIDE											BOC 153				152
BISMUTH PENTAFLUORIDE					507	168			48						152
BISMUTH SUBCHLORIDE Synonym: BISMUTH OXYCHLORIDE											BOC 153				153
BISMUTHYL CHLORIDE Synonym: BISMUTH OXYCHLORIDE											BOC 153				152
BISPHENOL A					511	169			48		BPA 154				154
BISPHENOL A DIGLYCIDYL ETHER					511	169					BDE 133				
BISPHENOL A EPICHLOROHYDRIN CONDENSATE Synonym: BISPHENOL A DIGLYCIDYL ETHER					511	169					BDE 133				
BISULFITE	2693/1760	60							48			130			
BITUMEN Synonym: ASPHALT	1999	27	105	12	307	97	17	−17	36		ASP 93	99	A12.0		155
BIVINYL Synonym: BUTADIENE	1010	17	139	12				40 –	53	48	BDI 134	148	B4.0	B 4	
BLACK LEAF 40 (40% WATER SOLUTION) Synonym: NICOTINE SULFATE	1658	55	623		2501	854			210		NCS 825	681			1658

HAZARDOUS MATERIALS REFERENCE BOOKS INDEX

CHEMICAL OR MATERIAL NAME	UN/NA Number	Guide Number (DOT)	Firefighter's Hazardous Materials Reference Book	First Aid Manual for Chemical Accidents	Sax's Dangerous Properties of Industrial Materials	Hazardous Chemicals Desk Reference	Rapid Guide to Hazardous Chemicals in the Workplace	Fire Protection Guide on Hazardous Materials (NFPA)	Firefighter's Handbook of Hazardous Materials	Pocket Guide to Chemical Hazards (NIOSH)	Chemical Hazard Response Information System (CHRIS)	Emergency Handling of Hazardous Materials in Surface Transportation (AAR)	Emergency Action Guides (EAG)	Chemical Data Notebook: A User's Manual	Condensed Chemical Dictionary
BLADAN	1705	15			3220	1101			262		TEP1104	890			1130
Synonym: TETRAETHYL PYROPHOSPHATE															
BLADEX					524	173			48						
BLASTING AGENT		46							48						156
BLEACHING POWDER	2208	35										131			156
BLENDED GASOLINE	1203	27	412	18	1779	601	104	–56	153		GAK 602	473	G1.0	G 1	554
Synonym: GASOLINE															
BLUE ASBESTOS	2212	31		12					48						
BLUE OIL	1547	57	85	12	254	72	13	29 –16	32	40	ANL 64	79	A10.0	A12	79
Synonym: ANILINE															
BLUE VERDIGRIS							55				CST 315				158
Synonym: COPPER SUBACETATE															
BLUE VITRIOL	9109		229	14	950	322	55				CSF 311				158
Synonym: COPPER SULFATE															
BOILER COMPOUND	1760	60			527				48		BCP 129				159
BOLETIC ACID	9126/9188		407		1760	595			152		FUM 599	465			543
Synonym: FUMARIC ACID															
BOMB SMOKE NONEXPLOSIVE	2028	60							49						
BORACIC ACID				12	529	174					BAC 114				161
Synonym: BORIC ACID															
BORANES					528				49						161
BORATE AND CHLORATE MIXTURE	1458	35							49						
BORAX ANHYDROUS			763	24	3066	1046	198				SDB1028				161
Synonym: SODIUM BORATE															
BORAZINE					529	173			49						
BORDEAUX ARSENITE	2759	55			529	173						135			

CHEMICAL OR MATERIAL NAME	UN/NA Number	Guide Number (DOT)	Firefighter's Hazardous Materials Reference Book	First Aid Manual for Chemical Accidents	Sax's Dangerous Properties of Industrial Materials	Hazardous Chemicals Desk Reference	Rapid Guide to Hazardous Chemicals in the Workplace	Fire Protection Guide on Hazardous Materials (NFPA)	Firefighter's Handbook of Hazardous Materials	Pocket Guide to Chemical Hazards (NIOSH)	Chemical Hazard Response Information System (CHRIS)	Emergency Handling of Hazardous Materials in Surface Transportation (AAR)	Emergency Action Guides (EAG)	Chemical Data Notebook: A User's Manual	Condensed Chemical Dictionary
BORIC ACID				12	529	174					BAC 114				162
BORNEOL	1312/1325	32			530	174		−19	49			135			163
BOROETHANE Synonym: DIBORANE	1911	18	263	15	1099	367	64	62 −33	49	82		323	D1.0		163
BOROHYDRIDE Synonym: SODIUM BOROHYDRIDE	1426	32	7		3066	1047			252		SBH1016	833			1051
BORON					531	174									163
BORON BROMIDE	2692	59							49						164
BORON CHLORIDE Synonym: BORON TRICHLORIDE	1741	15	133		532	175			49		BRT 168	136	B2.9		164
BORON COMPOUNDS					531	175			49						163
BORON FLUORIDE	1008	15							49						164
BORON HYDRIDE Synonym: DIBORANE	1911	18	263	15	1099	367	64	62 −33	49	82		323	D1.0		164
BORON OXIDE					531	175	26		49	46					164
BORON TRIBROMIDE	2692	59	132		532	175	26	37 –	49		BTB 173	136			164
BORON TRICHLORIDE	1741	15	133		532	175			49		BRT 168	136	B2.9		164
BORON TRIFLUORIDE	1008	15		12	532	176	27	38 –	50	46		136			164
BORON TRIFLUORIDE ACETIC ACID COMPLEX	1742	59			533	176			50			138			
BORON TRIFLUORIDE DIETHYL ETHERATE	2604	29							50			137			
BORON TRIFLUORIDE ETHERATE	2604	29			533			−19	50						165
BORON TRIFLUORIDE PROPIONIC ACID COMPLEX	1743	59							50			138			
BOTTLE GAS Synonym: PROPANE	1978/1075	22			2899		186		50		PRP 988	148			966

HAZARDOUS MATERIALS REFERENCE BOOKS INDEX

CHEMICAL OR MATERIAL NAME	UN/NA Number	Guide Number (DOT)	Firefighter's Hazardous Materials Reference Book	First Aid Manual for Chemical Accidents	Sax's Dangerous Properties of Industrial Materials	Hazardous Chemicals Desk Reference	Rapid Guide to Hazardous Chemicals in the Workplace	Fire Protection Guide on Hazardous Materials (NFPA)	Firefighter's Handbook of Hazardous Materials	Pocket Guide to Chemical Hazards (NIOSH)	Chemical Hazard Response Information System (CHRIS)	Emergency Handling of Hazardous Materials in Surface Transportation (AAR)	Emergency Action Guides (EAG)	Chemical Data Notebook: A User's Manual	Condensed Chemical Dictionary
BOTTLE GAS Synonym: BUTANE	1011	22			574		28		50		BUT 188	148	B4.0.5		177
BOTTLED GAS Synonym: LIQUEFIED PETROLEUM GAS	1075	22	517	19	2123	722	128		180		LPG 704	789	L1.0	L 1	703
BOV Synonym: SULFURIC ACID	1830	39	802	24	3163	1082	204	164–	259	200	SFA1037	871	S4.2	S 8	1104
BOX TOE GUM Synonym: COLLODION	2059/1993	56	220		940	319		–28	82		CLD 257	249			300
BOX TOE GUM	1133	26										138			
BP Synonym: DIBENZOYL PEROXIDE	2085/2087	49	262						97		DPO 469	320			368
BP Synonym: BENZOYL PEROXIDE	2085	49	122		392	130	23		42	44			B2.0	B 2	133
BPO Synonym: BENZOYL PEROXIDE	2085	49	122		392		23		42	44			B2.0	B 2	133
BPO Synonym: DIBENZOYL PEROXIDE	2085/2087	49	262			130			97		DPO 469	320			368
BRAKE FLUID HYDRAULIC	1118	27							50			138			166
BRIMSTONE Synonym: SULFUR	2448	32			3161	1080		–84			SXX1072	871	S3.5		168
BROCIDE Synonym: ETHYLENE DICHLORIDE	1184	26	372	17	1602	540	94	86 –50	143	110	EDC 519	430	E2.4.5	E 5	487
BROM O GAS Synonym: METHYL BROMIDE	1062	55	559	20	2274	776	137	110–67	192	146	MTB 805	619	M3.0	M 3	759
BROMALLYLENE Synonym: ALLYL BROMIDE	1099	57	39		105	34		22 –13	22		ABR 10	35			39

CHEMICAL OR MATERIAL NAME	UN/NA Number	Guide Number (DOT)	Firefighter's Hazardous Materials Reference Book	First Aid Manual for Chemical Accidents	Sax's Dangerous Properties of Industrial Materials	Hazardous Chemicals Desk Reference	Rapid Guide to Hazardous Chemicals in the Workplace	Fire Protection Guide on Hazardous Materials (NFPA)	Firefighter's Handbook of Hazardous Materials	Pocket Guide to Chemical Hazards (NIOSH)	Chemical Hazard Response Information System (CHRIS)	Emergency Handling of Hazardous Materials in Surface Transportation (AAR)	Emergency Action Guides (EAG)	Chemical Data Notebook: A User's Manual	Condensed Chemical Dictionary
BROMATE	1450	42			538	177			50			139			
BROMBENZYL CYANIDE LIQUID	1693	58													
BROMELITE *Synonym:* BERYLLIUM OXIDE	1566	53	130		421	143				46	BEO 141				141
BROMEX *Synonym:* NALED	2783/3082	55	600		2461	835	152		207		NLD 838	673			804
BROMIC ETHER	1891	58							50						
BROMINE	1744	59	134	12	539	178	27	38 –	50	48	BRX 170	139	B3.0	B 3	169
BROMINE AZIDE					540	178			50						169
BROMINE CHLORIDE	2901/1955	20							50			140			169
BROMINE CYANIDE	1889	55							50						169
BROMINE PENTAFLUORIDE	1745	44	135		540	179		39 –	50		BPF 158	140			170
BROMINE SOLUTION	1744	59	134		539	178	27		50		BRX 170	139		B 3	169
BROMINE TRIFLUORIDE	1746	44	136		541	179		39 –	50		BTF 176	141			170
BROMO-2-PROPANONE *Synonym:* BROMOACETONE	1569	55			562				51		BRE 166	141	B3.1		170
BROMOACETIC ACID SOLID	1938	60			542	180			51			141			170
BROMOACETIC ACID SOLUTION	1938	60			542	180			51			141			
BROMOACETONE	1569	55							51		BRE 166	141	B3.1		170
BOX TOE GUM	1133	60							51			138			
BROMOACETYLENE					542	180			51						
BROMOBENZENE	2514	26	137					–19	51		BBZ 125	142			170
BROMOBENZOL *Synonym:* BROMOBENZENE	2514	26	137					–19	51		BBZ 125	142			170
BROMOBENZYL CYANIDE	1694/1693	58							51			142			

CHEMICAL OR MATERIAL NAME	UN/NA Number	Guide Number (DOT)	Firefighter's Hazardous Materials Reference Book	First Aid Manual for Chemical Accidents	Sax's Dangerous Properties of Industrial Materials	Hazardous Chemicals Desk Reference	Rapid Guide to Hazardous Chemicals in the Workplace	Fire Protection Guide on Hazardous Materials (NFPA)	Firefighter's Handbook of Hazardous Materials	Pocket Guide to Chemical Hazards (NIOSH)	Chemical Hazard Response Information System (CHRIS)	Emergency Handling of Hazardous Materials in Surface Transportation (AAR)	Emergency Action Guides (EAG)	Chemical Data Notebook: A User's Manual	Condensed Chemical Dictionary
BROMOBENZYL NITRILE	1694	58				180			52						
BROMOBUTANE	1126	29							52		BBU 124				
BROMOCHLORODIFLUOROMETHANE	1974	12			546	181			52						171
BROMOCHLOROMETHANE	1887	58							52			143			171
BROMOETHANE	1891	58							52						171
BROMOETHYL ETHYL ETHER	2340	27							52						
BROMOFORM	2515	58		12	552	183	28		52	48	BRO 167	144			172
BROMOFUME Synonym: ETHYLENE DIBROMIDE	1605	55	371					85 –	143	110	EDB 518	430	E2.4.1		487
BROMOMETHANE Synonym: METHYL BROMIDE	1062	55	559	20	2274	776	137	110–67	52	146	MTB 805	619	M3.0	M 3	172
BROMOMETHYLBUTANE	2341	27							52						
BROMOMETHYLPROPANE	2342/1993	27							52			145			
BROMOPENTANE	2343	27							52						
BROMOPROPANE	2344/1993	29							53			145			
BROMOPROPYNE	2345	29							53						
BROMOSOL Synonym: METHYL BROMIDE	1062	55	559	20	2274	776	137	110–67	192	146	MTB 805	619	M3.0	M 3	759
BROMOSUCCINIMIDE (-n)					563	186			53						173
BROMOTOLUENE-alpha Synonym: BENZYL BROMIDE	1737	59	123		398	132			53		BBR 122	119			173
BROMOTOLUENE-meta															
BROMOTOLUENE-omega Synonym: BENZYL BROMIDE	1737	59	123		398	132			42		BBR 122	119			135
BROMOTOLUENE-ortho								–20	53						

CHEMICAL OR MATERIAL NAME	UN/NA Number	Guide Number (DOT)	Firefighter's Hazardous Materials Reference Book	First Aid Manual for Chemical Accidents	Sax's Dangerous Properties of Industrial Materials	Hazardous Chemicals Desk Reference	Rapid Guide to Hazardous Chemicals in the Workplace	Fire Protection Guide on Hazardous Materials (NFPA)	Firefighter's Handbook of Hazardous Materials	Pocket Guide to Chemical Hazards (NIOSH)	Chemical Hazard Response Information System (CHRIS)	Emergency Handling of Hazardous Materials in Surface Transportation (AAR)	Emergency Action Guides (EAG)	Chemical Data Notebook: A User's Manual	Condensed Chemical Dictionary
BROMOTOLUENE-para								-20	53						
BROMOTRICHLORMETHANE					564	186									
BROMOTRIFLUOROETHYLENE	2419/1954	17			565	186			53			146			173
BROMOTRIFLUOROMETHANE	1009	12							53			146			173
BROWN OIL Synonym: SULFURIC ACID	1830	39	802	24	3163	1082	204	164-	259	200	SFA1037	871	S4.2	S 8	1104
BROZONE Synonym: METHYL BROMIDE	1062	55	559	20	2274	776	137	110-67	192	146	MTB 805	619	M3.0	M 3	759
BRUCINE	1570	53	138		566	186			53		BRU 169	147			174
BRUCINE COMPOUNDS	1570	53			566				53		BRU 169				
BRUCINE DIHYDRATE Synonym: BRUCINE	1570	53	138		566	186			53		BRU 169	147			174
BTMAC Synonym: BENZYLTRIMETHYLAMMONIUM CHLORIDE					413						BMA 146				138
BUNKER C Synonym: FUEL OIL NO. 5	1993	27	406					-55	151			464	F3.3		
BUNKER C OIL Synonym: OILS FUEL NO. 6	1223	27									OSX 907				
BURNT LIME	1910	60							53						177
BUTA-1,3-DIENE Synonym: BUTADIENE	1010	17	139	12				40 -	53	48	BDI 134	148	B4.0	B 4	
BUTADIENE	1010	17	139	12				40 -	53	48	BDI 134	148	B4.0	B 4	
BUTADIENE MONOXIDE								-20	53						
BUTADIENE-alpha Synonym: BUTADIENE	1010	17	139	12				40 -	53	48	BDI 134	148	B4.0	B 4	

CHEMICAL OR MATERIAL NAME	UN/NA Number	Guide Number (DOT)	Firefighter's Hazardous Materials Reference Book	First Aid Manual for Chemical Accidents	Sax's Dangerous Properties of Industrial Materials	Hazardous Chemicals Desk Reference	Rapid Guide to Hazardous Chemicals in the Workplace	Fire Protection Guide on Hazardous Materials (NFPA)	Firefighter's Handbook of Hazardous Materials	Pocket Guide to Chemical Hazards (NIOSH)	Chemical Hazard Response Information System (CHRIS)	Emergency Handling of Hazardous Materials in Surface Transportation (AAR)	Emergency Action Guides (EAG)	Chemical Data Notebook: A User's Manual	Condensed Chemical Dictionary
BUTADIENE-gamma **Synonym: BUTADIENE**	1010	17	139	12				40 –	53	48	BDI 134	148	B4.0	B 4	
BUTADIENE-trans **Synonym: BUTADIENE**	1010	17	139	12				40 –	53	48	BDI 134	148	B4.0	B 4	
BUTAL **Synonym: BUTYRALDEHYDE (-n)**	1129	26	152	13	642	216		43 –25	61		BTR 182	172	B8.0		192
BUTALDEHYDE **Synonym: BUTYRALDEHYDE-n**	1129	26	152	13	642	216		43 –25	53		BTR 182	172	B8.0		177
BUTALYDE **Synonym: BUTYRALDEHYDE-n**	1129	26	152	13	642	216		43 –25	61		BTR 182	172	B8.0		192
BUTAN-2-ONE **Synonym: METHYL ETHYL KETONE**	1193	26	566	20	2319	793	143	–69	194		MEK 751	627	M3.2	M 5	768
BUTANAL **Synonym: BUTYRALDEHYDE (-n)**	1129	26	152	13	642	216		43 –25	53		BTR 182	172	B8.0		177
BUTANALOXIME	2840	26							53						
BUTANE	1011/1075	22	140	12	574	189	28	–20	54		BUT 188	148	B4.05		177
BUTANE MIXTURE	1011/1075	22	140	12	574	189	28	–20	54		BUT 188	148			
BUTANE SULTONE						190	29								
BUTANE-1,4-EPOXY **Synonym: TETRAHYDROFURAN**	2056	26	818		3227	1103	210	167–86	263	210	THF1116	894	T1.0	T 2	1131
BUTANE-n **Synonym: BUTANE**	1011/1075	22	140	12	574	189	28	–20	53		BUT 188	148	B4.05		177
BUTANEDIOL	1987	26							54						
BUTANEDIONE	2346	26							54						178
BUTANENITRILE	2411	28							54						178
BUTANETHIOL	2347	27							54						

CHEMICAL OR MATERIAL NAME	UN/NA Number	Guide Number (DOT)	Firefighter's Hazardous Materials Reference Book	First Aid Manual for Chemical Accidents	Sax's Dangerous Properties of Industrial Materials	Hazardous Chemicals Desk Reference	Rapid Guide to Hazardous Chemicals in the Workplace	Fire Protection Guide on Hazardous Materials (NFPA)	Firefighter's Handbook of Hazardous Materials	Pocket Guide to Chemical Hazards (NIOSH)	Chemical Hazard Response Information System (CHRIS)	Emergency Handling of Hazardous Materials in Surface Transportation (AAR)	Emergency Action Guides (EAG)	Chemical Data Notebook: A User's Manual	Condensed Chemical Dictionary
BUTANIC ACID Synonym: BUTYRIC ACID (-n)	2820	60	154		643	216		44 -25	61		BRA 164	173			192
BUTANOIC ACID Synonym: BUTYRIC ACID (-n)	2820	60	154		643	216		44 -25	54		BRA 164	173			178
BUTANOIC ANHYDRIDE	2739	60							54						
BUTANOL Synonym: BUTYL ALCOHOL (-n)	1120	26		12	593	196	31	-21	54	52	BAN 119	148	B7.1		181
BUTANOL-sec Synonym: BUTYL ALCOHOL (-sec)	1120	26			593	197	31	-21	56	52	BAS 120	148	B7.1		181
BUTANOL-tert Synonym: BUTYL ALCOHOL (-tert)	1120	26			594	197	31	-21	56	52	BAT 121	150	B7.1		182
BUTANOLAL	2839	55													
BUTANONE Synonym: METHYL ETHYL KETONE	1193	26	566	20	2319	793	143	-69	54		MEK 751	627	M3.2	M 5	768
BUTANOYL CHLORIDE	2353	29							54						178
BUTENE	1012	22	141						55		BTN 179		B5.0		
BUTENE RESINS Synonym: POLYBUTENE			697								PLB 962				930
BUTENE-alpha Synonym: BUTENE	1012	22	141						55		BTN 179		B5.0		
BUTENEDIOIC ACID-cis Synonym: MALEIC ACID	2215	60	527		2154	736			183		MLI 771	591			724
BUTENEDIOIC ACID-trans Synonym: FUMARIC ACID	9126/9188		407		1760	595			152		FUM 599	465			179
BUTENEDIOIC ANHYDRIDE-cis Synonym: MALEIC ANHYDRIDE	2215	60	528	19	2155	736	130	106-64	55	138	MLA 769	592	M1.0		724

CHEMICAL OR MATERIAL NAME	UN/NA Number	Guide Number (DOT)	Firefighter's Hazardous Materials Reference Book	First Aid Manual for Chemical Accidents	Sax's Dangerous Properties of Industrial Materials	Hazardous Chemicals Desk Reference	Rapid Guide to Hazardous Chemicals in the Workplace	Fire Protection Guide on Hazardous Materials (NFPA)	Firefighter's Handbook of Hazardous Materials	Pocket Guide to Chemical Hazards (NIOSH)	Chemical Hazard Response Information System (CHRIS)	Emergency Handling of Hazardous Materials in Surface Transportation (AAR)	Emergency Action Guides (EAG)	Chemical Data Notebook: A User's Manual	Condensed Chemical Dictionary
BUTENEDIOL	1987	26						−20							
BUTOXYDIETHYLENE GLYCOL Synonym: DIETHYLENE GLYCOL MONOBUTYL ETHER					1221			−38	111		DME 430				392
BUTOXYDIGLYCOL Synonym: DIETHYLENE GLYCOL MONOBUTYL ETHER					1221			−38	111		DME 430				392
BUTOXYL	2708/1993	26							55			149			
BUTOXYPROPYL TRICHLOROPHENOXYACETATE Synonym: 2,4,5-T ESTERS	2765	55			3176				259		TES1105				
BUTTER OF ANTIMONY Synonym: ANTIMONY TRICHLORIDE	1733	60	92						56		ATM 103	85			89
BUTTER OF ARSENIC Synonym: ARSENIC TRICHLORIDE	1560	55	99	12			16		35		AST 95	92	A11.0		97
BUTTERCUP YELLOW Synonym: ZINC CHROMATE			895		3539	1193	234				ZCR1201				1244
BUTYL 2,4,5-TRICHLOROPHENOXYACETATE Synonym: 2,4,5-T ESTERS	2765	55			639	215			259		TES1105				
BUTYL 2,4-DICHLOROPHENOXYACETATE Synonym: 2,4-D ESTERS	2765/3077	55	10		611				90		DES 393	336			
BUTYL 2-METHACRYLATE Synonym: BUTYL METHACRYLATE-n	2227/1993	26							58		BMN 148	156			187
BUTYL 2-METHYL-2-PROPENOATE Synonym: BUTYL METHACRYLATE-n	2227/1993	26							58		BMN 148	156			187
BUTYL 2-PROPENOATE-n Synonym: BUTYL ACRYLATE (-n)	2348/1993	26	146		592	196	30	41 −21	56		BTC 174	150	B7.0		
BUTYL ACETATE-n Synonym: BUTYL ACETATE	1123	26	142	13	591	195	29	−21	56	50	BCN 128	149	B6.0		180

CHEMICAL OR MATERIAL NAME	UN/NA Number	Guide Number (DOT)	Firefighter's Hazardous Materials Reference Book	First Aid Manual for Chemical Accidents	Sax's Dangerous Properties of Industrial Materials	Hazardous Chemicals Desk Reference	Rapid Guide to Hazardous Chemicals in the Workplace	Fire Protection Guide on Hazardous Materials (NFPA)	Firefighter's Handbook of Hazardous Materials	Pocket Guide to Chemical Hazards (NIOSH)	Chemical Hazard Response Information System (CHRIS)	Emergency Handling of Hazardous Materials in Surface Transportation (AAR)	Emergency Action Guides (EAG)	Chemical Data Notebook: A User's Manual	Condensed Chemical Dictionary
BUTYL ACETATE-sec	1123	26			591	195	30	-21	56	50	BTA 172				180
BUTYL ACETATE-tert	1123	26			591	195	30		56	50	BYA 189				180
BUTYL ACETOACETATE					591	196		-21	56						181
BUTYL ACRYLATE Synonym: BUTYL ACRYLATE-n	2348/1993	26	146		592	196	30	41 -21	56		BTC 174	150	B7.0		181
BUTYL ACRYLATE-iso	2527	27	145								BAI 116				
BUTYL ACRYLATE-n Synonym: BUTYL ACRYLATE	2348/1993	26	146		592	196	30	41 -21	56		BTC 174	150	B7.0		181
BUTYL ALCOHOL Synonym: BUTYL ALCOHOL-n	1120	26			593	196	31	-21	56	52	BAN 119	150	B7.1		181
BUTYL ALCOHOL-n Synonym: BUTYL ALCOHOL	1120	26			593	196	31	-21	56	52	BAN 119	150	B7.1		181
BUTYL ALCOHOL-sec	1120	26			593	197	31	-21	56	52	BAS 120		B7.1		181
BUTYL ALCOHOL-tert	1120	26			594	197	31	-21	56	52	BAT 121	150	B7.1		182
BUTYL ALDEHYDE Synonym: BUTYRALDEHYDE (-n)	1129	26	152	13	642	216		43 -25	61		BTR 182	172	B8.0		182
BUTYL ALDEHYDE (-n)	2045	26							56						182
BUTYL ALDEHYDE (-n) Synonym: BUTYRALDEHYDE (-n)	1129	26	152	13	642	216		43 -25	61		BTR 182	172	B8.0		182
BUTYL alpha-METHYLACRYLATE-n Synonym: BUTYL METHACRYLATE (-n)	2227/1993	26							58		BMN 148	156			187
BUTYL BENZENE (-n)	2709/1993	27			602	200		-21	56			151			183
BUTYL BENZOATE					603	201		-22	56						183
BUTYL BENZYL PHTHALATE	3082/9188	31									BPH 159	151			183
BUTYL BROMIDE (-n)	1126	29						-22	56			151			183

CHEMICAL OR MATERIAL NAME	UN/NA Number	Guide Number (DOT)	Firefighter's Hazardous Materials Reference Book	First Aid Manual for Chemical Accidents	Sax's Dangerous Properties of Industrial Materials	Hazardous Chemicals Desk Reference	Rapid Guide to Hazardous Chemicals in the Workplace	Fire Protection Guide on Hazardous Materials (NFPA)	Firefighter's Handbook of Hazardous Materials	Pocket Guide to Chemical Hazards (NIOSH)	Chemical Hazard Response Information System (CHRIS)	Emergency Handling of Hazardous Materials in Surface Transportation (AAR)	Emergency Action Guides (EAG)	Chemical Data Notebook: A User's Manual	Condensed Chemical Dictionary
BUTYL BROMIDE-sec	2339	27							56						183
BUTYL BUTYRATE (-n)								−22	56		BUB 184				183
BUTYL CARBINOL (-n)	1105	26							56						184
BUTYL CARBINOL-sec	1105	26							56						184
BUTYL CARBINOL-tert	1150	26						−22	56						
BUTYL CARBITOL Synonym: DIETHYLENE GLYCOL MONOBUTYL ETHER					1221			−38	56		DME 430				184
BUTYL CARBITOL ACETATE Synonym: DIETHYLENE GLYCOL MONOBUTYL ETHER ACETATE					606	202		−38			DEM 389				184
BUTYL CELLOSOLVE Synonym: ETHYLENE GLYCOL MONOBUTYL ETHER	2369	26						−51	56		EGM 532	432			184
BUTYL CELLOSOLVE ACETATE Synonym: ETHYLENE GLYCOL MONOBUTYL ETHER ACETATE		26						−51	56		EMA 542				184
BUTYL CHLORIDE (-n)	1127	27			607	203		−22	56		BYC 190	152			184
BUTYL CHLORIDE-sec	1127	27						−22	56						184
BUTYL CHLORIDE-tert	1127	27			607	203		−22	56						
BUTYL CHLOROFORMATE (-n)	2743	57									BFO 143				
BUTYL CHROMATE-tert					609	203			57	54					184
BUTYL CUMENE PEROXIDE-tert	2091	48							57						
BUTYL CUMYL PEROXIDE-tert	2091/3105	48							57			152			
BUTYL ETHANOATE Synonym: BUTYL ACETATE (-n)	1123	26	142	13	591	195	29	−21	57	50	BCN 128	149	B6.0		180
BUTYL ETHER Synonym: DI-n-BUTYL ETHER	1149	26			616	205		42 −42	57		DBE 344	153			186

CHEMICAL OR MATERIAL NAME	UN/NA Number	Guide Number (DOT)	Firefighter's Hazardous Materials Reference Book	First Aid Manual for Chemical Accidents	Sax's Dangerous Properties of Industrial Materials	Hazardous Chemicals Desk Reference	Rapid Guide to Hazardous Chemicals in the Workplace	Fire Protection Guide on Hazardous Materials (NFPA)	Firefighter's Handbook of Hazardous Materials	Pocket Guide to Chemical Hazards (NIOSH)	Chemical Hazard Response Information System (CHRIS)	Emergency Handling of Hazardous Materials in Surface Transportation (AAR)	Emergency Action Guides (EAG)	Chemical Data Notebook: A User's Manual	Condensed Chemical Dictionary
BUTYL ETHER-n Synonym: DI-n-BUTYL ETHER	1149	26			616	205		42 –42	57		DBE 344	153			186
BUTYL ETHYL ACETALDEHYDE	1191	26			616	205			57						186
BUTYL ETHYL ETHER (-n)	1179	26							57						186
BUTYL ETHYLENE	2370	27							57						
BUTYL FORMATE (-n)	1128	26			618	206		–23	57			153			186
BUTYL GLYCIDYL ETHER-n			147	13	618	206	32		57	54					186
BUTYL GLYCOLATE								–23	57						
BUTYL HYDRIDE	1011	22							57						
BUTYL HYDROPEROXIDE-tert	2094	51			619	206	33	–23	57						186
BUTYL HYDROPEROXIDE-tert	2092/3103	48				206		–23	57			153			186
BUTYL HYDROPEROXIDE-tert	3075	49							57						186
BUTYL HYDROPEROXIDE-tert	2093/3109	51			619	206	33	–23	57		BHP 145	154			186
BUTYL IMIDAZOLE	2690	55							57						
BUTYL ISOCYANATE-n	2485	57			622	207		–23				154			187
BUTYL ISOCYANATE-tert	2484	57							57						
BUTYL ISOPROPYL BENZENE HYDROPEROXIDE-tert	2091	48			622	207			57			155			187
BUTYL ISOVALERATE								–23	57						
BUTYL LACTATE	1993	27			622	207	33	–23	57			155			187
BUTYL LACTATE-n	1993	27			622	207	33	–23	57			155			187
BUTYL LITHIUM	2445	40			623										187
BUTYL MERCAPTAN Synonym: BUTYL MERCAPTAN-n	2347	27			623	208	33		57	54	BTM 178	155			187

CHEMICAL OR MATERIAL NAME	UN/NA Number	Guide Number (DOT)	Firefighter's Hazardous Materials Reference Book	First Aid Manual for Chemical Accidents	Sax's Dangerous Properties of Industrial Materials	Hazardous Chemicals Desk Reference	Rapid Guide to Hazardous Chemicals in the Workplace	Fire Protection Guide on Hazardous Materials (NFPA)	Firefighter's Handbook of Hazardous Materials	Pocket Guide to Chemical Hazards (NIOSH)	Chemical Hazard Response Information System (CHRIS)	Emergency Handling of Hazardous Materials in Surface Transportation (AAR)	Emergency Action Guides (EAG)	Chemical Data Notebook: A User's Manual	Condensed Chemical Dictionary
BUTYL MERCAPTAN-n Synonym: BUTYL MERCAPTAN	2347	27			623	208	33		57	54	BTM 178	155			187
BUTYL METHACRYLATE Synonym: BUTYL METHACRYLATE-n	2227/1993	26						-23	58		BMN 148	156			187
BUTYL METHACRYLATE-n Synonym: BUTYL METHACRYLATE	2227/1993	26						-23	58		BMN 148	156			187
BUTYL METHANOATE	1128	26							58						
BUTYL METHYL ETHER	2350/1993	26							58			156			
BUTYL METHYL KETONE-n Synonym: METHYL n-BUTYL KETONE	1993	27	576	20					58		MBK 729				760
BUTYL MONOPEROXYMALEATE-tert	2101	48							58						
BUTYL MONOPEROXYMALEATE-tert	2100	48							58						
BUTYL MONOPEROXYMALEATE-tert	2099	49							58						
BUTYL MONOPEROXYPHTHALATE-tert	2105	48							58						
BUTYL NITRATE (-n)					625			-23	58						188
BUTYL NITRITE Synonym: BUTYL NITRITE-n	2351/1993	26			626	209			58			156			
BUTYL NITRITE-n	2351/1993	26			626	209			58			156			
BUTYL NITRITE-sec	2351	26			626	209			58						
BUTYL NITRITE-tert	2351	26			626				58						188
BUTYL OLEATE					629	210		-24							188
BUTYL PERACETATE-tert	2096	48			630	210	34	-24							189
BUTYL PERACETATE-tert	2095				630		34	-24	58						189
BUTYL PERBENZOATE-tert	2890	48			630	211			58						189
BUTYL PERBENZOATE-tert	2097	48			630	211		-24	58						189

CHEMICAL OR MATERIAL NAME	UN/NA Number	Guide Number (DOT)	Firefighter's Hazardous Materials Reference Book	First Aid Manual for Chemical Accidents	Sax's Dangerous Properties of Industrial Materials	Hazardous Chemicals Desk Reference	Rapid Guide to Hazardous Chemicals in the Workplace	Fire Protection Guide on Hazardous Materials (NFPA)	Firefighter's Handbook of Hazardous Materials	Pocket Guide to Chemical Hazards (NIOSH)	Chemical Hazard Response Information System (CHRIS)	Emergency Handling of Hazardous Materials in Surface Transportation (AAR)	Emergency Action Guides (EAG)	Chemical Data Notebook: A User's Manual	Condensed Chemical Dictionary
BUTYL PERBENZOATE-tert	2098	48			630				58						189
BUTYL PEROXIDE-tert	2102	48			631	211	34		58						189
BUTYL PEROXY-2-ETHYLHEXANOATE-tert	2887/3106	48										158			189
BUTYL PEROXY-2-ETHYLHEXANOATE-tert	2886/3115	48										157			189
BUTYL PEROXY-2-ETHYLHEXANOATE-tert	2888/3117	52										157			189
BUTYL PEROXY-2-ETHYLHEXANOATE-tert	2143/3113	52							58			157			189
BUTYL PEROXY-3,5,5-TRIMETHYLHEXANOATE-tert	2104/3105	48							58			159			
BUTYL PEROXY-3-PHENYLPHTHALIDE-tert	2596	48							58						
BUTYL PEROXYACETATE-tert	2095/3101	49							58			159			189
BUTYL PEROXYACETATE-tert	2096/3103	48							58			159			189
BUTYL PEROXYBENZOATE-tert	2097/3103	48										160			189
BUTYL PEROXYBENZOATE-tert	2890/3106	48							58			160			189
BUTYL PEROXYBENZOATE-tert	2098	48													189
BUTYL PEROXYCROTONATE-tert	2183	48							58						
BUTYL PEROXYDICARBONATE	2169	52													
BUTYL PEROXYDICARBONATE (-n)	2170	52													
BUTYL PEROXYDICARBONATE-tert	2169	52							58						
BUTYL PEROXYDIETHYLACETATE-tert	2551/3105	48										161			
BUTYL PEROXYDIETHYLACETATE-tert	2144/3113	52							58			162			
BUTYL PEROXYISOBUTYRATE-tert	2142/3111	52							59			162			189
BUTYL PEROXYISOBUTYRATE-tert	2562/3115	52							59			162			189
BUTYL PEROXYISONONANOATE-tert	2104	48							59						
BUTYL PEROXYISOPROPYL CARBONATE-tert	2103/3103	49							59			163			189
BUTYL PEROXYMALEATE-tert	2100/3103	48							59			164			

CHEMICAL OR MATERIAL NAME	UN/NA Number	Guide Number (DOT)	Firefighter's Hazardous Materials Reference Book	First Aid Manual for Chemical Accidents	Sax's Dangerous Properties of Industrial Materials	Hazardous Chemicals Desk Reference	Rapid Guide to Hazardous Chemicals in the Workplace	Fire Protection Guide on Hazardous Materials (NFPA)	Firefighter's Handbook of Hazardous Materials	Pocket Guide to Chemical Hazards (NIOSH)	Chemical Hazard Response Information System (CHRIS)	Emergency Handling of Hazardous Materials in Surface Transportation (AAR)	Emergency Action Guides (EAG)	Chemical Data Notebook: A User's Manual	Condensed Chemical Dictionary
BUTYL PEROXYMALEATE-tert	2099/3102	49							59			163			
BUTYL PEROXYMALEATE-tert	2101/3108	48							59			164			
BUTYL PEROXYNEODECANOATE-tert	2177/3115	52							59			165			
BUTYL PEROXYNEODECANOATE-tert	2594/3115	52							59			164			
BUTYL PEROXYPHTHALATE-tert	2105/3102	48							59			165			
BUTYL PEROXYPIVALATE-tert	2110/3113	52			631	211		−24	59			165			189
BUTYL PEROXYPIVALATE-tert	3047	52							59						189
BUTYL PHENOL	1759	60										166			
BUTYL PHENOL	1760	60										166			
BUTYL PHENOL	2228	55							59			166			
BUTYL PHENYL ETHER (-n)					633	213		−24	59						190
BUTYL PHOSPHATE								−24	59						
BUTYL PHOSPHORIC ACID	1718	60							59						190
BUTYL PHTHALATE Synonym: DIBUTYL PHTHALATE	9095		265	15	1119	375	66	−34	59	84	DPA 458				190
BUTYL PHTHALATE-n	9188/3082			15								167			190
BUTYL PROPANOATE	1914	27			634	213			59						
BUTYL PROPIONATE (-n)	1914/1993	27						−24	59			167			190
BUTYL SULFIDE					636										191
BUTYL TITANATE Synonym: TETRABUTYL TITANATE			806		638	214			59		TBT1078				191
BUTYL TITANATE MONOMER Synonym: TETRABUTYL TITANATE			806								TBT1078				1126
BUTYL TOLUENE	2667	27		13					59		BUE 187	168			

CHEMICAL OR MATERIAL NAME	UN/NA Number	Guide Number (DOT)	Firefighter's Hazardous Materials Reference Book	First Aid Manual for Chemical Accidents	Sax's Dangerous Properties of Industrial Materials	Hazardous Chemicals Desk Reference	Rapid Guide to Hazardous Chemicals in the Workplace	Fire Protection Guide on Hazardous Materials (NFPA)	Firefighter's Handbook of Hazardous Materials	Pocket Guide to Chemical Hazards (NIOSH)	Chemical Hazard Response Information System (CHRIS)	Emergency Handling of Hazardous Materials in Surface Transportation (AAR)	Emergency Action Guides (EAG)	Chemical Data Notebook: A User's Manual	Condensed Chemical Dictionary
BUTYL TRICHLOROSILANE (-n)	1747	29	149		640	215		-24	59		BCS 131	168			191
BUTYL VINYL ETHER (-n)	2352/1993	26							59			168			192
BUTYL-1-BUTANAMINE-n Synonym: DI-n-BUTYLAMINE	2248/1993	68			1108	370					DBA 342	312	D3.2		371
BUTYL-2,4,6-TRINITRO-m-XYLENE-tert	2956	32							59						
BUTYL-2-PROPENOATE Synonym: BUTYL ACRYLATE (-n)	2348/1993	26	146		592	196	30	41 -21	59		BTC 174	150	B7.0		181
BUTYL-4,4-DI(tert-BUTYLPEROXY) VALERATE-n	2141/3106	48							59			169			
BUTYL-4,4-DI(tert-BUTYLPEROXY) VALERATE-n	2140/3103	48							60			169			
BUTYL-m-CRESOL-tert								-22	60						
BUTYLACID PHOSPHATE-n	1718	60							60			150			181
BUTYLALDECHYDE Synonym: BUTYRALDEHYDE (-n)	1129	26	152	13	642	216			61		BTR 182	172	B8.0		192
BUTYLAMINE Synonym: BUTYLAMINE-n	1125	68		13	594	197	32	41 -21	60	52	BAM 118	169	B7.3		182
BUTYLAMINE OLEATE								-21	60						
BUTYLAMINE-iso Synonym: ISOBUTYLAMINE	1214	68	465		2017	676	122	-61	172		IAM 659	547			652
BUTYLAMINE-n Synonym: BUTYLAMINE	1125	68		13	594	197	32	41 -21	60	52	BAM 118	169	B7.3		182
BUTYLAMINE-sec	1125	68			595	198	32	-21	60		BTL 177				182
BUTYLAMINE-tert	1125	68			595	198	32	-21	60		BUA 183	170			182
BUTYLAMINOETHYL METHACRYLATE-tert					598			-21	60						182
BUTYLANILINE (-n)	2738	55			601	199		-21	60						182
BUTYLBENZENE-sec	2709/1993	27			602	200		-22	56			151			183

HAZARDOUS MATERIALS REFERENCE BOOKS INDEX

CHEMICAL OR MATERIAL NAME	UN/NA Number	Guide Number (DOT)	Firefighter's Hazardous Materials Reference Book	First Aid Manual for Chemical Accidents	Sax's Dangerous Properties of Industrial Materials	Hazardous Chemicals Desk Reference	Rapid Guide to Hazardous Chemicals in the Workplace	Fire Protection Guide on Hazardous Materials (NFPA)	Firefighter's Handbook of Hazardous Materials	Pocket Guide to Chemical Hazards (NIOSH)	Chemical Hazard Response Information System (CHRIS)	Emergency Handling of Hazardous Materials in Surface Transportation (AAR)	Emergency Action Guides (EAG)	Chemical Data Notebook: A User's Manual	Condensed Chemical Dictionary
BUTYLBENZENE-tert	2709/1993	27			602	200		−22	56			151			183
BUTYLCARBINOL-n Synonym: AMYL ALCOHOL (-n)	1105/1987	26	78		242	69			56		AAN 4	74			184
BUTYLCARBINYL CHLORIDE-n Synonym: AMYL CHLORIDE (-n)	1107	26	79						31		AMY 61	75			74
BUTYLCYCLOHEXYL CHLOROFORMATE-tert	2747	55							60						
BUTYLCYCLOHEXYLAMINE-n					610			−22	57						
BUTYLENE Synonym: BUTENE	1012	22	150						60		BTN 179		B5.0		185
BUTYLENE GLYCOL-alpha								−23	61						
BUTYLENE GLYCOL-beta								−23	60						
BUTYLENE HYDRATE Synonym: BUTYL ALCOHOL (-sec)	1120	26			593	197	31	−21	56	52	BAS 120	150	B7.1		181
BUTYLENE OXIDE	3022/1993	26	151					42	61		BTO 180	171			
BUTYLENE OXIDE Synonym: TETRAHYDROFURAN	2056	26	818		3227	1103	210	167–86	263	210	THF1116	894	T1.0	T 2	1131
BUTYLENE-alpha Synonym: BUTENE	1012	22	141						60		BTN 179		B5.0		
BUTYLENE-beta	1012	22							60						
BUTYLETHYLACETALDEHYDE Synonym: ETHYLHEXALDEHYDE	1191	26	356		616	205			57		EHA 535	438			186
BUTYLETHYLAMINE Synonym: ETHYL-n-BUTYLAMINE-n	2733	29									EBA 507				481
BUTYLLITHIUM	2445	40			623			43 –	61						
BUTYLPHENOL-o-sec					632	212	34		61						
BUTYLPHENOL-p-sec	2229	53			632	212			61						

CHEMICAL OR MATERIAL NAME	UN/NA Number	Guide Number (DOT)	Firefighter's Hazardous Materials Reference Book	First Aid Manual for Chemical Accidents	Sax's Dangerous Properties of Industrial Materials	Hazardous Chemicals Desk Reference	Rapid Guide to Hazardous Chemicals in the Workplace	Fire Protection Guide on Hazardous Materials (NFPA)	Firefighter's Handbook of Hazardous Materials	Pocket Guide to Chemical Hazards (NIOSH)	Chemical Hazard Response Information System (CHRIS)	Emergency Handling of Hazardous Materials in Surface Transportation (AAR)	Emergency Action Guides (EAG)	Chemical Data Notebook: A User's Manual	Condensed Chemical Dictionary
BUTYLPHENOL-p-tert											BTP 181				
BUTYLTOLUENE-p-tert	2667	27					35		61	54		168			191
BUTYLTRICHLOROSILANE-n Synonym: BUTYL TRICHLOROSILANE	1747	29	149		640	215		−24	59		BCS 131	168			191
BUTYNEDIOL	2716	55							61						192
BUTYRAL Synonym: BUTYRALDEHYDE	1129	26	152	13	642	216		43 −25	61		BTR 182	172	B8.0		192
BUTYRALDEHYDE Synonym: BUTYRALDEHYDE-n	1129	26	152	13	642	216		43 −25	61		BTR 182	172	B8.0		192
BUTYRALDEHYDE-iso Synonym: BUTYRALDEHYDE	1129	26	153	13	642	216		43 −25	61		BAD 115	172	B8.0		192
BUTYRALDEHYDE-n Synonym: BUTYRALDEHYDE	1129	26	152	13	642	216		43 −25	61		BTR 182	172	B8.0		192
BUTYRALDOL								−25	61						
BUTYRALDOXIME	2840	26						−25	61			173			
BUTYRIC ACID Synonym: BUTYRIC ACID-n	2820	60	154		643	216		44 −25	61		BRA 164	173			192
BUTYRIC ACID ETHYL ESTER Synonym: ETHYL BUTYRATE	1180	26	349		1584	531		−49	61		EBR 508	412			482
BUTYRIC ACID-n Synonym: BUTYRIC ACID	2820	60	154		643	216		44 −25	61		BRA 164	173			192
BUTYRIC ALDEHYDE Synonym: BUTYRALDEHYDE (-n)	1129	26	152	13	642	216		43 −25	61		BTR 182	172	B8.0		192
BUTYRIC ANHYDRIDE	1760	60										173			192
BUTYRIC ANHYDRIDE	2739	60						−25	61						192

CHEMICAL OR MATERIAL NAME	UN/NA Number	Guide Number (DOT)	Firefighter's Hazardous Materials Reference Book	First Aid Manual for Chemical Accidents	Sax's Dangerous Properties of Industrial Materials	Hazardous Chemicals Desk Reference	Rapid Guide to Hazardous Chemicals in the Workplace	Fire Protection Guide on Hazardous Materials (NFPA)	Firefighter's Handbook of Hazardous Materials	Pocket Guide to Chemical Hazards (NIOSH)	Chemical Hazard Response Information System (CHRIS)	Emergency Handling of Hazardous Materials in Surface Transportation (AAR)	Emergency Action Guides (EAG)	Chemical Data Notebook: A User's Manual	Condensed Chemical Dictionary
BUTYRIC ETHER Synonym: ETHYL BUTYRATE	1180	26	349		1584	531		−49	61		EBR 508	412			482
BUTYROLACTONE								−25	61						192
BUTYRONE	2710	26							61						192
BUTYRONITRILE	2411/1993	28	155		643	217	35	−25	62		BNI 149	174			192
BUTYRONITRILE-n Synonym: BUTYRONITRILE	2411/1993	28	155		643		35		61		BNI 149	174			192
BUTYROYL CHLORIDE	2353	29							62						193
BUTYRYL CHLORIDE	2353/2358	29			644	218			62		BUC 185	174			193
BUTYRYLALDEHYDE Synonym: BUTYRALDEHYDE (-n)	1129	26	152	13	642	216		43 −25	61		BTR 182	172	B8.0		192
CACODYL									62						
CACODYL SULFIDE					647	219			62						
CACODYLIC ACID	1572/2811	53	156						62		CDA 235	175			194
CADMIUM				13	648	219	35		62						194
CADMIUM ACETATE	2570/3077	53	157		648	219	36		62		CAT 213	175			195
CADMIUM ACETATE DIHYDRATE Synonym: CADMIUM ACETATE	2570/3077	53	157		649	219	36		62		CAT 213	175			195
CADMIUM BROMIDE	2570/3077	53	158		649		36		62		CMB 263	175			195
CADMIUM BROMIDE TETRAHYDRATE Synonym: CADMIUM BROMIDE	2570/3077	53	158		649		36		62		CMB 263	175			195
CADMIUM CHLORATE					650	220	36		62						195
CADMIUM CHLORIDE	2570/3077	53	159		650	221	36		62		CDC 236	175			195
CADMIUM CHLORIDE 2.5 HYDRATE					651		36								
CADMIUM COMPOUND	2570	53			651	222	35		62			176			194

CHEMICAL OR MATERIAL NAME	UN/NA Number	Guide Number (DOT)	Firefighter's Hazardous Materials Reference Book	First Aid Manual for Chemical Accidents	Sax's Dangerous Properties of Industrial Materials	Hazardous Chemicals Desk Reference	Rapid Guide to Hazardous Chemicals in the Workplace	Fire Protection Guide on Hazardous Materials (NFPA)	Firefighter's Handbook of Hazardous Materials	Pocket Guide to Chemical Hazards (NIOSH)	Chemical Hazard Response Information System (CHRIS)	Emergency Handling of Hazardous Materials in Surface Transportation (AAR)	Emergency Action Guides (EAG)	Chemical Data Notebook: A User's Manual	Condensed Chemical Dictionary
CADMIUM DUST	2570	53		13	651		35			56					194
CADMIUM FLUOBORATE Synonym: CADMIUM FLUOROBORATE	2570	53			652	222	36				CFB 240				
CADMIUM FLUOBORATE SOLUTION Synonym: CADMIUM FLUOROBORATE	2570	53			652	222	36				CFB 240				
CADMIUM FLUOROBORATE	2570	53	160				36				CFB 240				
CADMIUM FUME Synonym: CADMIUM OXIDE	2570	53		13	654		36				COX 282				196
CADMIUM NITRATE	2570	53			653	223	36		62		CMN 267				196
CADMIUM NITRATE TETRAHYDRATE Synonym: CADMIUM NITRATE	2570	53			654	223	36		62		CMN 267				196
CADMIUM OXIDE	2570	53	162		654	224	36				COX 282				196
CADMIUM SULFATE	2570	53	163		656	225	36				CMS 270				197
CADOX HDP Synonym: CYCLOHEXANONE PEROXIDE	2896/3106	51							88		CHP 248	282			337
CADOX PS Synonym: DI(p-CHLOROBENZOYL) PEROXIDE	0149										DZP 498				
CAESIUM	1407/1383	40							62			176			197
CAESIUM HYDROXIDE	2682/1759	60							62			176			
CAESIUM NITRATE	1451/1479	35							62			177			
CAKE ALUMINUM Synonym: ALUMINUM SULFATE	9078/3077		46		132	43	9		24		ALM 43	47	A7.1.01		48
CALAMINE Synonym: ZINC CARBONATE	9157/3077		893		3538		234		288		ZCB1197	965			198
CALCIUM	1855	37			660	228			62			177			199
CALCIUM	1401	40	164		660	228		45 –	62		CAM 208	185			199

HAZARDOUS MATERIALS REFERENCE BOOKS INDEX

CHEMICAL OR MATERIAL NAME	UN/NA Number	Guide Number (DOT)	Firefighter's Hazardous Materials Reference Book	First Aid Manual for Chemical Accidents	Sax's Dangerous Properties of Industrial Materials	Hazardous Chemicals Desk Reference	Rapid Guide to Hazardous Chemicals in the Workplace	Fire Protection Guide on Hazardous Materials (NFPA)	Firefighter's Handbook of Hazardous Materials	Pocket Guide to Chemical Hazards (NIOSH)	Chemical Hazard Response Information System (CHRIS)	Emergency Handling of Hazardous Materials in Surface Transportation (AAR)	Emergency Action Guides (EAG)	Chemical Data Notebook: A User's Manual	Condensed Chemical Dictionary
CALCIUM ABIETATE Synonym: CALCIUM RESINATE	1313	32	177		671	234			63		CRE 300	184			200
CALCIUM ACETYLIDE Synonym: CALCIUM CARBIDE	1402	40	167	13	662	229		45 –	62		CCB 226	179	C1.0	C 1	201
CALCIUM ALKYLAROMATIC SULFONATE Synonym: DODECYLBENZENESULFONIC ACID CALCIUM SALT	2584	60									DCS 368				
CALCIUM ALKYLBENZENESULFONATE Synonym: DODECYLBENZENESULFONIC ACID CALCIUM SALT	2584	60									DCS 368				
CALCIUM ARSENATE	1573	53	165						62	56	CCA 225	178			200
CALCIUM ARSENITE	1574	53	166		661	228			63		CAS 212	178			200
CALCIUM ARSENITE SOLID Synonym: CALCIUM ARSENITE	1574	53	166		661	228			63		CAS 212	178			200
CALCIUM BIPHOSPHATE Synonym: CALCIUM PHOSPHATE					671	234					CAL 207				200
CALCIUM BISULFITE SOLUTION	2693	60			662	229			63						201
CALCIUM CARBIDE	1402	40	167	13	662	229		45 –	63		CCB 226	179	C1.0	C 1	201
CALCIUM CARBIMIDE	1403	40							63						
CALCIUM CARBONATE				13	663	229	36								201
CALCIUM CHLORATE	1452	35	168		663	230		45 –	63		CCC 227	179			201
CALCIUM CHLORATE SOLUTION	2429	35			663	230			63						
CALCIUM CHLORIDE				13	663	230					CLC 256				201
CALCIUM CHLORIDE ANHYDROUS Synonym: CALCIUM CHLORIDE				13	663	230					CLC 256				201

CHEMICAL OR MATERIAL NAME	UN/NA Number	Guide Number (DOT)	Firefighter's Hazardous Materials Reference Book	First Aid Manual for Chemical Accidents	Sax's Dangerous Properties of Industrial Materials	Hazardous Chemicals Desk Reference	Rapid Guide to Hazardous Chemicals in the Workplace	Fire Protection Guide on Hazardous Materials (NFPA)	Firefighter's Handbook of Hazardous Materials	Pocket Guide to Chemical Hazards (NIOSH)	Chemical Hazard Response Information System (CHRIS)	Emergency Handling of Hazardous Materials in Surface Transportation (AAR)	Emergency Action Guides (EAG)	Chemical Data Notebook: A User's Manual	Condensed Chemical Dictionary
CALCIUM CHLORIDE HYDRATES Synonym: CALCIUM CHLORIDE				13	663	230					CLC 256				201
CALCIUM CHLORITE	1453	35			664	230			63			179			202
CALCIUM CHLOROHYDROCHLORITE Synonym: CALCIUM HYPOCHLORITE	1748	45	172	13				46 –	63		CHY 252	182	C1.1		204
CALCIUM CHROMATE	9096/3077		169		664	230			63		CCR 232	180			202
CALCIUM CHROMATE (VI) Synonym: CALCIUM CHROMATE	9096/3077		169		664	230			63		CCR 232	180			202
CALCIUM CHROMATE DIHYDRATE Synonym: CALCIUM CHROMATE	9096/3077		169		664	231			63		CCR 232	180			202
CALCIUM CYANAMIDE	1403	40			665	231	37		63			180			202
CALCIUM CYANIDE	1575	55	170		665	231	37	46 –	63		CCN 230	180			202
CALCIUM DICARBIDE Synonym: CALCIUM CARBIDE	1402	40	167	13	662	229		45 –	63		CCB 226	179	C1.0	C 1	201
CALCIUM DIOXIDE Synonym: CALCIUM PEROXIDE	1457	35	175		670	234			63		CCP 231	183			202
CALCIUM DITHIONITE	1923	37							63			181			
CALCIUM DODECYLBENZENESULFONATE	9188/3077											181			
CALCIUM FLUORIDE					667						CAF 205				203
CALCIUM HYDRIDE	1404	40			668				63			181			203
CALCIUM HYDROGEN SULFITE SOLUTION	2693	60							63						203
CALCIUM HYDROSULFITE	1923	37							63			181			
CALCIUM HYDROXIDE			171	13	668	232	37		63		CAH 206				204
CALCIUM HYPOCHLORIDE Synonym: CALCIUM HYPOCHLORITE	1748	45	172	13				46 –	63		CHY 252	182	C1.1		204

CHEMICAL OR MATERIAL NAME	UN/NA Number	Guide Number (DOT)	Firefighter's Hazardous Materials Reference Book	First Aid Manual for Chemical Accidents	Sax's Dangerous Properties of Industrial Materials	Hazardous Chemicals Desk Reference	Rapid Guide to Hazardous Chemicals in the Workplace	Fire Protection Guide on Hazardous Materials (NFPA)	Firefighter's Handbook of Hazardous Materials	Pocket Guide to Chemical Hazards (NIOSH)	Chemical Hazard Response Information System (CHRIS)	Emergency Handling of Hazardous Materials in Surface Transportation (AAR)	Emergency Action Guides (EAG)	Chemical Data Notebook: A User's Manual	Condensed Chemical Dictionary
CALCIUM HYPOCHLORITE	1748	45	172	13				46 –	63		CHY 252	182	C1.1	C 2	204
CALCIUM HYPOCHLORITE	2208	35		13					63				C1.1	C 2	204
CALCIUM HYPOCHLORITE	2880	45		13					63			181	C1.1	C 2	204
CALCIUM HYPOCHLORITE MIXTURE	2208	35		13					63				C1.1	C 2	
CALCIUM MANGANESE SILICON	2844	40							63			182			
CALCIUM MANGANESE SILICON ALLOYS	1325	32													
CALCIUM NITRATE	1454	35	173		669	233			63		CNT 274	183			204
CALCIUM NITRATE TETRAHYDRATE Synonym: CALCIUM NITRATE	1454	35	173		669	233			63		CNT 274	183			204
CALCIUM OXIDE	1910	60	174	13	669	233	38	47 –	63	56	CAO 209	183			205
CALCIUM OXYCHLORIDE Synonym: CALCIUM HYPOCHLORITE	1748	45	172	13				46 –	63		CHY 252	182	C1.1	C 2	205
CALCIUM PERCHLORATE	1455	35							63						206
CALCIUM PERMANGANATE	1456	35			670	234			63			183			206
CALCIUM PEROXIDE	1457	35	175		670	234			63		CCP 231	183			206
CALCIUM PHOSPHATE					671	234					CAL 207				206
CALCIUM PHOSPHIDE	1360	41	176		671	234			63		CPP 293	184			207
CALCIUM PYROPHOSPHATE Synonym: CALCIUM PHOSPHATE					671	234					CAL 207				207
CALCIUM RESINATE	1313	32	177		671	234			63		CRE 300	184			207
CALCIUM RESINATE	1314	32			671	234			63						207
CALCIUM RESINATE FUSED Synonym: CALCIUM RESINATE	1313	32	177		671	234			63		CRE 300	184			207
CALCIUM ROSIN Synonym: CALCIUM RESINATE	1313	32	177		671	234			63		CRE 300	184			207

CHEMICAL OR MATERIAL NAME	UN/NA Number	Guide Number (DOT)	Firefighter's Hazardous Materials Reference Book	First Aid Manual for Chemical Accidents	Sax's Dangerous Properties of Industrial Materials	Hazardous Chemicals Desk Reference	Rapid Guide to Hazardous Chemicals in the Workplace	Fire Protection Guide on Hazardous Materials (NFPA)	Firefighter's Handbook of Hazardous Materials	Pocket Guide to Chemical Hazards (NIOSH)	Chemical Hazard Response Information System (CHRIS)	Emergency Handling of Hazardous Materials in Surface Transportation (AAR)	Emergency Action Guides (EAG)	Chemical Data Notebook: A User's Manual	Condensed Chemical Dictionary
CALCIUM SELENATE	2630	53							63						207
CALCIUM SILICIDE	1405	40			672				63			184			208
CALCIUM SILICON	1406	40							63			184			
CALCIUM SULFATE					672	235	38								208
CALCIUM SUPERPHOSPATE Synonym: CALCIUM PHOSPHATE					671	234					CAL 207				206
CALICHE Synonym: SODIUM NITRATE	1498	35	780		3097	1060			254		SDN1033	845	S2.6	S 5	210
CALOCHLOR Synonym: MERCURIC CHLORIDE	1624	53	532	19	674				186		MRC 792	599			740
CALOMEL Synonym: MERCUROUS CHLORIDE	2025	53	541	19	2192	749			64		MRR 795				210
CAMPHANOL-alpha	1312	32							64						
CAMPHENE	9011/3077	58	178		675	236			64		CPH 289	185			211
CAMPHOR	2717	32		13	675	236	39	-25	64	58		185			211
CAMPHOR OIL	1130	27	179		676	237		-25	64		CPO 292	185			
CANE SUGAR Synonym: SUCROSE			796		3151						SRS1056				212
CANNABIS					678				64						212
CANTHARIDES					679	237			64						
CAPRALDEHYDE Synonym: DECALDEHYDE											DAL 336				213
CAPRIC ALCOHOL Synonym: DECYL ALCOHOL (-n)	1987	26	258		1032	347			91		DAN 338	291			350
CAPRIC ALDEHYDE Synonym: DECALDEHYDE											DAL 336				

CHEMICAL OR MATERIAL NAME	UN/NA Number	Guide Number (DOT)	Firefighter's Hazardous Materials Reference Book	First Aid Manual for Chemical Accidents	Sax's Dangerous Properties of Industrial Materials	Hazardous Chemicals Desk Reference	Rapid Guide to Hazardous Chemicals in the Workplace	Fire Protection Guide on Hazardous Materials (NFPA)	Firefighter's Handbook of Hazardous Materials	Pocket Guide to Chemical Hazards (NIOSH)	Chemical Hazard Response Information System (CHRIS)	Emergency Handling of Hazardous Materials in Surface Transportation (AAR)	Emergency Action Guides (EAG)	Chemical Data Notebook: A User's Manual	Condensed Chemical Dictionary
CAPROALDEHYDE Synonym: HEXALDEHYDE (-n)	1207	26	431						64		HAL 618	513			598
CAPROIC ACID	2829	60						-25	64						213
CAPROLACTAM				13	680	238	39				CLS 259		C1.5		213
CAPROLACTAM-epsilon Synonym: CAPROLACTAM				13	680	238	39				CLS 259		C1.5		213
CAPRONALDEHYDE Synonym: HEXALDEHYDE (-n)	1207	26	431						159		HAL 618	513			598
CAPRONIC ALDEHYDE Synonym: HEXALDEHYDE (-n)	1207	26	431						159		HAL 618	513			598
CAPROYLALDEHYDE-n Synonym: HEXALDEHYDE (-n)	1207	26	431						159		HAL 618	513			598
CAPRYL ALCOHOL Synonym: OCTANOL	1987	26	650		2620				64		OTA 909	706	O1.0		847
CAPRYLALDEHYDE	1191	26						-25	64						
CAPRYLENE Synonym: 1-OCTENE	1993	27						-75	218		OTE 913				847
CAPRYLIC ALCOHOL Synonym: OCTANOL	1987	26	650		2620						OTA 909	706	O1.0		847
CAPRYLYL CHLORIDE	1191	26						-25	64						
CAPRYLYL PEROXIDE	2129	52							64						214
CAPRYLYL PEROXIDE SOLUTION	2129	52							64						
CAPTAN LIQUID	9099/3082	31	181		682	238	40		64		CPT 296	186			
CAPTAN SOLID	9099/3082		180		682	238	40		64		CPT 296	186			214
CARBACRYL Synonym: ACRYLONITRILE	1093	30	34	11	68	27	6	19 -12	20	34	ACN 23	13	A5.1	A 7	19

CHEMICAL OR MATERIAL NAME	UN/NA Number	Guide Number (DOT)	Firefighter's Hazardous Materials Reference Book	First Aid Manual for Chemical Accidents	Sax's Dangerous Properties of Industrial Materials	Hazardous Chemicals Desk Reference	Rapid Guide to Hazardous Chemicals in the Workplace	Fire Protection Guide on Hazardous Materials (NFPA)	Firefighter's Handbook of Hazardous Materials	Pocket Guide to Chemical Hazards (NIOSH)	Chemical Hazard Response Information System (CHRIS)	Emergency Handling of Hazardous Materials in Surface Transportation (AAR)	Emergency Action Guides (EAG)	Chemical Data Notebook: A User's Manual	Condensed Chemical Dictionary
CARBAMATE PESTICIDE	2757	55		13		239			64			187			
CARBAMATE PESTICIDE	2991	28		13		239			64			186			
CARBAMATE PESTICIDE	2758	28		13		239			64			187			
CARBAMATE PESTICIDE	2392/2757	55		13		239			64			187			
CARBAMIC ACID AMMONIUM SALT Synonym: AMMONIUM CARBAMATE	9083/3077	31	54		223	59			28		ACM 22	57			64
CARBAMIDE Synonym: UREA			865		3470	1166			283		URE1168		U3.0		214
CARBAMIDE PEROXIDE Synonym: UREA PEROXIDE	1511	35	866						283		UPO1166	945			214
CARBARYL	2757/3077	55	182		688	241	40		64	58	CBY 224	187			215
CARBARYL SOLID	2757	55	183		688	241	40		64		CBY 224				215
CARBIDE Synonym: CALCIUM CARBIDE	1402	40	167	13	662	229		45 –	65		CCB 226	179	C1.0	C 1	215
CARBINAMINE Synonym: MONOMETHYLAMINE	1061	19	596						206				M6.0		796
CARBINOL Synonym: METHANOL	1230	28	549				136		190			613	M2.0		216
CARBITOL Synonym: DIETHYLENE GLYCOL MONOETHYL ETHER	1993	27						–38	65		DGE 401	351			216
CARBOBENZOXY CHLORIDE Synonym: BENZYL CHLOROFORMATE	1739	39	125		400	133			42		BCF 126	120			135
CARBOFURAN	2757	55	184		695	242	40		65		CBF 218	188			216
CARBOLIC ACID Synonym: PHENOL	1671	55	185	22	2729		175	134–78	65	176	PHN 958	743	C2.0	P 1	217

HAZARDOUS MATERIALS REFERENCE BOOKS INDEX

CHEMICAL OR MATERIAL NAME	UN/NA Number	Guide Number (DOT)	Firefighter's Hazardous Materials Reference Book	First Aid Manual for Chemical Accidents	Sax's Dangerous Properties of Industrial Materials	Hazardous Chemicals Desk Reference	Rapid Guide to Hazardous Chemicals in the Workplace	Fire Protection Guide on Hazardous Materials (NFPA)	Firefighter's Handbook of Hazardous Materials	Pocket Guide to Chemical Hazards (NIOSH)	Chemical Hazard Response Information System (CHRIS)	Emergency Handling of Hazardous Materials in Surface Transportation (AAR)	Emergency Action Guides (EAG)	Chemical Data Notebook: A User's Manual	Condensed Chemical Dictionary
CARBOLIC ACID	2821	55	186			928					CBO 220			P 1	217
Synonym: CARBOLIC OIL															
CARBOLIC OIL	2821	55	186								CBO 220			P 1	217
CARBON	1362/1325	32	187	13	696	243	41		65			194			217
CARBON	1361	32		13	696	243	41		65						217
CARBON BISULFIDE	1131	28	188	13	698	244	42	47 -25	65	60	CBB 215	189	C2.1	C 3	218
Synonym: CARBON DISULFIDE															
CARBON BISULPHIDE	1131	28	188	13	698	244	42	47 -25	65	60	CBB 215	189	C2.1		218
Synonym: CARBON BISULFIDE															
CARBON BLACK	1993	27		13	697	243	41			58		189			218
CARBON CHLORIDE	1846	55	192	13	701	246	43	48 -	65	60	CBT 223	193	C2.3	C 5	221
Synonym: CARBON TETRACHLORIDE															
CARBON DICHLORIDE	1897	74							65						
CARBON DIOXIDE	2187	21		13	697	243	41		65	58		191	C2.2		219
CARBON DIOXIDE	1845	21		13	697		41		65	58		192	C2.2		219
CARBON DIOXIDE	1013	21	189	13	697	243	41		65	58	CDO 238	190	C2.2		219
CARBON DIOXIDE AIR MIXTURE	1956	12			698							191			
CARBON DIOXIDE ETHYLENE OXIDE MIXTURE	1952	12							65						
CARBON DIOXIDE ETHYLENE OXIDE MIXTURE	1041	17							65			190			
CARBON DIOXIDE NITROGEN GAS MIXTURE	1956	12										191			
CARBON DIOXIDE NITROUS OXIDE MIXTURE	1015	12			698	244			65			190			
CARBON DIOXIDE OXYGEN MIXTURE	1014	14			698				65			191			
CARBON DIOXIDE PROPYLENE OXIDE GAS MIXTURE	1956/1954	12										191			
CARBON DISULFIDE	1131	28	190	13	698	244	42	47 -25	65	60	CBB 215	189	C2.1	C 3	220
Synonym: CARBON BISULFIDE															

CHEMICAL OR MATERIAL NAME	UN/NA Number	Guide Number (DOT)	Firefighter's Hazardous Materials Reference Book	First Aid Manual for Chemical Accidents	Sax's Dangerous Properties of Industrial Materials	Hazardous Chemicals Desk Reference	Rapid Guide to Hazardous Chemicals in the Workplace	Fire Protection Guide on Hazardous Materials (NFPA)	Firefighter's Handbook of Hazardous Materials	Pocket Guide to Chemical Hazards (NIOSH)	Chemical Hazard Response Information System (CHRIS)	Emergency Handling of Hazardous Materials in Surface Transportation (AAR)	Emergency Action Guides (EAG)	Chemical Data Notebook: A User's Manual	Condensed Chemical Dictionary
CARBON DISULPHIDE Synonym: CARBON DISULFIDE	1131	28	190	13	698	244	42	47 –25	65	60	CBB 215	189	C2.1	C 3	220
CARBON HEXACHLORIDE	9037	53							65						220
CARBON HYDRIDE NITRIDE Synonym: HYDROCYANIC ACID	1051	13	442	18	1896	641	114	–59	164				H4.1	H 2	613
CARBON MONOXIDE	1016	18	191	13	700	245	42	48 –25	66	60	CMO 268	192	C2.2.1	C 4	221
CARBON MONOXIDE	9202	67		13	700	245	42		66			193	C2.2.1	C 4	221
CARBON MONOXIDE HYDROGEN MIXTURE	2600	18							66			192			
CARBON NITRIDE Synonym: CYANOGEN	1026	18	241		979	331	57	–29	66		CYG 330	277	C8.0		221
CARBON OXIDE Synonym: CARBON MONOXIDE	1016	18	191	13	700	245	42	48 –25	66	60	CMO 268	192	C2.2.1	C 4	221
CARBON OXYCHLORIDE Synonym: PHOSGENE	1076	15	682	22	2782	944	179	136–	66	180	PHG 956	752	P3.0	P 2	221
CARBON OXYSULFIDE	2204	18							66						221
CARBON REMOVER	1132	26			701	246			66						
CARBON SUBOXIDE									66						221
CARBON SULFIDE Synonym: CARBON BISULFIDE	1131	28	188				42		65		CBB 215	189	C2.1		218
CARBON TET Synonym: CARBON TETRACHLORIDE	1846	55	192	13	701	246	43	48 –	66	60	CBT 223	193	C2.3	C 5	221
CARBON TETRABROMIDE	2516	53			701	246	43					193			221
CARBON TETRACHLORIDE	1846	55	192	13	701	246	43	48 –	66	60	CBT 223	193	C2.3	C 5	221
CARBON TETRAFLUORIDE	1982	12			703	248			66						222
CARBON TRICHLORIDE	9037	53							66						222

CHEMICAL OR MATERIAL NAME	UN/NA Number	Guide Number (DOT)	Firefighter's Hazardous Materials Reference Book	First Aid Manual for Chemical Accidents	Sax's Dangerous Properties of Industrial Materials	Hazardous Chemicals Desk Reference	Rapid Guide to Hazardous Chemicals in the Workplace	Fire Protection Guide on Hazardous Materials (NFPA)	Firefighter's Handbook of Hazardous Materials	Pocket Guide to Chemical Hazards (NIOSH)	Chemical Hazard Response Information System (CHRIS)	Emergency Handling of Hazardous Materials in Surface Transportation (AAR)	Emergency Action Guides (EAG)	Chemical Data Notebook: A User's Manual	Condensed Chemical Dictionary
CARBONA Synonym: CARBON TETRACHLORIDE	1846	55	192	13	701	246	43	48 –	66	60	CBT 223	193	C2.3	C 5	221
CARBONIC ACID Synonym: AMMONIUM BICARBONATE	9081/3077	31	51		221	58			66		ABC 7	56			220
CARBONIC ACID DICHLORIDE Synonym: PHOSGENE	1076	15	682	22	2782	944	179	136–	231	180	PHG 956	752	P3.0	P 2	906
CARBONIC ACID DIETHYL ESTER Synonym: DIETHYL CARBONATE	2366/1993	26	287		1213	406		–37	108		DEC 383	345			390
CARBONIC ACID GAS Synonym: CARBON DIOXIDE	2187	21		13	697	243	41			58		191	C2.2		219
CARBONIC ACID GAS Synonym: CARBON DIOXIDE	1013	21	189	13	697	243	41			58	CDO 238	190	C2.2		219
CARBONIC ANHYDRIDE Synonym: CARBON DIOXIDE	1013	21	189	13	697	243	41		66	58	CDO 238	190	C2.2		219
CARBONIC ANHYDRIDE CARBON OXIDE Synonym: CARBON DIOXIDE	2187	21		13	697	243	41		65	58		191	C2.2		219
CARBONIC DICHLORIDE Synonym: PHOSGENE	1076	15	682	22	2782		179	136–	231	180	PHG 956	752	P3.0	P 2	906
CARBONIC OXIDE Synonym: CARBON MONOXIDE	1016	18	191	13	700	245	42	48 –25	66	60	CMO 268	192	C2.2.1	C 4	221
CARBONICE	1845	21													
CARBONOCHLORIDE ACID 1-METHYLETHYL ESTER Synonym: ISOPROPYL CHLOROFORMATE	2407/1993	57							175			561	15.0		661
CARBONOCHLORIDIC ACID 1-METHYLETHYL ESTER Synonym: ISOPROPYL CHLOROFORMATE	2407/1993	57							175			561	15.0		661
CARBONOCHLORIDIC ACID ETHYL ESTER Synonym: ETHYL CHLOROFORMATE	1182	28	352	16	1587	533		82 –50	137		ECF 512	413	E2.3.1		484

CHEMICAL OR MATERIAL NAME	UN/NA Number	Guide Number (DOT)	Firefighter's Hazardous Materials Reference Book	First Aid Manual for Chemical Accidents	Sax's Dangerous Properties of Industrial Materials	Hazardous Chemicals Desk Reference	Rapid Guide to Hazardous Chemicals In the Workplace	Fire Protection Guide on Hazardous Materials (NFPA)	Firefighter's Handbook of Hazardous Materials	Pocket Guide to Chemical Hazards (NIOSH)	Chemical Hazard Response Information System (CHRIS)	Emergency Handling of Hazardous Materials In Surface Transportation (AAR)	Emergency Action Guides (EAG)	Chemical Data Notebook: A User's Manual	Condensed Chemical Dictionary
CARBONOCHLORIDIC ACID METHYL ESTER Synonym: METHYL CHLOROFORMATE	1238	28	562	20					193		MHC 760	624	M3.1.1		762
CARBONYL CHLORIDE Synonym: PHOSGENE	1076	15	682	22	2782	944	179	136–	66	180	PHG 956	752	P3.0	P 2	222
CARBONYL DIAMINE PEROXIDE Synonym: UREA PEROXIDE	1511	35	866						283		UPO1166	945			1204
CARBONYL FLUORIDE	2417/1955	15			703	248	43		66			194			222
CARBONYL SULFIDE	2204/2676	18			704	249			66			195			222
CARBONYLDIAMIDE Synonym: UREA			865		3470	1166			283		URE1168		U3.0		1203
CARBOXIDE	1041	17							67						223
CARBOXYETHANE Synonym: PROPIONIC ACID	1848	60	721		2909	997	188	148–81	243		PNA 968	791	P7.1		968
CARBOXYLBENZENE Synonym: BENZOIC ACID	9094/3077	31	119		375	124		–17	41		BZA 191	115			131
CARENE											CAR 211				
CARPETING MEDIUM Synonym: ASPHALT BLENDING STOCKS STRAIGHT RUN RESIDUE	1999	27	105	12	307	97	17		36		ASR 94	99	A12.0		
CARTHAMUS TINCTORIUS OIL Synonym: OILS EDIBLE SAFFLOWER											OSF 905				
CARWINATE 125 M Synonym: DIPHENYLMETHANE DIISOCYANATE	2489	55	321						127		DPM 467				
CASING HEAD GASOLINE Synonym: GASOLINE	1203	27	412	18	1779	601	104	–56	153		GAK 602	473	G1.0	G 1	228

HAZARDOUS MATERIALS REFERENCE BOOKS INDEX

CHEMICAL OR MATERIAL NAME	UN/NA Number	Guide Number (DOT)	Firefighter's Hazardous Materials Reference Book	First Aid Manual for Chemical Accidents	Sax's Dangerous Properties of Industrial Materials	Hazardous Chemicals Desk Reference	Rapid Guide to Hazardous Chemicals in the Workplace	Fire Protection Guide on Hazardous Materials (NFPA)	Firefighter's Handbook of Hazardous Materials	Pocket Guide to Chemical Hazards (NIOSH)	Chemical Hazard Response Information System (CHRIS)	Emergency Handling of Hazardous Materials in Surface Transportation (AAR)	Emergency Action Guides (EAG)	Chemical Data Notebook: A User's Manual	Condensed Chemical Dictionary
CASINGHEAD GASOLINE Synonym: GASOLINE	1257	27	412		1779		104		67				G1.0	G 1	228
CASTOR BEANS MEAL POMACE OR FLAKE	2969	31			717	251			67			199			
CASTOR OIL					717	252		-25							229
CASTOR OIL HYDROGENATED								-25							230
CAT CRACKER FEEDSTOCK Synonym: FUEL OIL NO. 4	1993	27	405					-55	151			464	F3.2		
CATECHIN Synonym: CATECHOL					718	252	44		67		CTC 318				232
CATECHOL					718	252	44		67		CTC 318				232
CAUSTIC ALKALI LIQUIDS	1719	60							67			200			
CAUSTIC ARSENIC CHLORIDE Synonym: ARSENIC TRICHLORIDE	1560	55	99	12			16		35		AST 95	92	A11.0		97
CAUSTIC OIL OF ARSENIC Synonym: ARSENIC TRICHLORIDE	1560	55	99	12			16		35		AST 95	92	A11.0		97
CAUSTIC POTASH Synonym: POTASSIUM HYDROXIDE	1814	60	193	23	2871	984	186	145-	67		CPS 295	781	P6.0	P 5	233
CAUSTIC POTASH Synonym: POTASSIUM HYDROXIDE	1813	60	712	23	2871	984	186	145-	67			781	P6.0	P 5	233
CAUSTIC SODA Synonym: SODIUM HYDROXIDE	1823	60	777	24	3087	1056	200	159-	67	198	SHD1044	842	S2.5	S 4	233
CAUSTIC SODA	1824	60	194						67		CSS 314				233
CAUSTIC SODA SOLUTION SPENT	1824	60	194						67		CSS 314		C3.0		
CELFUME Synonym: METHYL BROMIDE	1062	55	559	20	2274	776	137	110-67	192	146	MTB 805	619	M3.0	M 3	759
CELLOIDIN	2000	32													235

CHEMICAL OR MATERIAL NAME	UN/NA Number	Guide Number (DOT)	Firefighter's Hazardous Materials Reference Book	First Aid Manual for Chemical Accidents	Sax's Dangerous Properties of Industrial Materials	Hazardous Chemicals Desk Reference	Rapid Guide to Hazardous Chemicals in the Workplace	Fire Protection Guide on Hazardous Materials (NFPA)	Firefighter's Handbook of Hazardous Materials	Pocket Guide to Chemical Hazards (NIOSH)	Chemical Hazard Response Information System (CHRIS)	Emergency Handling of Hazardous Materials in Surface Transportation (AAR)	Emergency Action Guides (EAG)	Chemical Data Notebook: A User's Manual	Condensed Chemical Dictionary
CELLOSOLVE Synonym: ETHYLENE GLYCOL MONOETHYL ETHER	1171	26	375					-51	67		EGE 529	432	E2.5		235
CELLOSOLVE ACETATE Synonym: ETHYLENE GLYCOL MONOETHYL ETHER ACE-TATE	1172	26	376					-51	67		EGA 526	432	E2.6		489
CELLULOID	2000/1325	32			726	253						200			236
CELLULOID SCRAP Synonym: CELLULOID	2002/2023	33			726	253			67			200			236
CELLULOSE FILM NEC PRINTED	1325	32													
CELLULOSE FILM SCRAP OR WASTE	1324	32										201			
CELLULOSE NITRATE SOLUTION Synonym: COLLODION	2059/1993	26	220		940	319		-28	82		CLD 257	249			300
CELLULOSE TETRANITRATE	2060	26			727	253			67						
CELPHOS Synonym: PHOSPHINE	2199	18	683	22	2783	945	180	136-80	67	180		752	P3.1		
CEMENT	1133	26		13	727	254			67			201			
CERIUM	1333/3078	32			730	255			68			202			242
CERIUM CRUDE	1333/1325	32			730				68			202			
CESIUM HYDROXIDE	2682	60			735		44		68						245
CESIUM HYDROXIDE SOLUTION	2681	60			735		44		68						
CESIUM METAL	1407	40			733	256			68						
CESIUM NITRATE	1451	35			735	257			68						245
CETYL SODIUM SULFATE Synonym: HEXADECYL SULFATE SODIUM SALT											HSS 645				
CETYLTRIMETHYLAMMONIUM CHLORIDE SOLUTION Synonym: HEXADECYLTRIMETHYLAMMONIUM CHLORIDE			430								HAC 615				246

HAZARDOUS MATERIALS REFERENCE BOOKS INDEX

CHEMICAL OR MATERIAL NAME	UN/NA Number	Guide Number (DOT)	Firefighter's Hazardous Materials Reference Book	First Aid Manual for Chemical Accidents	Sax's Dangerous Properties of Industrial Materials	Hazardous Chemicals Desk Reference	Rapid Guide to Hazardous Chemicals in the Workplace	Fire Protection Guide on Hazardous Materials (NFPA)	Firefighter's Handbook of Hazardous Materials	Pocket Guide to Chemical Hazards (NIOSH)	Chemical Hazard Response Information System (CHRIS)	Emergency Handling of Hazardous Materials in Surface Transportation (AAR)	Emergency Action Guides (EAG)	Chemical Data Notebook: A User's Manual	Condensed Chemical Dictionary
CHAMBER ACID Synonym: SULFURIC ACID	1830	39	802	24	3163	1082	204	164–	259	200	SFA1037	871	S4.2	S 8	248
CHARCOAL	1361	32	195		739	258			68		CHC 244	203			248
CHELEN Synonym: ETHYL CHLORIDE	1037	27	350		1586	532	92	82 –49	136	108	ECL 514	412	E2.3		483
CHEM BAM Synonym: NABAM	1609		599						207		NAB 816				803
CHEMICAL AMMUNITION	2017	58							68			208			
CHEMICAL AMMUNITION	2016	15							68						
CHEMICAL KIT	1760	60							68			208			
CHILE NITER Synonym: SODIUM NITRATE	1498	35	780		3097	1060			254		SDN1033	845	S2.6	S 5	1063
CHILE NITRATE Synonym: SODIUM NITRATE	1498	35	780		3097	1060			254		SDN1033	845	S2.6	S 5	1063
CHILE SALTPETER Synonym: SODIUM NITRATE	1498	35	780		3097	1060			254		SDN1033	845	S2.6	S 5	1063
CHINESE RED Synonym: MERCURIC SULFIDE	2025	53	539						186		MSF 800				743
CHINESE TANNIN Synonym: TANNIC ACID			805		3180	1089		–84			TNA1130				1114
CHINOLINE Synonym: QUINOLINE	2656	29	736		2968	1014		–83	68		QNL1005	805			987
CHINONE	2587	55							68						
CHLOR O PIC Synonym: CHLOROPICRIN	1580	56	210	14	863	296	50	54 –	77	70	CPL 290	229	C4.3		271

CHEMICAL OR MATERIAL NAME	UN/NA Number	Guide Number (DOT)	Firefighter's Hazardous Materials Reference Book	First Aid Manual for Chemical Accidents	Sax's Dangerous Properties of Industrial Materials	Hazardous Chemicals Desk Reference	Rapid Guide to Hazardous Chemicals in the Workplace	Fire Protection Guide on Hazardous Materials (NFPA)	Firefighter's Handbook of Hazardous Materials	Pocket Guide to Chemical Hazards (NIOSH)	Chemical Hazard Response Information System (CHRIS)	Emergency Handling of Hazardous Materials in Surface Transportation (AAR)	Emergency Action Guides (EAG)	Chemical Data Notebook: A User's Manual	Condensed Chemical Dictionary
CHLORACETIC ACID Synonym: CHLOROACETIC ACID	1751	60	199	13	761	270		50 –26	72		MCA 734	216	C3.9		261
CHLORACETONE	1695	59			743	260	44								
CHLORACETYL CHLORIDE Synonym: CHLOROACETYL CHLORIDE	1752	59	201		763	271	47	51 –	68		CAC 204	217			256
CHLORAL	2075	55							68			210			256
CHLORAL HYDRATE					743	260			68						257
CHLORALLYLENE Synonym: ALLYL CHLORIDE	1100	57	40	11	106	35	8	22 –13	22	36	ALC 40	36	A7.2		39
CHLORATE	1461	35			747	262			69			211			
CHLORATE AND BORATE MIXTURE	1458	35							69			210			
CHLORATE AND MAGNESIUM CHLORIDE MIXTURE	1459	35							69			210			
CHLORATE OF POTASH Synonym: POTASSIUM CHLORATE	1485	35	705	23	2861	979		143–	69		PCR 941	777			950
CHLORATE OF POTASSIUM Synonym: POTASSIUM CHLORATE	1485	35	705	23	2861	979		143–	238		PCR 941	777			950
CHLORATE OF SODA Synonym: SODIUM CHLORATE	1495	35	766	24	3070	1048		156–	69		SDC1029	834	S2.1	S 2	1052
CHLORBENZENE Synonym: CHLOROBENZENE	1134	27	203	13	768	273	47	51 –26	73	64	CRB 298	219	C4.1		263
CHLORBENZOL Synonym: CHLOROBENZENE	1134	27	203	13	768	273	47	51 –26	73	64	CRB 298	219	C4.1		263
CHLORCYAN Synonym: CYANOGEN CHLORIDE	1589	15	243	14	979	332	58		87		CCL 229	277	C8.0.1		334
CHLORDAN Synonym: CHLORDANE	2762	28	196		749	263	44		69	60	CDN 237	211			258

HAZARDOUS MATERIALS REFERENCE BOOKS INDEX

CHEMICAL OR MATERIAL NAME	UN/NA Number	Guide Number (DOT)	Firefighter's Hazardous Materials Reference Book	First Aid Manual for Chemical Accidents	Sax's Dangerous Properties of Industrial Materials	Hazardous Chemicals Desk Reference	Rapid Guide to Hazardous Chemicals in the Workplace	Fire Protection Guide on Hazardous Materials (NFPA)	Firefighter's Handbook of Hazardous Materials	Pocket Guide to Chemical Hazards (NIOSH)	Chemical Hazard Response Information System (CHRIS)	Emergency Handling of Hazardous Materials in Surface Transportation (AAR)	Emergency Action Guides (EAG)	Chemical Data Notebook: A User's Manual	Condensed Chemical Dictionary
CHLORDANE	2762	28	196		749	263	44		69	60	CDN 237	211			258
CHLORDECONE Synonym: KEPONE	2761/9189	55	493		2077	701			177		KPE 688	566			670
CHLORENE Synonym: ETHYL CHLORIDE	1037	27	350		1586	532	92	82 -49	136	108	ECL 514	412	E2.3		483
CHLORETHENE Synonym: VINYL CHLORIDE	1086	17	872	26	3495	1174	227	180–93	285	224	VCM1174	949	V1.1	V 1	1215
CHLORETHYL Synonym: ETHYL CHLORIDE	1037	27	350		1586	532	92	82 -49	136	108	ECL 514	412	E2.3		483
CHLORETHYLENE Synonym: VINYL CHLORIDE	1086	17	872	26	3495	1174	227	180–93	285	224	VCM1174	949	V1.1	V 1	1215
CHLOREX Synonym: 2,2-DICHLOROETHYL ETHER	1916/2810	55			1145	384		66 -35	69		DEE 385	334			258
CHLORFENVINFOS					752	265			69						
CHLORIC ACID	2626	35			753	265			69			212			259
CHLORIC ACID SODIUM SALT Synonym: SODIUM CHLORATE	1495	35	766	24	3070	1048		156–	253		SDC1029	834	S2.1	S 2	1052
CHLORIC ACID SOLUTION	2626	35			753				69			212			
CHLORIDE OF AMYL Synonym: AMYL CHLORIDE (-n)	1107	26	79						31		AMY 61	75			74
CHLORIDE OF PHOSPHORUS Synonym: PHOSPHORUS TRICHLORIDE	1809	39	690	22	2796	953	182	140–	69	184	PPT 981	758	P4.3	P 3	910
CHLORIDE OF SULFUR	1828	39							69						
CHLORIDUM Synonym: ETHYL CHLORIDE	1037	27	350		1586	532	92	82 -49	136	108	ECL 514	412	E2.3		483

CHEMICAL OR MATERIAL NAME	UN/NA Number	Guide Number (DOT)	Firefighter's Hazardous Materials Reference Book	First Aid Manual for Chemical Accidents	Sax's Dangerous Properties of Industrial Materials	Hazardous Chemicals Desk Reference	Rapid Guide to Hazardous Chemicals in the Workplace	Fire Protection Guide on Hazardous Materials (NFPA)	Firefighter's Handbook of Hazardous Materials	Pocket Guide to Chemical Hazards (NIOSH)	Chemical Hazard Response Information System (CHRIS)	Emergency Handling of Hazardous Materials in Surface Transportation (AAR)	Emergency Action Guides (EAG)	Chemical Data Notebook: A User's Manual	Condensed Chemical Dictionary
CHLORINATED BIPHENYL Synonym: POLYCHLORINATED BIPHENYL	2315/3151	31	698	23	2831	964	185	142–	69		PCB 934	773			931
CHLORINATED CAMPHENE	2761/3077	55			754	266	45		69	62		212			259
CHLORINATED DIPHENYL OXIDE					755	266	45		69	62					
CHLORINATED DIPHENYLS	2315	31							69						
CHLORINATED HYDROCARBON RESIDUE	1993	27										212			
CHLORINATED HYDROCHLORIC ETHER Synonym: 1,1-DICHLOROETHANE	2362/1993	27	1		1141	382	69		69	86	DCH 361	301			380
CHLORINATED LIME	2208/1760	35		13								213			259
CHLORINATED PHENOL PETROLEUM SOLUTION	2810	55										213			
CHLORINE	1017	20	197	13	756	267	45	49 –	69	62	CLX 261	214	C4.0	C 6	259
CHLORINE AZIDE					758	269			69						
CHLORINE CYANIDE Synonym: CYANOGEN CHLORIDE	1589	15	243	14	979	332	58		69		CCL 229	277	C8.0.1		334
CHLORINE DIOXIDE	9191	47		13	758	269	46		70	62					260
CHLORINE DIOXIDE HYDRATE	9191	47					46		70			214			
CHLORINE MONOXIDE								–26	70						260
CHLORINE PENTAFLUORIDE	2548/1955	44			758	269			70			215			
CHLORINE TRIFLUORIDE	1749	44	198	13	759	269	46	49 –	70	64	CTF 320	215			260
CHLORITE	1462	43			760	270			70						
CHLORO-2-a-TRIFLUORO TOLUENE-ortho	2234	27							71						
CHLORO-4-ETHYLBENZENE								–26							
CHLORO-a-a-a-TRIFLUOROTOLUENE-ortho	2234	27							72						
CHLORO-m-CRESOL-para	2669	55							72						

CHEMICAL OR MATERIAL NAME	UN/NA Number	Guide Number (DOT)	Firefighter's Hazardous Materials Reference Book	First Aid Manual for Chemical Accidents	Sax's Dangerous Properties of Industrial Materials	Hazardous Chemicals Desk Reference	Rapid Guide to Hazardous Chemicals in the Workplace	Fire Protection Guide on Hazardous Materials (NFPA)	Firefighter's Handbook of Hazardous Materials	Pocket Guide to Chemical Hazards (NIOSH)	Chemical Hazard Response Information System (CHRIS)	Emergency Handling of Hazardous Materials in Surface Transportation (AAR)	Emergency Action Guides (EAG)	Chemical Data Notebook: A User's Manual	Condensed Chemical Dictionary
CHLORO-m-NITROBENZENE Synonym: NITROCHLOROBENZENE-meta	1578	55	632						212			690	N3.2		
CHLORO-o-NITROBENZENE Synonym: NITROCHLOROBENZENE-ortho	1578	55	633		841	289			212			690	N3.3		
CHLOROACETALDEHYDE	2232/2810	55		13	760	270	46		72	64	CHO 247	216			261
CHLOROACETIC ACID	1750	59		13	761	270		50 –26	72		MCA 734	216	C3.9		261
CHLOROACETIC ACID	1751	60	199	13	761	270		50 –26	72		MCA 734	216	C3.9		261
CHLOROACETIC ACID ETHYL ESTER Synonym: ETHYL CHLOROACETATE	1181/1993	55	351					–49	136		ECA 510	412			483
CHLOROACETONE Synonym: CHLORACETONE	1695	59			743	260	44		72			217			261
CHLOROACETONITRILE	2668	55						50 –	72			217			261
CHLOROACETOPHENONE	1697	55	200	13	762			–26	73	64	CRA 297	217			261
CHLOROACETOPHENONE-alpha Synonym: CHLOROACETOPHENONE	1697	55	200	13	762	271	47	–26	73	64	CRA 297	217			261
CHLOROACETOPHENONE-omega Synonym: CHLOROACETOPHENONE	1697	55	200	13	762		47	–26	73	64	CRA 297	217			261
CHLOROACETYL CHLORIDE	1752	59	201		763	271	47	51 –	73		CAC 204	217			262
CHLOROANILINE	2018/2811	53			766				73			218			
CHLOROANILINE-meta Synonym: 3-CHLOROANILINE	2019/2810	55			766	272						218			262
CHLOROANILINE-ortho Synonym: 2-CHOLROANILINE	2019/2810	55			766	272			73			218			262
CHLOROANILINE-para Synonym: 4-CHLOROANILINE	2018/2811	53	202		767	272			73		CAP 210	218			262
CHLOROANISIDINE	2233	53							73			218			

CHEMICAL OR MATERIAL NAME	UN/NA Number	Guide Number (DOT)	Firefighter's Hazardous Materials Reference Book	First Aid Manual for Chemical Accidents	Sax's Dangerous Properties of Industrial Materials	Hazardous Chemicals Desk Reference	Rapid Guide to Hazardous Chemicals in the Workplace	Fire Protection Guide on Hazardous Materials (NFPA)	Firefighter's Handbook of Hazardous Materials	Pocket Guide to Chemical Hazards (NIOSH)	Chemical Hazard Response Information System (CHRIS)	Emergency Handling of Hazardous Materials in Surface Transportation (AAR)	Emergency Action Guides (EAG)	Chemical Data Notebook: A User's Manual	Condensed Chemical Dictionary
CHLOROBENZALDEHYDE-para								-26							
CHLOROBENZENE	1134	27	203	13	768	273	47	51 -26	73	64	CRB 298	219	C4.1		263
CHLOROBENZOL Synonym: CHLOROBENZENE	1134	27	203	13	768	273	47	51 -26	73	64	CRB 298	219	C4.1		263
CHLOROBENZONITRILE-ortho					771	273			73						
CHLOROBENZONITRILE-para					771				73						
CHLOROBENZOTRIFLUORIDE-meta	2234	27							73						
CHLOROBENZOTRIFLUORIDE-ortho	2234	27						-26							
CHLOROBENZOTRIFLUORIDE-para	1993	27													
CHLOROBENZOTRIFLUORIDE-para	2234	27			771	274			73						
CHLOROBENZOYL PEROXIDE-p,p Synonym: DI(p-CHLOROBENZOYL) PEROXIDE	0149										DZP 498				264
CHLOROBENZOYL PEROXIDE-para Synonym: DI(p-CHLOROBENZOYL) PEROXIDE	0149										DZP 498				264
CHLOROBENZOYL PEROXIDE-para	2113	48							74						264
CHLOROBENZOYL PEROXIDE-para	2114	48							74			219			264
CHLOROBENZOYL PEROXIDE-para	2115	48							73						264
CHLOROBENZYL CHLORIDE-ortho	2235	55							74			220			
CHLOROBENZYL CHLORIDE-para	2235	55							74			220			
CHLOROBENZYLIDENE MALONONITRILE-ortho				14	772	274	48			66					
CHLOROBROMOMETHANE	1887	58	204	14	774	274	48		74	66					
CHLOROBROMOPROPANE	2688	58							74						
CHLOROBUTADIENE Synonym: CHLOROPRENE (-beta)	1991	30	211						74	70	CRP 306	230	C4.4		272
CHLOROBUTYRONITRILE	1993	27										220			

HAZARDOUS MATERIALS REFERENCE BOOKS INDEX

CHEMICAL OR MATERIAL NAME	UN/NA Number	Guide Number (DOT)	Firefighter's Hazardous Materials Reference Book	First Aid Manual for Chemical Accidents	Sax's Dangerous Properties of Industrial Materials	Hazardous Chemicals Desk Reference	Rapid Guide to Hazardous Chemicals in the Workplace	Fire Protection Guide on Hazardous Materials (NFPA)	Firefighter's Handbook of Hazardous Materials	Pocket Guide to Chemical Hazards (NIOSH)	Chemical Hazard Response Information System (CHRIS)	Emergency Handling of Hazardous Materials in Surface Transportation (AAR)	Emergency Action Guides (EAG)	Chemical Data Notebook: A User's Manual	Condensed Chemical Dictionary
CHLOROCARBONIC ACID METHYL ESTER	1238	28	562	20					193		MHC 760	624	M3.1.1		762
Synonym: METHYL CHLOROFORMATE															
CHLOROCRESOL	2669/2811	55							74			221			
CHLOROCYAN	1589	15	243	14	979	332	58		87		CCL 229	277	C8.0.1		334
Synonym: CYANOGEN CHLORIDE															
CHLOROCYANIDE	1589	15	243	14	979	332	58		87		CCL 229	277	C8.0.1		334
Synonym: CYANOGEN CHLORIDE															
CHLOROCYANOGEN	1589	15	243	14	979	332	58		87		CCL 229	277	C8.0.1		334
Synonym: CYANOGEN CHLORIDE															
CHLORODIETHYALUMINUM	1101								74						
CHLORODIFLUOROBROMOMETHANE	1974	12							74			221			
CHLORODIFLUOROETHANE	2517	22		14					74			221			
CHLORODIFLUOROMETHANE	1018	12	593	14	791	278	48		74		MCF 737	222	M4.0		265
Synonym: MONOCHLORODIFLUOROMETHANE															
CHLORODIFLUOROMETHANE & CHLOROPENTAFLUORO-ETHANE MIX	1078	12							74			222			
CHLORODIMETHYL ETHER	1239	57					49	111–	193		CME 265	643	M3.1.2		762
Synonym: METHYL CHLOROMETHYL ETHER															
CHLORODINITROBENZENE	1577	56			799	279			75						
CHLORODINITROBENZOL	1577	56							75						
CHLORODIPHENYL	2315	31			800				75	66					265
CHLORODIPHENYL ARSINE	1699	55			800	279			75						
CHLOROETHANE	1037	27	350	14	1586	532	92	82 –49	75	108	ECL 514	412	E2.3		266
Synonym: ETHYL CHLORIDE															
CHLOROETHANOIC ACID	1751	60	199	13	761	270		50 –26	72		MCA 734	216	C3.9		261
Synonym: CHLOROACETIC ACID															

CHEMICAL OR MATERIAL NAME	UN/NA Number	Guide Number (DOT)	Firefighter's Hazardous Materials Reference Book	First Aid Manual for Chemical Accidents	Sax's Dangerous Properties of Industrial Materials	Hazardous Chemicals Desk Reference	Rapid Guide to Hazardous Chemicals in the Workplace	Fire Protection Guide on Hazardous Materials (NFPA)	Firefighter's Handbook of Hazardous Materials	Pocket Guide to Chemical Hazards (NIOSH)	Chemical Hazard Response Information System (CHRIS)	Emergency Handling of Hazardous Materials in Surface Transportation (AAR)	Emergency Action Guides (EAG)	Chemical Data Notebook: A User's Manual	Condensed Chemical Dictionary
CHLOROETHANOIC ACID	1750	59							75						
CHLOROETHENE Synonym: VINYL CHLORIDE	1086	17	872	26	3495	1174	227	180–93	75	224	VCM1174	949	V1.1	V 1	266
CHLOROETHYL ACETATE	1181	55						–26	75						
CHLOROETHYLENE Synonym: VINYL CHLORIDE	1086	17	872	26	3495	1174	227	180–93	75	224	VCM1174	949	V1.1	V 1	266
CHLOROFLUOROMETHANE				14	814	282	49		76						
CHLOROFORM	1888	55	205	14	815	282	49	52 –	76	68	CRF 301	223	C4.2	C 7	266
CHLOROFORMATE	2742	55							76			224			
CHLOROFORMIC ACID BENZYL ESTER Synonym: BENZYL CHLOROFORMATE	1739	39	125		400	133			42		BCF 126	120			135
CHLOROFORMIC ACID ETHYL ESTER Synonym: ETHYL CHLOROFORMATE	1182	28	352	16	1587	533		82 –50	137		ECF 512	413	E2.3.1		484
CHLOROFORMIC ACID ISOPROPYL ESTER Synonym: ISOPROPYL CHLOROFORMATE	2407/1993	57							175			561	15.0		661
CHLOROFORMIC ACID METHYL ESTER Synonym: METHYL CHLOROFORMATE	1238	28	562	20					193		MHC 760	624	M3.1.1		762
CHLOROFORMYL CHLORIDE Synonym: PHOSGENE	1076	15	682	22	2782	944	179	136–	231	180	PHG 956	752	P3.0	P 2	267
CHLOROHYDRIC ACID Synonym: HYDROGEN CHLORIDE	1050	15	447	18	1900	643	115	98 –	76	126	HDC 630	528	H4.3	H 3	616
CHLOROHYDRINS	2023	30	206								CHD 245				267
CHLOROISOPROPYL ALCOHOL	2611	57							76						267
CHLOROMETHANE Synonym: METHYL CHLORIDE	1063	18	561	14		781	138	111–67	76	146	MTC 806	623	M3.1	M 4	268

CHEMICAL OR MATERIAL NAME	UN/NA Number	Guide Number (DOT)	Firefighter's Hazardous Materials Reference Book	First Aid Manual for Chemical Accidents	Sax's Dangerous Properties of Industrial Materials	Hazardous Chemicals Desk Reference	Rapid Guide to Hazardous Chemicals in the Workplace	Fire Protection Guide on Hazardous Materials (NFPA)	Firefighter's Handbook of Hazardous Materials	Pocket Guide to Chemical Hazards (NIOSH)	Chemical Hazard Response Information System (CHRIS)	Emergency Handling of Hazardous Materials in Surface Transportation (AAR)	Emergency Action Guides (EAG)	Chemical Data Notebook: A User's Manual	Condensed Chemical Dictionary
CHLOROMETHOXYMETHANE Synonym: METHYL CHLOROMETHYL ETHER	1239	57					49	111–	193		CME 265	643	M3.1.2		762
CHLOROMETHYL CHLOROSULFONATE	2745	55							76						268
CHLOROMETHYL ETHER-bis	2249	55								68					
CHLOROMETHYL ETHYL ETHER	2354/1993	28			831	286			76			224			
CHLOROMETHYL ETHYLENE OXIDE Synonym: EPICHLOROHYDRIN	2023	30	336	16	1525	509	87	79 –46	132	102	EPC 552	401	E1.0	E 1	467
CHLOROMETHYL METHYL ETHER	1239	57	207		832	286	49		76	68	CME 265				
CHLOROMETHYL OXIRANE	2023	30							76						
CHLOROMETHYL PHENYL KETONE Synonym: CHLOROACETOPHENONE (-alpha)	1697	55	200		762	271	47	–26	73	64	CRA 297	217			261
CHLOROMETHYLCHLOROFORMATE	2745	55							76			224			268
CHLOROMETHYLMETHYL ETHER Synonym: METHYL CHLOROMETHYL ETHER	1239	57			832	286	49	111–	76		CME 265	643	M3.1.2		762
CHLOROMETHYLOXIRANE Synonym: EPICHLOROHYDRIN	2023	30	336	16	1525	509	87	79 –46	76	102	EPC 552	401	E1.0	E 1	467
CHLOROMETHYLPHENYLISOCYANATE	2236	55							76						
CHLORONAPHTHALENE				14					76						268
CHLORONITROANILINE	2237/2019	53							76			225			
CHLORONITROBENZENE	1578	55			840				76			225			
CHLORONITROBENZENE-meta Synonym: NITROCHLOROBENZENE	1578/2811	55							76			225	N3.2		269
CHLORONITROBENZENE-ortho Synonym: NITROCHLOROBENZENE	1578/2811	55	208		841	289			76		CNO 273	225	N3.3		269
CHLORONITROBENZENE-para Synonym: NITROCHLOROBENZENE	1578/2811	55		21	2525	863	159	–74	76	164		225	N3.4		269

CHEMICAL OR MATERIAL NAME	UN/NA Number	Guide Number (DOT)	Firefighter's Hazardous Materials Reference Book	First Aid Manual for Chemical Accidents	Sax's Dangerous Properties of Industrial Materials	Hazardous Chemicals Desk Reference	Rapid Guide to Hazardous Chemicals in the Workplace	Fire Protection Guide on Hazardous Materials (NFPA)	Firefighter's Handbook of Hazardous Materials	Pocket Guide to Chemical Hazards (NIOSH)	Chemical Hazard Response Information System (CHRIS)	Emergency Handling of Hazardous Materials in Surface Transportation (AAR)	Emergency Action Guides (EAG)	Chemical Data Notebook: A User's Manual	Condensed Chemical Dictionary
CHLORONITROTOLUENE	2433	53							77			225			
CHLOROPENTAFLUOROETHANE	1020	12		14	846	290	50		77			226			270
CHLOROPENTANES	1993	27										226			
CHLOROPHENATE	2905	53							77			227			
CHLOROPHENATE	2904	55							77			227			
CHLOROPHENOL	2020	53			849	291		53 –				228			
CHLOROPHENOL	2021	55			849	291		53 –				227			
CHLOROPHENOL-meta Synonym: 3-CHLOROPHENOL	2020	53			848	291			77			228			270
CHLOROPHENOL-ortho Synonym: 2-CHLOROPHENOL	2021	55			848	291		–27	77		CRH 302	227			270
CHLOROPHENOL-para Synonym: 4-CHLOROPHENOL	2020	53	209		848	291		–27	77		CPN 291	228			270
CHLOROPHENOTHANE	2761	55							77						
CHLOROPHENYL ISOCYANATE-meta					856				77						270
CHLOROPHENYL ISOCYANATE-para					856	294			77						270
CHLOROPHENYL TRICHLOROSILANE	1753	60			865	295			77			228			271
CHLOROPHOS Synonym: TRICHLORFON	2783/3018	55	835						272		TRC1146	917			1169
CHLOROPICRIN	1580	56	210	14	865	296	50	54 –	77	70	CPL 290	229	C4.3		271
CHLOROPICRIN AND METHYL BROMIDE MIXTURE	1581	55							77			228			
CHLOROPICRIN AND METHYL CHLORIDE MIXTURE	1582	18							77			228			
CHLOROPICRIN AND NONFLAMMABLE GAS MIXTURE	1955	15					50		77						
CHLOROPICRIN MIXTURE	1583	56		14	865		50		77	70		229	C4.3		
CHLOROPICRIN MIXTURE FLAMMABLE	2929	57		14	866	296	50		77	70					

CHEMICAL OR MATERIAL NAME	UN/NA Number	Guide Number (DOT)	Firefighter's Hazardous Materials Reference Book	First Aid Manual for Chemical Accidents	Sax's Dangerous Properties of Industrial Materials	Hazardous Chemicals Desk Reference	Rapid Guide to Hazardous Chemicals in the Workplace	Fire Protection Guide on Hazardous Materials (NFPA)	Firefighter's Handbook of Hazardous Materials	Pocket Guide to Chemical Hazards (NIOSH)	Chemical Hazard Response Information System (CHRIS)	Emergency Handling of Hazardous Materials in Surface Transportation (AAR)	Emergency Action Guides (EAG)	Chemical Data Notebook: A User's Manual	Condensed Chemical Dictionary
CHLOROPLATINIC ACID	2507	60			867	297			77			230			272
CHLOROPRENE	1991	30	211						77	70	CRP 306	230	C4.4		272
CHLOROPRENE-beta	1991	30	211						77	70	CRP 306	230	C4.4		272
Synonym: CHLOROPRENE															
CHLOROPROPANE	2356	26		14											
CHLOROPROPANOL	2611	57							78						
CHLOROPROPENE	2456	27		14					78						
CHLOROPROPIONIC ACID	2511	60							78						
CHLOROPROPIONIC ACID-alpha	2511	60			870			−27	78		CLA 255				272
Synonym: 2-CHLOROPROPIONIC ACID															
CHLOROPROPYL ALCOHOL-beta	2611	57							78						
CHLOROPROPYLENE OXIDE-gamma	2023	30	336	16	1525	509	87	79 −46	78	102	EPC 552	401	E1.0	E 1	467
Synonym: EPICHLOROHYDRIN															
CHLOROPROPYLENE-alpha	1100	57	40	11	106	35	8	22 −13	22	36	ALC 40	36	A7.2		272
Synonym: ALLYL CHLORIDE															
CHLOROPYRIDINE	2822	54							78						
CHLOROSILANE	2985	29			875	298		54 –	78			232			
CHLOROSILANE	2986	29			875	298		54 –	78			232			
CHLOROSILANE	2987	60			875	298		54 –	78						
CHLOROSILANE	2988	40			875	298		54 –	79			232			
CHLOROSULFONIC ACID	1754	39	212					55 –	79		CSA 309	233	C4.5	C 8	273
CHLOROSULFONIC ACID AND SULFUR TRIOXIDE MIXTURE	1754	39							79			234			
CHLOROSULFURIC ACID	1754	39	212		876	298		55 –	79		CSA 309	233	C4.5	C 8	273
Synonym: CHLOROSULFONIC ACID															
CHLOROTETRAFLUOROETHANE	1021	12			877	299			79			234			273

CHEMICAL OR MATERIAL NAME	UN/NA Number	Guide Number (DOT)	Firefighter's Hazardous Materials Reference Book	First Aid Manual for Chemical Accidents	Sax's Dangerous Properties of Industrial Materials	Hazardous Chemicals Desk Reference	Rapid Guide to Hazardous Chemicals in the Workplace	Fire Protection Guide on Hazardous Materials (NFPA)	Firefighter's Handbook of Hazardous Materials	Pocket Guide to Chemical Hazards (NIOSH)	Chemical Hazard Response Information System (CHRIS)	Emergency Handling of Hazardous Materials in Surface Transportation (AAR)	Emergency Action Guides (EAG)	Chemical Data Notebook: A User's Manual	Condensed Chemical Dictionary
CHLOROTHENE Synonym: TRICHLOROETHANE	2831	74		25							TCE1082		T4.2		273
CHLOROTOLUENE	2238/1993	27						-28				234			
CHLOROTOLUENE-alpha Synonym: BENZYL CHLORIDE	1738	59	124	12	400	133	23	-18	79	46	BCL 127	119			274
CHLOROTOLUENE-meta	2238	27									CTM 321				274
CHLOROTOLUENE-omega Synonym: BENZYL CHLORIDE	1738	59	124	12	400	133	23	-18	42	46	BCL 127	119			135
CHLOROTOLUENE-ortho	2238	27			878	299	51		79		CTO 322				274
CHLOROTOLUENE-para	2238	27	213								CRN 304				274
CHLOROTOLUIDINE	2239	55							79			234			
CHLOROTOLUOL	1738	59							79						
CHLOROTRIFLUOROETHYLENE Synonym: TRIFLUOROCHLOROETHYLENE	1082	17	850	14	885	301		-89	79		TFC1108	925			274
CHLOROTRIFLUOROMETHANE	1022	12		14	885	302			79			235			274
CHLOROTRIFLUOROMETHANE AND TRIFLUOROMETHANE MIXTURE	2599	12							79						
CHLOROTRIFLUOROMETHANE AND TRIFLUOROMETHANE MIXTURE	1078	12							79						
CHLOROTRIFLUOROMETHYLBENZENE	2234	27							79						275
CHLOROTRIFLUOROMETHYLBENZENE-meta	2234	27							79						
CHLOROTRIFLUOROMETHYLBENZENE-ortho	2234	27							79						
CHLOROTRIFLUOROMETHYLBENZENE-para	2234	27							79						
CHLOROTRIMETHYLSILANE Synonym: TRIMETHYLCHLOROSILANE	1298	29	854		886	1148		-90	79		TMC1127	931			1182

HAZARDOUS MATERIALS REFERENCE BOOKS INDEX

CHEMICAL OR MATERIAL NAME	UN/NA Number	Guide Number (DOT)	Firefighter's Hazardous Materials Reference Book	First Aid Manual for Chemical Accidents	Sax's Dangerous Properties of Industrial Materials	Hazardous Chemicals Desk Reference	Rapid Guide to Hazardous Chemicals in the Workplace	Fire Protection Guide on Hazardous Materials (NFPA)	Firefighter's Handbook of Hazardous Materials	Pocket Guide to Chemical Hazards (NIOSH)	Chemical Hazard Response Information System (CHRIS)	Emergency Handling of Hazardous Materials in Surface Transportation (AAR)	Emergency Action Guides (EAG)	Chemical Data Notebook: A User's Manual	Condensed Chemical Dictionary
CHLORPICRIN	1580	56							80						
CHLORPICRIN & NONFLAMMABLE NONLIQUEFIED COMPRESSED GAS	1955	15										236			
CHLORPYRIFOS	2783/3082	55			891	304	52		80			236			275
CHLORPYRIFOS Synonym: DURSBAN	1615										DUR 495				275
CHLORSULFONIC ACID Synonym: CHLOROSULFONIC ACID	1754	39	212					55 –	80		CSA 309	233	C4.5	C 8	273
CHLORTHEPIN Synonym: ENDOSULFAN	2761	55	332		1519	505	86		131		ESF 559	398			463
CHLORTHION				14					80						276
CHLORYL Synonym: ETHYL CHLORIDE	1037	27	350		1586	532	92	82 –49	136	108	ECL 514	412	E2.3		483
CHLORYL ANESTHETIC Synonym: ETHYL CHLORIDE	1037	27	350		1586	532	92	82 –49	136	108	ECL 514	412	E2.3		483
CHLORYLEN Synonym: TRICHLOROETHYLENE	1710	74	837	25	3352	1137	218	174–88		216	TCL1085	919	T5.0	T 6	1170
CHP Synonym: CUMENE HYDROPEROXIDE	2116	51	235					–29	85		CMH 266				329
CHROME CHLORIDE	9188/3082	31										236			
CHROME CHLORIDE	1760	60										237			
CHROME SULPHATE	9188/3082	31										237			
CHROMIC (III) ACETATE Synonym: CHROMIC ACETATE	9101/3077	31	214		898	305			80		CRT 308	237			279
CHROMIC ACETATE	9101/3077	31	214		898	305			80		CRT 308	237			279
CHROMIC ACID	1755	60		14	898	306	52		80	70		238	C5.0	C 9	279

CHEMICAL OR MATERIAL NAME	UN/NA Number	Guide Number (DOT)	Firefighter's Hazardous Materials Reference Book	First Aid Manual for Chemical Accidents	Sax's Dangerous Properties of Industrial Materials	Hazardous Chemicals Desk Reference	Rapid Guide to Hazardous Chemicals in the Workplace	Fire Protection Guide on Hazardous Materials (NFPA)	Firefighter's Handbook of Hazardous Materials	Pocket Guide to Chemical Hazards (NIOSH)	Chemical Hazard Response Information System (CHRIS)	Emergency Handling of Hazardous Materials in Surface Transportation (AAR)	Emergency Action Guides (EAG)	Chemical Data Notebook: A User's Manual	Condensed Chemical Dictionary
CHROMIC ACID Synonym: CHROMIC ANHYDRIDE	1463	42	215	14	898	306	52	56 –	80	70	CMA 262	238	C5.0	C 9	279
CHROMIC ACID DILITHIUM SALT Synonym: LITHIUM CHROMATE	9134/3077	31	522		2127	725			181		LCR 696	582			707
CHROMIC ACID STRONTIUM SALT (1:1) Synonym: STRONTIUM CHROMATE	9149/3077	45	793		3140	1072	201		257		SCM1023	860			1095
CHROMIC ANHYDRIDE Synonym: CHROMIC ACID	1463	42	216	14	898	306	52	56 –	80	70	CMA 262	238	C5.0	C 9	279
CHROMIC FLUORIDE	1756	60							80			239			279
CHROMIC FLUORIDE	1757	60							80			239			279
CHROMIC NITRATE	2720	35							80						279
CHROMIC OXIDE Synonym: CHROMIC ANHYDRIDE	1463	42	216						80		CMA 262	238			280
CHROMIC SULFATE	9100/3082	31	217						80		CHS 249	239			280
CHROMIC TRIOXIDE	1463	42													
CHROMIUM					899	307	52								280
CHROMIUM (II) COMPOUNDS					900	307				72					
CHROMIUM (III) COMPOUNDS					900	307				72					
CHROMIUM (VI) DIOXYCHLORIDE Synonym: CHROMYL CHLORIDE	1758	39	219			309		56 –	80		CMC 264				283
CHROMIUM (VI) OXIDE Synonym: CHROMIC ACID	1463	42	215	14	901	308	52	56 –	80	70		238	C5.0	C 9	279
CHROMIUM ACETATE Synonym: CHROMIC ACETATE	9101/3077	31	214		898	305			80		CRT 308	237			281
CHROMIUM CHLORIDE Synonym: CHROMOUS CHLORIDE	9102/3077		218	14		307			80		CRC 299	241			281

CHEMICAL OR MATERIAL NAME	UN/NA Number	Guide Number (DOT)	Firefighter's Hazardous Materials Reference Book	First Aid Manual for Chemical Accidents	Sax's Dangerous Properties of Industrial Materials	Hazardous Chemicals Desk Reference	Rapid Guide to Hazardous Chemicals in the Workplace	Fire Protection Guide on Hazardous Materials (NFPA)	Firefighter's Handbook of Hazardous Materials	Pocket Guide to Chemical Hazards (NIOSH)	Chemical Hazard Response Information System (CHRIS)	Emergency Handling of Hazardous Materials in Surface Transportation (AAR)	Emergency Action Guides (EAG)	Chemical Data Notebook: A User's Manual	Condensed Chemical Dictionary
CHROMIUM DICHLORIDE Synonym: CHROMOUS CHLORIDE	9102/3077		218						80		CRC 299	241			282
CHROMIUM III SULFATE Synonym: CHROMIC SULFATE	9100/3082	31	217						80		CHS 249	239			280
CHROMIUM LITHIUM OXIDE Synonym: LITHIUM CHROMATE	9134/3077	31	522		2127	725			181		LCR 696	582			707
CHROMIUM NITRATE	2720/1479	35			900				80			240			281
CHROMIUM OXIDE	1463	42			901	308									281
CHROMIUM OXYCHLORIDE Synonym: CHROMYL CHLORIDE	1758	39	219		904	309		56 –	80		CMC 264	240			281
CHROMIUM RESINATE	1325	32													
CHROMIUM SULFATE Synonym: CHROMIC SULFATE	9100/3077	31	217						80		CHS 249	240			282
CHROMIUM SULPHATE BASIC DRY	9188/3077	31										240			
CHROMIUM TRIACETATE Synonym: CHROMIC ACETATE	9101/3077	31	214		898	305			80		CRT 308	237			279
CHROMIUM TRIOXIDE Synonym: CHROMIC ANHYDRIDE	1463	42	216						80		CMA 262	238			282
CHROMIUM TRIOXIDE Synonym: CHROMIC ACID	1463	42	215	14	898	306	52	56 –	80	70		238	C5.0	C 9	282
CHROMOSULFURIC ACID	2240	39							80			241			
CHROMOUS CHLORIDE	9102/3077	31	218						80		CRC 299	241			282
CHROMYL CHLORIDE	1758	39	219		904	309		56 –	80		CMC 264				283
CHRYSENE	3082/9188	31			905	310	52					241			283
CIANURINA Synonym: MERCURIC CYANIDE	1636	53	533					107–	186		MCN 741	599			741

CHEMICAL OR MATERIAL NAME	UN/NA Number	Guide Number (DOT)	Firefighter's Hazardous Materials Reference Book	First Aid Manual for Chemical Accidents	Sax's Dangerous Properties of Industrial Materials	Hazardous Chemicals Desk Reference	Rapid Guide to Hazardous Chemicals in the Workplace	Fire Protection Guide on Hazardous Materials (NFPA)	Firefighter's Handbook of Hazardous Materials	Pocket Guide to Chemical Hazards (NIOSH)	Chemical Hazard Response Information System (CHRIS)	Emergency Handling of Hazardous Materials in Surface Transportation (AAR)	Emergency Action Guides (EAG)	Chemical Data Notebook: A User's Manual	Condensed Chemical Dictionary
CIGARETTE	1867	32				311			80						
CIGARETTE LIGHTER	1057	17							80			577			
CIGARETTE LIGHTER	1226	26							80						
CINENE	2052	27							80						284
CINNAMENE Synonym: STYRENE MONOMER	2055	27	795		3145	1074	202		80		STY1064	862	S3.0		1097
CINNAMENOL Synonym: STYRENE MONOMER	2055	27	795		3145	1074	202		257		STY1064	862	S3.0		1097
CINNAMOL Synonym: STYRENE MONOMER	2055	27	795		3145	1074	202		257		STY1064	862	S3.0		1097
CITRAL								−28	81						286
CITRIC ACID					916	313					CIT 254				286
CITRIC ACID DIAMMONIUM SALT Synonym: AMMONIUM CITRATE	9087/3077	31	58		225	61			28		ACI 20	58			65
CITRONELLEL								−28	81						
CITRONELLOL						313		−28	81						286
CLAUDETITE Synonym: ARSENIC TRIOXIDE	1561	53	100		299		16	33 –	35		ATO 105	93	A11.1	A14	98
CLEANING COMPOUND	1142	26							81						
CLEANING COMPOUND	1760	60							81						
CLEANING FLUID	1142	26							81						
CLEANING SOLVENTS 140 (60) CLASS								−28							
CLORETILO Synonym: ETHYL CHLORIDE	1037	27	350		1586	532	92	82 −49	136	108	ECL 514	412	E2.3		483

CHEMICAL OR MATERIAL NAME	UN/NA Number	Guide Number (DOT)	Firefighter's Hazardous Materials Reference Book	First Aid Manual for Chemical Accidents	Sax's Dangerous Properties of Industrial Materials	Hazardous Chemicals Desk Reference	Rapid Guide to Hazardous Chemicals in the Workplace	Fire Protection Guide on Hazardous Materials (NFPA)	Firefighter's Handbook of Hazardous Materials	Pocket Guide to Chemical Hazards (NIOSH)	Chemical Hazard Response Information System (CHRIS)	Emergency Handling of Hazardous Materials in Surface Transportation (AAR)	Emergency Action Guides (EAG)	Chemical Data Notebook: A User's Manual	Condensed Chemical Dictionary
CLOROX Synonym: SODIUM HYPOCHLORITE	1791	60	778	24	3088	1057			254		SHC1043	843			1059
CMME Synonym: METHYL CHLOROMETHYL ETHER	1239	57						111–	193			643	M3.1.2		762
CO RAL Synonym: COUMAPHOS	2783/3027	55	230		957	323					COU 281	268			320
COAL	1361	32							81			242			290
COAL DUST	1361	32			929	315			81						
COAL GAS	1023	18							81			242			290
COAL NAPHTHA Synonym: BENZENE	1114	27	115	12	356	115	21	35 –17	81	44	BNZ 152	112	B1.0	B 1	128
COAL OIL	1223	27							81						290
COAL OIL Synonym: FUEL OIL NO. 1	1993	27	403					–55				464	F3.0		290
COAL SPRAYING OIL PETROLEUM	1270/1993	27										242			
COAL TAR	1999				930	316	53		81						290
COAL TAR CREOSOTE	1136	27			930	316	53		81						291
COAL TAR CREOSOTE Synonym: CREOSOTE COAL TAR	1993	27	231						84		CCT 233		C6.0		291
COAL TAR DISTILLATE Synonym: CREOSOTE COAL TAR DISTILLATE	1137	27	231						81						291
COAL TAR DISTILLATE Synonym: CREOSOTE COAL TAR DISTILLATE	1136	27							81		CCT 233	243	C6.0		291
COAL TAR DYE	2801/1760	60				316						243			291
COAL TAR FIBRE SATURANT	1993	27										243			
COAL TAR LIGHT OIL	1137	27						–28	81						291

CHEMICAL OR MATERIAL NAME	UN/NA Number	Guide Number (DOT)	Firefighter's Hazardous Materials Reference Book	First Aid Manual for Chemical Accidents	Sax's Dangerous Properties of Industrial Materials	Hazardous Chemicals Desk Reference	Rapid Guide to Hazardous Chemicals in the Workplace	Fire Protection Guide on Hazardous Materials (NFPA)	Firefighter's Handbook of Hazardous Materials	Pocket Guide to Chemical Hazards (NIOSH)	Chemical Hazard Response Information System (CHRIS)	Emergency Handling of Hazardous Materials in Surface Transportation (AAR)	Emergency Action Guides (EAG)	Chemical Data Notebook: A User's Manual	Condensed Chemical Dictionary
COAL TAR NAPHTHA	2553	27							81						291
COAL TAR NAPHTHA SOLVENT	1136	27										244			
COAL TAR OIL	1137/1136	27							81			245			291
COAL TAR OIL	1136/1137	27							81			245			291
COAL TAR PITCH								−28							291
COAL TAR PITCH VOLATILES					930	316	53			72					
COATING SOLUTION	1139	26							81			245			291
COBALT (II) BROMIDE Synonym: COBALT BROMIDE (OUS)	9103/3077				932						COB 275	246			292
COBALT (II) CHLORIDE Synonym: COBALT CHLORIDE					932	317			81		CBC 216				293
COBALT ACETATE											CBA 214				292
COBALT ACETATE TETRAHYDRATE Synonym: COBALT ACETATE					931	317					CBA 214				292
COBALT AND COMPOUNDS					931	316	54								291
COBALT BROMIDE (OUS)	9103/3077				932						COB 275	246			292
COBALT CARBONYL					932	317	54		82						292
COBALT CHLORIDE					932	317					CBC 216				293
COBALT DIBROMIDE Synonym: COBALT BROMIDE (OUS)	9103/3077				932						COB 275	246			292
COBALT DIFORMATE Synonym: COBALT FORMATE	9104/3082										CFM 241	246			
COBALT FORMATE	9104/3082										CFM 241	246			
COBALT HYDROCARBONYL					934	318	54		82						293

HAZARDOUS MATERIALS REFERENCE BOOKS INDEX

CHEMICAL OR MATERIAL NAME	UN/NA Number	Guide Number (DOT)	Firefighter's Hazardous Materials Reference Book	First Aid Manual for Chemical Accidents	Sax's Dangerous Properties of Industrial Materials	Hazardous Chemicals Desk Reference	Rapid Guide to Hazardous Chemicals in the Workplace	Fire Protection Guide on Hazardous Materials (NFPA)	Firefighter's Handbook of Hazardous Materials	Pocket Guide to Chemical Hazards (NIOSH)	Chemical Hazard Response Information System (CHRIS)	Emergency Handling of Hazardous Materials in Surface Transportation (AAR)	Emergency Action Guides (EAG)	Chemical Data Notebook: A User's Manual	Condensed Chemical Dictionary
COBALT II ACETATE Synonym: COBALT ACETATE											CBA 214				292
COBALT II NITRATE Synonym: COBALT NITRATE					934				81		CON 278				292
COBALT II SULFATE Synonym: COBALT SULFATE					935	318					CBS 222				296
COBALT NAPHTHA	2001	32						−28	82						
COBALT NAPHTHENATE	2001/1325	32							82			245			293
COBALT NITRATE					934						CON 278				293
COBALT RESINATE	1318	32			935				82			246			296
COBALT SULFATE					935	318					CBS 222				296
COBALTOUS ACETATE Synonym: COBALT ACETATE											CBA 214				294
COBALTOUS BROMIDE Synonym: COBALT BROMIDE (OUS)	9103/3077				932				82		COB 275	246			294
COBALTOUS CHLORIDE Synonym: COBALT CHLORIDE					933	317					CBC 216				294
COBALTOUS CHLORIDE DIHYDRATE Synonym: COBALT CHLORIDE						317					CBC 216				293
COBALTOUS CHLORIDE HEXAHYDRATE Synonym: COBALT CHLORIDE					933	217					CBC 216				293
COBALTOUS FORMATE Synonym: COBALT FORMATE	9104/3082								82		CFM 241	246			295
COBALTOUS NITRATE Synonym: COBALT NITRATE					934				82		CON 278				295

CHEMICAL OR MATERIAL NAME	UN/NA Number	Guide Number (DOT)	Firefighter's Hazardous Materials' Reference Book	First Aid Manual for Chemical Accidents	Sax's Dangerous Properties of Industrial Materials	Hazardous Chemicals Desk Reference	Rapid Guide to Hazardous Chemicals in the Workplace	Fire Protection Guide on Hazardous Materials (NFPA)	Firefighter's Handbook of Hazardous Materials	Pocket Guide to Chemical Hazards (NIOSH)	Chemical Hazard Response Information System (CHRIS)	Emergency Handling of Hazardous Materials in Surface Transportation (AAR)	Emergency Action Guides (EAG)	Chemical Data Notebook: A User's Manual	Condensed Chemical Dictionary
COBALTOUS NITRATE HEXAHYDRATE Synonym: COBALT NITRATE					934						CON 278				293
COBALTOUS SULFAMATE	9105/3077								82			246			
COBALTOUS SULFATE HEPTAHYDRATE Synonym: COBALT SULFATE					935	318					CBS 222				296
COCAINE					936				82						297
COCCULUS	1584	53							82			247			297
COCONUT BUTTER Synonym: OILS EDIBLE COCONUT											OCC 871				298
COCONUT OIL Synonym: OILS EDIBLE COCONUT											OCC 871				298
CODOIL Synonym: OILS MISCELLANEOUS ROSIN	1286	26						-83			ORN 901				
CODOIL Synonym: OILS MISCELLANEOUS RESIN	1286	26									ORS 902				
COLAMINE Synonym: MONOETHANOLAMINE	2491	60	594					118-	206		MEA 749		M5.0		795
COLCHICINE					940				82						299
COLLIDIENE	1993	27										248			
COLLODION	2059/1993	26	220		940	319		-28	82		CLD 257	249			300
COLOGNE SPIRITS Synonym: ETHYL ALCOHOL	1170	26	344	16	1572	525	90	-48	82		EAL 504	408	E2.2	E 2	478
COLONIAL SPIRIT Synonym: METHYL ALCOHOL	1230	28	554	20	2251	770	136	-66	82	144	MAL 722	613		M 2	755
COLONIAL SPIRIT Synonym: METHANOL	1230	28	549						82			613	M2.0		751

HAZARDOUS MATERIALS REFERENCE BOOKS INDEX

CHEMICAL OR MATERIAL NAME	UN/NA Number	Guide Number (DOT)	Firefighter's Hazardous Materials Reference Book	First Aid Manual for Chemical Accidents	Sax's Dangerous Properties of Industrial Materials	Hazardous Chemicals Desk Reference	Rapid Guide to Hazardous Chemicals in the Workplace	Fire Protection Guide on Hazardous Materials (NFPA)	Firefighter's Handbook of Hazardous Materials	Pocket Guide to Chemical Hazards (NIOSH)	Chemical Hazard Response Information System (CHRIS)	Emergency Handling of Hazardous Materials in Surface Transportation (AAR)	Emergency Action Guides (EAG)	Chemical Data Notebook: A User's Manual	Condensed Chemical Dictionary
COLOPHONY SOLUTION Synonym: ROSIN SOLUTION	1993	27	739									818	R1.0		
COLUMBIAN SPIRIT Synonym: METHYL ALCOHOL	1230	28	554	20	2251	770	136	-66	82	144	MAL 722	613		M 2	755
COLUMBIAN SPIRITS Synonym: METHANOL	1230	28	549						82			613	M2.0		751
COMBUSTIBLE LIQUID	1993	27							82			249			
COMBUSTION IMPROVER C 12 Synonym: METHYLCYCLOPENTADIENYLMANGANESE TRI-CARBONYL	2810	55									MCT 746	645			764
COMMON VERDIGRIS Synonym: COPPER SUBACETATE							55				CST 315				315
COMPOUND POLISHING	1142	26							83			251			
COMPOUND TREE OR WEED KILLING	1993	27							83			251			
COMPOUND TREE OR WEED KILLING	1760	60							83			251			
COMPRESSED GAS	1956	12							83			256			304
COMPRESSED GAS	1955	15							83			255			304
COMPRESSED GAS	1954	22							83						304
COMPRESSED GAS	1953	18							83			255			304
CONDENSED PHOSPHORIC ACID Synonym: POLYPHOSPHORIC ACID	1760	60	700						237		PPA 972				940
CONOCO SA 597 Synonym: DODECYLBENZENESULFONIC ACID	2584	60	330						130		DSA 474	390	D5.0		441
COPPER					944	319	54								309
COPPER ACETATE	9106/3077		221		944	319	55				COP 279	257			310
COPPER ACETOARSENITE	1585	53	222				55				CAA 203	257			310

CHEMICAL OR MATERIAL NAME	UN/NA Number	Guide Number (DOT)	Firefighter's Hazardous Materials Reference Book	First Aid Manual for Chemical Accidents	Sax's Dangerous Properties of Industrial Materials	Hazardous Chemicals Desk Reference	Rapid Guide to Hazardous Chemicals in the Workplace	Fire Protection Guide on Hazardous Materials (NFPA)	Firefighter's Handbook of Hazardous Materials	Pocket Guide to Chemical Hazards (NIOSH)	Chemical Hazard Response Information System (CHRIS)	Emergency Handling of Hazardous Materials in Surface Transportation (AAR)	Emergency Action Guides (EAG)	Chemical Data Notebook: A User's Manual	Condensed Chemical Dictionary
COPPER ACETYLIDE					944	320	55		83						
COPPER AMMONIUM SULFATE Synonym: COPPER SULFATE AMMONIATED	9110						55				CSN 312				315
COPPER ARSENITE	1586	53	223				55		83		CPA 283	257			311
COPPER BASED PESTICIDE	2775	53					55		84			258			
COPPER BASED PESTICIDE	3010/2775	55					55		84			258			
COPPER BASED PESTICIDE	2776	28					55		83			257			
COPPER BOROFLUORIDE SOLUTION Synonym: COPPER FLUOROBORATE							55				CPF 287				
COPPER BROMIDE							55				CPB 284				311
COPPER BROMIDE (OUS)							55				CBD 217				
COPPER CHLORATE	2721	35			945		55		83			258			
COPPER CHLORIDE	2802	60	224	14	945	320	55		83		CPC 285	259			311
COPPER CYANIDE	1587	53	225		947	321	55		83		CCY 234	259			311
COPPER CYANIDE (OUS)	1587	53	225		947	321	55		83		CCY 234	259			
COPPER FLUOROBORATE							55				CPF 287				
COPPER FUME					948		55			74					
COPPER GLYCINATE							55				CPG 288				312
COPPER II FLUOBORATE SOLUTION Synonym: COPPER FLUOROBORATE							55				CPF 287				
COPPER IODIDE							55				CID 253				312
COPPER LACTATE (IC)							55				CLT 260				312
COPPER MONOBROMIDE Synonym: COPPER BROMIDE (OUS)							55				CBD 217				311
COPPER NAPHTHENATE	168/1993	26	226				55				CNN 272	260			313

CHEMICAL OR MATERIAL NAME	UN/NA Number	Guide Number (DOT)	Firefighter's Hazardous Materials Reference Book	First Aid Manual for Chemical Accidents	Sax's Dangerous Properties of Industrial Materials	Hazardous Chemicals Desk Reference	Rapid Guide to Hazardous Chemicals in the Workplace	Fire Protection Guide on Hazardous Materials (NFPA)	Firefighter's Handbook of Hazardous Materials	Pocket Guide to Chemical Hazards (NIOSH)	Chemical Hazard Response Information System (CHRIS)	Emergency Handling of Hazardous Materials in Surface Transportation (AAR)	Emergency Action Guides (EAG)	Chemical Data Notebook: A User's Manual	Condensed Chemical Dictionary
COPPER NITRATE	1479	35	227		948	321	55		83		CNI 271				313
COPPER ORTHOARSENITE Synonym: COPPER ARSENITE	1586	53	223		949	322	55		83		CPA 283	257			311
COPPER OXALATE	2449	54	228		949		55		83		COL 277				313
COPPER PLATING SOLUTION	2810	55					55					260			
COPPER RESINATE	1325	32					55								314
COPPER SELENATE	2630	53					55		83						314
COPPER SELENITE	2630	53					55		83						
COPPER SUBACETATE (IC)							55				CST 315				315
COPPER SULFATE	9109		229	14	950	322	55				CSF 311				315
COPPER SULFATE AMMONIATED	9110						55				CSN 312				315
COPPER SULFATE AND LIME COMBINED	9188/3077						55					260			
COPPER SULFATE PENTAHYDRATE Synonym: COPPER SULFATE	9109		229	14	951	322	55				CSF 311				315
COPPER SULPHATE AND SULPHUR	9188/3077						55					261			
COPPER TARTRATE (IC)	9111						55				CTT 323				
COPPERAS Synonym: FERROUS SULFATE	9125/3077		397		1702	579			147		FRS 593	450			311
COPRA	1363	37			952	322			84						
COPRA OIL Synonym: OILS EDIBLE COCONUT	1363	37			952	322			84		OCC 871				316
CORN OIL					955	322		-28							316
CORN SUGAR SOLUTION Synonym: DEXTROSE SOLUTION											DTS 492				317
CORN SYRUP											CSY 316				317

CHEMICAL OR MATERIAL NAME	UN/NA Number	Guide Number (DOT)	Firefighter's Hazardous Materials Reference Book	First Aid Manual for Chemical Accidents	Sax's Dangerous Properties of Industrial Materials	Hazardous Chemicals Desk Reference	Rapid Guide to Hazardous Chemicals in the Workplace	Fire Protection Guide on Hazardous Materials (NFPA)	Firefighter's Handbook of Hazardous Materials	Pocket Guide to Chemical Hazards (NIOSH)	Chemical Hazard Response Information System (CHRIS)	Emergency Handling of Hazardous Materials in Surface Transportation (AAR)	Emergency Action Guides (EAG)	Chemical Data Notebook: A User's Manual	Condensed Chemical Dictionary
CORROSIVE LIQUID	2922	59							84			263			317
CORROSIVE LIQUID	2920	29							84			263			317
CORROSIVE LIQUID	1760	60							84			263			317
CORROSIVE MERCURY CHLORIDE Synonym: MERCURIC CHLORIDE	1624	53	532	19					186		MRC 792	599			740
CORROSIVE SOLID	2923	59							84			265			317
CORROSIVE SOLID	1759	60							84			265			317
CORROSIVE SOLID	2921	34							84			265			317
COSMETICS	1760	60							84			266			318
COSMETICS	1759	60							84			267			318
COSMETICS	1479	35							84			266			318
COSMETICS	1325	32							84			267			318
COSMETICS	1993	27							84			266			318
COTTON	1365	32							84			268			318
COTTON DUST					957	323	55			74					
COTTON WASTE	1364	32							84			267			
COTTON WASTE NEC NOT FILTER PACKING FOR MANU-FACTURED	1325	32										267			
COTTONSEED OIL REFINED								–29							
COUMAPHOS	2783/3027	55	230		957	323					COU 281	268			320
COUMARIN DERIVATIVE PESTICIDE	3025	28							84						
COUMARIN DERIVATIVE PESTICIDE	3024	28							84			269			
COUMARIN DERIVATIVE PESTICIDE	3027/2783	55							84			268			
COUMARIN DERIVATIVE PESTICIDE	3026	55							84			268			

CHEMICAL OR MATERIAL NAME	UN/NA Number	Guide Number (DOT)	Firefighter's Hazardous Materials Reference Book	First Aid Manual for Chemical Accidents	Sax's Dangerous Properties of Industrial Materials	Hazardous Chemicals Desk Reference	Rapid Guide to Hazardous Chemicals in the Workplace	Fire Protection Guide on Hazardous Materials (NFPA)	Firefighter's Handbook of Hazardous Materials	Pocket Guide to Chemical Hazards (NIOSH)	Chemical Hazard Response Information System (CHRIS)	Emergency Handling of Hazardous Materials in Surface Transportation (AAR)	Emergency Action Guides (EAG)	Chemical Data Notebook: A User's Manual	Condensed Chemical Dictionary
CRAG HERBICIDE					959	324	55		84	74					
CRANKCASE OIL Synonym: OILS MISCELLANEOUS LUBRICATING	1270	27									OLB 884				
CRANKCASE OIL Synonym: OILS MISCELLANEOUS MOTOR	1270	27									OMT 890				
CREOSOL	2022	55		14	959										
CREOSOTE COAL TAR	1993	27	231	14					84		CCT 233		C6.0		322
CREOSOTE OIL Synonym: CREOSOTE COAL TAR	1993	27	231	14				−29	84		CCT 233		C6.0		322
CREOSOTE SALTS	1334	32							84						
CRESOL	2076	55	232		959	324	55	57 –	85	74	CRS 307	270	C7.0		322
CRESOL EPOXYPROPYL ETHER Synonym: CRESYL GLYCIDYL ETHER											CGE 242				
CRESOL-meta Synonym: CRESOL	2076	55	232		960	324	55	57 −29	85	74	CRL 303	270	C7.0		322
CRESOL-ortho Synonym: CRESOL	2076	55	232		960	325	55	57 −29	84	74	CRO 305	270	C7.0		322
CRESOL-para Synonym: CRESOL	2076	55	232		961	325	55	57 −29	85	74	CSO 313	270	C7.0		322
CRESOLE Synonym: CRESOL	2076	55	232		959	324	55	57 –	85	74	CRS 307	270	C7.0		322
CRESYL ACETATE-para								−29	85						323
CRESYL GLYCIDYL ETHER											CGE 242				
CRESYLIC ACID Synonym: CRESOL	2076	55	232		959	324	55	57 –	85	74	CRS 307	270	C7.0		322
CRESYLIC ACID	2922	59										271			323

CHEMICAL OR MATERIAL NAME	UN/NA Number	Guide Number (DOT)	Firefighter's Hazardous Materials Reference Book	First Aid Manual for Chemical Accidents	Sax's Dangerous Properties of Industrial Materials	Hazardous Chemicals Desk Reference	Rapid Guide to Hazardous Chemicals in the Workplace	Fire Protection Guide on Hazardous Materials (NFPA)	Firefighter's Handbook of Hazardous Materials	Pocket Guide to Chemical Hazards (NIOSH)	Chemical Hazard Response Information System (CHRIS)	Emergency Handling of Hazardous Materials in Surface Transportation (AAR)	Emergency Action Guides (EAG)	Chemical Data Notebook: A User's Manual	Condensed Chemical Dictionary
CRESYLIC ACID Synonym: XYLENOL	2261	55	883		3523	1184			288		XYL1190	962			1235
CRESYLIC ACID	2022	55							85			271			323
CRESYLIC ACID-meta Synonym: CRESOL (-meta)	2076	55	232		960	324	55	57 -29	85	74	CRL 303	270	C7.0		322
CRESYLIC ACIDS Synonym: CRESOL	2076	55	232		959	324	55	57 -	85	74	CRS 307	270	C7.0		323
CROLEAN Synonym: ACROLEIN	1092	30	31	11	63	25	5	17 -12	20	32	ARL 85	12	A4.0	A 5	18
CROPLAS EH Synonym: ETHYL HEXYL TALLATE											EHT 539				
CROTENALDEHYDE Synonym: CROTONALDEHYDE	1143	28	233	14	964	325	56	57 -29	85	76	CTA 317	271	C7.1		325
CROTON OIL Synonym: OILS MISCELLANEOUS CROTON									85		OCR 874				325
CROTON TIGLIUM OIL Synonym: OILS MISCELLANEOUS CROTON											OCR 874				325
CROTONAL Synonym: CROTONALDEHYDE	1143	28	233	14	964	325	56	57 -29	85	76	CTA 317	271	C7.1		325
CROTONALDEHYDE	1143	28	233	14	964	325	56	57 -29	85	76	CTA 317	271	C7.1		325
CROTONIC ACID	2823	60			965	326		-29	85			272			325
CROTONIC ALDEHYDE Synonym: CROTONALDEHYDE	1143	28	233	14	964	325	56	57 -29	85	76	CTA 317	271	C7.1		325
CROTONOEL Synonym: OILS MISCELLANEOUS CROTON											OCR 874				
CROTONONITRILE								-29							

CHEMICAL OR MATERIAL NAME	UN/NA Number	Guide Number (DOT)	Firefighter's Hazardous Materials Reference Book	First Aid Manual for Chemical Accidents	Sax's Dangerous Properties of Industrial Materials	Hazardous Chemicals Desk Reference	Rapid Guide to Hazardous Chemicals in the Workplace	Fire Protection Guide on Hazardous Materials (NFPA)	Firefighter's Handbook of Hazardous Materials	Pocket Guide to Chemical Hazards (NIOSH)	Chemical Hazard Response Information System (CHRIS)	Emergency Handling of Hazardous Materials in Surface Transportation (AAR)	Emergency Action Guides (EAG)	Chemical Data Notebook: A User's Manual	Condensed Chemical Dictionary
CROTONYL ALCOHOL	2614	26						−29	85						
CROTONYLENE	1144	27			966	327			85			272			326
CRUDE ARSENIC Synonym: ARSENIC TRIOXIDE	1561	53	100		299	92	16	33 –	35		ATO 105	93	A11.1	A14	98
CRUDE EPICHLOROHYDRIN Synonym: CHLOROHYDRINS	2023	30	206		818						CHD 245				267
CRUDE LIGHT OIL OF COAL TAR	1136/1137	27										272			
CRUDE OIL	1270	27			966				85						
CRUDE OIL PETROLEUM	1267	27										273			
CRYSTALLIZED VERDIGRIS Synonym: COPPER ACETATE	9106/3077		221		944	319	55				COP 279	257			310
CSA Synonym: CHLOROSULFONIC ACID	1754	39	212					55 –			CSA 309	233	C4.5	C 8	273
CTF Synonym: CHLORINE TRIFLUORIDE	1749	44	198	13	759	269	46	49 –	70	64	CTF 320	215			260
CTFE Synonym: TRIFLUOROCHLOROETHYLENE	1082	17	850					−89	276		TFC1108	925			328
CUBIC NITER Synonym: SODIUM NITRATE	1498	35	780		3097	1060			254		SDN1033	845	S2.6	S 5	1063
CUCUMBER DUST Synonym: CALCIUM ARSENATE	1573	53	165						62	56	CCA 225	178			200
CUMENE	1918/1993	28	234	14	969	328	56	58 −29	85	76	CUM 325	273			328
CUMENE HYDROPEROXIDE	2116	51	235					−29	85		CMH 266				329
CUMOL Synonym: CUMENE	1918/1993	28	234	14	969	328	56	58 −29	85	76	CUM 325	273			328

CHEMICAL OR MATERIAL NAME	UN/NA Number	Guide Number (DOT)	Firefighter's Hazardous Materials Reference Book	First Aid Manual for Chemical Accidents	Sax's Dangerous Properties of Industrial Materials	Hazardous Chemicals Desk Reference	Rapid Guide to Hazardous Chemicals in the Workplace	Fire Protection Guide on Hazardous Materials (NFPA)	Firefighter's Handbook of Hazardous Materials	Pocket Guide to Chemical Hazards (NIOSH)	Chemical Hazard Response Information System (CHRIS)	Emergency Handling of Hazardous Materials in Surface Transportation (AAR)	Emergency Action Guides (EAG)	Chemical Data Notebook: A User's Manual	Condensed Chemical Dictionary
CUMYL HYDROPEROXIDE Synonym: CUMENE HYDROPEROXIDE	2116/3109	51	235					-29	85		CMH 266	273			329
CUMYL PEROXY-neo-DECANOATE	2963	52							85						
CUMYL PEROXYPIVALATE	2964	52							85						
CUPRAMMONIUM SULFATE Synonym: COPPER SULFATE AMMONIATED	9110						55				CSN 312				315
CUPRIC ACETATE	9106								85						
CUPRIC ACETATE BASIC Synonym: COPPER SUBACETATE							55				CST 315				315
CUPRIC ACETATE MONOHYDRATE Synonym: COPPER ACETATE	9106/3077		221		944	319	55				COP 279	257			310
CUPRIC ACETATE-m-ARSENATE	1585	53							85						
CUPRIC AMINO ACETATE Synonym: COPPER GLYCINATE							55				CPG 288				312
CUPRIC AMMINE SULFATE Synonym: COPPER SULFATE AMMONIATED	9110						55				CSN 312				315
CUPRIC ARSENITE Synonym: COPPER ARSENITE	1586	53	223				55		83		CPA 283	257			311
CUPRIC BROMIDE ANHYDROUS Synonym: COPPER BROMIDE							55				CPB 284				311
CUPRIC CHLORIDE DIHYDRATE Synonym: COPPER CHLORIDE	2802	60	224	14	945	320	55		83		CPC 285	259			311
CUPRIC CHLORIDE SOLUTION AMMONIATED	1760	60										274			
CUPRIC FLUOBORATE SOLUTION Synonym: COPPER FLUOROBORATE							55				CPF 287				

CHEMICAL OR MATERIAL NAME	UN/NA Number	Guide Number (DOT)	Firefighter's Hazardous Materials Reference Book	First Aid Manual for Chemical Accidents	Sax's Dangerous Properties of Industrial Materials	Hazardous Chemicals Desk Reference	Rapid Guide to Hazardous Chemicals in the Workplace	Fire Protection Guide on Hazardous Materials (NFPA)	Firefighter's Handbook of Hazardous Materials	Pocket Guide to Chemical Hazards (NIOSH)	Chemical Hazard Response Information System (CHRIS)	Emergency Handling of Hazardous Materials in Surface Transportation (AAR)	Emergency Action Guides (EAG)	Chemical Data Notebook: A User's Manual	Condensed Chemical Dictionary
CUPRIC GREEN Synonym: COPPER ARSENITE	1586	53	223				55		83		CPA 283	257			311
CUPRIC NITRATE	1479	35	236						86			274			
CUPRIC NITRATE TRIHYDRATE Synonym: COPPER NITRATE	1479	35	227		948	321	55		83		CNI 271				313
CUPRIC OXALATE	2449/3077	54							86			274			
CUPRIC OXALATE HEMIHYDRATE Synonym: COPPER OXALATE	2449	54	228		949		55		83		COL 277				313
CUPRIC PHOSPHIDE									86						
CUPRIC SULFATE	9110/3077								86			275			
CUPRIC SULFATE Synonym: COPPER SULFATE	9109/3077	31	237	14	950	322	55		86		CSF 311	275			315
CUPRIC TARTRATE	9111/3077								86			275			
CUPRICIN Synonym: COPPER CYANIDE (OUS)	1587	53	225		947	321	55		83		CCY 234	259			311
CUPRIETHYLENEDIAMINE HYDROXIDE SOLUTION Synonym: CUPRIETHYLENEDIAMINE SOLUTION	1761	59	238						86		CES 239	275			330
CUPRIETHYLENEDIAMINE SOLUTION	1761	59	238						86		CES 239	275			330
CUPROUS CYANIDE Synonym: COPPER CYANIDE (OUS)	1587	53	225		947	321	55		83		CCY 234	259			330
CUPROUS IODIDE Synonym: COPPER IODIDE							55				CID 253				330
CURARE					971				86						331
CURATERR Synonym: CARBOFURAN	2757/2929	55	184		695	242	40				CBF 218	188			216

CHEMICAL OR MATERIAL NAME	UN/NA Number	Guide Number (DOT)	Firefighter's Hazardous Materials Reference Book	First Aid Manual for Chemical Accidents	Sax's Dangerous Properties of Industrial Materials	Hazardous Chemicals Desk Reference	Rapid Guide to Hazardous Chemicals in the Workplace	Fire Protection Guide on Hazardous Materials (NFPA)	Firefighter's Handbook of Hazardous Materials	Pocket Guide to Chemical Hazards (NIOSH)	Chemical Hazard Response Information System (CHRIS)	Emergency Handling of Hazardous Materials in Surface Transportation (AAR)	Emergency Action Guides (EAG)	Chemical Data Notebook: A User's Manual	Condensed Chemical Dictionary
CUT BACK ASPHALT Synonym: ASPHALT	1999	27	105	12	307	97	17	−17	36		ASP 93	99	A12.0		100
CYANACETIC ACID Synonym: CYANOACETIC ACID	1935	55	240		975			59 −	86		CYA 327				332
CYANAMIDE					973	329	57	−29	86						332
CYANIDE	1588	55	239		974	330	57		86			276			
CYANIDE Synonym: POTASSIUM CYANIDE	1680	55	707	23	974	330	57	144−	238		PTC 992	778			951
CYANIDE	1689	55			974	330	57			76					
CYANIDE MIXTURE DRY	1588	55					55		86			276			
CYANIDE OF CALCIUM Synonym: CALCIUM CYANIDE	1575	55	170		665	231	37	46 −	63		CCN 230	180			202
CYANIDE OF SODIUM Synonym: SODIUM CYANIDE	1689	55	768	24	3074	1050	199	157−	253		SCN1024	837	S2.2	S 3	1053
CYANIDE OF ZINC Synonym: ZINC CYANIDE	1713	53	896		3540	1194	234		288		ZCN1199	966			1244
CYANIDE SOLUTION	1935	55			974		57		86			276			
CYANIDES	1680	55			974	330	57			76					
CYANOACETIC ACID	1935	55	240		975			59 −	86		CYA 327				332
CYANOACETONITRILE	2647	53							86						
CYANOBENZENE Synonym: BENZONITRILE	2224	55	120		377	125			41		BZN 198	117			132
CYANOETHYLENE Synonym: ACRYLONITRILE	1093	30	34	11	68	27	6	19 −12	20	34	ACN 23	13	A5.1	A 7	19
CYANOGAS A DUST Synonym: CALCIUM CYANIDE	1575	55	170		665	232	37	46 −	63		CCN 230	180			202

CHEMICAL OR MATERIAL NAME	UN/NA Number	Guide Number (DOT)	Firefighter's Hazardous Materials Reference Book	First Aid Manual for Chemical Accidents	Sax's Dangerous Properties of Industrial Materials	Hazardous Chemicals Desk Reference	Rapid Guide to Hazardous Chemicals in the Workplace	Fire Protection Guide on Hazardous Materials (NFPA)	Firefighter's Handbook of Hazardous Materials	Pocket Guide to Chemical Hazards (NIOSH)	Chemical Hazard Response Information System (CHRIS)	Emergency Handling of Hazardous Materials in Surface Transportation (AAR)	Emergency Action Guides (EAG)	Chemical Data Notebook: A User's Manual	Condensed Chemical Dictionary
CYANOGAS G FUMIGANT Synonym: CALCIUM CYANIDE	1575	55	170		665	232	37	46 –	63		CCN 230	180			202
CYANOGEN	1026	18	241		979	331	57	–29	87		CYG 330	277	C8.0		333
CYANOGEN BROMIDE	1889	55	242	14	979	331		60 –	87		CBR 221	277			334
CYANOGEN CHLORIDE	1589	15	243	14	979	332	58		87		CCL 229	277	C8.0.1		334
CYANOGEN GAS Synonym: CYANOGEN	1026	18	241		979	331	57	59 –29	87		CYG 330	277	C8.0		333
CYANOMETHANE Synonym: ACETONITRILE	1648	28	22	11	24	12	3	15 –12	87	32	ATN 104	5			9
CYANOMETHYL ACETATE					981				87						334
CYANURIC CHLORIDE	2670/1759	60							87			278			335
CYCLAMEN ALDEHYDE					984	332		–30	87						335
CYCLOBUTANE	2601/1954	22			986	332		–30	87			278			336
CYCLOBUTYLCHLOROFORMATE	2744	55							87			278			
CYCLOBUTYLENE	2601	22							87						
CYCLODAN Synonym: ENDOSULFAN	2761/2996	55	332		1519	505	86		131		ESF 559	398			463
CYCLODODECATRIENE	2518	59							87						
CYCLOHEPTANE	2241/1993	27			988	333		–30	87		CYE 329	279			336
CYCLOHEPTATRIENE Synonym: 1,3,5-CYCLOHEPTATRIENE	2603	28			988	333			87						
CYCLOHEPTENE	2242/1993	27			989				87			279			
CYCLOHEXANE	1145	26	244	14	990	333	58	–30	88	76	CHX 251	280	C8.1		336
CYCLOHEXANETHIOL					990	333	58	–30	88						
CYCLOHEXANOL	1993	27	245	14	991	333	59	–30	88	78	CHN 246	280			337

CHEMICAL OR MATERIAL NAME	UN/NA Number	Guide Number (DOT)	Firefighter's Hazardous Materials Reference Book	First Aid Manual for Chemical Accidents	Sax's Dangerous Properties of Industrial Materials	Hazardous Chemicals Desk Reference	Rapid Guide to Hazardous Chemicals in the Workplace	Fire Protection Guide on Hazardous Materials (NFPA)	Firefighter's Handbook of Hazardous Materials	Pocket Guide to Chemical Hazards (NIOSH)	Chemical Hazard Response Information System (CHRIS)	Emergency Handling of Hazardous Materials in Surface Transportation (AAR)	Emergency Action Guides (EAG)	Chemical Data Notebook: A User's Manual	Condensed Chemical Dictionary
CYCLOHEXANOL ACETATE	2243/1993	27							88			281			337
CYCLOHEXANONE	1915/1993	26	246	14	991	334	59	−30	88	78	CCH 228	281	C8.2		337
CYCLOHEXANONE PEROXIDE	2119/3105	51	247						88			282			337
CYCLOHEXANONE PEROXIDE	2896	51							88		CHP 248	283			337
CYCLOHEXANONE PEROXIDE	2117	49							88						337
CYCLOHEXANONE PEROXIDE	2118/3104	51							88			281			337
CYCLOHEXANONE PEROXIDE AND DI(1-HYDROXYCYCLO-HEXYL)	2896	51										282			
CYCLOHEXATRIENE Synonym: BENZENE	1114	27	115	12	356	115	21	35 −17	40	44	BNZ 152	112	B1.0	B 1	128
CYCLOHEXENE	2256/1993	29			992	334	59	−30	88	78		283			337
CYCLOHEXENE OXIDE					992	335			88						338
CYCLOHEXENYL TRICHLOROSILANE	1762	29	248		994	335			88		CHT 250	284			338
CYCLOHEXYL ACETATE	2243/1993	27			994	335		−30	88		CYC 328	284			
CYCLOHEXYL ALCOHOL Synonym: CYCLOHEXANOL	1993	27	245	14	991	333	59	−30	88	78	CHN 246	280			337
CYCLOHEXYL FORMATE								−30	88						
CYCLOHEXYL ISOCYANATE	2488/2810	57			998	336			88			284			338
CYCLOHEXYL KETONE Synonym: CYCLOHEXANONE	1915/1993	26	246	14	991	334	59	−30	88	78	CCH 228	281	C8.2		337
CYCLOHEXYL MERCAPTAN	3054	28							88			285			
CYCLOHEXYL TRICHLOROSILANE	1763	60			1000	337		−30	88			285			339
CYCLOHEXYLAMINE	2357	68	249		994	336	60	60 −30	88		CHA 243	285			338
CYCLOHEXYLBENZENE								−30	88						338
CYCLOHEXYLMETHANE	2296	27							88						

CHEMICAL OR MATERIAL NAME	UN/NA Number	Guide Number (DOT)	Firefighter's Hazardous Materials Reference Book	First Aid Manual for Chemical Accidents	Sax's Dangerous Properties of Industrial Materials	Hazardous Chemicals Desk Reference	Rapid Guide to Hazardous Chemicals in the Workplace	Fire Protection Guide on Hazardous Materials (NFPA)	Firefighter's Handbook of Hazardous Materials	Pocket Guide to Chemical Hazards (NIOSH)	Chemical Hazard Response Information System (CHRIS)	Emergency Handling of Hazardous Materials in Surface Transportation (AAR)	Emergency Action Guides (EAG)	Chemical Data Notebook: A User's Manual	Condensed Chemical Dictionary
CYCLON Synonym: HYDROCYANIC ACID	1051	13	442	18	1896	641	114	–59	164				H4.1	H 2	613
CYCLONITE	0072				1001	337	60		89						339
CYCLONITE	0118				1001	337	60		89						339
CYCLOOCTADIENE	2520/1993	27							89			286			
CYCLOOCTADIENE PHOSPHINE	2940	37							89						
CYCLOOCTATETRAENE	2358	27							89						
CYCLOPENTADIENYLMANGANESE TRICARBONYL					1003	338	61		89						
CYCLOPENTANE	1146	27	250		1003	338	61	–31	89		CYP 331	286	C8.3		340
CYCLOPENTANE METHYL Synonym: METHYL CYCLOPENTANE	2298	26	564		2297	786		112–	193		MCP 743	626			764
CYCLOPENTANOL	2244	26						–31	89			286			340
CYCLOPENTANONE	2245/1993	26			1004	338		–31	89			287			340
CYCLOPENTENE	2246/1993	27			1005	339		–31	89		CPE 286	287			341
CYCLOPROPANE	1027	22	251		1008	339		–31	89		CPR 294	287			341
CYCLOPROPYL METHYL ETHER					1008				89						
CYCLOTETRAMETHYLENE OXIDE	3022	26				1103									
CYCLOTETRAMETHYLENE OXIDE Synonym: TETRAHYDROFURAN	2056	26	818		3227		210	167–86	89	210	THF1116	894	T1.0	T 2	1131
CYCLOTETRAMETHYLENE TETRANITRAMINE WET	0226				1010	340						288			
CYCLOTRIMETHYLENE TRINITRAMINE	0072								89			288			342
CYCLOTRIMETHYLENE TRINITRAMINE WET WITH NOT LESS 10%															
CYLOHEXANOL ACETATE	1993	27													
CYLOPENTANE METHYL	2298	26													

CHEMICAL OR MATERIAL NAME	UN/NA Number	Guide Number (DOT)	Firefighter's Hazardous Materials Reference Book	First Aid Manual for Chemical Accidents	Sax's Dangerous Properties of Industrial Materials	Hazardous Chemicals Desk Reference	Rapid Guide to Hazardous Chemicals in the Workplace	Fire Protection Guide on Hazardous Materials (NFPA)	Firefighter's Handbook of Hazardous Materials	Pocket Guide to Chemical Hazards (NIOSH)	Chemical Hazard Response Information System (CHRIS)	Emergency Handling of Hazardous Materials in Surface Transportation (AAR)	Emergency Action Guides (EAG)	Chemical Data Notebook: A User's Manual	Condensed Chemical Dictionary
CYMENE	2046/1993	27			1011				90		CMP 269	288			342
CYMENE-para	2046/1993	27	252		1011	340		-31	90		CMP 269	288			
CYMOGEN	1011	22													
CYMOL Synonym: CYMENE-para	2046	27	252		1011	340		-31	90		CMP 269				
CYTHION INSECTICIDE Synonym: MALATHION	2783/3082	55	526	19	2153	734	129		183	138	MLT 773	591			724
D D TURPENTINE Synonym: TURPENTINE	1299	27	856	26	3458	1162	225	-92	282	222	TPT1144	939			1194
DALAPON	1760	60	253						90		DLP 426				345
DALMATION INSECT POWDER Synonym: PYRETHRINS	9184/3082	31	732	23	2944	1007	191		245		PRR 989	800			979
DAMMONIUM OXALATE Synonym: AMMONIUM OXALATE	2449/3077	54	65		230	64			29		AOX 70	65			67
DBE Synonym: ETHYLENE DIBROMIDE	1605	55	371					85 –	143	110	EDB 518	430	E2.4.1		487
DBP Synonym: DIBUTYL PHTHALATE	9095		265	15	1119	375	66	-34		84	DPA 458				3647
DCE Synonym: VINYLIDENE CHLORIDE	1303	26	878	26	3498	1175	229	181-93	286		VCI1173	952	V1.2		1217
DCEE Synonym: 2,2-DICHLOROETHYL ETHER	1916/2810	55			1145	384		66 -35	103		DEE 385	334			
DCP DICALCIUM PHOSPHATE Synonym: CALCIUM PHOSPHATE					671	234					CAL 207				206
DDD	2761	55	254								DDD 373				347
DDT	2761/2810	55	255		1023	343	62		90	80	DDT 379	289			347

CHEMICAL OR MATERIAL NAME	UN/NA Number	Guide Number (DOT)	Firefighter's Hazardous Materials Reference Book	First Aid Manual for Chemical Accidents	Sax's Dangerous Properties of Industrial Materials	Hazardous Chemicals Desk Reference	Rapid Guide to Hazardous Chemicals in the Workplace	Fire Protection Guide on Hazardous Materials (NFPA)	Firefighter's Handbook of Hazardous Materials	Pocket Guide to Chemical Hazards (NIOSH)	Chemical Hazard Response Information System (CHRIS)	Emergency Handling of Hazardous Materials in Surface Transportation (AAR)	Emergency Action Guides (EAG)	Chemical Data Notebook: A User's Manual	Condensed Chemical Dictionary
DDT-p,p Synonym: **DDT**	2761/2810	55	255		1023	343	62		90	80	DDT 379	289			
DDVP	2783	55		14					90						348
DE FOL ATE Synonym: **SODIUM CHLORATE**	1495	35	766	24	3070	1048		156–	253		SDC1029	834	S2.1	S 2	1052
DE KALIN Synonym: **DECAHYDRONAPHTHALENE**	1147	27	257		1027	345		−31	90		DHN 405	289			349
DEA Synonym: **DIETHANOLAMINE**			286		1184		73	−36			DEA 381		D1.1.2		348
DEAC	1101								90						348
DEAD OIL Synonym: **CREOSOTE COAL TAR**	1993	27	231	14					84		CCT 233		C6.0		322
DEAE Synonym: **DIETHYLETHANOLAMINE**	2686	29							111		DAE 334				348
DEAK	1101								90						
DEC Synonym: **DECAHYDRONAPHTHALENE**	1147	27	257		1027	345		−31	90		DHN 405	289			348
DECABORANE	1868	34	256	15	1026	345	62	61 −31	90	80	DBR 352	289			348
DECACHLOROKETONE Synonym: **KEPONE**	2761/9189	55	493		2077	701					KPE 688	566			670
DECAHYDRONAPHTHALENE	1147	27	257		1027	345		−31	90		DHN 405	289			349
DECAHYDRONAPHTHALENE-trans	1147	27						−31	90						
DECALDEHYDE											DAL 336				
DECALIN Synonym: **DECAHYDRONAPHTHALENE**	1147	27	257		1027	345		−31	90		DHN 405	289			349

CHEMICAL OR MATERIAL NAME	UN/NA Number	Guide Number (DOT)	Firefighter's Hazardous Materials Reference Book	First Aid Manual for Chemical Accidents	Sax's Dangerous Properties of Industrial Materials	Hazardous Chemicals Desk Reference	Rapid Guide to Hazardous Chemicals in the Workplace	Fire Protection Guide on Hazardous Materials (NFPA)	Firefighter's Handbook of Hazardous Materials	Pocket Guide to Chemical Hazards (NIOSH)	Chemical Hazard Response Information System (CHRIS)	Emergency Handling of Hazardous Materials in Surface Transportation (AAR)	Emergency Action Guides (EAG)	Chemical Data Notebook: A User's Manual	Condensed Chemical Dictionary
DECANAL Synonym: DECALDEHYDE											DAL 336				349
DECANE	2247	27		15	1030	346		-31	90						349
DECANOL				15				-31	90						
DECANOYL PEROXIDE	2120/3114	52							91			290			349
DECENE-alpha Synonym: 1-DECENE								-31	91		DCE 359				350
DECHLORANE Synonym: MIREX	1615		589		2431	828			205		MRX 798				
DECYL ACRYLATE					1032			-31			DAR 340				
DECYL ALCOHOL Synonym: DECYL ALCOHOL-n	1987	26	258		1032	347			91		DAN 338	291			350
DECYL ALCOHOL-n Synonym: DECYL ALCOHOL	1987	26	258		1032	347			91		DAN 338	291			350
DECYL ALDEHYDE	1993	27										291			350
DECYL ALDEHYDE-n Synonym: DECALDEHYDE											DAL 336				350
DECYL HYDRIDE	2247	27							91						350
DECYL NITRATE								-31							
DECYLAMINE								-31	91						350
DECYLBENZENE Synonym: DECYLBENZENE-n								-31			DBZ 356				
DECYLBENZENE-n								-31			DBZ 356				
DECYLBENZENESULFONIC ACID Synonym: ALKYBENZENESULFONIC ACID											ABS 11				
DECYLMERCAPTAN-tert								-31							

CHEMICAL OR MATERIAL NAME	UN/NA Number	Guide Number (DOT)	Firefighter's Hazardous Materials Reference Book	First Aid Manual for Chemical Accidents	Sax's Dangerous Properties of Industrial Materials	Hazardous Chemicals Desk Reference	Rapid Guide to Hazardous Chemicals in the Workplace	Fire Protection Guide on Hazardous Materials (NFPA)	Firefighter's Handbook of Hazardous Materials	Pocket Guide to Chemical Hazards (NIOSH)	Chemical Hazard Response Information System (CHRIS)	Emergency Handling of Hazardous Materials in Surface Transportation (AAR)	Emergency Action Guides (EAG)	Chemical Data Notebook: A User's Manual	Condensed Chemical Dictionary
DEEP LEMON YELLOW **Synonym: STRONTIUM CHROMATE**	9149/3077	45	793		3140	1072	201		257		SCM1023	860			1095
DEG **Synonym: DIETHYLENE GLYCOL**			292	15	1219			−37	110		DEG 386				351
DEHPA **Synonym: DI(2-ETHYLHEXYL) PHOSPHORIC ACID**	1902	60	259						92		DEP 391				
DEHYDRITE **Synonym: MAGNESIUM PERCHLORATE**	1475	35	525						91		MPC 783	590			352
DELICIA **Synonym: PHOSPHINE**	2199	18	683	22	2783	945	180	136–80	231	180		752	P3.1		907
DEMETON				15	1039				91	80	DTN 490				353
DEMETON O					1039	349	62								
DEMETON S					1040	349	62								
DEN **Synonym: DIETHYLAMINE**	1154	68	290	15	1188	400	73	68 −36	110	90	DEN 390	349	D1.1.2.1		354
DENATURED ALCOHOL **Synonym: ETHYL ALCOHOL**	1170	26	344	16	1572	525	90	−32	91		EAL 504	408	E2.2	E 2	354
DETERGENT ALKYLATE **Synonym: DODECYLBENZENE**											DDB 371				441
DETIA **Synonym: PHOSPHINE**	2199	18	683	22	2783	945	180	136–80	231	180		752	P3.1		907
DEUMABIETIS OIL OF FIVE SIBERIAN **Synonym: PINE OIL**	1272	26	695		2810	959		−80	235			764	P5.0		918
DEUTERIUM	1957/1954	22				353		−32	91			294			357
DEXTROSE SOLUTION											DTS 492				358

CHEMICAL OR MATERIAL NAME	UN/NA Number	Guide Number (DOT)	Firefighter's Hazardous Materials Reference Book	First Aid Manual for Chemical Accidents	Sax's Dangerous Properties of Industrial Materials	Hazardous Chemicals Desk Reference	Rapid Guide to Hazardous Chemicals in the Workplace	Fire Protection Guide on Hazardous Materials (NFPA)	Firefighter's Handbook of Hazardous Materials	Pocket Guide to Chemical Hazards (NIOSH)	Chemical Hazard Response Information System (CHRIS)	Emergency Handling of Hazardous Materials in Surface Transportation (AAR)	Emergency Action Guides (EAG)	Chemical Data Notebook: A User's Manual	Condensed Chemical Dictionary
DI SYSTON Synonym: DISULFOTON	2783	55	325		1484	494	85		129		DIS 421	387			436
DI(1-HYDROXYCYCLOHEXYL) PEROXIDE	2 48/3106	48							91			303			
DI(2-CHLOROETHYL) ETHER Synonym: 2,2-DICHLOROETHYL ETHER	1916/2810	55			1145	384		66 –35	103		DEE 385	334			
DI(2-ETHYLHEXYL) ADIPATE Synonym: DIOCTYL ADIPATE								–44			DOA 453				424
DI(2-ETHYLHEXYL) PEROXYDICARBONATE	2360/3117	52			1226	411			91			304			
DI(2-ETHYLHEXYL) PEROXYDICARBONATE	2122/3113	52			1226	411			92			304			
DI(2-ETHYLHEXYL) PEROXYDICARBONATE	2123/3117	52			1226	411			91			304			
DI(2-ETHYLHEXYL) PHOSPHATE Synonym: DI(2-ETHYLHEXYL) PHOSPHORIC ACID	1902	60	259						92		DEP 391				
DI(2-ETHYLHEXYL) PHOSPHORIC ACID	1902	60	259						92		DEP 391				
DI(2-ETHYLHEXYL) PHTHALATE Synonym: DIOCTYL PHTHALATE	3082/9188				1445			–45			DOP 455	315			425
DI(2-ETHYLHEXYL) SULFOSUCCINATE SODIUM SALT Synonym: DIOCTYL SODIUM SULFOSUCCINATE											DSS 481				425
DI(2-HYDROXYETHYL) AMINE Synonym: DIETHANOLAMINE			286		1184	399	73	–36			DEA 381		D1.1.2		388
DI(2-METHYLBENZOYL) PEROXIDE	2593/3112	52							92			305			
DI(3,5,5-TRIMETHYL-1,2-DIOXOLANYL-3) PEROXIDE	2597/3116	52							92			309			
DI(3,5,5-TRIMETHYLHEXANOYL) PEROXIDE	2128/3115	52							92			310			
DI(4-CHLOROBENZOYL) PEROXIDE	2115	48							93			310			
DI(4-CHLOROBENZOYL) PEROXIDE Synonym: DI(p-CHLOROBENZOYL) PEROXIDE	0149										DZP 498				
DI(4-CHLOROBENZOYL) PEROXIDE	2113/3102	48							93			315			

HAZARDOUS MATERIALS REFERENCE BOOKS INDEX

CHEMICAL OR MATERIAL NAME	UN/NA Number	Guide Number (DOT)	Firefighter's Hazardous Materials Reference Book	First Aid Manual for Chemical Accidents	Sax's Dangerous Properties of Industrial Materials	Hazardous Chemicals Desk Reference	Rapid Guide to Hazardous Chemicals in the Workplace	Fire Protection Guide on Hazardous Materials (NFPA)	Firefighter's Handbook of Hazardous Materials	Pocket Guide to Chemical Hazards (NIOSH)	Chemical Hazard Response Information System (CHRIS)	Emergency Handling of Hazardous Materials in Surface Transportation (AAR)	Emergency Action Guides (EAG)	Chemical Data Notebook: A User's Manual	Condensed Chemical Dictionary
DI(4-CHLOROBENZOYL) PEROXIDE	2114	48							93						
DI(4-tert-BUTYLCYCLOHEXYL) PEROXYDICARBONATE	2894/3119	52							92			311			
DI(4-tert-BUTYLCYCLOHEXYL) PEROXYDICARBONATE	2154/3114	52							92			310			
DI(ETHYLENE OXIDE) Synonym: 1,4-DIOXANE	1165	26			1449	482	83	78 –	126		DOX 457	380			426
DI(p-CHLOROBENZOYL) PEROXIDE	1531										DZP 498				
DI(p-CHLOROBENZOYL) PEROXIDE	0149										DZP 498				
DI(p-CHLOROPHENYL) TRICHLOROMETHYLCARBINOL Synonym: 4,4-DICHLORO-alpha-TRICHLOROMETHYLBENZHY-DROL											DTM 489				
DI-2,4-DICHLOROBENZOYL PEROXIDE	2139	48							93						
DI-2,4-DICHLOROBENZOYL PEROXIDE	2137	48							93						
DI-2,4-DICHLOROBENZOYL PEROXIDE	2138	48							93						
DI-beta-HYDROXYETHOXYETHANE Synonym: TRIETHYLENE GLYCOL			848	25	3370	1142		-89			TEG1101				1175
DI-n-AMYL PHTHALATE											DAP 339				364
DI-n-AMYLAMINE Synonym: DIAMYL AMINE	2841	68			1077	361			93			318			364
DI-n-BUTYL ETHER	1149	26									DBE 344				373
DI-n-BUTYL KETONE	1224	26									DBK 346				
DI-n-BUTYL PEROXYDICARBONATE	2169/3115	52							93			311			
DI-n-BUTYLAMINE Synonym: DIBUTYLAMINE-n	2248/1993	68			1108	370			99		DBA 342	312	D3.2		371
DI-n-NONANOYL PEROXIDE	2130/3116	52							93			312			
DI-n-OCTANOYL PEROXIDE	2129	52							93						

CHEMICAL OR MATERIAL NAME	UN/NA Number	Guide Number (DOT)	Firefighter's Hazardous Materials Reference Book	First Aid Manual for Chemical Accidents	Sax's Dangerous Properties of Industrial Materials	Hazardous Chemicals Desk Reference	Rapid Guide to Hazardous Chemicals in the Workplace	Fire Protection Guide on Hazardous Materials (NFPA)	Firefighter's Handbook of Hazardous Materials	Pocket Guide to Chemical Hazards (NIOSH)	Chemical Hazard Response Information System (CHRIS)	Emergency Handling of Hazardous Materials in Surface Transportation (AAR)	Emergency Action Guides (EAG)	Chemical Data Notebook: A User's Manual	Condensed Chemical Dictionary
DI-n-OCTYL PHTHALATE	9 88/3082				1445							313			
DI-n-PROPYL PEROXYDICARBONATE	2 76/3113	52				489			93			313			
DI-n-PROPYLAMINE Synonym: DIPROPYLAMINE	2383/1993	68	322		1470	488			93		DNA 442	314	D4.0		432
DI-n-PROPYLAMINE	1993/2383	27									DNA 442	314	D4.0		432
DI-sec-BUTYL PEROXYDICARBONATE	2151/3115	52							94			314			
DI-sec-BUTYL PEROXYDICARBONATE	2150/3113	52										314			
DI-sec-BUTYL-p-PHENYLENEDIAMINE-n,n					1118	374		−33	94						
DI-sec-BUTYLAMINE	2248	68			1108			−33	94						
DI-tert-BUTYL PEROXIDE	2102/3107	48						−34	94			315			
DI-tert-BUTYL-p-CRESOL								−33	94						
DI-tert-BUTYLPEROXYPHTHALATE	2107/3107	48							94			299			
DI-tert-BUTYLPEROXYPHTHALATE	2106/3105	48							94			298			
DI-tert-BYTYLPEROXYPHTHALATE	2108/3106	48							94			299			
DIACETIC ETHER Synonym: ETHYL ACETOACETATE	1993	27	341					−47	94		EAA 499	406			478
DIACETONE Synonym: DIACETONE ALCOHOL	1148	26	260	15	1059	354	62	−32	94	82	DAA 332	316			359
DIACETONE ALCOHOL	1148	26	260	15	1059	354	62	−32	94	82	DAA 332	316			359
DIACETONE ALCOHOL PEROXIDE	2163/3115	52							94			316			
DIACETYL	2346	26										317			359
DIACETYL PEROXIDE	2084/3115	49							94			317			
DIACETYL PEROXIDE SOLUTION Synonym: ACETYL PEROXIDE	2084	49	26		51	23	4	−12	94		APS 80				13

HAZARDOUS MATERIALS REFERENCE BOOKS INDEX

CHEMICAL OR MATERIAL NAME	UN/NA Number	Guide Number (DOT)	Firefighter's Hazardous Materials Reference Book	First Aid Manual for Chemical Accidents	Sax's Dangerous Properties of Industrial Materials	Hazardous Chemicals Desk Reference	Rapid Guide to Hazardous Chemicals in the Workplace	Fire Protection Guide on Hazardous Materials (NFPA)	Firefighter's Handbook of Hazardous Materials	Pocket Guide to Chemical Hazards (NIOSH)	Chemical Hazard Response Information System (CHRIS)	Emergency Handling of Hazardous Materials in Surface Transportation (AAR)	Emergency Action Guides (EAG)	Chemical Data Notebook: A User's Manual	Condensed Chemical Dictionary
DIACETYLMETHANE Synonym: ACETYLACETONE	2310	26	28		37	17			94		ATA 97				360
DIALLYL MALEATE					1063	356			95						361
DIALLYL PHTHALATE					1064			−32	95						361
DIALLYLAMINE	2359/1993	29			1062	355			95			317			360
DIALLYLCYANAMIDE					1062				95						360
DIALLYLETHER	2360/1993	28			1063	356						318			
DIAMIDE Synonym: HYDRAZINE ANHYDROUS	2029	28	439	18	1885	635	111	95 −59	95	124		521	H3.0	H 1	
DIAMINE Synonym: HYDRAZINE ANHYDROUS	2029	28	439	18	1885	635	111	95 −59	95	124		521	H3.0	H 1	361
DIAMINE HYDRATE Synonym: HYDRAZINE SOLUTION	2030	59		18	1885	635	111		163	124	HDZ 633		H3.0	H 1	610
DIAMINOBENZENE-para	1993	27													
DIAMINODIPHENYL METHANE	2651	53							95						
DIAMINODIPHENYL-para	1885	53							95						362
DIAMMONIUM CHROMATE Synonym: AMMONIUM CHROMATE	9086/3077		57		224	61			28		ACH 19	58			65
DIAMMONIUM CITRATE Synonym: AMMONIUM CITRATE	9087/3077		58		225	61			28		ACI 20	58			65
DIAMMONIUM HYDROGEN PHOSPHATE Synonym: AMMONIUM PHOSPHATE					232	66					APP 78				363
DIAMMONIUM ORTHOPHOSPHATE Synonym: AMMONIUM PHOSPHATE					232	66					APP 78				68
DIAMMONIUM SULFIDE Synonym: AMMONIUM SULFIDE SOLUTION	2683/1993	28	70	12					30		ASF 90	68	A8.2		

CHEMICAL OR MATERIAL NAME	UN/NA Number	Guide Number (DOT)	Firefighter's Hazardous Materials Reference Book	First Aid Manual for Chemical Accidents	Sax's Dangerous Properties of Industrial Materials	Hazardous Chemicals Desk Reference	Rapid Guide to Hazardous Chemicals in the Workplace	Fire Protection Guide on Hazardous Materials (NFPA)	Firefighter's Handbook of Hazardous Materials	Pocket Guide to Chemical Hazards (NIOSH)	Chemical Hazard Response Information System (CHRIS)	Emergency Handling of Hazardous Materials in Surface Transportation (AAR)	Emergency Action Guides (EAG)	Chemical Data Notebook: A User's Manual	Condensed Chemical Dictionary
DIAMYL PHENOL	1993	27			1078				96						364
DIAMYL PHTHALATE Synonym: DI-n-AMYL PHTHALATE											DAP 339				364
DIAMYL SULFIDE	1993	27						-33	96			318			364
DIAMYLAMINE	2841	68			1077	361		61 -32	96			318			364
DIAMYLENE	1993	27						-32	96			319			364
DIANISIDINE DIISOCYANATE					1079	362	63								365
DIANISIDINE-ortho					1079	361	63	-33							
DIANTIMONY TRIOXIDE Synonym: ANTIMONY TRIOXIDE	1549/3077	60	94	12							ATX 109	87			90
DIAREX HF 77 Synonym: STYRENE MONOMER	2055	27	795		3145	1074	202		257		STY1064	862	S3.0		1097
DIARSENIC TRIOXIDE Synonym: ARSENIC TRIOXIDE	1561	53	100		299	92	16	33 –	35		ATO 105	93	A11.1	A14	98
DIATOMACEOUS EARTH					1080	362	64								365
DIAZINON	2783/2810	55	261	15	1083	363	64		96		DZN 497	319			366
DIAZOACETIC ESTER					1084	363			96						
DIAZOMETHANE					1085	363	64		97	82					367
DIBENZ (a,h) ANTHRACENE					1089	365									
DIBENZO [b,e] PYRININE Synonym: ACRIDINE	2713	32	30		59	25	5		19		ACD 16				
DIBENZOYL PEROXIDE	2089	49							97						
DIBENZOYL PEROXIDE	3074	49							97						
DIBENZOYL PEROXIDE	2088	49							97						

HAZARDOUS MATERIALS REFERENCE BOOKS INDEX

CHEMICAL OR MATERIAL NAME	UN/NA Number	Guide Number (DOT)	Firefighter's Hazardous Materials Reference Book	First Aid Manual for Chemical Accidents	Sax's Dangerous Properties of Industrial Materials	Hazardous Chemicals Desk Reference	Rapid Guide to Hazardous Chemicals in the Workplace	Fire Protection Guide on Hazardous Materials (NFPA)	Firefighter's Handbook of Hazardous Materials	Pocket Guide to Chemical Hazards (NIOSH)	Chemical Hazard Response Information System (CHRIS)	Emergency Handling of Hazardous Materials in Surface Transportation (AAR)	Emergency Action Guides (EAG)	Chemical Data Notebook: A User's Manual	Condensed Chemical Dictionary
DIBENZOYL PEROXIDE Synonym: BENZOYL PEROXIDE	2085/2087	49	262		392	130	23		97	44	DPO 469	320	B2.0	B 2	368
DIBENZOYL PEROXIDE	2090	49							97						
DIBENZOYL PEROXIDE	2087/2085	49							97			320			
DIBENZOYL PEROXIDE	2085/2087	49							97		DPO 469	320			
DIBENZYL ETHER								−33	97		DBN 349				368
DIBENZYL PEROXYDICARBONATE	2149/3112	52							97			322			
DIBENZYLDICHLOROSILANE	2434	60							97			323			
DIBK Synonym: DIISOBUTYL KETONE	1157	26	296		1300	433	75	−40		92	DIK 414	355			405
DIBORANE	1911	18	263	15	1099	367	64	62 −33	97	82		323	D1.0		368
DIBORANE 6 Synonym: DIBORANE	1911	18	263	15	1099	367	64	62 −33	97	82		323	D1.0		368
DIBORANE MIXTURE	1911	18	263	15	1099	367	64		97			323			
DIBORON HEXAHYDRIDE Synonym: DIBORANE	1911	18	263	15	1099	367	64	62 −33	97	82		323	D1.0		368
DIBORON TETRACHLORIDE					1099										
DIBROM Synonym: NALED	2783/3082	55	600		2461	835	152		207		NLD 838	673			804
DIBROMOACETYLENE					1100	368			98						369
DIBROMOBENZENE	2711	26							98			324			
DIBROMOBUTANONE	2648	55							98						
DIBROMOCHLOROPROPANE	2872	58			1101	369			98			324			
DIBROMODIFLUOROMETHANE	1941	58							98			324			369
DIBROMOETHANE	1605	55							98						

CHEMICAL OR MATERIAL NAME	UN/NA Number	Guide Number (DOT)	Firefighter's Hazardous Materials Reference Book	First Aid Manual for Chemical Accidents	Sax's Dangerous Properties of Industrial Materials	Hazardous Chemicals Desk Reference	Rapid Guide to Hazardous Chemicals in the Workplace	Fire Protection Guide on Hazardous Materials (NFPA)	Firefighter's Handbook of Hazardous Materials	Pocket Guide to Chemical Hazards (NIOSH)	Chemical Hazard Response Information System (CHRIS)	Emergency Handling of Hazardous Materials in Surface Transportation (AAR)	Emergency Action Guides (EAG)	Chemical Data Notebook: A User's Manual	Condensed Chemical Dictionary
DIBROMOETHANE-sym Synonym: ETHYLENE DIBROMIDE	1605	55	371					85 –	143	110	EDB 518	430	E2.4.1		487
DIBROMOMETHANE	2664	74			1105				98		DBH 345	324			370
DIBUTOXYMETHANE								–33							371
DIBUTYL ETHANOLAMINE-n,n	2873	55							98						
DIBUTYL ETHER Synonym: DI-n-BUTYL ETHER	1149	26	264					–33	99		DBE 344				373
DIBUTYL ETHER-n Synonym: DI-n-BUTYL ETHER	1149	26	264					–33	99		DBE 344				373
DIBUTYL MALEATE					1116	373		–34							373
DIBUTYL MERCURY					1116	373			99						
DIBUTYL OXALATE								–34	99						373
DIBUTYL OXIDE Synonym: DI-n-BUTYL ETHER	1149	26							99		DBE 344				373
DIBUTYL PEROXIDE-tert	2102	48							99						373
DIBUTYL PHOSPHATE					1119	374	66		99	84					374
DIBUTYL PHOSPHITE	1760	60			1119			–34	99			325			374
DIBUTYL PHTHALATE	9095		265	15	1119	375	66	–34		84	DPA 458				374
DIBUTYL SEBACATE					1120	375		–34							374
DIBUTYLAMINE Synonym: DI-n-BUTYLAMINE	2248/1993	68		15	1108	370		62 –33	99		DBA 342	312	D3.2		371
DIBUTYLAMINE-n Synonym: DI-n-BUTYLAMINE	2248/1993	68		15	1108	370		62 –33	99		DBA 342	312	D3.2		371
DIBUTYLAMINOETHANOL Synonym: 2-N-DIBUTYLAMINO ETHANOL	2873	55			1109	371		–33	99			325			

HAZARDOUS MATERIALS REFERENCE BOOKS INDEX

CHEMICAL OR MATERIAL NAME	UN/NA Number	Guide Number (DOT)	Firefighter's Hazardous Materials Reference Book	First Aid Manual for Chemical Accidents	Sax's Dangerous Properties of Industrial Materials	Hazardous Chemicals Desk Reference	Rapid Guide to Hazardous Chemicals in the Workplace	Fire Protection Guide on Hazardous Materials (NFPA)	Firefighter's Handbook of Hazardous Materials	Pocket Guide to Chemical Hazards (NIOSH)	Chemical Hazard Response Information System (CHRIS)	Emergency Handling of Hazardous Materials in Surface Transportation (AAR)	Emergency Action Guides (EAG)	Chemical Data Notebook: A User's Manual	Condensed Chemical Dictionary
DIBUTYLANILINE-n,n								-33	99						372
DIBUTYLMETHYLAMINE-n,n					1117				99						
DIBUTYLPHENOL											DBT 354				
DICAMBA	2769/3082	55	266						99		DIC 409	326			376
DICAMBA LIQUID	2769/3082	55	267						99		DIC 409	326			
DICARBOMETHOXYZINC	9153/3077		886		3536	1191	234		288		ZNA1211	964			1242
Synonym: ZINC ACETATE															
DICARBON HEXACHLORIDE	9037	53							99						
DICETYL PEROXYDICARBONATE	2895/3119	52							99			326			
DICETYL PEROXYDICARBONATE	2164/3116	52							99			326			
DICHLOBENIL SOLID	2769/3077	55	268						100		DIB 408	327			376
DICHLOBENIL LIQUID	2769/3077	55	269						100		DIB 408	327			
DICHLONE LIQUID	2761/3077	55	271						100		DCL 363	327			
DICHLONE SOLID	2761/3077	55	270						100		DCL 363	327			376
DICHLORICIDE	1592	58	273	15	1128	377	67	-34	101	84	DBP 351	330			378
Synonym: DICHLOROBENZENE (-para)															
DICHLOROACETIC ACID	1764	60							101			328			377
DICHLOROACETONITRILE					1124				101						
DICHLOROACETYL CHLORIDE	1765	60			1125	376		63 -34	101			328			377
DICHLOROACETYLENE					1125	376	66		101						377
DICHLOROANILINE	2811/1590	53										328			
DICHLOROANILINE	1590/2811	55						63 –	101			328			
DICHLOROBENZENE-meta	1592	58		15	1127	377					DBM 348				377
DICHLOROBENZENE-ortho	1591	58	272	15	1127	376	66	64 -34	101	84	DBO 350	329	D1.0.1		377

CHEMICAL OR MATERIAL NAME	UN/NA Number	Guide Number (DOT)	Firefighter's Hazardous Materials Reference Book	First Aid Manual for Chemical Accidents	Sax's Dangerous Properties of Industrial Materials	Hazardous Chemicals Desk Reference	Rapid Guide to Hazardous Chemicals in the Workplace	Fire Protection Guide on Hazardous Materials (NFPA)	Firefighter's Handbook of Hazardous Materials	Pocket Guide to Chemical Hazards (NIOSH)	Chemical Hazard Response Information System (CHRIS)	Emergency Handling of Hazardous Materials in Surface Transportation (AAR)	Emergency Action Guides (EAG)	Chemical Data Notebook: A User's Manual	Condensed Chemical Dictionary
DICHLOROBENZENE-para	1592	58	273	15	1128	377	67	-34	101	84	DBP 351	330			378
DICHLOROBENZOL-ortho Synonym: DICHLOROBENZENE-ortho	1591	58					66		101		DBO 350		D1.0.1		377
DICHLOROBENZOYL PEROXIDE-p,p Synonym: DI-(p-CHLOROBENZOYL) PEROXIDE	0149										DZP 498				
DICHLOROBUTENE	2924/2920	29	274						102			330			
DICHLOROBUTENE	1760	60									DCB 358				
DICHLORODIETHYL ETHER Synonym: 2,2-DICHLOROETHYL ETHER	1916/2810	55			1145	384		66 -35	103		DEE 385	334			380
DICHLORODIETHYL SILANE	1767	29			1136	379			102						
DICHLORODIFLUOROETHYLENE	9018/2810	74			1136	380			102			330			
DICHLORODIFLUOROMETHANE	1028	12	275	15	1137	380	68		102	86	DCF 360	331	D1.1		379
DICHLORODIFLUOROMETHANE & CHLORODIFLUOROMETH-ANE MIX	1078	12			1137	380			102						
DICHLORODIFLUOROMETHANE & DICHLOROTETRAFLUO-ROETHANE	1078	12							102			332			
DICHLORODIFLUOROMETHANE & DIFLUOROETHANE AZEO-TROPIC	2602/1078	12							102			332			
DICHLORODIFLUOROMETHANE & MONOCHLORODIFLUORO-METHANE	1078	12										332			
DICHLORODIFLUOROMETHANE & MONOCHLOROTRIFLUO-ROMETHANE	1078	12										333			
DICHLORODIFLUOROMETHANE & TRICHLOROFLUOROMETH-ANE	1078	12			1137				102			332			
DICHLORODIFLUOROMETHANE & TRICHLOROTRIFLUORO-ETHANE	1078	12			1138	381			102			333			

CHEMICAL OR MATERIAL NAME	UN/NA Number	Guide Number (DOT)	Firefighter's Hazardous Materials Reference Book	First Aid Manual for Chemical Accidents	Sax's Dangerous Properties of Industrial Materials	Hazardous Chemicals Desk Reference	Rapid Guide to Hazardous Chemicals in the Workplace	Fire Protection Guide on Hazardous Materials (NFPA)	Firefighter's Handbook of Hazardous Materials	Pocket Guide to Chemical Hazards (NIOSH)	Chemical Hazard Response Information System (CHRIS)	Emergency Handling of Hazardous Materials in Surface Transportation (AAR)	Emergency Action Guides (EAG)	Chemical Data Notebook: A User's Manual	Condensed Chemical Dictionary
DICHLORODIFLUOROMETHANE AND DIFLUOROETHANE MIXTURE	1078	12													
DICHLORODIMETHYL ETHER SYMMETRICAL	2249/2810	55							102			333			
DICHLORODIMETHYLSILANE Synonym: DIMETHYLDICHLOROSILANE	1162	29	305		1140	382		-42	102		DMD 429	370	D3.1		380
DICHLORODIMETHYLSILICON Synonym: DIMETHYLDICHLOROSILANE	1162	29	305					-42	122		DMD 429	370	D3.1		415
DICHLORODIPHENYLDICHLORO ETHANE Synonym: DDD	2761	55	254								DDD 373				380
DICHLORODIPHENYLSILANE Synonym: DIPHENYLDICHLOROSILANE	1769	29	320		1140	382		-45	103		DPD 461	382			429
DICHLORODIPHENYLSILICANE Synonym: DIPHENYLDICHLOROSILANE	1769	29	320					-45	127		DPD 461	382			429
DICHLORODIPHENYLTRICHLOROETHANE Synonym: DDT	2761/2810	55	255				62		103	80	DDT 379	289			380
DICHLOROETHANE	2362	27		15	1141				103						
DICHLOROETHANOYL CHLORIDE	1765	60							103						
DICHLOROETHER Synonym: 2,2-DICHLOROETHYL ETHER	1916/2810	55			1145			66 -35	103		DEE 385	334			380
DICHLOROETHYL ETHER	1916/2810	55	276		1145	384	69		103	88	DEE 385	334			380
DICHLOROETHYL OXIDE	1916	55							103						380
DICHLOROETHYLARSINE	1892	55			1143	383									380
DICHLOROETHYLENE	1150	29		15	1144	383		65 –				334			380
DICHLOROETHYLENE-cis	1150	29		15	1144	384	69		103						
DICHLOROETHYLENE-sym Synonym: 1,2-DICHLOROETHYLENE	1150/3082	29	3	15	1144	384	69	-35	103	88	DEL 388	301			380

CHEMICAL OR MATERIAL NAME	UN/NA Number	Guide Number (DOT)	Firefighter's Hazardous Materials Reference Book	First Aid Manual for Chemical Accidents	Sax's Dangerous Properties of Industrial Materials	Hazardous Chemicals Desk Reference	Rapid Guide to Hazardous Chemicals in the Workplace	Fire Protection Guide on Hazardous Materials (NFPA)	Firefighter's Handbook of Hazardous Materials	Pocket Guide to Chemical Hazards (NIOSH)	Chemical Hazard Response Information System (CHRIS)	Emergency Handling of Hazardous Materials in Surface Transportation (AAR)	Emergency Action Guides (EAG)	Chemical Data Notebook: A User's Manual	Condensed Chemical Dictionary
DICHLOROETHYLENE-trans	1150	29		15					103						
DICHLOROETHYLENE-unsym Synonym: VINYLIDENE CHLORIDE	1303	26	878	26	3498	1175	229	181–93	286		VCI1173	952	V1.2		1217
DICHLOROFLUOROMETHANE	1029	12		15	1147	386	70		103			334			
DICHLOROHYDRIN-alpha	2750	55							103						381
DICHLOROISOCYANURIC ACID	2465	45							103			335			381
DICHLOROISOCYANURIC ACID DRY	2465	45							103			335			
DICHLOROISOCYANURIC ACID SALTS	2465	45							103			335			
DICHLOROISOPROPYL ALCOHOL	2750	55							103						381
DICHLOROISOPROPYL ETHER	2490	59						−35	103			335			381
DICHLOROMETHANE	1593	74	277						104		DCM 364	335	D1.1.15		381
DICHLOROMETHYL ARSINE	1556	55			1150	386			104						
DICHLOROMETHYL ETHER Synonym: DICHLORODIMETHYL ETHER	2249/2810	55						66 –	104			333			381
DICHLOROMONOFLUOROMETHANE	1029	12							104	88	DFM 399				381
DICHLORONITROETHANE	2650	57			1154	388			104						
DICHLOROPENTANE	1152	27			1156	389		−35	104			336			382
DICHLOROPHENOXYACETIC ACID	2765	55	278						104						
DICHLOROPHENYL PHOSPHINE Synonym: BENZENE PHOSPHORUS DICHLORIDE	2798	39	117		1162	392			104		BPD 156				129
DICHLOROPHENYL TRICHLOROSILANE	1766	60			1162	392			104			337			383
DICHLOROPHENYLISOCYANATE	2250/2810	53							104			337			
DICHLOROPHENYLPHOSPHINE Synonym: BENZENE PHOSPHORUS DICHLORIDE	2798	39	117		1162	392			104		BPD 156				129

HAZARDOUS MATERIALS REFERENCE BOOKS INDEX

CHEMICAL OR MATERIAL NAME	UN/NA Number	Guide Number (DOT)	Firefighter's Hazardous Materials Reference Book	First Aid Manual for Chemical Accidents	Sax's Dangerous Properties of Industrial Materials	Hazardous Chemicals Desk Reference	Rapid Guide to Hazardous Chemicals in the Workplace	Fire Protection Guide on Hazardous Materials (NFPA)	Firefighter's Handbook of Hazardous Materials	Pocket Guide to Chemical Hazards (NIOSH)	Chemical Hazard Response Information System (CHRIS)	Emergency Handling of Hazardous Materials in Surface Transportation (AAR)	Emergency Action Guides (EAG)	Chemical Data Notebook: A User's Manual	Condensed Chemical Dictionary
DICHLOROPROPANE Synonym: 1,2-DICHLOROPROPANE	1279	27	4	15	2919	1001	183	150–82	105	188	DPP 470	798	P8.0.3		383
DICHLOROPROPANOL	2750	55							105						
DICHLOROPROPANONE	2649	55							105						
DICHLOROPROPENE Synonym: 1,3-DICHLOROPROPENE	2047	29	279		1164	393	70	67 –35	105		DPU 472	337			383
DICHLOROPROPENE AND PROPYLENE DICHLORIDE MIXTURE	2047	29							105			338			
DICHLOROPROPIONIC ACID	1760	60							105						
DICHLOROSILANE	2189/1953	19			1167	395		67 –35	105			339		D 1	
DICHLOROSTYRENE					1168				105						
DICHLOROSTYRENE-alpha-beta								–35	105						
DICHLOROTETRAFLUOROETHANE	1958/1078	12	280	15	1168	395	71		105	90	DTE 486	339			384
DICHLOROTRIAZINETRIONE	2465	45							105						
DICHLOROVOS Synonym: DICHLORVOS	2783	55	281		1172	396	71		105		DCV 369	340			384
DICHLORVOS LIQUID	2783	55	281		1172	396	71			90	DCV 369	340			384
DICHLORVOS SOLID	2783	55	282		1172	396	71				DCV 369	340			
DICHROMIUM SULFATE Synonym: CHROMIC SULFATE	9100/3082	31	217						80		CHS 249	239			280
DICHROMIUM TRISULFATE Synonym: CHROMIC SULFATE	9100/3082	31	217						80		CHS 249	239			280
DICOFOL Synonym: 4,4-DICHLORO-alpha-TRICHLOROMETHYLBENZHYDROL											DTM 489				384
DICOPHANE	2761	55							105						

CHEMICAL OR MATERIAL NAME	UN/NA Number	Guide Number (DOT)	Firefighter's Hazardous Materials Reference Book	First Aid Manual for Chemical Accidents	Sax's Dangerous Properties of Industrial Materials	Hazardous Chemicals Desk Reference	Rapid Guide to Hazardous Chemicals in the Workplace	Fire Protection Guide on Hazardous Materials (NFPA)	Firefighter's Handbook of Hazardous Materials	Pocket Guide to Chemical Hazards (NIOSH)	Chemical Hazard Response Information System (CHRIS)	Emergency Handling of Hazardous Materials in Surface Transportation (AAR)	Emergency Action Guides (EAG)	Chemical Data Notebook: A User's Manual	Condensed Chemical Dictionary
DICUMYL PEROXIDE	2121/3110	48							106			341			385
DICY Synonym: DICYCLOPENTADIENE	2048/1993	26	283		1178	398	72	-36	106		DPT 471	343			385
DICYAN Synonym: CYANOGEN	1026	18	241		979	331	57	-29	106		CYG 330	277	C8.0		385
DICYANOGEN Synonym: CYANOGEN	1026	18	241		979	331	57	-29	106		CYG 330	277	C8.0		333
DICYCLOHEPTADIENE	2251	26							106						
DICYCLOHEXANONE DIPEROXIDE Synonym: CYCLOHEXANONE PEROXIDE	2896	51							88		CHP 248	283			337
DICYCLOHEXYL	2565	60							106						385
DICYCLOHEXYL PEROXYDICARBONATE	2153/3114	52							106			342			
DICYCLOHEXYL PEROXYDICARBONATE	2152/3112	52							106			341			
DICYCLOHEXYLAMINE Synonym: DICYCLOHEXYLAMINE-N,N	2565/1760	60			1176	397		-35	106			342			385
DICYCLOHEXYLAMINE NITRITE	2687	53			1176				106						
DICYCLOPENTADIENE	2048/1993	26	283		1178	398	72	-36	106		DPT 471	343			385
DIDECANOLY PEROXIDE	2120	52							106						
DIELDRIN LIQUID	2761	55	284		1181	398	72		106	90	DED 384	344			
DIELDRIN SOLID	2761	55	285		1181	398	72		106		DED 384	344			387
DIESEL FUEL OIL NO. 1	1993	27						-36	107						
DIESEL FUEL OIL NO. 2	1993	27						-36	107						
DIESEL FUEL OIL NO. 4	1993	27						-36	107						
DIESEL IGNITION IMPROVER Synonym: AMYL NITRATE (-n)	1112/1993	26	82		246	71		29 -15	31		ANT 67	76			387

HAZARDOUS MATERIALS REFERENCE BOOKS INDEX

CHEMICAL OR MATERIAL NAME	UN/NA Number	Guide Number (DOT)	Firefighter's Hazardous Materials Reference Book	First Aid Manual for Chemical Accidents	Sax's Dangerous Properties of Industrial Materials	Hazardous Chemicals Desk Reference	Rapid Guide to Hazardous Chemicals in the Workplace	Fire Protection Guide on Hazardous Materials (NFPA)	Firefighter's Handbook of Hazardous Materials	Pocket Guide to Chemical Hazards (NIOSH)	Chemical Hazard Response Information System (CHRIS)	Emergency Handling of Hazardous Materials in Surface Transportation (AAR)	Emergency Action Guides (EAG)	Chemical Data Notebook: A User's Manual	Condensed Chemical Dictionary
DIESEL OIL Synonym: FUEL OIL NO. 2	1993	27	404					-55	151			464	F3.1		
DIESEL OIL LIGHT Synonym: OILS FUEL 1-D	1270	27	653								OOD 892				
DIESEL OIL MEDIUM Synonym: OILS FUEL 2-D	1270	27	653								OTD 912				
DIETHANOLAMINE			286		1184	399	73	-36			DEA 381		D1.1.2		388
DIETHANOLAMINE LAURYL SULFATE SOLUTION Synonym: DODECYL SULFATE DIETHANOLAMINE SALT											DSD 475				
DIETHION Synonym: ETHION	2783	55	339		1558	520	88		134		ETO 571	404			476
DIETHOXYCHLOROSILANE					1185				107						
DIETHOXYMETHANE	2373/1993	26							107			344			
DIETHOXYPROPENE	2374	26							107						
DIETHYL ACETOACETATE								-36	107						
DIETHYL ANILINE	2432	57			1209	404									390
DIETHYL BENZENE-meta	2049	29			1210	405		-36	107						
DIETHYL BENZENE-ortho	2049	29						-36	107						
DIETHYL BENZENE-para	2049	29						-36	107						
DIETHYL CARBAMYL CHLORIDE								-37	107						
DIETHYL CARBINOL	2706								107						390
DIETHYL CARBONATE	2366/1993	26	287		1213	406		-37	108		DEC 383	345			390
DIETHYL CELLOSOLVE Synonym: ETHYLENE GLYCOL DIETHYL ETHER	1153	26	374		1606	542		-51	108		EEE 522	431			488
DIETHYL CHLOROPHOSPHATE					1214	407			108						390

CHEMICAL OR MATERIAL NAME	UN/NA Number	Guide Number (DOT)	Firefighter's Hazardous Materials Reference Book	First Aid Manual for Chemical Accidents	Sax's Dangerous Properties of Industrial Materials	Hazardous Chemicals Desk Reference	Rapid Guide to Hazardous Chemicals in the Workplace	Fire Protection Guide on Hazardous Materials (NFPA)	Firefighter's Handbook of Hazardous Materials	Pocket Guide to Chemical Hazards (NIOSH)	Chemical Hazard Response Information System (CHRIS)	Emergency Handling of Hazardous Materials in Surface Transportation (AAR)	Emergency Action Guides (EAG)	Chemical Data Notebook: A User's Manual	Condensed Chemical Dictionary
DIETHYL CHLOROTHIOPHOSPHATE	1760	60										345			
DIETHYL DICHLOROSILANE	1767	29										346			391
DIETHYL ETHANEDIOATE	2525	54							108						
DIETHYL ETHER Synonym: ETHYL ETHER	1155	26	354	16	1614	546	96	87 -52	108	112	EET 524	346	E2.8	E 3	393
DIETHYL FLUOROPHOSPHATE					1224	410			108						
DIETHYL FUMARATE					1225			-39							
DIETHYL GLYCOL								-39	108						
DIETHYL KETONE	1156	26			1228		74	-39	108		DEK 387	347			395
DIETHYL MALEATE					1229				108						395
DIETHYL MALONATE								-39	108						396
DIETHYL MERCURY				15	1229	413			108						
DIETHYL OXALATE	2525	54			1232	414			108						396
DIETHYL OXIDE Synonym: ETHYL ETHER	1155	26	354	16	1614	546	96	87 -52	108	112	EET 524	346	E2.8	E 3	396
DIETHYL PEROXIDE					1233			-39	108						
DIETHYL PEROXYDICARBONATE	2175/3115	52			1233	415			108			347			
DIETHYL PHOSPHOROCHLORODITHIOATE-o,o	2751	59			1235	415			108						396
DIETHYL PHTHALATE	9188/3082		288		1236	415	74	-39	108		DPH 465	348			396
DIETHYL SELENIDE								-39	108						
DIETHYL SUCCINATE								-39	108						397
DIETHYL SULFATE	1594/2810	55	289		1240	418	75	70 -39	108		DSU 483	349			397
DIETHYL SULFIDE	2375/1993	28							108			348			397
DIETHYL SULPHATE	2810/1594	55										349			

HAZARDOUS MATERIALS REFERENCE BOOKS INDEX

CHEMICAL OR MATERIAL NAME	UN/NA Number	Guide Number (DOT)	Firefighter's Hazardous Materials Reference Book	First Aid Manual for Chemical Accidents	Sax's Dangerous Properties of Industrial Materials	Hazardous Chemicals Desk Reference	Rapid Guide to Hazardous Chemicals in the Workplace	Fire Protection Guide on Hazardous Materials (NFPA)	Firefighter's Handbook of Hazardous Materials	Pocket Guide to Chemical Hazards (NIOSH)	Chemical Hazard Response Information System (CHRIS)	Emergency Handling of Hazardous Materials in Surface Transportation (AAR)	Emergency Action Guides (EAG)	Chemical Data Notebook: A User's Manual	Condensed Chemical Dictionary
DIETHYL TARTRATE								−39	108						397
DIETHYL TELLURIDE					1241			70 –	108						
DIETHYL-1,3-BUTANEDIAMINE-n,n								−37	107						
DIETHYL-1,3-PROPANEDIAMINE-n,n	2684	29							109						
DIETHYL-O,O	2783/3027	55	230		957	323					COU 281	268			320
Synonym: COUMAPHOS															
DIETHYL-o-PHTHALATE	9188/3082	31	288		1236	415	74		109		DPH 465	348			396
Synonym: DIETHYL PHTHALATE															
DIETHYLACETALDEHYDE	1178	26			1186	400			110						388
DIETHYLACETAMIDE					1187										
DIETHYLACETIC ACID					1187	400			110						388
DIETHYLALUMINUM CHLORIDE	1101/3052			15	1188			68 −36	107			349			388
DIETHYLALUMINUM HYDRIDE				15				−36	107						388
DIETHYLAMINE	1154	68	290	15	1188	400	73	68 −36	110	90	DEN 390	349	D1.1.2.1		388
DIETHYLAMINOETHANOL	2686/1993	29		15	1192				110		DAE 334	350			389
Synonym: DIETHYLETHANOLAMINE															
DIETHYLAMINOETHOXYETHANOL	1993	27													389
DIETHYLAMINOETHOXYETHANOL	1987	26													389
DIETHYLAMINOPROPYLAMINE	2684	29							110						
DIETHYLANILINE	2432	57			1209										
DIETHYLANILINE-n,n	2432	57			1209	404		−36							390
DIETHYLBENZENE	2049/1993	29	291		1210	404			107		DEB 382	345			390
DIETHYLBERYLLIUM					1211	405			110						
DIETHYLCADMIUM					1212	405			110						390

CHEMICAL OR MATERIAL NAME	UN/NA Number	Guide Number (DOT)	Firefighter's Hazardous Materials Reference Book	First Aid Manual for Chemical Accidents	Sax's Dangerous Properties of Industrial Materials	Hazardous Chemicals Desk Reference	Rapid Guide to Hazardous Chemicals in the Workplace	Fire Protection Guide on Hazardous Materials (NFPA)	Firefighter's Handbook of Hazardous Materials	Pocket Guide to Chemical Hazards (NIOSH)	Chemical Hazard Response Information System (CHRIS)	Emergency Handling of Hazardous Materials in Surface Transportation (AAR)	Emergency Action Guides (EAG)	Chemical Data Notebook: A User's Manual	Condensed Chemical Dictionary
DIETHYLCYCLOHEXANE					1214			−37	108						390
DIETHYLENE DIAMINE	2685	29						−37	110						391
DIETHYLENE DIOXIDE	1165	26							110						
DIETHYLENE GLYCOL			292	15	1219	409		−37	110		DEG 386				391
DIETHYLENE GLYCOL DIBENZOATE					1220			−37							391
DIETHYLENE GLYCOL DIETHYL ETHER								−37	111						392
DIETHYLENE GLYCOL DIMETHYL ETHER								−38	111		DGD 400				392
DIETHYLENE GLYCOL DINITRATE	0075				1220	409			111			352			392
DIETHYLENE GLYCOL ETHYL ETHER Synonym: DIETHYLENE GLYCOL MONOETHYL ETHER	1993	27						−38	111		DGE 401	351			392
DIETHYLENE GLYCOL METHYL ETHER Synonym: DIETHYLENE GLYCOL MONOMETHYL ETHER					1221			−38	111		DGM 402				393
DIETHYLENE GLYCOL METHYL ETHER ACETATE								−38	111						
DIETHYLENE GLYCOL MONOAMINE Synonym: 2-(2-AMINOETHOXY) ETHANOL	1760	60	13		170	51			26			50			56
DIETHYLENE GLYCOL MONOBUTYL ETHER					1221			−38	111		DME 430				392
DIETHYLENE GLYCOL MONOBUTYL ETHER ACETATE								−38			DEM 389				392
DIETHYLENE GLYCOL MONOETHYL ETHER	1993	27						−38	111		DGE 401	351			392
DIETHYLENE GLYCOL MONOISOBUTYL ETHER					1221			−38	111						
DIETHYLENE GLYCOL MONOMETHYL ETHER					1221			−38	111		DGM 402				393
DIETHYLENE IMIDOXIDE Synonym: MORPHOLINE	2054	29	597		2447	833	151	120–73	206	156	MPL 788	659			798
DIETHYLENE OXIDE Synonym: TETRAHYDROFURAN	2056	26	818		3227	1103	210	167–86	111	210	THF1116	894	T1.0	T 2	393

HAZARDOUS MATERIALS REFERENCE BOOKS INDEX

CHEMICAL OR MATERIAL NAME	UN/NA Number	Guide Number (DOT)	Firefighter's Hazardous Materials Reference Book	First Aid Manual for Chemical Accidents	Sax's Dangerous Properties of Industrial Materials	Hazardous Chemicals Desk Reference	Rapid Guide to Hazardous Chemicals in the Workplace	Fire Protection Guide on Hazardous Materials (NFPA)	Firefighter's Handbook of Hazardous Materials	Pocket Guide to Chemical Hazards (NIOSH)	Chemical Hazard Response Information System (CHRIS)	Emergency Handling of Hazardous Materials in Surface Transportation (AAR)	Emergency Action Guides (EAG)	Chemical Data Notebook: A User's Manual	Condensed Chemical Dictionary
DIETHYLENE OXIMIDE **Synonym: MORPHOLINE**	2054	29	597		2447	833	151	120–73	111	156	MPL 788	659			798
DIETHYLENEDIAMINE **Synonym: PIPERAZINE**	2579/1760	60	696		2811	959		–80			PPZ 983	766			391
DIETHYLENEIMIDE OXIDE **Synonym: MORPHOLINE**	2054	29	597		2447	833	151	120–73	206	156	MPL 788	659			798
DIETHYLENETRIAMINE	2079/1760	29	293		1222	409	74	69 –38	111		DET 394	352			393
DIETHYLENIMIDE OXIDE	2054	29							111						
DIETHYLETHANOLAMINE	2686	29							111		DAE 334				
DIETHYLETHANOLAMINE-n,n	2686	29						–38	111						393
DIETHYLETHYLENE DIAMINE	2685	29							111						391
DIETHYLETHYLENEDIAMINE-n,n	2685	29						–38	108						393
DIETHYLHYDROXYLAMINE	1993	27			1227				111			352			395
DIETHYLLAURAMIDE-n,n								–39	108						
DIETHYLMAGNESIUM	1367	40			1229	413			111						
DIETHYLOLAMINE **Synonym: DIETHANOLAMINE**			286		1184	399	73	–36			DEA 381		D1.1.2		388
DIETHYLPHOSPHINE					1235				111						
DIETHYLPHOSPHITE					1235				111						396
DIETHYLTHIOPHOSPHORYL CHLORIDE	2751	59							111			353			
DIETHYLZINC	1366/2845	40	294		1243	419		71 –39	109		DEZ 395	353			397
DIFLUOROCHLOROETHANE	2517	22							112			221			
DIFLUOROCHLOROMETHANE **Synonym: MONOCHLORODIFLUOROMETHANE**	1018	12	593						205		MCF 737		M4.0		399
DIFLUOROCHLOROMETHYLMETHANE	2517	22							112						

CHEMICAL OR MATERIAL NAME	UN/NA Number	Guide Number (DOT)	Firefighter's Hazardous Materials Reference Book	First Aid Manual for Chemical Accidents	Sax's Dangerous Properties of Industrial Materials	Hazardous Chemicals Desk Reference	Rapid Guide to Hazardous Chemicals in the Workplace	Fire Protection Guide on Hazardous Materials (NFPA)	Firefighter's Handbook of Hazardous Materials	Pocket Guide to Chemical Hazards (NIOSH)	Chemical Hazard Response Information System (CHRIS)	Emergency Handling of Hazardous Materials in Surface Transportation (AAR)	Emergency Action Guides (EAG)	Chemical Data Notebook: A User's Manual	Condensed Chemical Dictionary
DIFLUORODIBROMOMETHANE	1941	58			1245	420	72		112	92					399
DIFLUORODICHLOROMETHANE Synonym: DICHLORODIFLUOROMETHANE	1028	12	275	15	1137	380	68		102	86	DCF 360	331	D1.1		399
DIFLUOROETHANE	1030	22		15					112			353			
DIFLUOROETHYLENE	1959	22		15					112						
DIFLUOROMONOCHLOROETHANE	2517	22							112						399
DIFLUOROMONOCHLOROMETHANE Synonym: MONOCHLORODIFLUOROMETHANE	1018	12	593						112		MCF 737		M4.0		399
DIFLUOROPHOSPHORIC ACID	1768	59	295						112		DFA 396	354			399
DIFLUOROPHOSPHORUS ACID Synonym: DIFLUOROPHOSPHORIC ACID	1768	59	295						112		DFA 396	354			399
DIFORMYL Synonym: GLYOXAL			421		1802	609			155		GOS 610				570
DIGLYCIDYL ETHER				15	1250	421	75		112	92					400
DIGLYCOL Synonym: DIETHYLENE GLYCOL			292	15	1219	409		-37	110		DEG 386				400
DIGLYCOL CHLOROHYDRIN								-40	112						400
DIGLYCOL MONOBUTYL ETHER Synonym: DIETHYLENE GLYCOL MONOBUTYL ETHER					1221			-38	111		DME 430				392
DIGLYCOL MONOBUTYL ETHER ACETATE Synonym: DIETHYLENE GLYCOL MONOBUTYL ETHER ACE-TATE								-38			DEM 389				392
DIGLYME Synonym: DIETHYLENE GLYCOL DIMETHYL ETHER								-38	111		DGD 400				401
DIGLYCOLAMINE Synonym: 2-(2-AMINOETHOXY) ETHANOL	1760	60	13		170	51			26			50			56

HAZARDOUS MATERIALS REFERENCE BOOKS INDEX

CHEMICAL OR MATERIAL NAME	UN/NA Number	Guide Number (DOT)	Firefighter's Hazardous Materials Reference Book	First Aid Manual for Chemical Accidents	Sax's Dangerous Properties of Industrial Materials	Hazardous Chemicals Desk Reference	Rapid Guide to Hazardous Chemicals in the Workplace	Fire Protection Guide on Hazardous Materials (NFPA)	Firefighter's Handbook of Hazardous Materials	Pocket Guide to Chemical Hazards (NIOSH)	Chemical Hazard Response Information System (CHRIS)	Emergency Handling of Hazardous Materials in Surface Transportation (AAR)	Emergency Action Guides (EAG)	Chemical Data Notebook: A User's Manual	Condensed Chemical Dictionary
DIHEPTYL PHTHALATE											DHP 406				
DIHEXYL ETHER					1252			-40	113						
DIHEXYLAMINE					1252	422		-40	113						401
DIHYDRO-1H-AZIRINE Synonym: ETHYLENEIMINE	1185	30	380	17	1610	544	95	86 -51	144	110	ETI 567	437	E2.6.1		490
DIHYDRO-2,5-DIOXOFURAN Synonym: MALEIC ANHYDRIDE	2215	60	528	19	2155	736	130	106-64	184	138	MLA 769	592	M1.0		724
DIHYDROAZIRINE Synonym: ETHYLENEIMINE	1185	30	380	17	1610	544	95	86 -51	144	110	ETI 567	437	E2.6.1		490
DIHYDROGEN DIOXIDE Synonym: HYDROGEN PEROXIDE	2015	47	450	18		644	115	100–	165	126	HPO 643	532	H4.5	H 5	616
DIHYDROGEN DIOXIDE Synonym: HYDROGEN PEROXIDE	2014	45		18	1910	644	115	99 –	165	126		531	H4.5	H 5	616
DIHYDROGEN MONOSULFIDE Synonym: HYDROGEN SULFIDE	1053	13	452	18	1903	646	116	100–59	165	128	HDS 632	532	H4.7	H 6	617
DIHYDROGEN SELENIDE Synonym: HYDROGEN SELENIDE	2202	13	451	18	1903	645	116		165	128		532	H4.6		617
DIHYDROGEN SULFIDE Synonym: HYDROGEN SULFIDE	1053	13	452	18	1903	646	116	100–59	165	128	HDS 632	532	H4.7	H 6	617
DIHYDROOXIRENE Synonym: ETHYLENE OXIDE	1040	69	377	17	1611	544	96	87 -52	144	112	EOX 550	434	E2.7	E 6	491
DIHYDROPYRAN	2376	26			1278	428		-40	113			354			
DIHYDROXYBENZENE-meta Synonym: RESORCINOL	2876	55	738	23	2984	1019	192	-83	113		RSC1006	813			403
DIHYDROXYBENZENE-ortho								-40	113						403

CHEMICAL OR MATERIAL NAME	UN/NA Number	Guide Number (DOT)	Firefighter's Hazardous Materials Reference Book	First Aid Manual for Chemical Accidents	Sax's Dangerous Properties of Industrial Materials	Hazardous Chemicals Desk Reference	Rapid Guide to Hazardous Chemicals in the Workplace	Fire Protection Guide on Hazardous Materials (NFPA)	Firefighter's Handbook of Hazardous Materials	Pocket Guide to Chemical Hazards (NIOSH)	Chemical Hazard Response Information System (CHRIS)	Emergency Handling of Hazardous Materials in Surface Transportation (AAR)	Emergency Action Guides (EAG)	Chemical Data Notebook: A User's Manual	Condensed Chemical Dictionary
DIHYDROXYBENZENE-para Synonym: HYDROQUINONE	2662	53	453	18	1906	647	116	-40	113	128	HDQ 631	534			403
DIHYDROXYBENZOL Synonym: RESORCINOL	2876	55	738	23	2984	1019	192	-83	113		RSC1006	813			1003
DIHYDROXYDIETHYL ETHER-b,b Synonym: DIETHYLENE GLYCOL			292	15	1219	409		-37	110		DEG 386				391
DIHYDROXYDIPHENYLDIMETHYLMETHANE-p,p Synonym: BISPHENOL A					511	169			48		BPA 154				154
DIHYDROXYETHYLAMINE-b,b' Synonym: DIETHANOLAMINE			286		1184	399	73	-36			DEA 381		D1.1.2		388
DIISOBUTENE Synonym: DIISOBUTYLENE	2050/1993	26	298					-40	114		DBL 347	356	D2.0		405
DIISOBUTYL KETONE	1157	26	296		1300	433	75	-40		92	DIK 414	355			405
DIISOBUTYL PHTHALATE					1300	434		-40			DIT 422				405
DIISOBUTYLALUMINUM HYDRIDE								-40	113						405
DIISOBUTYLAMINE	2361/1993	68			1298	433		-40	114		DBU 355	355			405
DIISOBUTYLCARBINOL	1993/1987	45	297	15	1299	433		-40	113		DBC 343	355			405
DIISOBUTYLENE	2050/1993	26	298					-40	114		DBL 347	356	D2.0		405
DIISOBUTYLENE-alpha Synonym: DIISOBUTYLENE	2050/1993	26	298					-40	114		DBL 347	356	D2.0		405
DIISOBUTYLENE-beta Synonym: DIISOBUTYLENE	2050/1993	26	298					-40	114		DBL 347	356	D2.0		405
DIISOBUTYRYL PEROXIDE	2182/3111	52			1300				114			356			
DIISOCYANATOTOLUENE Synonym: TOLUENE DIISOCYANATE	2078	57	828			1127	215		269		TDI1095	910	T4.1	T 5	1157
DIISODECYL PHTHALATE								-40			DID 410				406

HAZARDOUS MATERIALS REFERENCE BOOKS INDEX

CHEMICAL OR MATERIAL NAME	UN/NA Number	Guide Number (DOT)	Firefighter's Hazardous Materials Reference Book	First Aid Manual for Chemical Accidents	Sax's Dangerous Properties of Industrial Materials	Hazardous Chemicals Desk Reference	Rapid Guide to Hazardous Chemicals in the Workplace	Fire Protection Guide on Hazardous Materials (NFPA)	Firefighter's Handbook of Hazardous Materials	Pocket Guide to Chemical Hazards (NIOSH)	Chemical Hazard Response Information System (CHRIS)	Emergency Handling of Hazardous Materials in Surface Transportation (AAR)	Emergency Action Guides (EAG)	Chemical Data Notebook: A User's Manual	Condensed Chemical Dictionary
DIISOOCTYL ACID PHOSPHATE	1902	60			1301	434			114			356			406
DIISOPROPANOLAMINE					1302			-41			DIP 419				406
DIISOPROPYL	2457	27							114						406
DIISOPROPYL BENZENE					1303			-41			DIX 424				
DIISOPROPYL ETHER Synonym: ISOPROPYL ETHER	1159	26	483		2050	691	125	72 -63	114	134	IPE 673	357	D2.1		407
DIISOPROPYL OXIDE Synonym: DIISOPROPYL ETHER	1159	26				691		72 -	114			357	D2.1		407
DIISOPROPYL OXIDE Synonym: ISOPROPYL ETHER	1159	26	483		2050		125	-63	114	134	IPE 673		D2.1		407
DIISOPROPYL PERCARBONATE Synonym: ISOPROPYL PERCARBONATE	2134	52	485						175		IPC 671				662
DIISOPROPYL PEROXYDICARBONATE Synonym: ISOPROPYL PERCARBONATE	2134/3115	52	485					-41	114		IPC 671	357			407
DIISOPROPYLAMINE	1158	68	299		1302	434	76	71 -41	114	94	DIA 407	358			406
DIISOPROPYLAMINOETHANOL	1987/2825	26										358			406
DIISOPROPYLBENZENE HYDROPEROXIDE	2171	48	300						114		DIH 413				407
DIISOPROPYLETHANOLAMINE	2825	68			1305										
DIISOPROPYLETHANOLAMINE-n,n	2825	68			1305			-41	114						407
DIISOPROPYLMERCURY					1305	435			114						
DIISOTRIDECYLPEROXYDICARBONATE	2889/3115	52							115			359			
DIKETENE	2521/1993	57						72 -41	115						408
DILAUROYL PEROXIDE	2893	48							115						
DILAUROYL PEROXIDE Synonym: LAUROYL PEROXIDE	2124/3106	48	498		2091	706	127		115		LPO 705	359			684

CHEMICAL OR MATERIAL NAME	UN/NA Number	Guide Number (DOT)	Firefighter's Hazardous Materials Reference Book	First Aid Manual for Chemical Accidents	Sax's Dangerous Properties of Industrial Materials	Hazardous Chemicals Desk Reference	Rapid Guide to Hazardous Chemicals in the Workplace	Fire Protection Guide on Hazardous Materials (NFPA)	Firefighter's Handbook of Hazardous Materials	Pocket Guide to Chemical Hazards (NIOSH)	Chemical Hazard Response Information System (CHRIS)	Emergency Handling of Hazardous Materials in Surface Transportation (AAR)	Emergency Action Guides (EAG)	Chemical Data Notebook: A User's Manual	Condensed Chemical Dictionary
DILITHIUM CHROMATE	9 34/3077	31	522		2127	725			181		LCR 696	582			707
Synonym: LITHIUM CHROMATE															
DILUTE SULFURIC ACID	1832	39	803	24	3163	1082	204			200	SAC1008	872	S4.3	S 8	1104
Synonym: SULFURIC ACID															
DIMAZINE	1163	28			1381	459	79	−43	115	96	DMH 432				417
Synonym: 1,1-DIMETHYLHYDRAZINE															
DIMECRON									115						
DIMETHOXY METHANE	1234	26							115						410
DIMETHOXY STRYCHNINE	1570	53							115						411
DIMETHOXYBENZENE-para					1314										
DIMETHOXYETHANE	2252	27			1316	439			115						
DIMETHOXYETHYL PHTHALATE					1317	439		−41							
DIMETHOXYMETHANE	1234	26	568						115		MTF 808				410
Synonym: METHYL FORMAL															
DIMETHYL ACETAMIDE	1993	27			1321	441	77		120	94	DAC 333	360			411
DIMETHYL ANTHRANILATE								−42	116						413
DIMETHYL BENZYL HYDROPEROXIDE	2116	51													
DIMETHYL CARBAMYL CHLORIDE	2262	60					78		116						
DIMETHYL CARBAMYL CHLORIDE-n,n	2262	60					78		116						
DIMETHYL CARBINOL	1219	26							116						414
DIMETHYL CARBONATE	1161	26							121			361			414
DIMETHYL CELLOSOLVE	2252	27						−51	143		EGD 528				488
Synonym: ETHYLENE GLYCOL DIMETHYL ETHER															
DIMETHYL CHLORACETAL								−42	121						
DIMETHYL CHLOROTHIOPHOSPHATE	2267	59													

HAZARDOUS MATERIALS REFERENCE BOOKS INDEX

CHEMICAL OR MATERIAL NAME	UN/NA Number	Guide Number (DOT)	Firefighter's Hazardous Materials Reference Book	First Aid Manual for Chemical Accidents	Sax's Dangerous Properties of Industrial Materials	Hazardous Chemicals Desk Reference	Rapid Guide to Hazardous Chemicals in the Workplace	Fire Protection Guide on Hazardous Materials (NFPA)	Firefighter's Handbook of Hazardous Materials	Pocket Guide to Chemical Hazards (NIOSH)	Chemical Hazard Response Information System (CHRIS)	Emergency Handling of Hazardous Materials in Surface Transportation (AAR)	Emergency Action Guides (EAG)	Chemical Data Notebook: A User's Manual	Condensed Chemical Dictionary
DIMETHYL DECALIN								–42	116						
DIMETHYL DITHIOPHOSPHATE OF DIETHYL MERCAPTOSUCCINATE	2783	55													
DIMETHYL ETHER	1033	22	301					74 –	116		DIM 416	361			416
DIMETHYL FORMAL Synonym: METHYL FORMAL	1234	26	568								MTF 808				
DIMETHYL FORMALDEHYDE Synonym: ACETONE	1090	26	20	11	22	10	2	–11	17	30	ACT 26	4	A3.0	A 3	9
DIMETHYL FORMOCARBOTHIALDINE					1378	459			116						
DIMETHYL KETONE Synonym: ACETONE	1090	26	20	11	22	10	2	–11	116	30	ACT 26	4	A3.0	A 3	418
DIMETHYL MALONATE					1388				116						418
DIMETHYL MERCURY				16	1388										
DIMETHYL METHYL AMINE-n,n Synonym: TRIMETHYLAMINE	1083	19	853	26	3399	1146	221	178–	278		TMA1126	930	T6.0	T 7	1181
DIMETHYL O-p-NITROPHENYL THIOPHOSPHATE-O,O Synonym: METHYL PARATHION	2783	55	577	20	2369	809	148		196		MPT 790	635			418
DIMETHYL PHOSPHOROCHLORIDOTHIOATE	2267	59			1411	470									
DIMETHYL PHTHALATE	9188				1411	470	80	–44	117	98	DTL 488				419
DIMETHYL S-(4-OXOBENXOTRIAZINO-3-METHYL) PHOSPHORODITHIO	2783	55							117						
DIMETHYL SELENIDE					1415				117						
DIMETHYL SILICONE FLUIDS Synonym: DIMETHYLPOLYSILOXANE					1412	470					DMP 437				420
DIMETHYL SILICONE OIL Synonym: DIMETHYLPOLYSILOXANE					1412	470					DMP 437				420

CHEMICAL OR MATERIAL NAME	UN/NA Number	Guide Number (DOT)	Firefighter's Hazardous Materials Reference Book	First Aid Manual for Chemical Accidents	Sax's Dangerous Properties of Industrial Materials	Hazardous Chemicals Desk Reference	Rapid Guide to Hazardous Chemicals in the Workplace	Fire Protection Guide on Hazardous Materials (NFPA)	Firefighter's Handbook of Hazardous Materials	Pocket Guide to Chemical Hazards (NIOSH)	Chemical Hazard Response Information System (CHRIS)	Emergency Handling of Hazardous Materials in Surface Transportation (AAR)	Emergency Action Guides (EAG)	Chemical Data Notebook: A User's Manual	Condensed Chemical Dictionary
DIMETHYL SULFATE	1595	57	302	16	1416	472	80	75 −44	117	98	DSF 477	362	D3.1.1		420
DIMETHYL SULFIDE	1164	27	303					75 −44	117		DSL 478	363			420
DIMETHYL SULFOXIDE					1417	472		−44	117		DMS 438				421
DIMETHYL TEREPHTHALATE						473		−44			DMT 439				421
DIMETHYL THIOPHOSHORYL CHLORIDE	2267	59							117			363			
DIMETHYL-1,2-DIBROMO-2,2-DICHLORETHYL PHOSPHATE	2783	55								96					
DIMETHYL-1-HEXANOLS Synonym: ISOOCTYL ALCOHOL	1987	26	475		2032	683	123	−62			IOA 668	552			657
DIMETHYL-o,o Synonym: TRICHLORFON	2783/3018	55	835						272		TRC1146	917			1169
DIMETHYL-o,o-DICHLOROVINYL-2,2-PHOSPHATE (TECHNICAL)	2783	55						−42							
DIMETHYL-p-NITROSOANILINE	1369	32							119						
DIMETHYLACETAL	2377	27			1321				120						411
DIMETHYLACETAMIDE	1993	27			1321		77	−41	120		DAC 333	360			411
DIMETHYLACETAMIDE-n,n Synonym: DIMETHYLACETAMIDE	1993	27			1321	441	77	−41	120		DAC 333	360			411
DIMETHYLACETYLENE	1144	27							120						411
DIMETHYLALDEHYDE Synonym: ISOOCTALDEHYDE	1993	27	474								IOC 669				
DIMETHYLAMINE	1032	19	304	16	1324	442	77	73 −41	120	94	DMA 427	367	D3.0		411
DIMETHYLAMINE SOLUTION	1160	26		16	1324	442	77	73 −41	120			368	D3.0		
DIMETHYLAMINOACETONITRILE	2378	28			1326	443			120						
DIMETHYLAMINOETHANOL	2051/1993	29			1331	445			120			369			
DIMETHYLAMINOETHYL METHACRYLATE	2522/1993	55			1334	446			120			368			412

HAZARDOUS MATERIALS REFERENCE BOOKS INDEX

CHEMICAL OR MATERIAL NAME	UN/NA Number	Guide Number (DOT)	Firefighter's Hazardous Materials Reference Book	First Aid Manual for Chemical Accidents	Sax's Dangerous Properties of Industrial Materials	Hazardous Chemicals Desk Reference	Rapid Guide to Hazardous Chemicals in the Workplace	Fire Protection Guide on Hazardous Materials (NFPA)	Firefighter's Handbook of Hazardous Materials	Pocket Guide to Chemical Hazards (NIOSH)	Chemical Hazard Response Information System (CHRIS)	Emergency Handling of Hazardous Materials in Surface Transportation (AAR)	Emergency Action Guides (EAG)	Chemical Data Notebook: A User's Manual	Condensed Chemical Dictionary
DIMETHYLANILINE	2253	57		16	1347		77			96					413
DIMETHYLANILINE-n,n	2253	57			1347	449	77	-42							
DIMETHYLARSINE					1348	449			121						
DIMETHYLARSINIC ACID Synonym: CACODYLIC ACID	1572/2811	53	156						121		CDA 235	175			413
DIMETHYLBENZENE HYDROPEROXIDE-alpha,alpha Synonym: CUMENE HYDROPEROXIDE	2116	51	235					-29			CMH 266				329
DIMETHYLBENZYL HYDROPEROXIDE Synonym: CUMENE HYDROPEROXIDE	2116	51	235					-29	85		CMH 266				329
DIMETHYLBERYLLIUM					1356	451			121						
DIMETHYLBUTANE	2457	27							116						
DIMETHYLCADMIUM					1359	452			121						414
DIMETHYLCARBAMOYL CHLORIDE	2262	60			1361	453	78		116		DCR 367				414
DIMETHYLCARBINOL Synonym: ISOPROPYL ALCOHOL	1219	26	482	18	2040		124	-63	116	132	IPA 670	555		1 2	414
DIMETHYLCARBINOL Synonym: ISOPROPANOL	1219	26	480			687			116			555	14.0		414
DIMETHYLCHLOROETHER Synonym: METHYL CHLOROMETHYL ETHER	1239	57						111-	193			643	M3.1.2		762
DIMETHYLCHLOROSILANE	2985	29													
DIMETHYLCYANAMIDE					1364	454		-42	116						415
DIMETHYLDICHLOROSILANE	1162	29	305					-42	122		DMD 429	370	D3.1		415
DIMETHYLDICHLOROVINYL PHOSPHATE	2783	55							122						415
DIMETHYLDIETHOXYSILANE	2380/1993	26							122			370			
DIMETHYLDIOXANE	2707/1993	27			1370			-43	122			371			415

CHEMICAL OR MATERIAL NAME	UN/NA Number	Guide Number (DOT)	Firefighter's Hazardous Materials Reference Book	First Aid Manual for Chemical Accidents	Sax's Dangerous Properties of Industrial Materials	Hazardous Chemicals Desk Reference	Rapid Guide to Hazardous Chemicals in the Workplace	Fire Protection Guide on Hazardous Materials (NFPA)	Firefighter's Handbook of Hazardous Materials	Pocket Guide to Chemical Hazards (NIOSH)	Chemical Hazard Response Information System (CHRIS)	Emergency Handling of Hazardous Materials in Surface Transportation (AAR)	Emergency Action Guides (EAG)	Chemical Data Notebook: A User's Manual	Condensed Chemical Dictionary
DIMETHYLDISULFIDE	2381/1993	27			1371	457			122			361			
DIMETHYLENE METHANE	2200	22							122						416
DIMETHYLENE OXIDE Synonym: ETHYLENE OXIDE	1040	69	377	17	1611	544	96	87 –52	122	112	EOX 550	434	E2.7	E 6	491
DIMETHYLENEIMINE Synonym: ETHYLENEIMINE	1185	30	380	17	1610	544	95	86 –51	122	110	ETI 567	437	E2.6.1		490
DIMETHYLENIMINE Synonym: ETHYLENEIMINE	1185	30	380	17	1610	544	95	86 –51	144	110	ETI 567	437	E2.6.1		490
DIMETHYLETHANOLAMINE	2051	29							122		DMB 428	372			
DIMETHYLETHANOLAMINE-n,n	2051	29							122						
DIMETHYLETHYL NITRITE-alpha,alpha	2351	26							122						
DIMETHYLETHYLAMINE	1993	27										372			
DIMETHYLFORMAMIDE	2265	26	306		1378	458	79		116	96	DMF 431				
DIMETHYLFORMAMIDE-n,n Synonym: DIMETHYLFORMAMIDE	2265	26	306		1378	458	79	–43	116	96	DMF 431				416
DIMETHYLHEXANE DIHYDROPEROXIDE	2174	49	307		1381				122		DDW 380				
DIMETHYLHYDRAZINE Synonym: 1,2-DIMETHYLHYDRAZINE	2382	57		16	1382	460	80		123		DML 433	372			
DIMETHYLHYDRAZINE Synonym: 1,1-DIMETHYLHYDRAZINE	1163	57	308	16	1381	459	79	74 –	122		DMH 432	362			417
DIMETHYLHYDRAZINE-unsym Synonym: 1,1-DIMETHYLHYDRAZINE	1163	57		16	1381	459	79	–43	122	96	DMH 432	362			417
DIMETHYLISOPROPANOLAMINE-n,n								–43	123						
DIMETHYLMAGNESIUM	1368	40			1387	461			116						
DIMETHYLMETHANE Synonym: PROPANE	1978	22		23	2899	993	186	–81	117	186	PRP 988		L1.0		418

HAZARDOUS MATERIALS REFERENCE BOOKS INDEX

CHEMICAL OR MATERIAL NAME	UN/NA Number	Guide Number (DOT)	Firefighter's Hazardous Materials Reference Book	First Aid Manual for Chemical Accidents	Sax's Dangerous Properties of Industrial Materials	Hazardous Chemicals Desk Reference	Rapid Guide to Hazardous Chemicals in the Workplace	Fire Protection Guide on Hazardous Materials (NFPA)	Firefighter's Handbook of Hazardous Materials	Pocket Guide to Chemical Hazards (NIOSH)	Chemical Hazard Response Information System (CHRIS)	Emergency Handling of Hazardous Materials in Surface Transportation (AAR)	Emergency Action Guides (EAG)	Chemical Data Notebook: A User's Manual	Condensed Chemical Dictionary
DIMETHYLPHENOL	2261	55	883			1184			288		XYL1190	962			419
Synonym: XYLENOL															
DIMETHYLPHOSPHINE					1410				123						
DIMETHYLPIPERAZINE-cis								−44	123		DMP 437				420
DIMETHYLPOLYSILOXANE					1412	470									
DIMETHYLPROPANE	2044	22							123						
DIMETHYLZINC	1370	40	309		1425				123		DMZ 441	373			
DIMYRISTYL PEROXYDICARBONATE	2892/3119	52							124			374			
DIMYRISTYL PEROXYDICARBONATE	2595/3116	52							124			374			
DINITRO-o-CRESOL	1598/2811	53			1430	477			124	98	DNC 444	374			
DINITROANILINE	1596/2811	56	310						124			375			422
DINITROBENZENE	1597	56		16	1428	475	81		124	98		375			
DINITROBENZENE SOLUTION	1597	56		16	1428	475	81		124	98		375			
DINITROBENZENE-meta	1597	56		16	1428	475	81		124	98	DNB 443				
DINITROBENZENE-ortho	1597	56	311	16	1428	476	81		124	98	DNO 448				
DINITROBENZENE-para	1597	56		16	1428	476	81		124	98	DNZ 452				
DINITROBENZOL	1597	56		16	1428	475	81		124	98	DNB 443				
Synonym: DINITROBENZENE-meta															
DINITROBENZOL-ortho	1597	56	311	16	1428	476	81		124	98	DNO 448				
Synonym: DINITROBENZENE-ortho															
DINITROCHLOROBENZENE	1577	56						77 −44	125			376			
DINITROCHLOROBENZOL	1577	56										376			
DINITROCRESOL	1598	53	313	16	1430	477					DNC 444				
DINITROCYCLOHEXYL PHENOL	9026/2810	53							125			376			423

CHEMICAL OR MATERIAL NAME	UN/NA Number	Guide Number (DOT)	Firefighter's Hazardous Materials Reference Book	First Aid Manual for Chemical Accidents	Sax's Dangerous Properties of Industrial Materials	Hazardous Chemicals Desk Reference	Rapid Guide to Hazardous Chemicals in the Workplace	Fire Protection Guide on Hazardous Materials (NFPA)	Firefighter's Handbook of Hazardous Materials	Pocket Guide to Chemical Hazards (NIOSH)	Chemical Hazard Response Information System (CHRIS)	Emergency Handling of Hazardous Materials in Surface Transportation (AAR)	Emergency Action Guides (EAG)	Chemical Data Notebook: A User's Manual	Condensed Chemical Dictionary
DINITROGEN MONOXIDE Synonym: **NITROUS OXIDE**	1070	14	644						216		NTO 858	700	N3.5		833
DINITROGEN OXIDE Synonym: **NITROUS OXIDE**	1070	14	644						216		NTO 858	700	N3.5		833
DINITROGEN TETROXIDE Synonym: **NITROGEN TETROXIDE**	1067	20	637		2537	867		128–	125		NOX 845	377	N3.4.6		423
DINITROPHENOL	1320/2926	36		16	1434	479			125			377			423
DINITROPHENOL SOLUTION	1599	57	314	16	1434	479			125						
DINITROPHENOL SOLUTION	2326/1320	34		16	1434							377			
DINITROPHENOL-alpha Synonym: **2,4-DINITROPHENOL**	0076			16	1435	479			125		DNP 449				
DINITROPHENOL-beta Synonym: **2,6-DINITROPHENOL**	1599	57		16	1435	479					DNH 446				
DINITROPHENOL-gamma Synonym: **2,5-DINITROPHENOL**	1599	57		16	1435						DNE 445				
DINITROPHENOL-o,o Synonym: **2,6-DINITROPHENOL**	1599	57		16	1435	479					DNH 446				
DINITROPHENOLATE	1321	36							125			378			
DINITROPHENYLAMINE	1596	56							125						
DINITRORESORCINOL	1322	36							125			378			
DINITROSO-DIMETHYL TEREPHTHALAMIDE	2973	71							125						
DINITROSOPENTAMETHYLENE TETRAMINE	2972	71							125						424
DINITROTOLUENE	1600	56	316	16	1440	480	82	77 –	125	100		380			424
DINITROTOLUENE	2038/2811	56	315	16	1440	480	82	77 –	126	100		379			424
DIOCTYL ADIPATE						481		-44			DOA 453				424

HAZARDOUS MATERIALS REFERENCE BOOKS INDEX

CHEMICAL OR MATERIAL NAME	UN/NA Number	Guide Number (DOT)	Firefighter's Hazardous Materials Reference Book	First Aid Manual for Chemical Accidents	Sax's Dangerous Properties of Industrial Materials	Hazardous Chemicals Desk Reference	Rapid Guide to Hazardous Chemicals in the Workplace	Fire Protection Guide on Hazardous Materials (NFPA)	Firefighter's Handbook of Hazardous Materials	Pocket Guide to Chemical Hazards (NIOSH)	Chemical Hazard Response Information System (CHRIS)	Emergency Handling of Hazardous Materials in Surface Transportation (AAR)	Emergency Action Guides (EAG)	Chemical Data Notebook: A User's Manual	Condensed Chemical Dictionary
DIOCTYL PHTHALATE					1445			−45			DOP 455				425
DIOCTYL SODIUM SULFOSUCCINATE											DSS 481				425
DIOCTYLAMINE					1443				126						424
DIOFORM Synonym: 1,2-DICHLOROETHYLENE	1150/3082	29	3		1144	384	69	−35	103	88	DEL 388	301			
DIOLAMINE Synonym: DIETHANOLAMINE			286		1484	399	73	−36			DEA 381		D1.1.2		388
DIOXANE Synonym: 1,4-DIOXANE	1165	26	317	16	1449	482	83	78 –	126	100	DOX 457	380			426
DIOXANE-para Synonym: 1,4-DIOXANE	1165	26	317	16	1449	482	83	78 −45	126		DOX 457	380			426
DIOXATHION					1450	483	83		126						426
DIOXOLANE Synonym: DIOXOLAN	1166	26			1451	483		−45	126						
DIOXONIUM PERCHLORATE SOLUTION Synonym: PERCHLORIC ACID	1873	47	672	22	2699	917			225		PCL 937	732			884
DIPENTENE	2052/1993	27	318					−45	126		DPN 468	381			427
DIPENTYL PHTHALATE Synonym: DI-n-AMYL PHTHALATE											DAP 339				364
DIPEROXY AZELAIC ACID	2958	52							126						
DIPHENYL				16					127	100	DIL 415				428
DIPHENYL DICHLOROSILANE	1769	29	320					−45	127		DPD 461	382			429
DIPHENYL ETHER	2489	55									DPE 462				430
DIPHENYL KETONE Synonym: BENZOPHENONE					379	125					BZP 200				430

CHEMICAL OR MATERIAL NAME	UN/NA Number	Guide Number (DOT)	Firefighter's Hazardous Materials Reference Book	First Aid Manual for Chemical Accidents	Sax's Dangerous Properties of Industrial Materials	Hazardous Chemicals Desk Reference	Rapid Guide to Hazardous Chemicals in the Workplace	Fire Protection Guide on Hazardous Materials (NFPA)	Firefighter's Handbook of Hazardous Materials	Pocket Guide to Chemical Hazards (NIOSH)	Chemical Hazard Response Information System (CHRIS)	Emergency Handling of Hazardous Materials in Surface Transportation (AAR)	Emergency Action Guides (EAG)	Chemical Data Notebook: A User's Manual	Condensed Chemical Dictionary
DIPHENYL METHANONE Synonym: BENZOPHENONE					379	125					BZP 200				132
DIPHENYL OXIDE Synonym: DIPHENYL ETHER	2489	55						–45	127		DPE 462				430
DIPHENYL OXIDE	1693	58						–45	127						431
DIPHENYLAMINE			319	16	1457	485	84	–45	127		DAM 337				428
DIPHENYLAMINECHLOROARSINE	1698	55							127			382			428
DIPHENYLCHLORARSINE	1699	55							127						
DIPHENYLCHLOROARSINE	1699/2810	55							127			382			429
DIPHENYLMETHANE								–45	127						430
DIPHENYLMETHANE DIISOCYANATE	2489	55	321						127		DPM 467	382			
DIPHENYLMETHANE-4,4'-DIISOCYANATE (MDI) Synonym: DIPHENYLMETHANE DIISOCYANATE	2489	55	321								DPM 467	382			430
DIPHENYLMETHANE-4,4-DIISOCYANATE Synonym: DIPHENYLMETHANE DIISOCYANATE	2489	55	321						127		DPM 467	382			430
DIPHENYLMETHYL BROMIDE	1770	60													430
DIPHENYLMETHYL DIISOCYANATE	2489	55							127						
DIPHENYLOXIDE-4,4'-DI-SULFOHYDRAZIDE	2951	72							127						
DIPHENYLSILICAON DICHLORIDE Synonym: DIPHENYLDICHLOROSILANE	1769	29	320					–45	127		DPD 461	382			429
DIPHOSGENE Synonym: PHOSGENE	1076	15	682	22	2782	944	179	136–	127	180	PHG 956	752	P3.0	P 2	432
DIPICRYL SULFIDE	2852/0401	33							127			383			432
DIPPING ACID Synonym: SULFURIC ACID	1830	39	802	24	3163	1082	204	164–	127	200	SFA1037	871	S4.2	S 8	1104

HAZARDOUS MATERIALS REFERENCE BOOKS INDEX

CHEMICAL OR MATERIAL NAME	UN/NA Number	Guide Number (DOT)	Firefighter's Hazardous Materials Reference Book	First Aid Manual for Chemical Accidents	Sax's Dangerous Properties of Industrial Materials	Hazardous Chemicals Desk Reference	Rapid Guide to Hazardous Chemicals in the Workplace	Fire Protection Guide on Hazardous Materials (NFPA)	Firefighter's Handbook of Hazardous Materials	Pocket Guide to Chemical Hazards (NIOSH)	Chemical Hazard Response Information System (CHRIS)	Emergency Handling of Hazardous Materials in Surface Transportation (AAR)	Emergency Action Guides (EAG)	Chemical Data Notebook: A User's Manual	Condensed Chemical Dictionary
DIPROPANOATE Synonym: CROTONALDEHYDE	1143	28	233	14	964	325	56	57 −29	85	76	CTA 317	271	C7.1		325
DIPROPIONYL PEROXIDE	2132/3117	52			1470	488			128			384			
DIPROPYL ETHER	2384/1993	26							128			384			
DIPROPYL KETONE	2710/1993	26			1472	488	84		128			385			433
DIPROPYLALUMINUM HYDRIDE								−45	128						
DIPROPYLAMINE Synonym: DI-n-PROPYLAMINE	2383/1993	68	322	16	1470	488		−45	128		DNA 442	384	D4.0		432
DIPROPYLANILINE-n,n Synonym: NITRALIN	1609		625								NTL 856				822
DIPROPYLENE GLYCOL								−45			DPG 464				432
DIPROPYLENE GLYCOL METHYL ETHER					1471	488	82	−45	128	102	DPY 473				
DIPROPYLENE GLYCOL MONOMETHYL ETHER									128						432
DIPROPYLENE TRIAMINE	2269	60							128						433
DIPROPYLMETHANE Synonym: HEPTANE (-n)	1206	27	425	18	1830	616	106	−57	128	120	HPT 644	505	H1.0.3		433
DIPTEREX Synonym: TRICHLORFON	2783/3018	55	835	16					128		TRC1146	917			433
DIQUAT LIQUID	2781	55	324	16	1475	490	85		128		DIQ 420				
DIQUAT SOLID	2781/3077	55	323	16	1475	490	85		128		DIQ 420	385			433
DISILANE					1476	490			128						
DISINFECTANT	1903	60							128			386			434
DISINFECTANT	1601	55							128			386			434
DISODIUM ARSENATE HEPTAHYDRATE Synonym: SODIUM ARSENATE	1685	53	756						252		SDA1027	831			1049

CHEMICAL OR MATERIAL NAME	UN/NA Number	Guide Number (DOT)	Firefighter's Hazardous Materials Reference Book	First Aid Manual for Chemical Accidents	Sax's Dangerous Properties of Industrial Materials	Hazardous Chemicals Desk Reference	Rapid Guide to Hazardous Chemicals in the Workplace	Fire Protection Guide on Hazardous Materials (NFPA)	Firefighter's Handbook of Hazardous Materials	Pocket Guide to Chemical Hazards (NIOSH)	Chemical Hazard Response Information System (CHRIS)	Emergency Handling of Hazardous Materials in Surface Transportation (AAR)	Emergency Action Guides (EAG)	Chemical Data Notebook: A User's Manual	Condensed Chemical Dictionary
DISODIUM DIHYDROGEN PYROPHOSPHATE Synonym: SODIUM PHOSPHATE	9147/3077				3106	1063			255		SPP1054	849			434
DISODIUM ETHYLENEBIS (DITHIOCABAMATE) Synonym: NABAM	1609		599						128		NAB 816				434
DISODIUM METHANE ARSONATE Synonym: METHANEARSONIC ACID SODIUM SALT	1557	53	548								MSA 799				
DISODIUM METHYL ARSONATE Synonym: METHANEARSONIC ACID SODIUM SALT	1557	53	548								MSA 799				434
DISODIUM NITRILOTRIACETATE Synonym: NITRILOTRIACETIC ACID AND SALTS			629		1481						NAA 815				
DISODIUM SELENITE Synonym: SODIUM SELENITE	2630	53	784		3112	1064			255		SSE1058	850			1068
DISPERSANT GAS	1954	22							129						
DISPERSANT GAS	1078	12							129						
DISTEARYL PEROXYDICARBONATE	2592/3106	48							129			386			
DISTILLATES FLASHED FEED STOCKS	1268	27									DFF 398				
DISTILLATES STRAIGHT RUN	1268	27									DSR 480				
DISUCCINIC ACID PEROXIDE	2962	52							129						
DISUCCINIC ACID PEROXIDE	2135	49							129						
DISULFATOZIRCONIC ACID Synonym: ZIRCONIUM SULFATE	9163	31	912		3551	1198	235		290		ZCS1202	973			1253
DISULFIRAM					1484	494	85								436
DISULFOTON LIQUID	2783/3018	55	325		1484	494	85		129		DIS 421	387			436
DISULFOTON SOLID	2783	55	326		1484	494	85		129			387			
DISULFUR DICHLORIDE Synonym: SULFUR CHLORIDE	1828	39	799					−84	258			869	54.0		1103

CHEMICAL OR MATERIAL NAME	UN/NA Number	Guide Number (DOT)	Firefighter's Hazardous Materials Reference Book	First Aid Manual for Chemical Accidents	Sax's Dangerous Properties of Industrial Materials	Hazardous Chemicals Desk Reference	Rapid Guide to Hazardous Chemicals in the Workplace	Fire Protection Guide on Hazardous Materials (NFPA)	Firefighter's Handbook of Hazardous Materials	Pocket Guide to Chemical Hazards (NIOSH)	Chemical Hazard Response Information System (CHRIS)	Emergency Handling of Hazardous Materials in Surface Transportation (AAR)	Emergency Action Guides (EAG)	Chemical Data Notebook: A User's Manual	Condensed Chemical Dictionary
DISULFURIC ACID	1831	39							129						
DISULFURYL CHLORIDE	1817	39							129						436
DITHANE Synonym: NABAM	1609		599		1484				207		NAB 816				803
DITHIOCARBAMATE PESTICIDE	2771	55		16					129			388			
DITHIOCARBAMATE PESTICIDE	3005	28		16					129						
DITHIOCARBAMATE PESTICIDE	3006/2771	55		16					129			388			
DITHIOCARBAMATE PESTICIDE	2772	28		16					129			388			
DITHIOCARBONIC ANHYDRIDE Synonym: CARBON BISULFIDE	1131	28	188						65			189	C2.1		218
DITHIONIC ACID Synonym: SULFURIC ACID	1830	39	802	24	3163		204	164–	259	200	SFA1037	871	S4.2	S 8	1104
DITHIONITE HYPOSULFITE Synonym: SODIUM HYDROSULFITE	1384	37	776		3086	1056			254			842	S2.4		1058
DITHIOPYROPHOSPHORIC ACID Synonym: TETRAETHYL DITHIOPYROPHOSPHATE	1704	55	811						262		TED1100	888			1129
DITHIOSYSTOX Synonym: DISULFOTON	2783	55	325		1484	494	85		129		DIS 421	387			436
DITHYL-5,2-(ETHYLTHIO) ETHYL PHOSPHODITHIOATE-O,O Synonym: DISULFOTON	2783	55	325		1484	494	85		129		DIS 421	387			436
DITRIDECYL PHTHALATE					1491			–46			DTP 491				438
DIURON	2767/3077	55	327		1491	496	85		129		DIU 423	389			438
DIVINYL Synonym: BUTADIENE	1010	17	139	12				40 –	129	48	BDI 134	148	B4.0	B 4	
DIVINYL ACETYLENE								–46	129						438
DIVINYL ETHER	1167	30						–46	129			389			438

CHEMICAL OR MATERIAL NAME	UN/NA Number	Guide Number (DOT)	Firefighter's Hazardous Materials Reference Book	First Aid Manual for Chemical Accidents	Sax's Dangerous Properties of Industrial Materials	Hazardous Chemicals Desk Reference	Rapid Guide to Hazardous Chemicals in the Workplace	Fire Protection Guide on Hazardous Materials (NFPA)	Firefighter's Handbook of Hazardous Materials	Pocket Guide to Chemical Hazards (NIOSH)	Chemical Hazard Response Information System (CHRIS)	Emergency Handling of Hazardous Materials in Surface Transportation (AAR)	Emergency Action Guides (EAG)	Chemical Data Notebook: A User's Manual	Condensed Chemical Dictionary
DIVINYL OXIDE	1167	30							129						438
DIVINYL-beta	1010	17							129						
DIVINYLBENZENE	1993				1491	497	86	79 -46	130			389			438
DIVINYLENE OXIDE Synonym: FURAN	2389	26			1763	596		-55	152		FUR 600		F3.3.1		545
DL LACTIC ACID AMMONIUM SALT Synonym: AMMONIUM LACTATE											ALT 46				
DM	1698	55							130						439
DMA Synonym: DIMETHYLAMINE	1032	19	304	16	1492	497	77	73 -41	120	94	DMA 427	367	D3.0		439
DMDT Synonym: METHOXYCHLOR	2761/3077	55	550		2224	763	133		191	140	MOC 780	615			439
DMF Synonym: DIMETHYLFORMAMIDE	2265	26	306		1378	458	79		116	96	DMF 431				439
DMS Synonym: DIMETHYL SULFIDE	1164	27	303					75 -44	117		DSL 478	363			420
DMSO Synonym: DIMETHYL SULFOXIDE					1417	472		-44	117		DMS 438				440
DNB-meta Synonym: DINITROBENZENE-meta	1597	56		16	1428	475	81		124	98	DNB 443				
DNBP				14											440
DNOC	1598	53		14											440
DNP Synonym: 2,6-DINITROPHENOL	1599	57			1435	479					DNH 446				440
DNPA Synonym: DI-n-PROPYLAMINE	2383/1993	68			1470	488			93		DNA 442	314	D4.0		432

HAZARDOUS MATERIALS REFERENCE BOOKS INDEX

CHEMICAL OR MATERIAL NAME	UN/NA Number	Guide Number (DOT)	Firefighter's Hazardous Materials Reference Book	First Aid Manual for Chemical Accidents	Sax's Dangerous Properties of Industrial Materials	Hazardous Chemicals Desk Reference	Rapid Guide to Hazardous Chemicals in the Workplace	Fire Protection Guide on Hazardous Materials (NFPA)	Firefighter's Handbook of Hazardous Materials	Pocket Guide to Chemical Hazards (NIOSH)	Chemical Hazard Response Information System (CHRIS)	Emergency Handling of Hazardous Materials in Surface Transportation (AAR)	Emergency Action Guides (EAG)	Chemical Data Notebook: A User's Manual	Condensed Chemical Dictionary
DNT Synonym: 2,4-DINITROTOLUENE	1600	56			1440	480		-44	125		DTT 493				440
DO 14 Synonym: PROPARGITE	2765/3082	55	718						242		PRG 986	790			967
DO BENZENESULFONIC ACID Synonym: ALKYLBENZENESULFONIC ACID											ABS 11				
DOA Synonym: DIOCTYL ADIPATE								-44			DOA 453				440
DODECANE	1993				1493			-46	130						440
DODECANOL Synonym: LINEAR ALCOHOL			515								LAL 692				441
DODECANOL											DDN 376				441
DODECANOYL PEROXIDE Synonym: LAUROYL PEROXIDE	2124	48	328		2091	706	127		130		LPO 705				684
DODECENE	1993	27									DOD 454	390			441
DODECENE NON LINEAR Synonym: DODECENE	1993	27									DOD 454	390			441
DODECENE NON LINEAR Synonym: PROPYLENE TETRAMER	2850/1993	27	730						244		PTT1004	799			
DODECYL ALCOHOL Synonym: DODECANOL					1495	498					DDN 376				441
DODECYL MERCAPTAN Synonym: LAURYL MERCAPTAN	2124	48	498		2091	706	127	-46			LRM 707				441
DODECYL MERCAPTAN-tert								-46	130						
DODECYL PHENOL					1498	499		-46							442

CHEMICAL OR MATERIAL NAME	UN/NA Number	Guide Number (DOT)	Firefighter's Hazardous Materials Reference Book	First Aid Manual for Chemical Accidents	Sax's Dangerous Properties of Industrial Materials	Hazardous Chemicals Desk Reference	Rapid Guide to Hazardous Chemicals in the Workplace	Fire Protection Guide on Hazardous Materials (NFPA)	Firefighter's Handbook of Hazardous Materials	Pocket Guide to Chemical Hazards (NIOSH)	Chemical Hazard Response Information System (CHRIS)	Emergency Handling of Hazardous Materials in Surface Transportation (AAR)	Emergency Action Guides (EAG)	Chemical Data Notebook: A User's Manual	Condensed Chemical Dictionary
DODECYL SULFATE AMMONIUM SALT Synonym: AMMONIUM LAURYL SULFATE											ALS 45				
DODECYL SULFATE DIETHANOLAMINE SALT											DSD 475				
DODECYL SULFATE MAGNESIUM SALT											DSM 479				
DODECYL SULFATE SODIUM SALT											DDS 378				
DODECYL SULFATE TRIETHANOLAMINE SALT											DST 482				
DODECYL TRICHLOROSILANE	1771	60	329			500			130		DTC 485	391			442
DODECYLBENZENE											DDB 371				441
DODECYLBENZENE-n Synonym: DODECYLBENZENE											DDB 371				441
DODECYLBENZENESULFONIC ACID	2584	60	330						130		DSA 474	390	D5.0		441
DODECYLBENZENESULFONIC ACID CALCIUM SALT	2584	60									DCS 368				
DODECYLBENZENESULFONIC ACID ISOPROPYLAMINE SALT											DAI 335				
DODECYLBENZENESULFONIC ACID TRIETHANOLAMINE SALT											DBS 353				
DODECYLENE-alpha Synonym: 1-DODECENE											DDC 372				
DODECYLETHYLENE Synonym: 1-TETRADECENE								-85			TTD1152				1128
DOLOCHLOR Synonym: CHLOROPICRIN	1580	56	210	14	865	296	50	54 –	77	70	CPL 290	229	C4.3		271
DOP Synonym: DIOCTYL PHTHALATE					1445	481		-45			DOP 455				442
DORMANT OIL Synonym: OILS MISCELLANEOUS SPRAY	1270	27									OSY 908				443
DOW FUME 40 Synonym: ETHYLENE DIBROMIDE	1605	55	371		1601			85 –	143	110	EDB 518	430	E2.4.1		487

HAZARDOUS MATERIALS REFERENCE BOOKS INDEX

CHEMICAL OR MATERIAL NAME	UN/NA Number	Guide Number (DOT)	Firefighter's Hazardous Materials Reference Book	First Aid Manual for Chemical Accidents	Sax's Dangerous Properties of Industrial Materials	Hazardous Chemicals Desk Reference	Rapid Guide to Hazardous Chemicals in the Workplace	Fire Protection Guide on Hazardous Materials (NFPA)	Firefighter's Handbook of Hazardous Materials	Pocket Guide to Chemical Hazards (NIOSH)	Chemical Hazard Response Information System (CHRIS)	Emergency Handling of Hazardous Materials in Surface Transportation (AAR)	Emergency Action Guides (EAG)	Chemical Data Notebook: A User's Manual	Condensed Chemical Dictionary
DOWANOL 33B **Synonym: PROPYLENE GLYCOL METHYL ETHER**								–82	244		PME 966				
DOWANOL DB **Synonym: DIETHYLENE GLYCOL MONOBUTYL ETHER**					1221			–38	111		DME 430				392
DOWANOL DE **Synonym: DIETHYLENE GLYCOL MONOETHYL ETHER**	1993	27						–38	111		DGE 401	351			392
DOWANOL DM **Synonym: DIETHYLENE GLYCOL MONOMETHYL ETHER**					1221			–38	111		DGM 402				393
DOWANOL EB **Synonym: ETHYLENE GLYCOL MONOBUTYL ETHER**	2369	26						–51	143		EGM 532	432			489
DOWANOL EE **Synonym: ETHYLENE GLYCOL MONOETHYL ETHER**	1171	26	375					–51	143		EGE 529	432	E2.5		489
DOWANOL EM **Synonym: ETHYLENE GLYCOL MONOMETHYL ETHER**	1188	26						–51	143		EME 544	433			489
DOWANOL PM **Synonym: PROPYLENE GLYCOL METHYL ETHER**								–82	244		PME 966				
DOWANOL TE **Synonym: ETHOXY TRIGLYCOL**					1568	523		–47	135		ETG 565				477
DOWCO 179 **Synonym: DURSBAN**	1615										DUR 495				448
DOWFUME **Synonym: ETHYLENE DIBROMIDE**	1605	55	371					85 –	143	110	EDB 518	430	E2.4.1		444
DOWFUME **Synonym: METHYL BROMIDE**	1062	55	559	20	2274	776	137	110–67	192	146	MTB 805	619	M3.0	M 3	444
DOWICIDE 2 **Synonym: TRICHLOROPHENOL**	2020/2902	53	839								TPH1139	921			

CHEMICAL OR MATERIAL NAME	UN/NA Number	Guide Number (DOT)	Firefighter's Hazardous Materials Reference Book	First Aid Manual for Chemical Accidents	Sax's Dangerous Properties of Industrial Materials	Hazardous Chemicals Desk Reference	Rapid Guide to Hazardous Chemicals in the Workplace	Fire Protection Guide on Hazardous Materials (NFPA)	Firefighter's Handbook of Hazardous Materials	Pocket Guide to Chemical Hazards (NIOSH)	Chemical Hazard Response Information System (CHRIS)	Emergency Handling of Hazardous Materials in Surface Transportation (AAR)	Emergency Action Guides (EAG)	Chemical Data Notebook: A User's Manual	Condensed Chemical Dictionary
DOWICIDE 7 Synonym: PENTACHLOROPHENOL	2020/3082	53		22	2679	912	172	133–	223	174	PCP 940	729			879
DOWTHERM											DTH 487				444
DOWTHERM E Synonym: DICHLOROBENZENE-ortho	1591	58	272	15	1127	376	66	64 –34	101	84	DBO 350	329	D1.0.1		377
DRACYCLIC ACID Synonym: BENZOIC ACID	9094/3077		119		375	124		–17	41		BZA 191	115			131
DRI TRI Synonym: SODIUM PHOSPHATE	9148/3077	31	783		3106	1063					SPH1053	849			445
DRIER	1168	26							130			391			445
DRIER	1371	32							130						445
DROP LEAF Synonym: SODIUM CHLORATE	1495	35	766	24	3070	1048		156–	253		SDC1029	834	S2.1	S 2	1052
DRUGS	1351/2810	11							130			392			445
DRY AND CLEAR Synonym: BENZOYL PEROXIDE	2085	49	122		392	130	23		42	44			B2.0	B 2	133
DRY ICE	1845	21							130			192			446
DRYCLEANER NAPHTHA Synonym: NAPHTHA STODDARD SOLVENT	1268	27	604								NSS 849				
DRYING OIL EPOXIDES Synonym: EPOXIDIZED VEGETABLE OILS											EVO 573				
DSMA Synonym: METHANEARSONIC ACID SODIUM SALT	1557	53	548								MSA 799				446
DUBLOFIX Synonym: ETHYL CHLORIDE	1037	27	350		1586	532	92	82 –49	136	108	ECL 514	412	E2.3		483
DUMASIN	2245	26							130						

HAZARDOUS MATERIALS REFERENCE BOOKS INDEX

CHEMICAL OR MATERIAL NAME	UN/NA Number	Guide Number (DOT)	Firefighter's Hazardous Materials Reference Book	First Aid Manual for Chemical Accidents	Sax's Dangerous Properties of Industrial Materials	Hazardous Chemicals Desk Reference	Rapid Guide to Hazardous Chemicals in the Workplace	Fire Protection Guide on Hazardous Materials (NFPA)	Firefighter's Handbook of Hazardous Materials	Pocket Guide to Chemical Hazards (NIOSH)	Chemical Hazard Response Information System (CHRIS)	Emergency Handling of Hazardous Materials in Surface Transportation (AAR)	Emergency Action Guides (EAG)	Chemical Data Notebook: A User's Manual	Condensed Chemical Dictionary
DURSBAN	1615										DUR 495				448
DUST LAYING OIL Synonym: ASPHALT BLENDING STOCKS ROOFERS FLUX	1999	27	105	12	307	97	17				ARF 84	99	A12.0		
DUTCH LIQUID Synonym: ETHYLENE DICHLORIDE	1184	26	372	17	1602	540	94	86 −50	143	110	EDC 519	430	E2.4.5	E 5	487
DYE	1602	55							130			395			
DYE	2801	60							130						
DYE INTERMEDIATE	2801	60							130			395			450
DYE INTERMEDIATE	1602	55							130			395			450
DYLOX Synonym: TRICHLORFON	2783/3018	55	835						130		TRC1146	917			450
DYNAMITE					1506	501			131						451
DYTOL M 83 Synonym: OCTANOL	1987	26	650		2620						OTA 909	706	O1.0		847
DYTOL S 91 Synonym: DECYL ALCOHOL (-n)	1987	26	258		1032	347			91		DAN 338	291			350
E 1059									131						
E 605 Synonym: PARATHION	2783	55	663	22	2667	907	170		131			727			872
E D BEE Synonym: ETHYLENE DIBROMIDE	1605	55	371		1601			85 –	131	110	EDB 518	430	E2.4.1		487
E 3314 Synonym: HEPTACHLOR	2761/3077	55	424		1826	615	106		156	120	HTC 646	505			591
EAA Synonym: ETHYL ACETOACETATE	1993	27	341					−47	135		EAA 499	406			478

CHEMICAL OR MATERIAL NAME	UN/NA Number	Guide Number (DOT)	Firefighter's Hazardous Materials Reference Book	First Aid Manual for Chemical Accidents	Sax's Dangerous Properties of Industrial Materials	Hazardous Chemicals Desk Reference	Rapid Guide to Hazardous Chemicals in the Workplace	Fire Protection Guide on Hazardous Materials (NFPA)	Firefighter's Handbook of Hazardous Materials	Pocket Guide to Chemical Hazards (NIOSH)	Chemical Hazard Response Information System (CHRIS)	Emergency Handling of Hazardous Materials in Surface Transportation (AAR)	Emergency Action Guides (EAG)	Chemical Data Notebook: A User's Manual	Condensed Chemical Dictionary
EADC Synonym: ETHYLALUMINUM DICHLORIDE	1924	40						−48	131		EAD 501				452
EASC Synonym: ETHYL ALUMINUM SESQUICHLORIDE	1925	40	345					−48	136		EAS 506				452
EB Synonym: ETHYLBENZENE	1175	26	367	16	1579	528	91	81 −48	141	106	ETB 561	409	E2.2.3		480
EBDC SODIUM SALT Synonym: NABAM	1609		599						207		NAB 816				803
ECH Synonym: EPICHLOROHYDRIN	2023	30	336	16	1525	509	87	79 −46	132	102	EPC 552	401	E1.0	E 1	467
ECRINITRIT Synonym: SODIUM NITRITE	1500	35	781		3098	1060			254		SNT1051	846			1063
EDB Synonym: ETHYLENE DIBROMIDE	1605	55	371					85 −	131	110	EDB 518	430	E2.4.1		452
EDC Synonym: ETHYLENE DICHLORIDE	1184	26	372	17	1602	540	94	86 −50	143	110	EDC 519	430	E2.4.5	E 5	487
EDCO Synonym: METHYL BROMIDE	1062	55	559	20	2274	776	137	110−67	192	146	MTB 805	619	M3.0	M 3	759
EDIBLE TALLOW Synonym: TALLOW								−84			TLO1125				1114
EDTA Synonym: ETHYLENEDIAMINE TETRAACETIC ACID	9117/3077		379		1601	539			131		EDT 521	405			453
EICOSANE								−46							453
EKTASOLVE DB ACETATE Synonym: DIETHYLENE GLYCOL MONOBUTYL ETHER ACETATE								−38			DEM 389				392

HAZARDOUS MATERIALS REFERENCE BOOKS INDEX

CHEMICAL OR MATERIAL NAME	UN/NA Number	Guide Number (DOT)	Firefighter's Hazardous Materials Reference Book	First Aid Manual for Chemical Accidents	Sax's Dangerous Properties of Industrial Materials	Hazardous Chemicals Desk Reference	Rapid Guide to Hazardous Chemicals in the Workplace	Fire Protection Guide on Hazardous Materials (NFPA)	Firefighter's Handbook of Hazardous Materials	Pocket Guide to Chemical Hazards (NIOSH)	Chemical Hazard Response Information System (CHRIS)	Emergency Handling of Hazardous Materials in Surface Transportation (AAR)	Emergency Action Guides (EAG)	Chemical Data Notebook: A User's Manual	Condensed Chemical Dictionary
EKTASOLVE EE Synonym: ETHYLENE GLYCOL MONOETHYL ETHER	1171	26	375					-51	143		EGE 529	432	E2.5		489
EKTASOLVE EE ACETATE Synonym: ETHYLENE GLYCOL MONOETHYL ETHER ACE-TATE	1172	26	376					-51	143		EGA 526	432	E2.6		489
ELAYL Synonym: ETHYLENE	1038	22		16	1597	536		83 -50	142		ETL 568	436	E2.4	E 4	485
ELAYL	1962	22							131						
ELECTRICAL INSULATING OIL Synonym: OILS MISCELLANEOUS TRANSFORMER	1270	27									OTF 914				
ELECTRO CF 12 Synonym: DICHLORODIFLUOROMETHANE	1028	12	275	15	1137	380	68		102	86	DCF 360	331	D1.1		379
ELECTRO CF 22 Synonym: MONOCHLORODIFLUOROMETHANE	1018	12	593						205				M4.0		795
ELECTROLYTE BATTERY FLUID ACID	2796	39							131						
ELEMENTAL PHOSPHORUS Synonym: PHOSPHORUS YELLOW & WHITE	1381	38	693	22	2791	949	180	137–	232	182	PPW 982	760	P4.1		
EMBAFUME Synonym: METHYL BROMIDE	1062	55	559	20	2274	776	137	110–67	192	146	MTB 805	619	M3.0	M 3	759
EMERALD GREEN Synonym: COPPER ACETOARSENITE	1585	53	222				55		83		CAA 203	257			460
EMKANOL Synonym: ETHYLENE GLYCOL MONOETHYL ETHER	1171	26	375					-51	143		EGE 529	432	E2.5		489
EMMATOS	2783	55							131						
ENAMEL	1263	26							131			397			462

CHEMICAL OR MATERIAL NAME	UN/NA Number	Guide Number (DOT)	Firefighter's Hazardous Materials Reference Book	First Aid Manual for Chemical Accidents	Sax's Dangerous Properties of Industrial Materials	Hazardous Chemicals Desk Reference	Rapid Guide to Hazardous Chemicals in the Workplace	Fire Protection Guide on Hazardous Materials (NFPA)	Firefighter's Handbook of Hazardous Materials	Pocket Guide to Chemical Hazards (NIOSH)	Chemical Hazard Response Information System (CHRIS)	Emergency Handling of Hazardous Materials in Surface Transportation (AAR)	Emergency Action Guides (EAG)	Chemical Data Notebook: A User's Manual	Condensed Chemical Dictionary
ENANTHIC ALCOHOL **Synonym: HEPTANOL**			426	18							HTN 648				
ENDOSULFAN	2761	55	332		1519	505	86		131		ESF 559	398			463
ENDRATE **Synonym: ETHYLENEDIAMINE TETRAACETIC ACID**	9117/3077		379		1601	539			144		EDT 521	405			486
ENDRIN	2761/2996	55	334		1520	507	86		131	102	EDR 520	399			463
ENGINE STARTING FLUID	1960	22							131			399			
ENT 16,391 **Synonym: KEPONE**	2761/9189	55	493		2077	701			177		KPE 688	566			670
ENT 25,719 **Synonym: MIREX**	1615		589		2431	828			205		MRX 798				
ENT 27,311 **Synonym: DURSBAN**	1615										DUR 495				448
EO **Synonym: ETHYLENE OXIDE**	1040	69	377	17	1611	544	96	87 -52	144	112	EOX 550	434	E2.7	E 6	491
EPHEDRINE					1523	508									467
EPHEDRINE HYDROCHLORIDE					1523	508									
EPI **Synonym: EPICHLOROHYDRIN**	2023	30	336	16	1525	509	87	79 -46	132	102	EPC 552	401	E1.0	E 1	467
EPI CLEAR **Synonym: BENZOYL PEROXIDE**	2085	49	122		392	130	23		42	44			B2.0	B 2	133
EPIBROMOHYDRIN	2558/2810	57							132			400			
EPICHLORHYDRIN **Synonym: EPICHLOROHYDRIN**	2023	30	336	16	1525	509	87	79 -46	132	102	EPC 552	401	E1.0	E 1	467
EPICHLOROHYDRIN	2023	30	336	16	1525	509	87	79 -46	132	102	EPC 552	401	E1.0	E 1	467

CHEMICAL OR MATERIAL NAME	UN/NA Number	Guide Number (DOT)	Firefighter's Hazardous Materials Reference Book	First Aid Manual for Chemical Accidents	Sax's Dangerous Properties of Industrial Materials	Hazardous Chemicals Desk Reference	Rapid Guide to Hazardous Chemicals in the Workplace	Fire Protection Guide on Hazardous Materials (NFPA)	Firefighter's Handbook of Hazardous Materials	Pocket Guide to Chemical Hazards (NIOSH)	Chemical Hazard Response Information System (CHRIS)	Emergency Handling of Hazardous Materials in Surface Transportation (AAR)	Emergency Action Guides (EAG)	Chemical Data Notebook: A User's Manual	Condensed Chemical Dictionary
EPICHLOROHYDRIN-alpha Synonym: EPICHLOROHYDRIN	2023	30	336	16	1525	509	87	79 –46	132	102	EPC 552	401	E1.0	E 1	467
EPICHLOROPHYDRIN Synonym: EPICHLOROHYDRIN	2023	30	336	16	1525	509	87	79 –46	132	102	EPC 552	401	E1.0	E 1	467
EPN				16	1527	509	87		132	102					468
EPOXIDIZED DRYING OILS Synonym: EPOXIDIZED VEGETABLE OILS											EVO 573				
EPOXIDIZED OILS Synonym: EPOXIDIZED VEGETABLE OILS											EVO 573				
EPOXIDIZED TALL OIL OCTYL ESTER Synonym: OCTYL EPOXY TALLATE			652								OET 878				
EPOXIDIZED VEGETABLE OILS											EVO 573				
EPOXY PROPANE Synonym: PROPYLENE OXIDE	1280	26	729	23	2922	1003	190	150–82	244	190	POX 971	798	P8.1		971
EPOXY RESINS UNCURED					1536	512			132						
EPOXYETHANE Synonym: ETHYLENE OXIDE	1040	69	377	17	1611	544	96	87 –52	132	112	EOX 550	434	E2.7	E 6	468
EPOXYETHOXYPROPANE	2752	26							132						
ERADICATOR	1850	27							133						
ERIOCHOLCITE Synonym: COPPER CHLORIDE	2802	60	224	14	946	320	55		83		CPC 285	259			311
ERYTHRENE Synonym: BUTADIENE	1010	17	139	12				40 –	133	48	BDI 134	148	B4.0	B 4	
ESKIMON 11 Synonym: TRICHLOROFLUOROMETHANE	3082/9188		838		3353	1138	219		273		TCF1083	919			1170

CHEMICAL OR MATERIAL NAME	UN/NA Number	Guide Number (DOT)	Firefighter's Hazardous Materials Reference Book	First Aid Manual for Chemical Accidents	Sax's Dangerous Properties of Industrial Materials	Hazardous Chemicals Desk Reference	Rapid Guide to Hazardous Chemicals in the Workplace	Fire Protection Guide on Hazardous Materials (NFPA)	Firefighter's Handbook of Hazardous Materials	Pocket Guide to Chemical Hazards (NIOSH)	Chemical Hazard Response Information System (CHRIS)	Emergency Handling of Hazardous Materials in Surface Transportation (AAR)	Emergency Action Guides (EAG)	Chemical Data Notebook: A User's Manual	Condensed Chemical Dictionary
ESKIMON 12 **Synonym: DICHLORODIFLUOROMETHANE**	1028	12	275	15	1137	380	68		102	86	DCF 360	331	D1.1		379
ESKIMON 22 **Synonym: CHLORODIFLUOROMETHANE**	1018	12		14	791		48		74		MCF 737	222			265
ESKIMON 22 **Synonym: MONOCHLORODIFLUOROMETHANE**	1018	12	593			278			205				M4.0		795
ESKINION 12 **Synonym: DICHLORODIFLUOROMETHANE**	1028	12	275	15	1137	380	68		102	86	DCF 360	331	D1.1		379
ESSENCE OF MIRBANE **Synonym: NITROBENZENE**	1662	55	631	21	2518	861	158	124–74	212	162	NTB 852		N3.1		825
ESSENCE OF MYRBANE **Synonym: NITROBENZENE**	1662	55	631	21	2518	861	158	124–74	212	162	NTB 852		N3.1		825
ESTER **Synonym: METHYL CHLOROFORMATE**	1238	28	562	20	1547	514			193		MHC 760	624	M3.1.1		473
ETCHING ACID	1790	59										401			
ETHANAL **Synonym: ACETALDEHYDE**	1089	26	16	11	5	3	1	13 –11	133	30	AAD 2	1	A1.0	A 1	474
ETHANE	1035	22	337	16	1553	517		–46	133		ETH 566	402			474
ETHANE	1961	22		16	1553	517			133		ETH 566	402			474
ETHANE CARBOXYLIC ACID **Synonym: PROPIONIC ACID**	1848	60	721		2909	997	188	148–81	243		PNA 968	791	P7.1		968
ETHANE DINITRILE **Synonym: CYANOGEN**	1026	18	241		979	331	57	–29	134		CYG 330	277	C8.0		333
ETHANE HEXAMERCARBIDE					1554				133						
ETHANE PROPANE MIXTURE	1961/1954	22							133			402			

CHEMICAL OR MATERIAL NAME	UN/NA Number	Guide Number (DOT)	Firefighter's Hazardous Materials Reference Book	First Aid Manual for Chemical Accidents	Sax's Dangerous Properties of Industrial Materials	Hazardous Chemicals Desk Reference	Rapid Guide to Hazardous Chemicals in the Workplace	Fire Protection Guide on Hazardous Materials (NFPA)	Firefighter's Handbook of Hazardous Materials	Pocket Guide to Chemical Hazards (NIOSH)	Chemical Hazard Response Information System (CHRIS)	Emergency Handling of Hazardous Materials in Surface Transportation (AAR)	Emergency Action Guides (EAG)	Chemical Data Notebook: A User's Manual	Condensed Chemical Dictionary
ETHANECARBOXYLIC ACID Synonym: PROPIONIC ACID	1848	60	721		2909	997	188	148–81	243		PNA 968	791	P7.1		968
ETHANEDIAL Synonym: GLYOXAL			421		1802	609			134		GOS 610				570
ETHANEDINITRILE Synonym: CYANOGEN	1026	18	241		979	331	57	−29	134		CYG 330	277	C8.0		333
ETHANEDIOIC ACID Synonym: OXALIC ACID	2449	54	658	21	2639	899	169	131–	134	170	OXA 919				860
ETHANEDIOIC ACID DISODIUM SALT Synonym: SODIUM OXALATE			782		3103						SOX1052				1063
ETHANEDIOYL CHLORIDE									134						474
ETHANENITRILE Synonym: ACETONITRILE	1648	28	22	11	24	12	3	15 −12	134	32	ATN 104	5			9
ETHANESULFONYL CHLORIDE					1555				134						
ETHANETHIOL Synonym: ETHYL MERCAPTAN	1228	28									EMC 543				474
ETHANETHIOL Synonym: ETHYL MERCAPTAN	2363	27	358		1631	550	97	−53	134	112		418	E2.9		474
ETHANETHIOLIC ACID	2436	26							134						475
ETHANOIC ACID Synonym: ACETIC ACID GLACIAL	2789	29	18	11	15	6	1	14 −11	134	30	AAC 1	3	A2.0	A 2	475
ETHANOIC ANHYDRIDE Synonym: ACETIC ANHYDRIDE	1715	39	19	11	18	8	2	14 −11	134	30	ACA 13	3	A2.1		7
ETHANOL Synonym: ETHYL ALCOHOL	1170	26	344	16	1572	525	88	−48	134		EAL 504	408	E2.2	E 2	475

CHEMICAL OR MATERIAL NAME	UN/NA Number	Guide Number (DOT)	Firefighter's Hazardous Materials Reference Book	First Aid Manual for Chemical Accidents	Sax's Dangerous Properties of Industrial Materials	Hazardous Chemicals Desk Reference	Rapid Guide to Hazardous Chemicals in the Workplace	Fire Protection Guide on Hazardous Materials (NFPA)	Firefighter's Handbook of Hazardous Materials	Pocket Guide to Chemical Hazards (NIOSH)	Chemical Hazard Response Information System (CHRIS)	Emergency Handling of Hazardous Materials in Surface Transportation (AAR)	Emergency Action Guides (EAG)	Chemical Data Notebook: A User's Manual	Condensed Chemical Dictionary
ETHANOLAMINE **Synonym: MONOETHANOLAMINE**	2491	60	594	16	1555	518	88	118–47	134	104	MEA 749	403	M5.0		475
ETHANOLAMINE-beta **Synonym: MONOETHANOLAMINE**	2491	60	594	16			88	118–			MEA 749		M5.0		795
ETHANOYL BROMIDE	1716	60							134						
ETHANOYL CHLORIDE **Synonym: ACETYL CHLORIDE**	1717	29	25		41	19		16 –12	134		ACC 15	7			12
ETHANOYL IODIDE	1898	60							134						
ETHANOYL PEROXIDE	2084	49							134						
ETHENE **Synonym: ETHYLENE**	1038	22		16	1597	536		83 –50	142		ETL 568	436	E2.4	E 4	475
ETHENE **Synonym: ETHYLENE**	1962	21	368	16	1597	536		83 –50	134		ETL 568	436	E2.4	E 4	475
ETHENEOXIDE **Synonym: ETHYLENE OXIDE**	1040	69	377	17	1611	544	96	87 –52	144	112	EOX 550	434	E2.7	E 6	491
ETHENYL ETHANOATE **Synonym: VINYL ACETATE**	1301	26	871	26	3492		227	179–92	134		VAM1172	948	V1.0		1215
ETHENYLBENZENE **Synonym: STYRENE MONOMER**	2055	27	795		3145	1074	202		257		STY1064	862	S3.0	S 6	1097
ETHENYLOXYETHENE	1167	30							134						
ETHER **Synonym: ETHYL ETHER**	1155	26	338	16	1556	519	96	87 –52	134	112	EET 524	346	E2.8	E 3	475
ETHER CHLORATUS **Synonym: ETHYL CHLORIDE**	1037	27	350		1586	532	92	82 –49		108	ECL 514	412	E2.3		483
ETHER HYDROCHLORIC **Synonym: ETHYL CHLORIDE**	1037	27	350		1586	532	92	82 –49		108	ECL 514	412	E2.3		483

CHEMICAL OR MATERIAL NAME	UN/NA Number	Guide Number (DOT)	Firefighter's Hazardous Materials Reference Book	First Aid Manual for Chemical Accidents	Sax's Dangerous Properties of Industrial Materials	Hazardous Chemicals Desk Reference	Rapid Guide to Hazardous Chemicals in the Workplace	Fire Protection Guide on Hazardous Materials (NFPA)	Firefighter's Handbook of Hazardous Materials	Pocket Guide to Chemical Hazards (NIOSH)	Chemical Hazard Response Information System (CHRIS)	Emergency Handling of Hazardous Materials in Surface Transportation (AAR)	Emergency Action Guides (EAG)	Chemical Data Notebook: A User's Manual	Condensed Chemical Dictionary
ETHER VINYL ETHYL Synonym: VINYL ETHYL ETHER	1302	26	873					-93			VEE1175	950			1216
ETHERIN Synonym: ETHYLENE	1038	22		16	1597	536		83 -50	142		ETL 568	436	E2.4	E 4	485
ETHIDE	2650	57							134						
ETHINE Synonym: ACETYLENE	1001	17	29	11	43	20	3	17 -12	134		ACE 17	9		A 4	12
ETHINYL TRICHLORIDE Synonym: TRICHLOROETHYLENE	1710	74	837	25	3352	1137	218	174-88	134	216	TCL1085	919	T5.0	T 6	1170
ETHION	2783/3018	55	339		1556	520	88		134		ETO 571	404			476
ETHIOPS MINERAL Synonym: MERCURIC SULFIDE	2025	53	539						186		MSF 800				
ETHIRYDINE Synonym: ETHYLENEIMINE	1185	30	380	17	1610	544	95	86 -51	144	110	ETI 567	437	E2.6.1		490
ETHOXY ACETATE Synonym: ETHYLENE GLYCOL MONOETHYL ETHER ACETATE	1172	26	376					-51	143		EGA 526	432	E2.6		489
ETHOXY DIGLYCOL Synonym: DIETHYLENE GLYCOL MONOETHYL ETHER	1993	27						-38	111		DGE 401	351			392
ETHOXY TRIGLYCOL					1568	523		-47	135		ETG 565				477
ETHOXYACETYLENE					1560	521		-47	135						
ETHOXYBENZENE								-47	135						
ETHOXYDIHYDROPYRAN											EHP 538				
ETHOXYETHANE Synonym: ETHYL ETHER	1155	26	354	16	1614	546	96	87 -52	137	112	EET 524	346	E2.8	E 3	492
ETHOXYETHANOL	1171	26			1563				135						

CHEMICAL OR MATERIAL NAME	UN/NA Number	Guide Number (DOT)	Firefighter's Hazardous Materials Reference Book	First Aid Manual for Chemical Accidents	Sax's Dangerous Properties of Industrial Materials	Hazardous Chemicals Desk Reference	Rapid Guide to Hazardous Chemicals in the Workplace	Fire Protection Guide on Hazardous Materials (NFPA)	Firefighter's Handbook of Hazardous Materials	Pocket Guide to Chemical Hazards (NIOSH)	Chemical Hazard Response Information System (CHRIS)	Emergency Handling of Hazardous Materials in Surface Transportation (AAR)	Emergency Action Guides (EAG)	Chemical Data Notebook: A User's Manual	Condensed Chemical Dictionary
ETHOXYETHYL ACETATE Synonym: ETHYLENE GLYCOL MONOETHYL ETHER ACETATE	1172	26	376		1563			–51	135		EGA 526	432	E2.6		489
ETHOXYETHYL ACETATE-beta Synonym: ETHYLENE GLYCOL MONOETHYL ETHER ACETATE	1172	26	376					–51	135		EGA 526	432	E2.6		489
ETHOXYLATED DODECANOL											EOD 547				
ETHOXYLATED DODECYL ALCOHOL Synonym: ETHOXYLATED DODECANOL											EOD 547				
ETHOXYLATED LAURYL ALCOHOL Synonym: ETHOXYLATED DODECANOL											EOD 547				
ETHOXYLATED MYRISTYL ALCOHOL Synonym: ETHOXYLATED TETRADECANOL											EOT 549				
ETHOXYLATED NONYLPHENOL											ENP 546				
ETHOXYLATED PENTADECANOL											EOP 548				
ETHOXYLATED PENTADECYLALCOHOL Synonym: ETHOXYLATED PENTADECANOL											EOP 548				
ETHOXYLATED TETRADECANOL											EOT 549				
ETHOXYLATED TETRADECYL ALCOHOL Synonym: ETHOXYLATED TETRADECANOL											EOT 549				
ETHOXYLATED TRIDECANOL											ETD 563				
ETHOXYLATED TRIDECYL ALCOHOL Synonym: ETHOXYLATED TRIDECANOL											ETD 563				
ETHOXYPROPANE	2615	26							135						
ETHOXYTRIETHYLENE GLYCOL Synonym: ETHOXY TRIGLYCOL					1568	523		–47	135		ETG 565				477

HAZARDOUS MATERIALS REFERENCE BOOKS INDEX

CHEMICAL OR MATERIAL NAME	UN/NA Number	Guide Number (DOT)	Firefighter's Hazardous Materials Reference Book	First Aid Manual for Chemical Accidents	Sax's Dangerous Properties of Industrial Materials	Hazardous Chemicals Desk Reference	Rapid Guide to Hazardous Chemicals in the Workplace	Fire Protection Guide on Hazardous Materials (NFPA)	Firefighter's Handbook of Hazardous Materials	Pocket Guide to Chemical Hazards (NIOSH)	Chemical Hazard Response Information System (CHRIS)	Emergency Handling of Hazardous Materials in Surface Transportation (AAR)	Emergency Action Guides (EAG)	Chemical Data Notebook: A User's Manual	Condensed Chemical Dictionary
ETHOXYTRIMETHYLSILANE					1569				135						
ETHYL 2-HYDROXYPROPANOATE Synonym: ETHYL LACTATE	1192	26	357					-53			ELT 541	418			495
ETHYL 2-HYDROXYPROPIONATE Synonym: ETHYL LACTATE	1192	26	357					-53	140		ELT 541	418			495
ETHYL 2-METHACRYLATE Synonym: ETHYL METHACRYLATE	2277/1993	26	359		1633	551		-53	138		ETM 569	419	E3.0		496
ETHYL 2-METHYL-2-PROPENOATE Synonym: ETHYL METHACRYLATE	2277/1993	26	359		1633	551		-53	138		ETM 569	419	E3.0		496
ETHYL 2-PROPENOATE Synonym: ETHYL ACRYLATE	1917	26	343	16	1571	524	90	80 -48	136	106	EAC 500	407	E2.1		478
ETHYL 3,3-DI(tert-BUTYLPEROXY) BUTYRATE	2184	49							140						
ETHYL 3,3-DI(tert-BUTYLPEROXY) BUTYRATE	2185	48							140						
ETHYL 3,3-DI(tert-BUTYLPEROXY) BUTYRATE	2598	48							140						
ETHYL 3-OXOBUTANOATE Synonym: ETHYL ACETOACETATE	1993	27	341					-47	140		EAA 499	406			478
ETHYL ACETANILIDE-n					1570	523		-47	135						477
ETHYL ACETATE	1173	26	340	16	1570	523	89	-47	135	104	ETA 560	406	E2.0		477
ETHYL ACETIC ACID	2820	60							135						478
ETHYL ACETOACETATE	1993	27	341					-47	135		EAA 499	406			478
ETHYL ACETONE	1249	26							135						478
ETHYL ACETYL GLYCOLATE								-47	135						
ETHYL ACETYLENE	2452/1954	17			1571	524			141	1		407			478
ETHYL ACRYLATE	1173	26	342	16											478
ETHYL ACRYLATE	1917	26	343	16	1571	524	90	80 -48	136	106	EAC 500	407	E2.1		478

CHEMICAL OR MATERIAL NAME	UN/NA Number	Guide Number (DOT)	Firefighter's Hazardous Materials Reference Book	First Aid Manual for Chemical Accidents	Sax's Dangerous Properties of Industrial Materials	Hazardous Chemicals Desk Reference	Rapid Guide to Hazardous Chemicals in the Workplace	Fire Protection Guide on Hazardous Materials (NFPA)	Firefighter's Handbook of Hazardous Materials	Pocket Guide to Chemical Hazards (NIOSH)	Chemical Hazard Response Information System (CHRIS)	Emergency Handling of Hazardous Materials in Surface Transportation (AAR)	Emergency Action Guides (EAG)	Chemical Data Notebook: A User's Manual	Condensed Chemical Dictionary
ETHYL ALCOHOL	1170	26	344	16	1572	525	90	−48	136		EAL 504	408	E2.2	E 2	478
ETHYL ALCOHOL ANHYDROUS	1986	28		16										E 2	
ETHYL ALDEHYDE Synonym: ACETALDEHYDE	1089	26	16	11	5	3	1	13 −11	136	30	AAD 2	1	A1.0	A 1	5
ETHYL alpha-HYDROXYPROPIONATE Synonym: ETHYL LACTATE	1192	26	357					−53	138		ELT 541	418			495
ETHYL alpha-METHYLMETHACRYLATE Synonym: ETHYL METHACRYLATE	2277/1993	26	359		1633	551		−53	138		ETM 569	419	E3.0		496
ETHYL ALUMINUM DICHLORIDE	1924	40						−48	136		EAD 501				479
ETHYL ALUMINUM SESQUICHLORIDE	1925	40	345					−48	136		EAS 506				479
ETHYL AMINO ETHANOL	1987	26						−48	136						479
ETHYL AMYL KETONE	2271/1993	26			1576	527	91		136		EAK 503	409			479
ETHYL BENZENE	1175	26	367	16	1579	528	91	81 −48	141	106	ETB 561	409	E2.2.3		480
ETHYL BENZOATE					1579	529		−48	136						480
ETHYL BENZOL Synonym: ETHYL BENZENE	1175	26	367		1579	528	91	81 −48	141	106	ETB 561	409	E2.2.3		480
ETHYL BORATE	1176	26						−48	136			410			481
ETHYL BROMIDE	1891/2810	58	346		1581	529	92	−48	136	106		410			481
ETHYL BROMOACETATE	1603/2810	55						−48	136			410			481
ETHYL BROMOETHANOATE	1603	55							136						
ETHYL BUTANOATE Synonym: ETHYL BUTYRATE	1180	26	349		1584			−49	136		EBR 508	412			481
ETHYL BUTYL ALCOHOL	1993	27													
ETHYL BUTYL ALCOHOL	1987/2275	26										411			
ETHYL BUTYL CARBONATE								−49	142						482

CHEMICAL OR MATERIAL NAME	UN/NA Number	Guide Number (DOT)	Firefighter's Hazardous Materials Reference Book	First Aid Manual for Chemical Accidents	Sax's Dangerous Properties of Industrial Materials	Hazardous Chemicals Desk Reference	Rapid Guide to Hazardous Chemicals in the Workplace	Fire Protection Guide on Hazardous Materials (NFPA)	Firefighter's Handbook of Hazardous Materials	Pocket Guide to Chemical Hazards (NIOSH)	Chemical Hazard Response Information System (CHRIS)	Emergency Handling of Hazardous Materials in Surface Transportation (AAR)	Emergency Action Guides (EAG)	Chemical Data Notebook: A User's Manual	Condensed Chemical Dictionary
ETHYL BUTYL ETHER	1179	26			1583	531		-49	136			411			482
ETHYL BUTYL KETONE			348		1583	531	92	-49	136	108					482
ETHYL BUTYRATE	1180	26	349		1584			-49	142		EBR 508	412			482
ETHYL CAPROATE	1177	26			1585	532		-49	136						482
ETHYL CAPRYLATE								-49	136						483
ETHYL CARBONATE Synonym: DIETHYL CARBONATE	2366/1993	26	287		1213	406		-37	136		DEC 383	345			483
ETHYL CELLOSOLVE Synonym: ETHYLENE GLYCOL MONOETHYL ETHER	1171	26	375					-51	143		EGE 529	432	E2.5		489
ETHYL CHLORACETATE Synonym: ETHYL CHLOROACETATE	1181/1993	55	351		1585	532		-49	136		ECA 510	412			483
ETHYL CHLORIDE	1037	27	350		1586	532	92	82 -49	136	108	ECL 514	412	E2.3		483
ETHYL CHLOROACETATE Synonym: ETHYL CHLORACETATE	1181/1993	55	351		1585	532		-49	136		ECA 510	412			483
ETHYL CHLOROCARBONATE Synonym: ETHYL CHLOROFORMATE	1182	28	352	16	1587	533		82 -50	136		ECF 512	413	E2.3.1		483
ETHYL CHLOROETHANOATE Synonym: ETHYL CHLOROACETATE	1181/1993	55	351					-49	137		ECA 510	412			483
ETHYL CHLOROFORMATE	1182	28	352	16	1587	533		82 -50	137		ECF 512	413	E2.3.1		484
ETHYL CHLOROPROPIONATE	2935	29							137						
ETHYL CHLOROTHIOFORMATE	2826	59							137			413			
ETHYL CHLOROTHIOLFORMATE	2826	59							137						
ETHYL CROTONATE	1862	26			1588			-50	137			413			484
ETHYL CYANIDE	2404	28							137						484
ETHYL CYANOACETATE	2666	55			1589	534		-50	137			414			484

CHEMICAL OR MATERIAL NAME	UN/NA Number	Guide Number (DOT)	Firefighter's Hazardous Materials Reference Book	First Aid Manual for Chemical Accidents	Sax's Dangerous Properties of Industrial Materials	Hazardous Chemicals Desk Reference	Rapid Guide to Hazardous Chemicals in the Workplace	Fire Protection Guide on Hazardous Materials (NFPA)	Firefighter's Handbook of Hazardous Materials	Pocket Guide to Chemical Hazards (NIOSH)	Chemical Hazard Response Information System (CHRIS)	Emergency Handling of Hazardous Materials in Surface Transportation (AAR)	Emergency Action Guides (EAG)	Chemical Data Notebook: A User's Manual	Condensed Chemical Dictionary
ETHYL DICHLORIDE	1993	27										414			
ETHYL DICHLOROPHOSPHATE Synonym: ETHYL PHOSPHORODICHLORIDATE	1760/2927	60	363		1654	558			139		EPP 556	422			
ETHYL DL LACTATE Synonym: ETHYL LACTATE	1192	26	357					-53	138		ELT 541	418			495
ETHYL ETHANOATE Synonym: ETHYL ACETATE	1173	26	340	16	1570	523	89	-47	137	104	ETA 560	406	E2.0		477
ETHYL ETHANOATE ACETIC ESTER Synonym: ETHYL ACETATE	1173	26	340	16	1570	523	89	-47	135	104	ETA 560	406	E2.0		477
ETHYL ETHER	1155	26	354	16	1614	546	96	87 -52	137	112	EET 524	346	E2.8	E 3	492
ETHYL FLUID	1649	56							137						
ETHYL FLUORIDE	2453	12		16				-52	137			415			
ETHYL FORMATE	1190	26	355		1617	547	97	-52	137	112	EFM 525	415			492
ETHYL FORMATE-ortho	1190	26						-52	137						
ETHYL FORMIC ESTER Synonym: ETHYL FORMATE	1190	26	355		1617	547	97	-52	137	112	EFM 525	415			492
ETHYL GLYCOL ACETATE Synonym: ETHYLENE GLYCOL MONOETHYL ETHER ACE-TATE	1172	26	376					-51	137		EGA 526	432	E2.6		490
ETHYL GUTHION	2783	55			1618	547			137						
ETHYL HEXALDEHYDE	1191	26	356						144		EHA 535	438			
ETHYL HEXYL TALLATE											EHT 539				
ETHYL HEXYLAMINE	2276	29				548			145						
ETHYL HEXYLCHLOROFORMATE	2748	55							137						
ETHYL HYDRATE Synonym: ETHYL ALCOHOL	1170	26	344	16	1572	525	90	-48	136		EAL 504	408	E2.2	E 2	478

HAZARDOUS MATERIALS REFERENCE BOOKS INDEX

CHEMICAL OR MATERIAL NAME	UN/NA Number	Guide Number (DOT)	Firefighter's Hazardous Materials Reference Book	First Aid Manual for Chemical Accidents	Sax's Dangerous Properties of Industrial Materials	Hazardous Chemicals Desk Reference	Rapid Guide to Hazardous Chemicals in the Workplace	Fire Protection Guide on Hazardous Materials (NFPA)	Firefighter's Handbook of Hazardous Materials	Pocket Guide to Chemical Hazards (NIOSH)	Chemical Hazard Response Information System (CHRIS)	Emergency Handling of Hazardous Materials in Surface Transportation (AAR)	Emergency Action Guides (EAG)	Chemical Data Notebook: A User's Manual	Condensed Chemical Dictionary
ETHYL HYDRIDE	1035	22							138						
ETHYL HYDROXIDE Synonym: ETHYL ALCOHOL	1170	26	344	16	1572	525	90	-48	136		EAL 504	408	E2.2	E 2	478
ETHYL IODIDE					1628				138						495
ETHYL IODOACETATE					1628				138						495
ETHYL ISOBUTYRATE	2385/1993	26			1628	549		-53	138			439			495
ETHYL ISOCYANATE	2481/1993	28			1629	549			138			417			495
ETHYL ISOCYANIDE					1629				138						
ETHYL LACTATE	1192	26	357					-53	138		ELT 541	418			495
ETHYL MALONATE					1631	550			138						496
ETHYL MERCAPTAN	1228	28									EMC 543				496
ETHYL MERCAPTAN	2363	27	358		1631	550	97	-53	138	112		418	E2.9		496
ETHYL META TOLUIDINE-n	1993	27													501
ETHYL METHACRYLATE	2277/1993	26	359		1633	551		-53	138		ETM 569	419	E3.0		496
ETHYL METHACRYLATE INHIBITED Synonym: ETHYL METHACRYLATE	2277/1993	26	359		1633	551		-53	138		ETM 569	419	E3.0		496
ETHYL METHANOATE Synonym: ETHYL FORMATE	1190	26	355		1617	547	97	-52	138	112	EFM 525	415			492
ETHYL METHYL ACRYLATE Synonym: ETHYL METHACRYLATE	2277/1993	26	359		1633	551		-53	138		ETM 569	419	E3.0		496
ETHYL METHYL ETHER	1039	26			1637	552		89 –	138			419			497
ETHYL METHYL KETONE Synonym: METHYL ETHYL KETONE	1193	26	566	20	2319	793	143	-69	138	8	MEK 751	627	M3.2	M 5	497
ETHYL METHYL KETONE PEROXIDE	2550	51							138						
ETHYL METHYL KETONE PEROXIDE	2127	49							138						

CHEMICAL OR MATERIAL NAME	UN/NA Number	Guide Number (DOT)	Firefighter's Hazardous Materials Reference Book	First Aid Manual for Chemical Accidents	Sax's Dangerous Properties of Industrial Materials	Hazardous Chemicals Desk Reference	Rapid Guide to Hazardous Chemicals in the Workplace	Fire Protection Guide on Hazardous Materials (NFPA)	Firefighter's Handbook of Hazardous Materials	Pocket Guide to Chemical Hazards (NIOSH)	Chemical Hazard Response Information System (CHRIS)	Emergency Handling of Hazardous Materials in Surface Transportation (AAR)	Emergency Action Guides (EAG)	Chemical Data Notebook: A User's Manual	Condensed Chemical Dictionary
ETHYL NITRATE	1993	27		16	1642	554		−53	138						497
ETHYL NITRILE Synonym: ACETONITRILE	1648	28	22	11	24	12	3	15 −12	138	32	ATN 104	5			9
ETHYL NITRITE	1194	30	360		1643	554		89 −53	138		ETN 570	420			497
ETHYL ORTHOFORMATE	2324/1993	26			1648	556			139			420			
ETHYL ORTHOSILICATE Synonym: ETHYL SILICATE	1292	29	365		1659	560	98	−54	139	114	ESC 558				500
ETHYL OXALATE	2525	54						−53	139			421			498
ETHYL OXIDE	1155	26							139						
ETHYL PARATHION Synonym: PARATHION	2784	28		22	2667	907			222		PTO1000				498
ETHYL PERCHLORATE					1649	557			139						
ETHYL PHENYL DICHLOROSILANE	2435	39	361						139		EPS 557	440			499
ETHYL PHENYL KETONE					1652			−54							499
ETHYL PHENYLACETATE					1650	557		−53							498
ETHYL PHENYLAMINE	2272	55							139						
ETHYL PHOSPHINE									139						
ETHYL PHOSPHONOTHIOIC DICHLORIDE	1760	60	362						139		EPD 553				
ETHYL PHOSPHONOUS DICHLORIDE	2845	40				558			139			422			
ETHYL PHOSPHORODICHLORIDATE	1760/2927	60	363			558			139		EPP 556	422			
ETHYL PHOSPHORODICHLORIDOTHIONATE Synonym: ETHYL PHOSPHONOTHIOIC DICHLORIDE	1760	60	362						139		EPD 553				
ETHYL PHTHALATE Synonym: DIETHYL PHTHALATE	9 88/3082		288		1236	415	74	−39	139		DPH 465	348			499
ETHYL PIPERIDINE	2386	26							139						

HAZARDOUS MATERIALS REFERENCE BOOKS INDEX

CHEMICAL OR MATERIAL NAME	UN/NA Number	Guide Number (DOT)	Firefighter's Hazardous Materials Reference Book	First Aid Manual for Chemical Accidents	Sax's Dangerous Properties of Industrial Materials	Hazardous Chemicals Desk Reference	Rapid Guide to Hazardous Chemicals in the Workplace	Fire Protection Guide on Hazardous Materials (NFPA)	Firefighter's Handbook of Hazardous Materials	Pocket Guide to Chemical Hazards (NIOSH)	Chemical Hazard Response Information System (CHRIS)	Emergency Handling of Hazardous Materials in Surface Transportation (AAR)	Emergency Action Guides (EAG)	Chemical Data Notebook: A User's Manual	Condensed Chemical Dictionary
ETHYL PROPENOATE	1917	26	343	16	1571	524	90	80 -48	139	106	EAC 500	407	E2.1		478
Synonym: ETHYL ACRYLATE															
ETHYL PROPENYL ETHER					1656			-54	139						
ETHYL PROPIONATE	1195	26	364		1656	559		-54	139			423			499
ETHYL PROPIONYL	1156	26							139						499
ETHYL PROPYL ETHER	2615	26			1657	559		-54	139			423			
ETHYL SILICATE	1292	29	365		1659	560	98	-54	139	114	ESC 558	891			500
ETHYL SILICATE 40	1292	29	365		1659	560	98	-54	139	114	ESC 558	891			500
Synonym: ETHYL SILICATE															
ETHYL SILICATE CONDENSED	1292	29	365		1659	560	98	-54	139	114	ESC 558	891			500
Synonym: ETHYL SILICATE															
ETHYL SULFATE	1594	55							139						500
ETHYL SULFHYDRATE	2363	27	358		1631		97	-53	139	112		418	E2.9		500
Synonym: ETHYL MERCAPTAN															
ETHYL SULFHYDRATE	1228	28			1631	550	97				EMC 543				500
Synonym: ETHYL MERCAPTAN															
ETHYL SULFIDE	2375	28			1660	560			139						500
ETHYL SULFURIC ACID	2571	60							146			440			500
ETHYL TETRAPHOSPHATE	1611	55							139						
ETHYL THIONOPHOSPHORYL DICHLORIDE	1760	60	362						139		EPD 553				
Synonym: ETHYL PHOSPHONOTHIOIC DICHLORIDE															
ETHYL TOLUIDINE	2754	55							140						
ETHYL VINYL ETHER	1302	26	873		1668	562		-93	140		VEE1175	950			501
Synonym: VINYL ETHYL ETHER															
ETHYL-2-HYDROXYPROPIONATE	1192	26							140						

CHEMICAL OR MATERIAL NAME	UN/NA Number	Guide Number (DOT)	Firefighter's Hazardous Materials Reference Book	First Aid Manual for Chemical Accidents	Sax's Dangerous Properties of Industrial Materials	Hazardous Chemicals Desk Reference	Rapid Guide to Hazardous Chemicals in the Workplace	Fire Protection Guide on Hazardous Materials (NFPA)	Firefighter's Handbook of Hazardous Materials	Pocket Guide to Chemical Hazards (NIOSH)	Chemical Hazard Response Information System (CHRIS)	Emergency Handling of Hazardous Materials in Surface Transportation (AAR)	Emergency Action Guides (EAG)	Chemical Data Notebook: A User's Manual	Condensed Chemical Dictionary
ETHYL-n-BUTYL ETHER	1179	26			1583				141						482
ETHYL-n-BUTYLAMINE	1993	27													481
ETHYL-n-BUTYLAMINE-n	2733	29									EBA 507				
ETHYL-n-BUTYRATE	1180	26			1584	531									
ETHYL-n-CYCLOHEXYLAMINE	1993	27			1590						ECC 511	424			484
ETHYL-o-FORMATE	1190	26							141						492
ETHYLACETIC ACID Synonym: BUTYRIC ACID (-n)	2820	60	154		643	216		44 -25	135		BRA 164	173			478
ETHYLAMINE	1036	68	366	16	1573	526	90	-48	141	106	EAM 505	427	E2.2.1		479
ETHYLAMINE SOLUTION	2270/1993	29		16	1573		90		141		EAM 505	427			479
ETHYLANILINE (-n)	2272	55			1577	528		80 -48	141						479
ETHYLANILINE-ortho	2273	55							141						480
ETHYLBENZENE	1175	26	367	16	1579	528	91	81 -48	141	106	ETB 561	409	E2.2.3		480
ETHYLBENZOL	1175	26							141						
ETHYLBENZYL TOLUIDINE	2753	53							141						
ETHYLBENZYLANILINE	2274	53						-48	136						481
ETHYLBUTANOL	2275	26	347						141		EBT 509				
ETHYLBUTYL ACETATE	1177	26							142			411			
ETHYLBUTYLAMINE Synonym: ETHYL-n-BUTYLAMINE-n	2733	29						-49	142		EBA 507				481
ETHYLBUTYRALDEHYDE	1178	26							142			411			
ETHYLCARBINOL Synonym: PROPYL ALCOHOL (-n)	1274	26	724	23	2915	1000	189		136	188	'PAL 925	794			969
ETHYLCYCLOBUTANE					1589			-50	137						

CHEMICAL OR MATERIAL NAME	UN/NA Number	Guide Number (DOT)	Firefighter's Hazardous Materials Reference Book	First Aid Manual for Chemical Accidents	Sax's Dangerous Properties of Industrial Materials	Hazardous Chemicals Desk Reference	Rapid Guide to Hazardous Chemicals in the Workplace	Fire Protection Guide on Hazardous Materials (NFPA)	Firefighter's Handbook of Hazardous Materials	Pocket Guide to Chemical Hazards (NIOSH)	Chemical Hazard Response Information System (CHRIS)	Emergency Handling of Hazardous Materials in Surface Transportation (AAR)	Emergency Action Guides (EAG)	Chemical Data Notebook: A User's Manual	Condensed Chemical Dictionary
ETHYLCYCLOHEXANE								-50	137		ECY 516				484
ETHYLCYCLOHEXYLAMINE-n	2920/1760	29			1590	534		-50	142		ECC 511	428			484
ETHYLCYCLOPENTANE					1590			-50	137						484
ETHYLDICHLOROARSINE	1892/2810	55							142			428			484
ETHYLDICHLOROSILANE	1183	29	353					-50	142		ECS 515	415			485
ETHYLDIMETHYLMETHANE Synonym: ISOPENTANE	1265	27	476		1595	536	93	-62	137		IPT 680		12.0		485
ETHYLENBIS (DITHIOCARBAMIC ACID) DISODIUM SALT Synonym: NABAM	1609		599						207		NAB 816				803
ETHYLENE	1038	22		16	1597	536		83 -50	142		ETL 568	436	E2.4	E 4	485
ETHYLENE	1962	21	368	16	1597	536		83 -50	142		ETL 568	436	E2.4	E 4	485
ETHYLENE ACETATE Synonym: ETHYLENE GLYCOL DIACETATE	1993	27			1605	542			142		EGY 534	430			488
ETHYLENE ALDEHYDE Synonym: ACROLEIN	1092	30	31	11	63	25	5	˙17 -12	142	32	ARL 85	12	A4.0	A 5	18
ETHYLENE BIS(IMINODIACETIC ACID) Synonym: ETHYLENEDIAMINE TETRAACETIC ACID	9117/3077		379			539			144		EDT 521	405			485
ETHYLENE BROMIDE Synonym: ETHYLENE DIBROMIDE	1605	55	371					85 -	142	110	EDB 518	430	E2.4.1		485
ETHYLENE CARBOXYLIC ACID Synonym: ACRYLIC ACID	2218	29	33		65	26	6	18 -12	20		ACR 25	13	A5.0	A 6	18
ETHYLENE CHLORHYDRIN Synonym: ETHYLENE CHLOROHYDRIN	1135	55	369	16	1599	538	93	83 -	142	108	ECH 513	429			486
ETHYLENE CHLORIDE Synonym: ETHYLENE DICHLORIDE	1184	26	372	17	1602	540	94	86 -50	142	110	EDC 519	430	E2.4.5	E 5	486
ETHYLENE CHLOROHYDRIN	1135	55	369	16	1599	538	93	83 -	142	108	ECH 513	429			486

CHEMICAL OR MATERIAL NAME	UN/NA Number	Guide Number (DOT)	Firefighter's Hazardous Materials Reference Book	First Aid Manual for Chemical Accidents	Sax's Dangerous Properties of Industrial Materials	Hazardous Chemicals Desk Reference	Rapid Guide to Hazardous Chemicals in the Workplace	Fire Protection Guide on Hazardous Materials (NFPA)	Firefighter's Handbook of Hazardous Materials	Pocket Guide to Chemical Hazards (NIOSH)	Chemical Hazard Response Information System (CHRIS)	Emergency Handling of Hazardous Materials in Surface Transportation (AAR)	Emergency Action Guides (EAG)	Chemical Data Notebook: A User's Manual	Condensed Chemical Dictionary
ETHYLENE CYANOHYDRIN			370					84 -50	142		ETC 562				486
ETHYLENE DIACETATE Synonym: ETHYLENE GLYCOL DIACETATE	1993	27			1605	542			143		EGY 534	430			488
ETHYLENE DIBROMIDE	1605	55	371					85 -	143	110	EDB 518	430	E2.4.1		487
ETHYLENE DICHLORIDE	1184	26	372	17	1602	540	94	86 -50	143	110	EDC 519	430	E2.4.5	E 5	487
ETHYLENE DIFLUORIDE	1030	22							143						
ETHYLENE DIHYDRATE Synonym: ETHYLENE GLYCOL			373	17	1604	541	94	-50	143		EGL 531		E2.4.7		487
ETHYLENE DIPROPIONATE Synonym: CROTONALDEHYDE	1143	28	233	14	964	325	56	57 -29	85	76	CTA 317	271	C7.1		325
ETHYLENE GAS	1962	21		16		536			142		ETL 568				
ETHYLENE GAS	1038	22		16		536			142		ETL 568				
ETHYLENE GLYCOL			373	17	1604	541	94	-50	143		EGL 531		E2.4.7		487
ETHYLENE GLYCOL BIS(2,3-EPOXY-2-METHYL-PROPYL) ETHER					1605				143						
ETHYLENE GLYCOL DIACETATE	1993	27			1605	542			143		EGY 534	430			488
ETHYLENE GLYCOL DIBUTYL ETHER								-50	143		EGB 527				488
ETHYLENE GLYCOL DIETHYL ETHER	1153	26	374		1606	542		-51	143		EEE 522	431			488
ETHYLENE GLYCOL DIFORMATE	1993	27			1606	542			143			431			488
ETHYLENE GLYCOL DIHYDROXYDIETHYL ETHER Synonym: TRIETHYLENE GLYCOL			848	25	3370	1142		-89			TEG1101				1175
ETHYLENE GLYCOL DIMETHYL ETHER	2252	27						-51	143		EGD 528				488
ETHYLENE GLYCOL DINITRATE				17	1606	542	94		143	110					488
ETHYLENE GLYCOL DIPROPIONATE Synonym: CROTONALDEHYDE	1143/1993	28	233	14	964	325	56	57 -29	85	76	CTA 317	431	C7.1		488

HAZARDOUS MATERIALS REFERENCE BOOKS INDEX

CHEMICAL OR MATERIAL NAME	UN/NA Number	Guide Number (DOT)	Firefighter's Hazardous Materials Reference Book	First Aid Manual for Chemical Accidents	Sax's Dangerous Properties of Industrial Materials	Hazardous Chemicals Desk Reference	Rapid Guide to Hazardous Chemicals in the Workplace	Fire Protection Guide on Hazardous Materials (NFPA)	Firefighter's Handbook of Hazardous Materials	Pocket Guide to Chemical Hazards (NIOSH)	Chemical Hazard Response Information System (CHRIS)	Emergency Handling of Hazardous Materials in Surface Transportation (AAR)	Emergency Action Guides (EAG)	Chemical Data Notebook: A User's Manual	Condensed Chemical Dictionary
ETHYLENE GLYCOL ETHYL ETHER Synonym: ETHYLENE GLYCOL MONOETHYL ETHER	1171	26	375					–51	143		EGE 529	432	E2.5		489
ETHYLENE GLYCOL ETHYLBUTYL ETHER								–51	143						
ETHYLENE GLYCOL ISOPROPYL ETHER								–51	143		EGI 530				
ETHYLENE GLYCOL METHYL ETHER	1188	26			1607	543	95		143						
ETHYLENE GLYCOL MONOACETATE					1608			–51							488
ETHYLENE GLYCOL MONOBENZYL ETHER					1608			–51	143						488
ETHYLENE GLYCOL MONOBUTYL ETHER	2369	26						–51	143		EGM 532	432			489
ETHYLENE GLYCOL MONOBUTYL ETHER ACETATE								–51	143		EMA 542				489
ETHYLENE GLYCOL MONOETHYL ETHER	1171	26	375					–51	143		EGE 529	432	E2.5		489
ETHYLENE GLYCOL MONOETHYL ETHER ACETATE	1172	26	376					–51	143		EGA 526	432	E2.6		489
ETHYLENE GLYCOL MONOISOBUTYL ETHER								–51	143						
ETHYLENE GLYCOL MONOMETHYL ETHER	1188	26		17				–51	143		EME 544	433			489
ETHYLENE GLYCOL MONOMETHYL ETHER ACETATE	1189	26			1608	543	95	–51	143			433			490
ETHYLENE GLYCOL MONOMETHYL ETHER FORMAL								–51	144						
ETHYLENE GLYCOL N-BUTYL ETHER	2369	26						–50	144						
ETHYLENE OXIDE	1040	69	377	17	1611	544	96	87 –52	144	112	EOX 550	434	E2.7	E 6	491
ETHYLENE OXIDE CARBON DIOXIDE MIXTURE	1041	17			1612	545			144						
ETHYLENE OXIDE CARBON DIOXIDE MIXTURE	1952	12			1612										
ETHYLENE OXIDE DICHLORODIFLUOROMETHANE GAS MIX-TURE	1956	12										434			
ETHYLENE OXIDE DICHLORODIFLUOROMETHANE MIXTURE	1954	22										435			
ETHYLENE OXIDE PROPYLENE OXIDE MIXTURE	2983	26							144			435			

CHEMICAL OR MATERIAL NAME	UN/NA Number	Guide Number (DOT)	Firefighter's Hazardous Materials Reference Book	First Aid Manual for Chemical Accidents	Sax's Dangerous Properties of Industrial Materials	Hazardous Chemicals Desk Reference	Rapid Guide to Hazardous Chemicals in the Workplace	Fire Protection Guide on Hazardous Materials (NFPA)	Firefighter's Handbook of Hazardous Materials	Pocket Guide to Chemical Hazards (NIOSH)	Chemical Hazard Response Information System (CHRIS)	Emergency Handling of Hazardous Materials in Surface Transportation (AAR)	Emergency Action Guides (EAG)	Chemical Data Notebook: A User's Manual	Condensed Chemical Dictionary
ETHYLENE PROPIONATE Synonym: CROTONALDEHYDE	1143	28	233	14	964	325	56	57 -29	85	76	CTA 317	271	C7.1		325
ETHYLENE SULFIDE					1612	545									
ETHYLENE TETRACHLORIDE	1897	74							144						
ETHYLENE TRICHLORIDE Synonym: TRICHLOROETHYLENE	1710	74	837	25	3352	1137	218	174–88	144	216	TCL1085	919	T5.0	T 6	1170
ETHYLENEDIAMINE	1604	29	378					84 -50	142	108	EDA 517	437			486
ETHYLENEDIAMINE TETRAACETIC ACID	9117/3077		379		1601	539			144		EDT 521	405			486
ETHYLENEDINITRILO TETRAACETIC ACID Synonym: ETHYLENEDIAMINE TETRAACETIC ACID	9117/3077		379		1601	539			144		EDT 521	405			486
ETHYLENEIMINE	1185	30	380	17	1610	544	95	86 -51	144	110	ETI 567	437	E2.6.1		490
ETHYLENIMINE Synonym: ETHYLENEIMINE	1185	30	380	17	1610	544	95	86 -51	144	110	ETI 567	437	E2.6.1		491
ETHYLETHANOLAMINE-n								-52	137						491
ETHYLETHYLENE Synonym: BUTENE	1012	22	141						55				B5.0		492
ETHYLFORMIC ACID Synonym: PROPIONIC ACID	1848	60	721		2909	997	188	148–81	243		PNA 968	791	P7.1		968
ETHYLHEXOIC ACID	1760	60										438			
ETHYLHEXYL ACETATE	1993	27		17								438			
ETHYLHEXYLACRYLATE	1993	27										439			
ETHYLHYDROSULFIDE Synonym: ETHYL MERCAPTAN	2363	27	358		1631	550	97	-53	138	112		418	E2.9		496
ETHYLIC ACID Synonym: ACETIC ACID	2789	29		11	15	6	1		16	30	AAC 1	3	A2.0	A 2	7

HAZARDOUS MATERIALS REFERENCE BOOKS INDEX

CHEMICAL OR MATERIAL NAME	UN/NA Number	Guide Number (DOT)	Firefighter's Hazardous Materials Reference Book	First Aid Manual for Chemical Accidents	Sax's Dangerous Properties of Industrial Materials	Hazardous Chemicals Desk Reference	Rapid Guide to Hazardous Chemicals in the Workplace	Fire Protection Guide on Hazardous Materials (NFPA)	Firefighter's Handbook of Hazardous Materials	Pocket Guide to Chemical Hazards (NIOSH)	Chemical Hazard Response Information System (CHRIS)	Emergency Handling of Hazardous Materials in Surface Transportation (AAR)	Emergency Action Guides (EAG)	Chemical Data Notebook: A User's Manual	Condensed Chemical Dictionary
ETHYLIDENE CHLORIDE Synonym: 1,1-DICHLOROETHANE	2362	27	1		1141	382	69		145	86	DCH 361	301			380
ETHYLIDENE DICHLORIDE Synonym: 1,1-DICHLOROETHANE	2362	27	1		1141	382	69		145	86	DCH 361	301			380
ETHYLIDENE DIFLUORIDE Synonym: 1,1-DIFLUOROETHANE	1030	22	2		1627	549			145		DFE 397				399
ETHYLIDENE FLUORIDE Synonym: 1,1-DIFLUOROETHANE	1030	22	2						112		DFE 397				399
ETHYLIDENE NORBORNENE					1627	549	97				ENB 545				
ETHYLIDENENORBORNYLENE Synonym: ETHYLIDENE NORBORNENE					1627	549	97				ENB 545				
ETHYLIDENENORCAMPHENE Synonym: ETHYLIDENE NORBORNENE					1627	549	97				ENB 545				
ETHYLLITHIUM					1630				145						495
ETHYLMONOBROMOACETATE	1603	55							145						
ETHYLMONOCHLOROACETATE	1181	55							145						
ETHYLMORPHOLINE (-n)						554	98		138	114					497
ETHYLOLAMINE Synonym: MONOETHANOLAMINE	2491	60	594					118–	145		MEA 749		M5.0		
ETHYLORTHOSILICATE	1292	29							145						
ETHYLOXIDE Synonym: ETHYLENE OXIDE	1040	69	377	17	1611	544	96	87 –52	144	112	EOX 550	434	E2.7	E 6	491
ETHYLPHENOL-para								–53	139						498
ETHYLPHENYLDICHLOROSILANE	2435	39							139		EPS 557	440			499
ETHYLPYROPHOSPHATE Synonym: TETRAETHYL PYROPHOSPHATE	1705	15				1101			262		TEP1104				499

CHEMICAL OR MATERIAL NAME	UN/NA Number	Guide Number (DOT)	Firefighter's Hazardous Materials Reference Book	First Aid Manual for Chemical Accidents	Sax's Dangerous Properties of Industrial Materials	Hazardous Chemicals Desk Reference	Rapid Guide to Hazardous Chemicals in the Workplace	Fire Protection Guide on Hazardous Materials (NFPA)	Firefighter's Handbook of Hazardous Materials	Pocket Guide to Chemical Hazards (NIOSH)	Chemical Hazard Response Information System (CHRIS)	Emergency Handling of Hazardous Materials in Surface Transportation (AAR)	Emergency Action Guides (EAG)	Chemical Data Notebook: A User's Manual	Condensed Chemical Dictionary
ETHYLSILICON TRICHLORIDE Synonym: ETHYLTRICHLOROSILANE	1196	29	381		1665	562		90 −54	140		ETS 572	440			501
ETHYLTHIOALCOHOL Synonym: ETHYL MERCAPTAN	2363	27	358		1631	550	97	−53	139	112		418	E2.9		496
ETHYLTOLUENE-meta								−54	146						
ETHYLTOLUENE-ortho								−54	146						
ETHYLTOLUENE-para					1664			−54	146						
ETHYLTRICHLOROSILANE	1196	29	381		1665	562		90 −54	140		ETS 572	440			501
ETHYLZINC Synonym: DIETHYLZINC	1366	40	294		1243	419		71 −39	140		DEZ 395	353			501
ETHYNE Synonym: ACETYLENE	1001	17	29	11	43	20	3	17 −12	146		ACE 17	9		A 4	501
ETHYNE CALCIUM DERIV Synonym: CALCIUM CARBIDE	1402	40	167	13	662	229		45 −	63		CCB 226	179	C1.0	C 1	201
ETIOLOGIC AGENT	2814	24	382						146			440			
ETO Synonym: ETHYLENE OXIDE	1040	69	377	17	1611	544	96	87 −52	144	112	EOX 550	434	E2.7	E 6	491
ETOH Synonym: ETHYL ALCOHOL	1170	26	344	16	1572	525	90	−48	136		EAL 504	408	E2.2	E 2	478
EUFIN Synonym: DIETHYL CARBONATE	2366/1993	26	287		1213	406		−37	108		DEC 383	345			390
EUNATROL Synonym: OLEIC ACID SODIUM SALT			656		2630	896					OAC 866				
EXITELITE Synonym: ANTIMONY TRIOXIDE	1549/3077	60	94	12							ATX 109	87			90
EXPLOSIVE A	0081	46	383		1678	564			146			441			

CHEMICAL OR MATERIAL NAME	UN/NA Number	Guide Number (DOT)	Firefighter's Hazardous Materials Reference Book	First Aid Manual for Chemical Accidents	Sax's Dangerous Properties of Industrial Materials	Hazardous Chemicals Desk Reference	Rapid Guide to Hazardous Chemicals in the Workplace	Fire Protection Guide on Hazardous Materials (NFPA)	Firefighter's Handbook of Hazardous Materials	Pocket Guide to Chemical Hazards (NIOSH)	Chemical Hazard Response Information System (CHRIS)	Emergency Handling of Hazardous Materials in Surface Transportation (AAR)	Emergency Action Guides (EAG)	Chemical Data Notebook: A User's Manual	Condensed Chemical Dictionary
EXPLOSIVE B	0082	46			1678	564			146			441			
EXPLOSIVE C	0083	50			1678	564			146			442			
EXTRACT	1197	26							146			442			
EXTRACT	1169	26							146			442			
F 11 Synonym: TRICHLOROFLUOROMETHANE	3082/9188	31	838	25	3353	1138	219		273		TCF1083	919			1170
F 12 Synonym: DICHLORODIFLUOROMETHANE	1028	12	275	15	1137	380	68		102	86	DCF 360	331	D1.1		379
F 22 Synonym: MONOCHLORODIFLUOROMETHANE	1018	12	593						205				M4.0		795
FA Synonym: FURFURYL ALCOHOL	2874	55	409		1766	597	104	-55	152	118	FAL 575		F4.0		546
FABRIC	1373	32							146						506
FACTITIOUS AIR Synonym: NITROUS OXIDE	1070	14	644						216		NTO 858	700	N3.5		833
FALL Synonym: SODIUM CHLORATE	1495	35	766	24	3070	1048		156-	253		SDC1029	834	S2.1	S 2	1052
FAMPHUR					1684	571			146						506
FANNOFORM Synonym: FORMALDEHYDE SOLUTION	1198/2209	29	401	17	1748	591	102	91 -54	150	116	FMS 590	462	F2.0	F 2	
FASCIOLIN Synonym: CARBON TETRACHLORIDE	1846	55	192	13	701	246	43	48 -	66	60	CBT 223	193	C2.3	C 5	221
FAST RED GG BASE Synonym: 4-NITROANILINE	1661	55			1661		158		211		NAL 818				
FAST RED IG BASE Synonym: 4-NITROANILINE	1661	55			1661		158		211		NAL 818				

CHEMICAL OR MATERIAL NAME	UN/NA Number	Guide Number (DOT)	Firefighter's Hazardous Materials Reference Book	First Aid Manual for Chemical Accidents	Sax's Dangerous Properties of Industrial Materials	Hazardous Chemicals Desk Reference	Rapid Guide to Hazardous Chemicals in the Workplace	Fire Protection Guide on Hazardous Materials (NFPA)	Firefighter's Handbook of Hazardous Materials	Pocket Guide to Chemical Hazards (NIOSH)	Chemical Hazard Response Information System (CHRIS)	Emergency Handling of Hazardous Materials in Surface Transportation (AAR)	Emergency Action Guides (EAG)	Chemical Data Notebook: A User's Manual	Condensed Chemical Dictionary
FAST RED TR BASE Synonym: 4-CHLORO-o-TOLUIDINE	2239	55			879	299	51				CTD 319				274
FAST WHITE Synonym: LEAD SULFATE	1794	60			2105	716	128		179		LSF 709	575			692
FC 12 Synonym: DICHLORODIFLUOROMETHANE	1028	12	275	15	1137	380	68		102	86	DCF 360	331	D1.1		379
FC 22 Synonym: MONOCHLORODIFLUOROMETHANE	1018	12	593						205				M4.0		795
FENAMIPHOS					1693	574	98		146						510
FENTHION					1696	575	99		147						510
FERBAM					1697	576	99		147	114					511
FERMEATATION BUTYL ALCOHOL Synonym: ISOBUTYL ALCOHOL	1212/1120	26			2016	676	121	-61	171	132	IAL 658	545			652
FERMENICIDE LIQUID Synonym: SULFUR DIOXIDE	1079	16	800	24	3162	1081	203	163-	258	200	SFD1039	870	S4.1	S 7	1104
FERMENICIDE POWDER Synonym: SULFUR DIOXIDE	1079	16	800	24	3162	1081	203	163-	258	200	SFD1039	870	S4.1	S 7	1104
FERMENTATION ALCOHOL Synonym: ETHYL ALCOHOL	1170	26	344	16	1572	525	90	-48	136		EAL 504	408	E2.2	E 2	511
FERMENTATION AMYL ALCOHOL Synonym: ISOAMYL ALCOHOL	1105	26	461		2011	673	120	-60	171	130	IAA 655				649
FERRIC AMMONIUM CITRATE	9118/3077	31	384						147			445			512
FERRIC AMMONIUM OXALATE	9119/9188	31	385						147		FAO 578	445			512
FERRIC ARSENATE SOLID	1606	53							147			445			512
FERRIC ARSENITE	1607	53							147			446			512
FERRIC CHLORIDE	1773	60	387		1698	577	99		147		FCL 580	447			512

HAZARDOUS MATERIALS REFERENCE BOOKS INDEX

CHEMICAL OR MATERIAL NAME	UN/NA Number	Guide Number (DOT)	Firefighter's Hazardous Materials Reference Book	First Aid Manual for Chemical Accidents	Sax's Dangerous Properties of Industrial Materials	Hazardous Chemicals Desk Reference	Rapid Guide to Hazardous Chemicals in the Workplace	Fire Protection Guide on Hazardous Materials (NFPA)	Firefighter's Handbook of Hazardous Materials	Pocket Guide to Chemical Hazards (NIOSH)	Chemical Hazard Response Information System (CHRIS)	Emergency Handling of Hazardous Materials in Surface Transportation (AAR)	Emergency Action Guides (EAG)	Chemical Data Notebook: A User's Manual	Condensed Chemical Dictionary
FERRIC CHLORIDE ANHYDROUS Synonym: FERRIC CHLORIDE	1773	60	387		1698	577	99		147		FCL 580	447			512
FERRIC CHLORIDE HEXAHYDRATE Synonym: FERRIC CHLORIDE	1773	60	387		1698	577	99		147		FCL 580	447			512
FERRIC CHLORIDE SOLUTION	2582	60	386		1698	577	99		147			446	F1.0		
FERRIC FLUORIDE	9120/3077		388		1698	577			147		FFX 585	447			513
FERRIC GLYCEROPHOSPHATE			389								FCP 581				513
FERRIC NITRATE	1466	35	390						147		FNT 591	447			513
FERRIC NITRATE NONAHYDRATE Synonym: FERRIC NITRATE	1466	35	390						147		FNT 591	447			513
FERRIC SULFATE	9121/3082	31	391		1699	577			147		FSF 595	448			514
FERROCENE					1699	578	100								515
FERROCERIUM	1323	32							147						515
FERROCHROME EXOTHERMIC					1700	578			147						
FERROMANGANESE EXOTHERMIC					1700	578									
FERROSILICON	1408	41			1700	578			147			448			516
FERROUS AMMONIUM SULFATE	9122/3077	31	392						147		FAS 579	449			517
FERROUS AMMONIUM SULFATE HEXAHYDRATE Synonym: FERROUS AMMONIUM SULFATE	9122/3077	31	392						147		FAS 579	449			517
FERROUS AMMONIUM SULPHATE FERTILIZER GRADE	9188														
FERROUS ARSENATE	1608	53							147			449			517
FERROUS BOROFLUORIDE Synonym: FERROUS FLUOROBORATE			395								FFB 584				
FERROUS CHLORIDE	1759	60	394		1701	579			147		FEC 582	449			517
FERROUS CHLORIDE SOLUTION	1760	60	393		1701	579			147			449			

CHEMICAL OR MATERIAL NAME	UN/NA Number	Guide Number (DOT)	Firefighter's Hazardous Materials Reference Book	First Aid Manual for Chemical Accidents	Sax's Dangerous Properties of Industrial Materials	Hazardous Chemicals Desk Reference	Rapid Guide to Hazardous Chemicals in the Workplace	Fire Protection Guide on Hazardous Materials (NFPA)	Firefighter's Handbook of Hazardous Materials	Pocket Guide to Chemical Hazards (NIOSH)	Chemical Hazard Response Information System (CHRIS)	Emergency Handling of Hazardous Materials in Surface Transportation (AAR)	Emergency Action Guides (EAG)	Chemical Data Notebook: A User's Manual	Condensed Chemical Dictionary
FERROUS CHLORIDE TETRAHYDRATE Synonym: FERROUS CHLORIDE	1759	60	394		1701	579			147		FEC 582	449			517
FERROUS FLUOROBORATE			395								FFB 584				
FERROUS METAL	2793	32							147			450			
FERROUS OXALATE			396								FOX 592				517
FERROUS OXALATE DIHYDRATE Synonym: FERROUS OXALATE			396								FOX 592				517
FERROUS SULFATE	9125/3077		397		1702	579			147		FRS 593	450			518
FERROUS SULPHATE LESS 40% WATER	9188/3082											450			
FERROUS SULPHATE NOT LESS 40% WATER	9188														
FERROUS-o-ARSENATE	1608	53							147						
FERROVANADIUM DUST					1703	580	100			114					
FERROX Synonym: FERROUS OXALATE			396								FOX 592				517
FERTILIZER ACID Synonym: SULFURIC ACID	1830	39	802	24	3163	1082	204	164–	259	200	SFA1037	871	S4.2	S 8	1104
FERTILIZER AMMONIATING SOLUTION	1043	16							147			451			
FERTILIZER SOLUTION CONSISTING OF WATER FREE AMMONIA	1760	60													
FERTILIZING COMPOUNDS NEC DRY	1479	35										451			
FIBER	1372	32							147						519
FIBER	1373	32							147						519
FIBROUS GLASS				17	1705	580	100								
FILM MOTION PICTURE	1324	32							147						
FILM NITROCELLULOSE BASE	1324	32							147						

HAZARDOUS MATERIALS REFERENCE BOOKS INDEX

CHEMICAL OR MATERIAL NAME	UN/NA Number	Guide Number (DOT)	Firefighter's Hazardous Materials Reference Book	First Aid Manual for Chemical Accidents	Sax's Dangerous Properties of Industrial Materials	Hazardous Chemicals Desk Reference	Rapid Guide to Hazardous Chemicals in the Workplace	Fire Protection Guide on Hazardous Materials (NFPA)	Firefighter's Handbook of Hazardous Materials	Pocket Guide to Chemical Hazards (NIOSH)	Chemical Hazard Response Information System (CHRIS)	Emergency Handling of Hazardous Materials in Surface Transportation (AAR)	Emergency Action Guides (EAG)	Chemical Data Notebook: A User's Manual	Condensed Chemical Dictionary
FILMCOL Synonym: ETHYL ALCOHOL	1170	26	344	16	1572	525	90	-48	136		EAL 504	408	E2.2	E 2	478
FILMERINE Synonym: SODIUM NITRITE	1500	35	781		3098	1060			254		SNT1051	846			1063
FIRE EXTINGUISHER	1044	12							147			451			
FIRE EXTINGUISHER CHARGE	1774	60							147			452			
FIRE LIGHTER	2623	32							147			452			
FIRE PROOFING COMPOUNDS LIQUID	1760	60													
FISH MEAL AND SCRAP	1374	32							148			453			523
FISH MEAL AND SCRAP	2216	31							148						523
FLAME COLORING COMPOUNDS	1479	35													
FLAME RETARDANT COMPOUND LIQUID	1760	60										453			
FLAMMABLE GAS	1954	22							148						
FLAMMABLE GAS IN LIGHTER	1057	17							148						
FLAMMABLE LIQUID	2924	29							148			454			
FLAMMABLE LIQUID	1993	27							148			454			
FLAMMABLE LIQUID	1992	28							148			454			
FLAMMABLE LIQUID PREPARATIONS	1142	26							148			455			
FLAMMABLE SOLID	1325	32							148			456			
FLAMMABLE SOLID	2925	34							148			455			
FLAMMABLE SOLID	2926	34							148			456			
FLARES HIGHWAY SAFETY SOLIDIFIED FUEL	1325	32										457			
FLAX FIBER	1372	32													

CHEMICAL OR MATERIAL NAME	UN/NA Number	Guide Number (DOT)	Firefighter's Hazardous Materials Reference Book	First Aid Manual for Chemical Accidents	Sax's Dangerous Properties of Industrial Materials	Hazardous Chemicals Desk Reference	Rapid Guide to Hazardous Chemicals in the Workplace	Fire Protection Guide on Hazardous Materials (NFPA)	Firefighter's Handbook of Hazardous Materials	Pocket Guide to Chemical Hazards (NIOSH)	Chemical Hazard Response Information System (CHRIS)	Emergency Handling of Hazardous Materials in Surface Transportation (AAR)	Emergency Action Guides (EAG)	Chemical Data Notebook: A User's Manual	Condensed Chemical Dictionary
FLAXSEED OIL Synonym: **OILS MISCELLANEOUS LINSEED**											OLS 887				527
FLOOR COVERING UNDER LAYMENT COMPOUNDS	1993	27													
FLOOR POLISH OR WAX	1142	26													
FLOUR BLEACHING COMPOUNDS NEC	1479	35										458			
FLOWERS OF ANTIMONY Synonym: **ANTIMONY TRIOXIDE**	1549/3077	60	94	12							ATX 109	87			90
FLUE DUST	2811	53							148						
FLUKOIDS Synonym: **CARBON TETRACHLORIDE**	1846	55	192	13	701	246	43	48 –	66	60	CBT 223	193	C2.3	C 5	221
FLUOBORIC ACID	1775	60						90 –	148			458			529
FLUOHYDRIC ACID Synonym: **HYDROFLUORIC ACID SOLUTION**	1790	59	443	18	1897	642	114		164		HFA 635	527	H4.1.1		614
FLUORHYDRIC ACID Synonym: **HYDROFLUORIC ACID SOLUTION**	1790	59	443	18	1897	642	114				HFA 635	527	H4.1.1		614
FLUORIC ACID Synonym: **HYDROFLUORIC ACID SOLUTION**	1790	59	443	18	1897	642	114		148		HFA 635	527	H4.1.1		614
FLUORIDE					1717	583	101								
FLUORIDES	1690	54			1717	583	101			116					
FLUORINE	1045	20	398	17	1717	583	101	91 –	148	116	FXX 601	458		F 1	530
FLUORINE	9192	25		17					148	116				F 1	530
FLUORINE MONOXIDE	2190	20							148						
FLUORINE NITRATE									148						530
FLUORISTAN Synonym: **STANNOUS FLUORIDE**			790								STF1061				1085

HAZARDOUS MATERIALS REFERENCE BOOKS INDEX

CHEMICAL OR MATERIAL NAME	UN/NA Number	Guide Number (DOT)	Firefighter's Hazardous Materials Reference Book	First Aid Manual for Chemical Accidents	Sax's Dangerous Properties of Industrial Materials	Hazardous Chemicals Desk Reference	Rapid Guide to Hazardous Chemicals in the Workplace	Fire Protection Guide on Hazardous Materials (NFPA)	Firefighter's Handbook of Hazardous Materials	Pocket Guide to Chemical Hazards (NIOSH)	Chemical Hazard Response Information System (CHRIS)	Emergency Handling of Hazardous Materials in Surface Transportation (AAR)	Emergency Action Guides (EAG)	Chemical Data Notebook: A User's Manual	Condensed Chemical Dictionary
FLUOROACETIC ACID	2642/2811	59							149			459			530
FLUOROANILINE	2944	55							149						
FLUOROANILINE	2941	55							149			459			
FLUOROBENZENE	2387/1993	27			1722	586		−54	149		FLB 587	459			531
FLUOROCARBON 12	1028	12	275	15	1137	380	68		102	86	DCF 360	331	D1.1		531
Synonym: DICHLORODIFLUOROMETHANE															
FLUOROETHANE	2453	12			1729	587			149						
FLUOROETHANE GASES NONFLAMMABLE	1956	12										460			
FLUOROETHANOIC ACID	2642	59			1729	587			149						
FLUOROETHYLENE	1860	17	874	26	3498	1175	228	−93	149		VFI1176	950			1217
Synonym: VINYL FLUORIDE															
FLUOROFORMYL FLUORIDE	2417	15							149						532
FLUOROHYDRIC ACID GAS	1052	15	449					99 –	164	126	HFX 636	530	H4.4	H 4	616
Synonym: HYDROGEN FLUORIDE															
FLUOROMETHANE	2454	22		17	1733	588			149			628	M3.4.1		532
Synonym: METHYL FLUORIDE															
FLUOROPHOSPHORIC ACID	1776	59							150			460			
FLUOROSILIC ACID	1778	60	399	17					150		FSL 596				532
Synonym: FLUOSILICIC ACID															
FLUOROSILICIC ACID	1778/1831	60							150			460	H4.2		532
Synonym: HYDROFLUOROSILICIC ACID															
FLUOROSULFONIC ACID	1777	39	400						150		FSA 594	461			532
Synonym: FLUOSULFONIC ACID															
FLUOROSULFURIC ACID	1777	39	400		1740	589			150		FSA 594				532
Synonym: FLUOSULFONIC ACID															
FLUOROTOLUENE	2388/1993	27							150			461			

CHEMICAL OR MATERIAL NAME	UN/NA Number	Guide Number (DOT)	Firefighter's Hazardous Materials Reference Book	First Aid Manual for Chemical Accidents	Sax's Dangerous Properties of Industrial Materials	Hazardous Chemicals Desk Reference	Rapid Guide to Hazardous Chemicals in the Workplace	Fire Protection Guide on Hazardous Materials (NFPA)	Firefighter's Handbook of Hazardous Materials	Pocket Guide to Chemical Hazards (NIOSH)	Chemical Hazard Response Information System (CHRIS)	Emergency Handling of Hazardous Materials in Surface Transportation (AAR)	Emergency Action Guides (EAG)	Chemical Data Notebook: A User's Manual	Condensed Chemical Dictionary
FLUOROTRICHLOROMETHANE	1078	12							150	116					532
FLUORSPAR Synonym: CALCIUM FLUORIDE					667						CAF 205				532
FLUOSILICIC ACID Synonym: HYDROFLUOROSILICIC ACID	1778	60	399	17					150		FSL 596		H4.2		532
FLUOSPAR Synonym: CALCIUM FLUORIDE					667						CAF 205				203
FLUOSULFONIC ACID	1777	39	400						150		FSA 594				533
FLUOSULFURIC ACID	1777	39							150						
FLUXING OIL Synonym: ASPHALT BLENDING STOCKS ROOFERS FLUX	1999	27	105	12	307	97	17				ARF 84	99	A12.0		
FOLIAGE OIL Synonym: OILS MISCELLANEOUS SPRAY	1270	27									OSY 908				
FOLIC ACID					1745	590									534
FONOFOS	1045	20			1746	590	101		150						535
FORMALDEHYDE DIMETHYLACETOL Synonym: METHYL FORMAL	1234	26	568								MTF 808				
FORMALDEHYDE GAS				17					150						
FORMALDEHYDE POLYMER Synonym: PARAFORMALDEHYDE	2213	32	661		2664	906		131–76	222		PFA 950	725			871
FORMALDEHYDE SOLUTION	1 98/2209	29	401	17	1748	591	102	91 –54	150	116	FMS 590	462	F2.0	F 2	
FORMALDEHYDE SOLUTION	2209	29		17	1748	591	102	91 –54	150	116	FMS 590	462	F2.0	F 2	
FORMALIN Synonym: FORMALDEHYDE SOLUTION	1 98/2209	29	401	17	1748	591	102	91 –54	150	116	FMS 590	462	F2.0	F 2	536
FORMALIN Synonym: FORMALDEHYDE	2209	29					102		150		FMS 590	462		F 2	536

CHEMICAL OR MATERIAL NAME	UN/NA Number	Guide Number (DOT)	Firefighter's Hazardous Materials Reference Book	First Aid Manual for Chemical Accidents	Sax's Dangerous Properties of Industrial Materials	Hazardous Chemicals Desk Reference	Rapid Guide to Hazardous Chemicals in the Workplace	Fire Protection Guide on Hazardous Materials (NFPA)	Firefighter's Handbook of Hazardous Materials	Pocket Guide to Chemical Hazards (NIOSH)	Chemical Hazard Response Information System (CHRIS)	Emergency Handling of Hazardous Materials in Surface Transportation (AAR)	Emergency Action Guides (EAG)	Chemical Data Notebook: A User's Manual	Condensed Chemical Dictionary
FORMALITH Synonym: FORMALDEHYDE SOLUTION	1198/2209	29	401	17	1748	591	102	91 −54	150	116	FMS 590	462	F2.0	F 2	
FORMAMIDE					1749	592	102	−54	150		FAM 576				536
FORMIC ACID	1779	60	402	17	1750	592	102	92 −55	150	118	FMA 589	463			537
FORMIC ACID AMMONIUM SALT Synonym: AMMONIUM FORMATE					226						AFM 34				66
FORMIC ACID BUTYL ESTER	1128	26							150						
FLUOROSILIC ACID Synonym: ETHYL FORMATE	1778	26	355		1617	547		−52	150	112	EFM 439				492
FORMIC ACID METHYL ESTER Synonym: METHYL FORMATE	1243	26	569		2323	796	144	114−69	150	150	MFM 757	628	M3.4.4		769
FORMIC ACID ZINC SALT Synonym: ZINC FORMATE	9159/3077		900				234		289		ZFM1207	967			1245
FORMIC ALDEHYDE Synonym: FORMALDEHYDE SOLUTION	1198/2209	29	401	17	1748	591	102	91 −54	150	116	FMS 590	462	F2.0	F 2	537
FORMIC ALDEHYDE SOLUTION Synonym: FORMALDEHYDE SOLUTION	1198/2209	29	401	17	1748	591	102	91 −54	150	116	FMS 590	462	F2.0	F 2	
FORMIC ANAMMONIDE Synonym: HYDROCYANIC ACID	1051	13	442	18	1896	641	114	−59	164				H4.1	H 2	613
FORMIC ETHER Synonym: ETHYL FORMATE	1190	26	355		1617	547	97	−52	150	112	EFM 525	415			492
FORMONITRILE Synonym: HYDROCYANIC ACID	1051	13	442	18	1896	641	114	−59	164				H4.1	H 2	537
FORMYL HYDRAZINE					1753										
FORMYL TRICHLORIDE Synonym: CHLOROFORM	1888	55	205	14	815	282	49	52 −	151	68	CRF 301	223	C4.2	C 7	266

CHEMICAL OR MATERIAL NAME	UN/NA Number	Guide Number (DOT)	Firefighter's Hazardous Materials Reference Book	First Aid Manual for Chemical Accidents	Sax's Dangerous Properties of Industrial Materials	Hazardous Chemicals Desk Reference	Rapid Guide to Hazardous Chemicals in the Workplace	Fire Protection Guide on Hazardous Materials (NFPA)	Firefighter's Handbook of Hazardous Materials	Pocket Guide to Chemical Hazards (NIOSH)	Chemical Hazard Response Information System (CHRIS)	Emergency Handling of Hazardous Materials in Surface Transportation (AAR)	Emergency Action Guides (EAG)	Chemical Data Notebook: A User's Manual	Condensed Chemical Dictionary
FORMYLIC ACID Synonym: FORMIC ACID	1779	60	402	17	1750	592	102	92 –55	150	118	FMA 589	463			537
FOWLERS SOLUTION Synonym: POTASSIUM ARSENITE	1678	54	704		1757	977			237		POA 969	776			949
FREEMANS WHITE LEAD Synonym: LEAD SULFATE	1794	60			2105	716	128		179		LSF 709	575			692
FRENCH VERDIGRIS Synonym: COPPER SUBACETATE							55				CST 315				315
FREON 10 Synonym: CARBON TETRACHLORIDE	1846	55	192	13	701	246	43	48 –	66	60	CBT 223	193	C2.3	C 5	221
FREON 11 Synonym: TRICHLOROFLUOROMETHANE	3082/9188	31	838	17	3353	1138	219		151		TCF1083	919			1170
FREON 112				17					151						541
FREON 113				17	1758	593	103		151						
FREON 114	1958	12		17	1758		103		151						
FREON 115	1020	12		17					151						
FREON 116	2193	12		17					151						
FREON 12 Synonym: DICHLORODIFLUOROMETHANE	1028	12	275	17	1137	380	68		151	86	DCF 360	331	D1.1		379
FREON 13	1022	52		17					151						
FREON 14	1982	12		17					151						
FREON 20 Synonym: CHLOROFORM	1888	55	205	14	815	282	49	52 –	76	68	CRF 301	223	C4.2	C 7	266
FREON 21	1029	12		17					151						
FREON 22 Synonym: CHLORODIFLUOROMETHANE	1018	12		17	791	278	48		151		MCF 737	222			541

HAZARDOUS MATERIALS REFERENCE BOOKS INDEX

CHEMICAL OR MATERIAL NAME	UN/NA Number	Guide Number (DOT)	Firefighter's Hazardous Materials Reference Book	First Aid Manual for Chemical Accidents	Sax's Dangerous Properties of Industrial Materials	Hazardous Chemicals Desk Reference	Rapid Guide to Hazardous Chemicals in the Workplace	Fire Protection Guide on Hazardous Materials (NFPA)	Firefighter's Handbook of Hazardous Materials	Pocket Guide to Chemical Hazards (NIOSH)	Chemical Hazard Response Information System (CHRIS)	Emergency Handling of Hazardous Materials in Surface Transportation (AAR)	Emergency Action Guides (EAG)	Chemical Data Notebook: A User's Manual	Condensed Chemical Dictionary
FREON 22	1028	12		17											541
FREON 22	1018	12	593	17					151				M4.0		541
Synonym: **MONOCHLORODIFLUOROMETHANE**															
FREON 23	1984	12							151						
FREON 40	1063	18	561	20	2284	781	138	111–67	193	146	MTC 806	623	M3.1	M 4	762
Synonym: **METHYL CHLORIDE**															
FREON 500									151						
FREON F 12	1028	12	275	15	1137	380	68		102	86	DCF 360	331	D1.1		379
Synonym: **DICHLORODIFLUOROMETHANE**															
FRIGEN 11	3082/9188	31	838	25	3353		219		273		TCF1083	919			1170
Synonym: **TRICHLOROFLUOROMETHANE**															
FRIGEN 12	1028	12	275	15	1137	380	68		102	86	DCF 360	331	D1.1		379
Synonym: **DICHLORODIFLUOROMETHANE**															
FRIGEN 22	1018	12	593						205				M4.0		795
Synonym: **MONOCHLORODIFLUOROMETHANE**															
FUEL	1375	37							151						542
FUEL AVIATION	1863	27							152			464			543
FUEL OIL	1993	27			1759	594						464			
FUEL OIL 1 D	1270	27	653								ODS 877				
Synonym: **OILS DIESEL**															
FUEL OIL 2 D	1270	27	653								ODS 877				
Synonym: **OILS DIESEL**															
FUEL OIL DIESEL	1993	27													
FUEL OIL NO. 1	1223	27	489		2071				151		JPO 685				671
Synonym: **JET FUELS JP-1**															
FUEL OIL NO. 1	1993	27	403					–55	151			464	F3.0		

CHEMICAL OR MATERIAL NAME	UN/NA Number	Guide Number (DOT)	Firefighter's Hazardous Materials Reference Book	First Aid Manual for Chemical Accidents	Sax's Dangerous Properties of Industrial Materials	Hazardous Chemicals Desk Reference	Rapid Guide to Hazardous Chemicals in the Workplace	Fire Protection Guide on Hazardous Materials (NFPA)	Firefighter's Handbook of Hazardous Materials	Pocket Guide to Chemical Hazards (NIOSH)	Chemical Hazard Response Information System (CHRIS)	Emergency Handling of Hazardous Materials in Surface Transportation (AAR)	Emergency Action Guides (EAG)	Chemical Data Notebook: A User's Manual	Condensed Chemical Dictionary
FUEL OIL NO. 1 Synonym: OILS MISCELLANEOUS RANGE	1223	27							151		ORG 900				
FUEL OIL NO. 1 Synonym: KEROSENE	1223	27	494		2078	701	126		151		KRS 689	566			
FUEL OIL NO. 2	1993	27	404					–55	151			464	F3.1		
FUEL OIL NO. 3	1993	27							151						
FUEL OIL NO. 4	1993	27	405					–55	151			464	F3.2		
FUEL OIL NO. 5	1993	27	406					–55	151			464	F3.3		
FUEL OIL NO. 6	1993	27						–55	152						
FUEL OIL TREATING COMPOUNDS	1760	60													
FUMARIC ACID	9126/9188		407		1760	595			152		FUM 599	465			543
FUMARYL CHLORIDE	1780	60			1761	595			152			465			543
FUMIGANT 1 Synonym: METHYL BROMIDE	1062	55	559	20	2274	776	137	110–67	192	146	MTB 805	619	M3.0	M 3	759
FUMIGRAIN Synonym: ACRYLONITRILE	1093	30	34	11	68	27	6	19 –12	20	34	ACN 23	13	A5.1	A 7	19
FUMING LIQUID ARSENIC Synonym: ARSENIC TRICHLORIDE	1560	55	99	12			16		152		AST 95	92	A11.0		97
FUMING SULFURIC ACID Synonym: OLEUM	1831	39	657						152		OLM 886		O2.0	S 8	852
FUMING SULPHURIC ACID Synonym: OLEUM	1831	39	657								OLM 886		O2.0	S 8	852
FUMO GAS Synonym: ETHYLENE DIBROMIDE	1605	55	371		1601			85 –	143	110	EDB 518	430	E2.4.1		487
FUNGICIDE	2902	55							152						544
FUNGICIDE	1759	60													544

HAZARDOUS MATERIALS REFERENCE BOOKS INDEX

CHEMICAL OR MATERIAL NAME	UN/NA Number	Guide Number (DOT)	Firefighter's Hazardous Materials Reference Book	First Aid Manual for Chemical Accidents	Sax's Dangerous Properties of Industrial Materials	Hazardous Chemicals Desk Reference	Rapid Guide to Hazardous Chemicals in the Workplace	Fire Protection Guide on Hazardous Materials (NFPA)	Firefighter's Handbook of Hazardous Materials	Pocket Guide to Chemical Hazards (NIOSH)	Chemical Hazard Response Information System (CHRIS)	Emergency Handling of Hazardous Materials in Surface Transportation (AAR)	Emergency Action Guides (EAG)	Chemical Data Notebook: A User's Manual	Condensed Chemical Dictionary
FURADAN Synonym: CARBOFURAN	2757	55	184		695	242	40		152		CBF 218	188			544
FURAL Synonym: FURFURAL	1199	29	408	17	1765	597	103	92 –55	152	118	FFA 583	467			545
FURAL PYROMUCIC ALDEHYDE Synonym: FURFURAL	1199	29	408	17	1765	597	103	92 –55	152	118	FFA 583	467			545
FURALE	1199	29							152						
FURAN	2389	26			1763	596		–55	152		FUR 600	467	F3.3.1		545
FURANIDINE Synonym: TETRAHYDROFURAN	2056	26	818		3227	1103	210	167–86	263	210	THF1116	894	T1.0	T 2	1131
FURFURAL	1199	29	408	17	1765	597	103	92 –55	152	118	FFA 583	467			545
FURFURAL ALCOHOL Synonym: FURFURYL ALCOHOL	2874	55	409	17	1766	597	104	–55	152	118	FAL 575		F4.0		546
FURFURALCOHOL Synonym: FURFURYL ALCOHOL	2874	55	409		1766	597	104	–55	152	118	FAL 575		F4.0		546
FURFURALDEHYDE Synonym: FURFURAL	1199	29	408	17	1765	597	103	92 –55	152	118	FFA 583	467			546
FURFURAMIDE					1766	597			152						546
FURFURAN Synonym: FURAN	2389	26			1763	596		–55	152		FUR 600	467	F3.3.1		546
FURFURANE	2389	26							152						
FURFUROLE Synonym: FURFURAL	1199	29	408	17	1765	597	103	92 –55	152	118	FFA 583	467			545
FURFURYL ACETATE								–55	152						546
FURFURYL ALCOHOL	2874	55	409		1766	597	104	–55	152	118	FAL 575	467	F4.0		546
FURFURYLAMINE	2526/1993	28			1767	598		–55	152			468			

CHEMICAL OR MATERIAL NAME	UN/NA Number	Guide Number (DOT)	Firefighter's Hazardous Materials Reference Book	First Aid Manual for Chemical Accidents	Sax's Dangerous Properties of Industrial Materials	Hazardous Chemicals Desk Reference	Rapid Guide to Hazardous Chemicals in the Workplace	Fire Protection Guide on Hazardous Materials (NFPA)	Firefighter's Handbook of Hazardous Materials	Pocket Guide to Chemical Hazards (NIOSH)	Chemical Hazard Response Information System (CHRIS)	Emergency Handling of Hazardous Materials in Surface Transportation (AAR)	Emergency Action Guides (EAG)	Chemical Data Notebook: A User's Manual	Condensed Chemical Dictionary
FUROLE	1199	29							152						
FUROYL CHLORIDE					1769				152						547
FURYL ALCOHOL Synonym: FURFURYL ALCOHOL	2874	55	409		1766	597	104	-55	153	118	FAL 575		F4.0		546
FURYLCARBINOL-alpha Synonym: FURFURYL ALCOHOL	2874	55	409		1766	597	104	-55	152	118	FAL 575		F4.0		546
FUSEL OIL	1201	26			1771	599			153			469			
FUSEL OIL Synonym: ISOAMYL ALCOHOL	1105	26	461		2011	673	120	-60	171	130	IAA 655				649
FYDE Synonym: FORMALDEHYDE SOLUTION	1198/2209	29	401	17	1748	591	102	91 -54	150	116	FMS 590	462	F2.0	F 2	
G 24,480	2783	55													
GAA Synonym: ACRYLIC ACID	2218	29	33		65	26	6	18 -12	20		ACR 25	13	A5.0	A 6	18
GALENA Synonym: LEAD SULFIDE	2811/3077	53	509	19	2105	716	128				LSU 710	576			550
GALENA ORE CRUDE	9138/3077											471			
GALLIC ACID			410		1776	600					GLA 608				550
GALLIC ACID MONOHYDRATE Synonym: GALLIC ACID			410		1776	600					GLA 608				550
GALLIUM	2803	60			1776	600			153			471			550
GALLIUM ARSENIDE					1776			93 -	153						550
GALLIUM METAL LIQUID Synonym: GALLIUM	2803	60			1776	600						471			550
GALLOTANNIC ACID Synonym: TANNIC ACID			805		3180	1089		-84			TNA1130				551

CHEMICAL OR MATERIAL NAME	UN/NA Number	Guide Number (DOT)	Firefighter's Hazardous Materials Reference Book	First Aid Manual for Chemical Accidents	Sax's Dangerous Properties of Industrial Materials	Hazardous Chemicals Desk Reference	Rapid Guide to Hazardous Chemicals in the Workplace	Fire Protection Guide on Hazardous Materials (NFPA)	Firefighter's Handbook of Hazardous Materials	Pocket Guide to Chemical Hazards (NIOSH)	Chemical Hazard Response Information System (CHRIS)	Emergency Handling of Hazardous Materials in Surface Transportation (AAR)	Emergency Action Guides (EAG)	Chemical Data Notebook: A User's Manual	Condensed Chemical Dictionary
GALLOTANNIN Synonym: **TANNIC ACID**			805		3180	1089		−84			TNA1130				1114
GAMBIER EXTRACTS LIQUID	1993	27													
GAMMEXANE Synonym: **BENZENE HEXACHLORIDE**	2761	55	116		360	117			153		BHC 144				129
GAS BLAST FURNACE								−55	153						
GAS COAL GAS								−55							
GAS DRIPS	1864	27							153			472			
GAS EX B Synonym: **PHOSPHINE**	2199	18	683	22	2783	945	180	136–80	231	180		752	P3.1		907
GAS IDENTIFICATION SET	9035	15							153			472			
GAS NATURAL								−55	153						553
GAS OIL CRACKED	1202	27	411						153		GOC 609				
GAS OIL GAS								−55	153						
GAS PRODUCER								−55	153						
GAS WATER								−55	153						
GAS WATER (CARBURETED)								−55	153						
GASOHOL	1203	27							153			472			553
GASOLINE	1203	27	412	18	1779	601	104	−56	153		GAK 602	473	G1.0	G 1	554
GASOLINE AUTOMOTIVE	1203	27	412	18	1779		104		153		GAT 603			G 1	
GASOLINE AVIATION	1203	27	413	18	1779			−56	153		GAV 604			G 1	
GASOLINE BLACK CONTAINING HEAVY DISTILLATES	1993	27													
GASOLINE BLENDING STOCKS ALKYLATES	1203	27									GAK 602				
GASOLINE BLENDING STOCKS REFORMATES	1203	27									GRF 612				

CHEMICAL OR MATERIAL NAME	UN/NA Number	Guide Number (DOT)	Firefighter's Hazardous Materials Reference Book	First Aid Manual for Chemical Accidents	Sax's Dangerous Properties of Industrial Materials	Hazardous Chemicals Desk Reference	Rapid Guide to Hazardous Chemicals in the Workplace	Fire Protection Guide on Hazardous Materials (NFPA)	Firefighter's Handbook of Hazardous Materials	Pocket Guide to Chemical Hazards (NIOSH)	Chemical Hazard Response Information System (CHRIS)	Emergency Handling of Hazardous Materials in Surface Transportation (AAR)	Emergency Action Guides (EAG)	Chemical Data Notebook: A User's Manual	Condensed Chemical Dictionary
GASOLINE CASINGHEAD	1257	27	414	18	1779			-56	153		GCS 607			G 1	
GASOLINE HEART CUT	1203	27													
GASOLINE POLYMER	1215										GPL 611				
GASOLINE STRAIGHT RUN	1203	27			1779						GSR 613				
GC 1,189 Synonym: KEPONE	2761	55	493		2077	701			177		KPE 688				670
GELATINE DYNAMITE	0081		415						154			473			555
GELBIN YELLOW ULTRAMARINE Synonym: CALCIUM CHROMATE	9096/3077		169		664	230			63		CCR 232	180			202
GEMALGENE Synonym: TRICHLOROETHYLENE	1710	74	837	25	3352	1137	218	174-88	273	216	TCL1085	919	T5.0		1170
GENETRON 11 Synonym: TRICHLOROFLUOROMETHANE	3082/9188	31	838	25	3353	1138	219		273		TCF1083	919			1170
GENETRON 1113 Synonym: TRIFLUOROCHLOROETHYLENE	1082	17	850					-89	276		TFC1108	925			1177
GENETRON 12 Synonym: DICHLORODIFLUOROMETHANE	1028	12	275	15		380	68		102	86	DCF 360	331	D1.1		379
GENETRON 22 Synonym: CHLORODIFLUOROMETHANE	1018	12		14	791	278	48		74		MCF 737	222			265
GENETRON 22 Synonym: MONOCHLORODIFLUOROMETHANE	1018	12	593										M4.0		795
GERANIOL								-56							558
GERANIOL FORMATE					1782	602			154						
GERANIOL NEROL MIXTURE	1993	27										474			
GERANYL FORMATE								-56	154						558

HAZARDOUS MATERIALS REFERENCE BOOKS INDEX

CHEMICAL OR MATERIAL NAME	UN/NA Number	Guide Number (DOT)	Firefighter's Hazardous Materials Reference Book	First Aid Manual for Chemical Accidents	Sax's Dangerous Properties of Industrial Materials	Hazardous Chemicals Desk Reference	Rapid Guide to Hazardous Chemicals in the Workplace	Fire Protection Guide on Hazardous Materials (NFPA)	Firefighter's Handbook of Hazardous Materials	Pocket Guide to Chemical Hazards (NIOSH)	Chemical Hazard Response Information System (CHRIS)	Emergency Handling of Hazardous Materials in Surface Transportation (AAR)	Emergency Action Guides (EAG)	Chemical Data Notebook: A User's Manual	Condensed Chemical Dictionary
GERHARDITE Synonym: COPPER NITRATE	1479	35	227		948	321	55		83		CNI 271				313
GERMANE	2192	18						94 –	154			474			559
GERMANIUM MONOHYDRIDE					1786				154						
GERMANIUM TETRAHYDRIDE	2192	18			1786	604	105		154						559
GETTYSOLVE C Synonym: HEPTANE (-n)	1206	27	425	18	1830	616	106	–57	157	120	HPT 644	505	H1.0.3		592
GILSONITE					1788	605			154						560
GLACIAL ACETIC ACID Synonym: ACETIC ACID GLACIAL	2789	29	18	11	15	6	1	14 –11	154	30	AAC 1	3	A2.0	A 2	7
GLACIAL ACRYLIC ACID Synonym: ACRYLIC ACID	2218	29	33		65	26	6	18 –12	20		ACR 25	13	A5.0	A 6	18
GLUCINUM	1567	32							154						
GLUCITOL D Synonym: SORBITOL			789		3124	1068					SBT1018				1077
GLUCOSE SOLUTION Synonym: DEXTROSE SOLUTION					1790						DTS 492				564
GLUE CATALYST NEC	1993	27										474			
GLUTARALDEHYDE			416	18	1793	606	105		154		GTA 614				565
GLYCERIN				18	1794	606	105								566
GLYCERIN DICHLOROHYDRIN-alpha-beta	2750	55						–56	154						
GLYCERINE	1760	60	417			606		–56			GCR 606	474			
GLYCERITE Synonym: TANNIC ACID			805		3180	1089		–84			TNA1130				1114
GLYCEROL Synonym: GLYCERINE	1760	60	417			606		–56			GCR 606	474			566

CHEMICAL OR MATERIAL NAME	UN/NA Number	Guide Number (DOT)	Firefighter's Hazardous Materials Reference Book	First Aid Manual for Chemical Accidents	Sax's Dangerous Properties of Industrial Materials	Hazardous Chemicals Desk Reference	Rapid Guide to Hazardous Chemicals in the Workplace	Fire Protection Guide on Hazardous Materials (NFPA)	Firefighter's Handbook of Hazardous Materials	Pocket Guide to Chemical Hazards (NIOSH)	Chemical Hazard Response Information System (CHRIS)	Emergency Handling of Hazardous Materials in Surface Transportation (AAR)	Emergency Action Guides (EAG)	Chemical Data Notebook: A User's Manual	Condensed Chemical Dictionary
GLYCEROL EPICHLOROHYDRIN Synonym: EPICHLOROHYDRIN	2023	30	336	16	1525	509	87	79 –46	132	102	EPC 552	401	E1.0	E 1	467
GLYCEROL-alpha-MONOCHLOROHYDRIN	2689	55							154			475			
GLYCERYL TRINITRATE SOLUTION	1204	26							154						568
GLYCIDALDEHYDE	2622/1993	28	418		1797	607			154			475			
GLYCIDOL			419	18	1797	608	105		154	118					568
GLYCIDYL ACRYLATE				18				–56	154						569
GLYCIDYL alpha-METHYL ACRYLATE Synonym: GLYCIDYL METHACRYLATE	1993	27	420								GCM 605				
GLYCIDYL CHLORIDE Synonym: EPICHLOROHYDRIN	2023	30	336	16	1525	509	87	79 –46	132	102	EPC 552	401	E1.0	E 1	467
GLYCIDYL METHACRYLATE	1993	27	420								GCM 605				
GLYCINE COPPER COMPLEX Synonym: COPPER GLYCINATE							55				CPG 288				312
GLYCINOL Synonym: MONOETHANOLAMINE	2491	60	594					118–	206		MEA 749		M5.0		795
GLYCOCOLL COPPER Synonym: COPPER GLYCINATE							55				CPG 288				312
GLYCOL Synonym: ETHYLENE GLYCOL	1993	27	373	17	1604	541	94	–50	155		EGL 531	434	E2.4.7		569
GLYCOL BOTTOMS	1693	58													
GLYCOL BROMIDE Synonym: ETHYLENE DIBROMIDE	1605	55	371		1601			85 –	143	110	EDB 518	430	E2.4.1		487
GLYCOL BUTYL ETHER Synonym: ETHYLENE GLYCOL MONOBUTYL ETHER	2369	26						–51	143		EGM 532	432			489

CHEMICAL OR MATERIAL NAME	UN/NA Number	Guide Number (DOT)	Firefighter's Hazardous Materials Reference Book	First Aid Manual for Chemical Accidents	Sax's Dangerous Properties of Industrial Materials	Hazardous Chemicals Desk Reference	Rapid Guide to Hazardous Chemicals in the Workplace	Fire Protection Guide on Hazardous Materials (NFPA)	Firefighter's Handbook of Hazardous Materials	Pocket Guide to Chemical Hazards (NIOSH)	Chemical Hazard Response Information System (CHRIS)	Emergency Handling of Hazardous Materials in Surface Transportation (AAR)	Emergency Action Guides (EAG)	Chemical Data Notebook: A User's Manual	Condensed Chemical Dictionary
GLYCOL CHLOROHYDRIN Synonym: ETHYLENE CHLOROHYDRIN	1135	55	369	16	1599	538	93	83 –	155	108	ECH 513	429			569
GLYCOL CYANOHYDRIN Synonym: ETHYLENE CYANOHYDRIN			370					84 –50	155		ETC 562				486
GLYCOL DIACETATE Synonym: ETHYLENE GLYCOL DIACETATE	1993	27				542		–57	155		EGY 534	430			569
GLYCOL DIBROMIDE Synonym: ETHYLENE DIBROMIDE	1605	55	371		1601			85 –	155	110	EDB 518	430	E2.4.1		487
GLYCOL DICHLORIDE Synonym: ETHYLENE DICHLORIDE	1184	26	372	17	1602	540	94	86 –50	155	110	EDC 519	430	E2.4.5	E 5	487
GLYCOL ETHER EE ACETATE Synonym: ETHYLENE GLYCOL MONOETHYL ETHER ACE- TATE	1172	26	376					–51	143		EGA 526	432	E2.6		489
GLYCOL ETHERS	1993	27			1799							475			
GLYCOL MONOBUTYL ETHER	2369	26							155						
GLYCOL MONOBUTYL ETHER ACETATE Synonym: ETHYLENE GLYCOL MONOBUTYL ETHER ACE- TATE								–51	155		EMA 542				489
GLYCOL MONOETHYL ETHER Synonym: ETHYLENE GLYCOL MONOETHYL ETHER	1171	26	375					–51	143		EGE 529		E2.5		489
GLYCOL MONOETHYL ETHER ACETATE Synonym: ETHYLENE GLYCOL MONOETHYL ETHER ACE- TATE	1172	26	376					–51	143		EGA 526	432	E2.6		489
GLYCOL MONOMETHYLETHER Synonym: ETHYLENE GLYCOL MONOMETHYL ETHER	1188	26						–51	143		EME 544	433			489
GLYOXAL			421		1802	609			155		GOS 610				570

CHEMICAL OR MATERIAL NAME	UN/NA Number	Guide Number (DOT)	Firefighter's Hazardous Materials Reference Book	First Aid Manual for Chemical Accidents	Sax's Dangerous Properties of Industrial Materials	Hazardous Chemicals Desk Reference	Rapid Guide to Hazardous Chemicals in the Workplace	Fire Protection Guide on Hazardous Materials (NFPA)	Firefighter's Handbook of Hazardous Materials	Pocket Guide to Chemical Hazards (NIOSH)	Chemical Hazard Response Information System (CHRIS)	Emergency Handling of Hazardous Materials in Surface Transportation (AAR)	Emergency Action Guides (EAG)	Chemical Data Notebook: A User's Manual	Condensed Chemical Dictionary
GRAIN ALCOHOL Synonym: ETHYL ALCOHOL	1170	26	344	16	1572	525	90	-48	155		EAL 504	408	E2.2	E 2	574
GRAPE SUGAR SOLUTION Synonym: DEXTROSE SOLUTION											DTS 492				574
GREEN NICKEL OXIDE Synonym: NICKEL HYDROXIDE	9140/3077	31	619		2494	849	156				NKH 836	680			818
GREEN OIL Synonym: ANTHRACENE			87		262	76	14	-17	155		ATH 101				83
GREEN VERDIGRIS Synonym: COPPER SUBACETATE							55				CST 315				315
GREEN VITRIOL Synonym: FERROUS SULFATE	9125/3077		397		1702	579			147		FRS 593	450			518
GRENADE	2017	58													
GRENADE	2016	15	422						155			476			
GUAIACOL					1810	611			155						577
GUANIDINE NITRATE	1467	43			1813				155			477			578
GUM CAMPHOR	2717	32							155						579
GUM TURPENTINE Synonym: TURPENTINE	1299	27	856	26	3458	1162	225	-92	282	222	TPT1144	939			580
GUSATHION INSECTICIDE Synonym: AZINPHOS METHYL	2783	55	106		323	100	18		37	42	AZM 112	103			109
GUTHION	2783	55							156						580
GUTHION INSECTICIDE Synonym: AZINPHOS METHYL	2783	55	106		323	100	18		37	42	AZM 112	103			109
GUTTA PERCHA SOLUTION	1205	27							156						580
HAFNIUM	2545	40			1818	613						478			581

CHEMICAL OR MATERIAL NAME	UN/NA Number	Guide Number (DOT)	Firefighter's Hazardous Materials Reference Book	First Aid Manual for Chemical Accidents	Sax's Dangerous Properties of Industrial Materials	Hazardous Chemicals Desk Reference	Rapid Guide to Hazardous Chemicals in the Workplace	Fire Protection Guide on Hazardous Materials (NFPA)	Firefighter's Handbook of Hazardous Materials	Pocket Guide to Chemical Hazards (NIOSH)	Chemical Hazard Response Information System (CHRIS)	Emergency Handling of Hazardous Materials in Surface Transportation (AAR)	Emergency Action Guides (EAG)	Chemical Data Notebook: A User's Manual	Condensed Chemical Dictionary
HAFNIUM METAL	1326	32			1818	613			156	120		478			
HAFNIUM METAL	2545	40			1818	613			156	120		478			
HAFNIUM SOLUTION	1326	32				613									
HAIR WET	1325	32													
HALOCARBON 12 Synonym: DICHLORODIFLUOROMETHANE	1028	12	275	15	1137	380	68		102	86	DCF 360	331	D1.1		379
HALOGENATED IRRITATING LIQUID	1610/1693	58							156			479			
HALOGENATED WAXES Synonym: POLYCHLORINATED BIPHENYL	2315/3151	31	698	23	2831	964	185	142–	236		PCB 934	773			931
HALON 1001 Synonym: METHYL BROMIDE	1062	55	559	20	2274	776	137	110–67	192	146	MTB 805	619	M3.0	M 3	759
HALON 104 Synonym: CARBON TETRACHLORIDE	1846	55	192	13	701	246	43	48 –	66	60	CBT 223	193	C2.3	C 5	221
HALON 122 Synonym: DICHLORODIFLUOROMETHANE	1028	12	275	15	1137	380	68		102	86	DCF 360	331	D1.1		379
HALON 1220 Synonym: DICHLORODIFLUOROMETHANE	1028	12	275	15	1137	380	68		102	86	DCF 360	331	D1.1		379
HALOTHANE	1326	32		18	1821	613			156						583
HALOTHANE	2545	40		18	1821	613			156						583
HARTSHORN Synonym: AMMONIUM CARBONATE	9084/3077	31	55	11	223	60			28		ACB 14	57			585
HAY	1327	32							156						
HAZARDOUS SUBSTANCE	9188	31							156						
HAZARDOUS WASTE	9189/3082	31	423						156			479			586
HCN Synonym: HYDROCYANIC ACID	1051	13	442	18	1896	641	114	–59	164				H4.1	H 2	613

CHEMICAL OR MATERIAL NAME	UN/NA Number	Guide Number (DOT)	Firefighter's Hazardous Materials Reference Book	First Aid Manual for Chemical Accidents	Sax's Dangerous Properties of Industrial Materials	Hazardous Chemicals Desk Reference	Rapid Guide to Hazardous Chemicals in the Workplace	Fire Protection Guide on Hazardous Materials (NFPA)	Firefighter's Handbook of Hazardous Materials	Pocket Guide to Chemical Hazards (NIOSH)	Chemical Hazard Response Information System (CHRIS)	Emergency Handling of Hazardous Materials in Surface Transportation (AAR)	Emergency Action Guides (EAG)	Chemical Data Notebook: A User's Manual	Condensed Chemical Dictionary
HEA Synonym: 2-HYDROXYETHYL ACRYLATE								-60	166		HAI 617				586
HEAVY HYDROGEN	1957	22							156						587
HELIUM	1046	12			1824	614			156			504	H1.0		588
HELIUM	1963	21			1824	614			156			504	H1.0		588
HELIUM ARGON GAS MIXTURE	1956	12										503			
HELIUM BUTANE GAS MIXTURE	1954	22										503			
HELIUM BUTANE GAS MIXTURE	1956	12										503			
HELIUM CARBON DIOXIDE NITROGEN MIXTURE	1956	12										504			
HELIUM ISOBUTANE GAS MIXTURE	1954	22													
HELIUM ISOBUTANE GAS MIXTURE	1956	12										504			
HELIUM NITROGEN MIXTURE	1956	12										504			
HELIUM OXYGEN MIXTURE	1980	14				614			156						
HELLEBOREIN					1824				156						
HEMITERPENE Synonym: ISOPRENE	1218	27	479		2037	685		102-62	174		IPR 679	554	13.0	1 1	658
HENDECANE	2330	27						-57	156						
HENDECANOIC ALCOHOL Synonym: UNDECANOL			857								UND1165				
HEOD Synonym: DIELDRIN	2761	55	284		1181	398	72		106	90	DED 384	344			590
HEPAR SULFURIS	1382	32													
HEPTACHLOR	2761/3077	55	424		1826	615	106		156	120	HTC 646	505			591
HEPTAFLUOROBUTYRIC ACID					1828	616			156						591
HEPTALDEHYDE	3056	26							157						591

HAZARDOUS MATERIALS REFERENCE BOOKS INDEX

CHEMICAL OR MATERIAL NAME	UN/NA Number	Guide Number (DOT)	Firefighter's Hazardous Materials Reference Book	First Aid Manual for Chemical Accidents	Sax's Dangerous Properties of Industrial Materials	Hazardous Chemicals Desk Reference	Rapid Guide to Hazardous Chemicals in the Workplace	Fire Protection Guide on Hazardous Materials (NFPA)	Firefighter's Handbook of Hazardous Materials	Pocket Guide to Chemical Hazards (NIOSH)	Chemical Hazard Response Information System (CHRIS)	Emergency Handling of Hazardous Materials in Surface Transportation (AAR)	Emergency Action Guides (EAG)	Chemical Data Notebook: A User's Manual	Condensed Chemical Dictionary
HEPTALDEHYDE-n	3056	26							157						591
HEPTANE Synonym: HEPTANE-n	1206	27	425	18	1830	616	106	−57	157	120	HPT 644	505	H1.0.3		592
HEPTANE-n Synonym: HEPTANE	1206	27	425	18	1830	616	106	−57	157	120	HPT 644	505	H1.0.3		592
HEPTANOL	1993	27	426	18							HTN 648				
HEPTENE (-n)	2278	27	427		1832	617			157						
HEPTYL ALCOHOL Synonym: HEPTANOL			426	18	1833	617			157		HTN 648				592
HEPTYL CARBINOL Synonym: OCTANOL	1987	26	650								OTA 909	706	O1.0		847
HEPTYL HYDRIDE Synonym: HEPTANE (-n)	1206	27	425	18	1830	616	106	−57	157	120	HPT 644	505	H1.0.3		592
HEPTYLAMINE								−57	157						593
HEPTYLCARBINOL Synonym: OCTANOL	1987	26	650								OTA 909	706	O1.0		847
HEPTYLENE Synonym: 1-HEPTENE	2278	27							157		HTE 647				592
HEPTYLENE-2-trans	2278	27						−57	157						
HEPTYLENE-alpha	2278	27							157						
HEXA Synonym: HEXAMETHYLENETETRAMINE	1328	32	434		1856	625			160		HMT 639				598
HEXA BENZENESULFONIC ACID Synonym: ALKYLBENZENESULFONIC ACID											ABS 11				
HEXACARBONYLCHROMIUM					1838	618	107								
HEXACHLORO DIPHENYL OXIDE								−57	158						595

CHEMICAL OR MATERIAL NAME	UN/NA Number	Guide Number (DOT)	Firefighter's Hazardous Materials Reference Book	First Aid Manual for Chemical Accidents	Sax's Dangerous Properties of Industrial Materials	Hazardous Chemicals Desk Reference	Rapid Guide to Hazardous Chemicals in the Workplace	Fire Protection Guide on Hazardous Materials (NFPA)	Firefighter's Handbook of Hazardous Materials	Pocket Guide to Chemical Hazards (NIOSH)	Chemical Hazard Response Information System (CHRIS)	Emergency Handling of Hazardous Materials in Surface Transportation (AAR)	Emergency Action Guides (EAG)	Chemical Data Notebook: A User's Manual	Condensed Chemical Dictionary
HEXACHLOROACETONE	2661	54							158			506			595
HEXACHLOROBENZENE	2729	53		18	1839	618	107		158		HCZ 628	507			595
HEXACHLOROBUTADIENE	2279	55			1840	618	107	-57	158		HCB 622	507			595
HEXACHLOROCYCLOPENTADIENE	2646	55	428		1841	619	108		158		HCC 623	507	H1.0.4		595
HEXACHLOROCYCLOPENTADIENE DIMER Synonym: MIREX	1615		589		2431	828			205		MRX 798				
HEXACHLORODISILANE					1842				158						
HEXACHLOROETHANE	9037/2811	53	429	18	1842	619	108		158	122	HCE 624	508			595
HEXACHLORONAPHTHALENE					1843	620	108		158	122					596
HEXACHLOROPHENE	2875	53			1843	620			158		HCP 627	508			596
HEXACHLOROPROPENE					1844	621			158						
HEXADECYL SULFATE SODIUM SALT											HSS 645				
HEXADECYL TRICHLOROSILANE	1781	60			1846	622			158			509			596
HEXADECYLTRIMETHYLAMMONIUM CHLORIDE			430								HAC 615				
HEXADIENE	2458	29							158			509			
HEXADRIN Synonym: ENDRIN	2761/2996	55	334		1520	507	86		131	102	EDR 520	399			463
HEXAETHYL TETRAPHOSPHATE	1611/2783	55			1848	622			159			510			597
HEXAETHYL TETRAPHOSPHATE AND COMPRESSED GAS MIXTURE	1612	15							159			510			
HEXAETHYL TETRAPHOSPHATE MIXTURE	2783	55							159			510			
HEXAFLUOROACETONE	2420	15			1849	623	109		159			511	H2.5		597
HEXAFLUOROACETONE	2810	55													597
HEXAFLUOROACETONE HYDRATE	2552/2810	55			1849	623			159			511			

HAZARDOUS MATERIALS REFERENCE BOOKS INDEX

CHEMICAL OR MATERIAL NAME	UN/NA Number	Guide Number (DOT)	Firefighter's Hazardous Materials Reference Book	First Aid Manual for Chemical Accidents	Sax's Dangerous Properties of Industrial Materials	Hazardous Chemicals Desk Reference	Rapid Guide to Hazardous Chemicals in the Workplace	Fire Protection Guide on Hazardous Materials (NFPA)	Firefighter's Handbook of Hazardous Materials	Pocket Guide to Chemical Hazards (NIOSH)	Chemical Hazard Response Information System (CHRIS)	Emergency Handling of Hazardous Materials in Surface Transportation (AAR)	Emergency Action Guides (EAG)	Chemical Data Notebook: A User's Manual	Condensed Chemical Dictionary
HEXAFLUOROACETONE HYDRATE	2810/2552	55										511			
HEXAFLUORODICHLOROBUTENE					1850	623									
HEXAFLUOROETHANE	2193	12		18					159			512			597
HEXAFLUOROISOPROPANOL	1760	60			1850	623						512			
HEXAFLUOROPHOSPHORIC ACID	1782	59			1850	623			159			512			597
HEXAFLUOROPROPYLENE	1858	12							159			512			597
HEXAFLUOROPROPYLENE OXIDE	1956	12							159			513			
HEXAFLUOSILICIC ACID Synonym: HYDROFLUOROSILICIC ACID	1778	60							164				H4.2		614
HEXAFLUOSILICIC ACID Synonym: FLUOSILICIC ACID	1778	60	399	17					150		FSL 596				532
HEXAHYDRIC ALCOHOL Synonym: SORBITOL			789		3124	1068					SBT1018				597
HEXAHYDRO-1,4-DIAZINE Synonym: PIPERAZINE	2579/1760	60	696		2811	959		−80			PPZ 983	766			919
HEXAHYDRO-2H-AZEPINE-2-ONE Synonym: CAPROLACTAM				13	680	238	39				CLS 259		C1.5		213
HEXAHYDROANILINE Synonym: CYCLOHEXYLAMINE	2357	68	249		994	336	60	60 −30	159		CHA 243	285			597
HEXAHYDROAZEPINE Synonym: HEXAMETHYLENIMINE	2493	29	433		1851				160		HMI 638				
HEXAHYDROBENZENE Synonym: CYCLOHEXANE	1145	26	244	14	990	333	58	−30	159	76	CHX 251	280	C8.1		597
HEXAHYDROCRESOL	2617	26							159						597
HEXAHYDROMETHYL PHENOL	2617	26							159						597

CHEMICAL OR MATERIAL NAME	UN/NA Number	Guide Number (DOT)	Firefighter's Hazardous Materials Reference Book	First Aid Manual for Chemical Accidents	Sax's Dangerous Properties of Industrial Materials	Hazardous Chemicals Desk Reference	Rapid Guide to Hazardous Chemicals in the Workplace	Fire Protection Guide on Hazardous Materials (NFPA)	Firefighter's Handbook of Hazardous Materials	Pocket Guide to Chemical Hazards (NIOSH)	Chemical Hazard Response Information System (CHRIS)	Emergency Handling of Hazardous Materials in Surface Transportation (AAR)	Emergency Action Guides (EAG)	Chemical Data Notebook: A User's Manual	Condensed Chemical Dictionary
HEXAHYDROPHENOL Synonym: CYCLOHEXANOL	1993	27	245		991	333	59	−30	159	78	CHN 246	280			597
HEXAHYDROPYRAZINE Synonym: PIPERAZINE	2579/1760	60	696		2811	959		−80			PPZ 983	766			919
HEXAHYDROPYRIDINE	2401	29							159						597
HEXAHYDROXYL	2263	27							159						
HEXALDEHYDE	1207	26	431						159		HAL 618	513			598
HEXALDEHYDE (-n)	1207	26	431						159		HAL 618	513			598
HEXALIN Synonym: CYCLOHEXANOL	1993	27	245	14	991	333	59	−30	159	78	CHN 246	280			598
HEXALIN ACETATE	2243	27							159						
HEXAMETHYL PHOSPHORAMIDE					1858	626	109		160						599
HEXAMETHYLDISILAZANE	1993	27						95 −	160			513			598
HEXAMETHYLENE Synonym: CYCLOHEXANE	1145	26	244	14	990	333	58	−30	160	76	CHX 251	280	C8.1		598
HEXAMETHYLENE DIAMINE	2280	60	432						160			515			598
HEXAMETHYLENE DIAMINE SOLUTION	1783	60							160		HMD 637	515	H1.1		598
HEXAMETHYLENE DIISOCYANATE	2281/2811	55							160			515			598
HEXAMETHYLENEIMINE	2493	29	433						160			516			598
HEXAMETHYLENETETRAMINE	1328	32	434		1856	625			160		HMT 639				598
HEXAMETHYLENIMINE	2493	29							160		HMI 638				
HEXAMETHYLPHOSPHORAMIDE					1858	626	109		160						599
HEXAMINE Synonym: HEXAMETHYLENETETRAMINE	1328	32	434		1857	625			160		HMT 639	516			599

HAZARDOUS MATERIALS REFERENCE BOOKS INDEX

CHEMICAL OR MATERIAL NAME	UN/NA Number	Guide Number (DOT)	Firefighter's Hazardous Materials Reference Book	First Aid Manual for Chemical Accidents	Sax's Dangerous Properties of Industrial Materials	Hazardous Chemicals Desk Reference	Rapid Guide to Hazardous Chemicals in the Workplace	Fire Protection Guide on Hazardous Materials (NFPA)	Firefighter's Handbook of Hazardous Materials	Pocket Guide to Chemical Hazards (NIOSH)	Chemical Hazard Response Information System (CHRIS)	Emergency Handling of Hazardous Materials in Surface Transportation (AAR)	Emergency Action Guides (EAG)	Chemical Data Notebook: A User's Manual	Condensed Chemical Dictionary
HEXANAL Synonym: HEXALDEHYDE (-n)	1207	26	431		1859			−58	160		HAL 618	513			598
HEXANAPHTHENE Synonym: CYCLOHEXANE	1145	26	244	14	990	333	58	−30	88	76	CHX 251	280	C8.1		599
HEXANDIOL	1693	58										516			
HEXANE Synonym: HEXANE-n	1208	27	435	18	1859	627	109	−58	160	122	HXA 649	517	H2.0		599
HEXANE-n Synonym: HEXANE	1208	27	435	18	1859	627	109	−58	160	122	HXA 649	517	H2.0		599
HEXANEDIOC ACID Synonym: ADIPIC ACID	9077/3077		35		79	28		12	20		ADA 28	15	A6.0	A 8	24
HEXANEDIOIC ACID Synonym: ADIPIC ACID	9077/3077		35		79	28		−12	20		ADA 28	15	A6.0	A 8	599
HEXANOIC ACID	1760/2829	60			1862	628					HXO 653	518			599
HEXANOL Synonym: HEXANOL-n	2282	26	436	18					161		HXN 652	519			
HEXANOL-n Synonym: HEXANOL	2282	26	436	18					161		HXN 652	519			
HEXANON Synonym: CYCLOHEXANONE	1915	26	246	14	991	334	59	−30	88	78	CCH 228	281	C8.2		337
HEXENE	2370	27	437		1865	629			161			519			
HEXONE Synonym: METHYL ISOBUTYL KETONE	1245/1993	26	571	20	1868	630	110	−70	162	124	MIK 765	630	M3.3		600
HEXENE-alpha Synonym: 1-HEXENE	2370	27			1865	629		−58	161		HXE 650				600

CHEMICAL OR MATERIAL NAME	UN/NA Number	Guide Number (DOT)	Firefighter's Hazardous Materials Reference Book	First Aid Manual for Chemical Accidents	Sax's Dangerous Properties of Industrial Materials	Hazardous Chemicals Desk Reference	Rapid Guide to Hazardous Chemicals in the Workplace	Fire Protection Guide on Hazardous Materials (NFPA)	Firefighter's Handbook of Hazardous Materials	Pocket Guide to Chemical Hazards (NIOSH)	Chemical Hazard Response Information System (CHRIS)	Emergency Handling of Hazardous Materials in Surface Transportation (AAR)	Emergency Action Guides (EAG)	Chemical Data Notebook: A User's Manual	Condensed Chemical Dictionary
HEXENE-Iso Synonym: 2-METHYL-1-PENTENE	2288	27			2372			-71	198		MPE 785				777
HEXYL ACETATE Synonym: METHYL AMYL ACETATE	1233	26	556		1870	630		-58	162		HAE 616	642			600
HEXYL ACETATE-sec	1233	26			1870	631	111		162	124					601
HEXYL ALCOHOL	2282	26			1870	631		-58	162						601
HEXYL ALCOHOL-n Synonym: HEXANOL (-n)	2282	26	436	18	1870	631			162		HXN 652	519			601
HEXYL ALCOHOL-sec	2282	26						-58							
HEXYL ALCOHOL-sec Synonym: ETHYLBUTANOL	2275	26	347					-58	162		EBT 509				
HEXYL ETHER								-59	162						601
HEXYL HYDRIDE Synonym: HEXANE (-n)	1208	27	435	18	1859	627	109	-58	162		HXA 649	517	H2.0		599
HEXYL METHACRYLATE								-59	162						601
HEXYL TRICHLOROSILANE	1784	29			1875	632			162			520			602
HEXYLAMINE	1993	27			1871	631		-59	162			520			601
HEXYLENE Synonym: 1-HEXENE	2370	27			1865	629		-58	162		HXE 650				601
HEXYLENE GLYCOL	2030	59	438		1873	632	111	-59	162		HXG 651				601
HF Synonym: HYDROGEN FLUORIDE	1052	15	449					99 –	164	126	HFX 636	530	H4.4	H 4	602
HF A Synonym: HYDROGEN FLUORIDE	1052	15	449					99 –	164	126	HFX 636	530	H4.4	H 4	616
HHDN Synonym: ALDRIN	2761	55	37	11	94	31	7		21	34	ALD 40	21			602

HAZARDOUS MATERIALS REFERENCE BOOKS INDEX

CHEMICAL OR MATERIAL NAME	UN/NA Number	Guide Number (DOT)	Firefighter's Hazardous Materials Reference Book	First Aid Manual for Chemical Accidents	Sax's Dangerous Properties of Industrial Materials	Hazardous Chemicals Desk Reference	Rapid Guide to Hazardous Chemicals in the Workplace	Fire Protection Guide on Hazardous Materials (NFPA)	Firefighter's Handbook of Hazardous Materials	Pocket Guide to Chemical Hazards (NIOSH)	Chemical Hazard Response Information System (CHRIS)	Emergency Handling of Hazardous Materials in Surface Transportation (AAR)	Emergency Action Guides (EAG)	Chemical Data Notebook: A User's Manual	Condensed Chemical Dictionary
HI DRY Synonym: TETRAETHYLENE GLYCOL	1993	27	815		3218	1100		–85			TTG1155	892			1129
HIGH SPEED BEARING OIL Synonym: OILS MISCELLANEOUS SPINDLE	1270	27						–83			OSD 904				
HIGHER FATTY ALCOHOL Synonym: TALLOW FATTY ALCOHOL											TFA1107				
HISTAMINE					1878	633			162						603
HMDA Synonym: HEXAMETHYLENEDIAMINE	1783	60							160		HMD 637	515	H1.1		598
HMDA Synonym: HEXAMETHYLENE DIAMINE SOLUTION	1783	60							160		HMD 637	515	H1.1		598
HOME HEATING OIL Synonym: FUEL OIL NO. 2	1993	27	404		1758			–55	151			464	F3.1		
HOME HEATING OIL Synonym: OILS FUEL. 2	1223	27						–55			OTW 917				
HOMOPIPERIDINE Synonym: HEXAMETHYLENIMINE	2493	29	433						160		HMI 638				
HOUSEHOLD AMMONIA Synonym: AMMONIUM HYDROXIDE	2672	60	62	12	228	62	11		29		AMH 52	60	A8.0	A10	66
HTH Synonym: CALCIUM HYPOCHLORITE	1748	45	172	13				46 –	63		CHY 252	182	C1.1	C 2	607
HTH DRY CHLORINE Synonym: CALCIUM HYPOCHLORITE	1748	45	172	13				46 –	63		CHY 252	182	C1.1	C 2	204
HYDRACRYLIC ACID beta LACTONE Synonym: PROPIOLACTONE-beta			719		2907	997	187		242	186	PLT 964				
HYDRACRYLONITRILE Synonym: ETHYLENE CYANOHYDRIN			370		1883	634		84 –50	162		ETC 562				486

CHEMICAL OR MATERIAL NAME	UN/NA Number	Guide Number (DOT)	Firefighter's Hazardous Materials Reference Book	First Aid Manual for Chemical Accidents	Sax's Dangerous Properties of Industrial Materials	Hazardous Chemicals Desk Reference	Rapid Guide to Hazardous Chemicals in the Workplace	Fire Protection Guide on Hazardous Materials (NFPA)	Firefighter's Handbook of Hazardous Materials	Pocket Guide to Chemical Hazards (NIOSH)	Chemical Hazard Response Information System (CHRIS)	Emergency Handling of Hazardous Materials in Surface Transportation (AAR)	Emergency Action Guides (EAG)	Chemical Data Notebook: A User's Manual	Condensed Chemical Dictionary
HYDRAZINE ANHYDROUS	2029	28	439	18	1885	635	111	95 –59	163	124		521	H3.0	H 1	
HYDRAZINE AQUEOUS	2029	28	439	18	1885	635	111	95 –	163	124		521	H3.0	H 1	
HYDRAZINE AQUEOUS	2030	59		18	1885	635	111		163	124	HDZ 633		H3.0	H 1	
HYDRAZINE BASE Synonym: HYDRAZINE ANHYDROUS	2029	28	439	18	1885	635	111	95 –59	163	124		521	H3.0	H 1	
HYDRAZINE HYDRATE Synonym: HYDRAZINE SOLUTION	2030	59		18	1886	636	112		163	124	HDZ 633		H3.0	H 1	611
HYDRAZINE HYDROCHLORIDE					1887	636	112								
HYDRAZINE SOLUTION	2030	59		18	1885	635	111		163	124	HDZ 633		H3.0		610
HYDRAZINE SULFATE (1:1)					1887	636	112								611
HYDRAZOIC ACID					1890	637	113		163						611
HYDRAZOIC ACID SODIUM SALT Synonym: SODIUM AZIDE	1687	56	159		3063	1045	198		252		SAZ1014	832			1049
HYDRAZOMETHANE Synonym: METHYL HYDRAZINE	1244	57	586		2327	797	144	114–	195	152	MHZ 762	648	M3.4.6		770
HYDRIDE	1409	40			1890	638			163			521			611
HYDRIODIC ACID	1787	60	440	18	1891	638		96 –	163			521			611
HYDROBROMIC ACID	1788	60			1891	639	113	96 –	163			522			612
HYDROBROMIC ACID ANHYDROUS Synonym: HYDROGEN BROMIDE	1048	15	446	18		639			164	124	HBR 621	528	H4.2.1		616
HYDROBROMIC ACID MONOAMMONIATE Synonym: AMMONIUM BROMIDE					222	59					ANB 62				64
HYDROBROMIC ETHER	1891	58							163						
HYDROCARBON GAS	1964	22			1891	639			163			522			
HYDROCARBON GAS	1965	22			1891	639			163			523			

HAZARDOUS MATERIALS REFERENCE BOOKS INDEX

CHEMICAL OR MATERIAL NAME	UN/NA Number	Guide Number (DOT)	Firefighter's Hazardous Materials Reference Book	First Aid Manual for Chemical Accidents	Sax's Dangerous Properties of Industrial Materials	Hazardous Chemicals Desk Reference	Rapid Guide to Hazardous Chemicals in the Workplace	Fire Protection Guide on Hazardous Materials (NFPA)	Firefighter's Handbook of Hazardous Materials	Pocket Guide to Chemical Hazards (NIOSH)	Chemical Hazard Response Information System (CHRIS)	Emergency Handling of Hazardous Materials in Surface Transportation (AAR)	Emergency Action Guides (EAG)	Chemical Data Notebook: A User's Manual	Condensed Chemical Dictionary
HYDROCHLORIC ACID	1050	15		18	1892	639	113		163						613
HYDROCHLORIC ACID ANHYDROUS Synonym: HYDROGEN CHLORIDE	1050	15	447	18	1892	639	113	98 –	163	126	HDC 630	528	H4.3	H 3	616
HYDROCHLORIC ACID GAS Synonym: HYDROGEN CHLORIDE	1050	15	447	18	1900	639	113	98 –		126	HDC 630	528	H4.3	H 3	616
HYDROCHLORIC ACID SOLUTION	1789	60	441		1892	639	113		163		HCL 625	524	H4.0		613
HYDROCHLORIC ETHER Synonym: ETHYL CHLORIDE	1037	27	350		1586	532	92	82 –49	164	108	ECL 514	412	E2.3		483
HYDROCHLORIDE Synonym: HYDROGEN CHLORIDE	1050	15	447	18	1900	643	115	98 –	164	126	HDC 630	528	H4.3	H 3	616
HYDROCYANIC ACID Synonym: HYDROGEN CYANIDE	1051	13	442	18	1896	641	114	98 –59	164	126	HCN 626	529	H4.1	H 2	613
HYDROCYANIC ACID	1613	55		18	1896	641	114	–59	164			525	H4.1	H 2	613
HYDROCYANIC ACID SODIUM SALT Synonym: SODIUM CYANIDE	1689	55	768	24	1897	1050	199	157–	253		SCN1024	837	S2.2	S 3	1053
HYDROFLUORIC ACID	1052	15		18	1897	642	114								614
HYDROFLUORIC ACID ANHYDROUS Synonym: HYDROGEN FLUORIDE	1052	15	449	18	1897	642	114	99 –		126	HFX 636	530	H4.4	H 4	616
HYDROFLUORIC ACID SOLUTION	1790	59	443	18	1897	642	114		164		HFA 635	527	H4.1.1		
HYDROFLUORIC AND SULFURIC ACID MIXTURE	1786	59			1898				164			527			
HYDROFLUOROSILICIC ACID	1778	60							164				H4.2		614
HYDROFLUOSILIC ACID Synonym: FLUOSILICIC ACID	1778	60	399	17					150		FSL 596				532
HYDROFLUOSILICIC ACID Synonym: HYDROFLUOROSILICIC ACID	1778	60							164				H4.2		614

CHEMICAL OR MATERIAL NAME	UN/NA Number	Guide Number (DOT)	Firefighter's Hazardous Materials Reference Book	First Aid Manual for Chemical Accidents	Sax's Dangerous Properties of Industrial Materials	Hazardous Chemicals Desk Reference	Rapid Guide to Hazardous Chemicals in the Workplace	Fire Protection Guide on Hazardous Materials (NFPA)	Firefighter's Handbook of Hazardous Materials	Pocket Guide to Chemical Hazards (NIOSH)	Chemical Hazard Response Information System (CHRIS)	Emergency Handling of Hazardous Materials in Surface Transportation (AAR)	Emergency Action Guides (EAG)	Chemical Data Notebook: A User's Manual	Condensed Chemical Dictionary
HYDROFURAN	2056	26	818		3227	1103	210	167–86	263	210	THF1116	894	T1.0	T 2	1131
Synonym: TETRAHYDROFURAN															
HYDROGEN	1966	22	445		1898	642		97 –59	164		HXX 654	533			614
HYDROGEN	1049	22	444		1898	642		97 –59	164		HXX 654	533			614
HYDROGEN AND METHANE MIXTURE	2034/1954	22							164			528			
HYDROGEN ANTIMONIDE	2676	18							164						
HYDROGEN ARSENIDE	2188	18							164						
HYDROGEN BROMIDE	1048	15	446	18					164	124	HBR 621	528	H4.2.1		616
HYDROGEN BROMIDE ANHYDROUS	1048	15	446	18					164	124	HBR 621	528	H4.2.1		616
Synonym: HYDROGEN BROMIDE															
HYDROGEN BROMIDE SOLUTION	1788	60		18					164						
HYDROGEN CARBOXYLIC ACID	1779	60							164						
HYDROGEN CHLORIDE	1050	15	447	18	1900	643	115	98 –	164	126	HDC 630	528	H4.3	H 3	616
HYDROGEN CHLORIDE	2186	15		18	1900	643	115	98 –	164			529	H4.3	H 3	616
HYDROGEN CHLORIDE SOLUTION	1789	60		18	1900	643	115		164						
HYDROGEN CYANIDE	1614	57							164			530			616
HYDROGEN CYANIDE	1051	13	448		1896	641	114	98 –59	164	126	HCN 626	529	H4.1	H 2	616
Synonym: HYDROCYANIC ACID															
HYDROGEN DIOXIDE	2014	45		18	1901	644	115	99 –	165	126		531	H4.5	H 5	616
Synonym: HYDROGEN PEROXIDE															
HYDROGEN DIOXIDE	2015	47	450	18	1901	644	115	100–	164	126	HPO 533	532	H4.5	H 5	616
Synonym: HYDROGEN PEROXIDE															
HYDROGEN DISULFIDE					1900				164						
HYDROGEN FLUORIDE	1052	15	449					99 –	164	126	HFX 636	530	H4.4	H 4	616

HAZARDOUS MATERIALS REFERENCE BOOKS INDEX

CHEMICAL OR MATERIAL NAME	UN/NA Number	Guide Number (DOT)	Firefighter's Hazardous Materials Reference Book	First Aid Manual for Chemical Accidents	Sax's Dangerous Properties of Industrial Materials	Hazardous Chemicals Desk Reference	Rapid Guide to Hazardous Chemicals in the Workplace	Fire Protection Guide on Hazardous Materials (NFPA)	Firefighter's Handbook of Hazardous Materials	Pocket Guide to Chemical Hazards (NIOSH)	Chemical Hazard Response Information System (CHRIS)	Emergency Handling of Hazardous Materials in Surface Transportation (AAR)	Emergency Action Guides (EAG)	Chemical Data Notebook: A User's Manual	Condensed Chemical Dictionary
HYDROGEN FLUORIDE AQUEOUS SOLUTION Synonym: HYDROFLUORIC ACID SOLUTION	1790	59	443		1897	642	114		164		HFA 635	527	H4.1.1		
HYDROGEN FLUORIDE SOLUTION	1790	59							164					H 4	616
HYDROGEN HEXAFLUOROSILICATE Synonym: HYDROFLUOROSILICIC ACID	1778	60							164				H4.2		616
HYDROGEN HEXAFLUOROSILICATE Synonym: FLUOSILICIC ACID	1778	60	399	17					150		FSL 596				616
HYDROGEN IODIDE	1956	12													616
HYDROGEN IODIDE	2197	15							164			531			616
HYDROGEN IODIDE SOLUTION	1787	60							164						
HYDROGEN NITRATE Synonym: NITRIC ACID FUMING	2032	44		21	2509	856		123–	210	160		684	N2.0	N 1	823
HYDROGEN OXIDE Synonym: HYDROGEN PEROXIDE	2014	45		18	1901	644	115	99 –	165	126		531	H4.5	H 5	616
HYDROGEN OXIDE Synonym: HYDROGEN PEROXIDE	2015	47	450	18	1901	644	115	100–	165	126	HPO 643	532	H4.5	H 5	616
HYDROGEN PEROXIDE	2015	47	450	18	1901	644	115	100–	165	126	HPO 643	532	H4.5	H 5	616
HYDROGEN PEROXIDE CARBAMIDE Synonym: UREA PEROXIDE	1511	35	866						283		UPO1166	945			1204
HYDROGEN PEROXIDE SOLUTION	2014	45		18	1901	644	115	99 –	165	126		531	H4.5	H 5	
HYDROGEN PEROXIDE SOLUTION	2984	60		18	1902				165	126			H4.5	H 5	
HYDROGEN PHOSPHIDE	2199	18							165						617
HYDROGEN SELENIDE	2202	13	451	18	1903	645	116		165	128		532	H4.6		617
HYDROGEN SULFIDE	1053	13	452	18	1903	646	116	100–59	165	128	HDS 632	532	H4.7	H 6	617
HYDROGEN-para Synonym: HYDROGEN	1966	22	445		1898	642		97 –59	164		HXX 654	533			615

CHEMICAL OR MATERIAL NAME	UN/NA Number	Guide Number (DOT)	Firefighter's Hazardous Materials Reference Book	First Aid Manual for Chemical Accidents	Sax's Dangerous Properties of Industrial Materials	Hazardous Chemicals Desk Reference	Rapid Guide to Hazardous Chemicals in the Workplace	Fire Protection Guide on Hazardous Materials (NFPA)	Firefighter's Handbook of Hazardous Materials	Pocket Guide to Chemical Hazards (NIOSH)	Chemical Hazard Response Information System (CHRIS)	Emergency Handling of Hazardous Materials in Surface Transportation (AAR)	Emergency Action Guides (EAG)	Chemical Data Notebook: A User's Manual	Condensed Chemical Dictionary
HYDROGENATED TERPHENYLS					1900	643	112								615
HYDROPEROXIDE Synonym: HYDROGEN PEROXIDE	2014	45		18	1901	644	115	99 –	165	126		531	H4.5	H 5	618
HYDROPEROXIDE Synonym: HYDROGEN PEROXIDE	2015	47	450	18	1901	644	115	100–	165	126	HPO 643	532	H4.5	H 5	618
HYDROQUINOL Synonym: HYDROQUINONE	2662	53	453	18	1906	647	116	-59	165	128	HDQ 631	534			618
HYDROQUINONE	2662	53	453	18	1906	647	116	-59	165	128	HDQ 631	534			618
HYDROQUINONE MONOMETHYL ETHER								-59	165						619
HYDROSELENIC ACID	2202	13							165						
HYDROSILICOFLUORIC ACID Synonym: HYDROFLUOROSILICIC ACID	1778	60	454						165				H4.2		619
HYDROSULFITE OF SODA Synonym: SODIUM HYDROSULFITE	1384	37	776		3086	1056			254			842	S2.4		1058
HYDROSULFURIC ACID Synonym: HYDROGEN SULFIDE	1053	13	452	18	1903	646	116	100-59	165	128	HDS 632	532	H4.7	H 6	617
HYDROXETHYL-1,2-ETHANEDIAMINE-n Synonym: AMINOETHYLETHANOLAMINE	1760	60	47		172						AEA 30	50			57
HYDROXY BIACETYL Synonym: ACETIC ANHYDRIDE	1715	39	19	11	18	8	2	14 -11	17	30	ACA 13	3	A2.1		7
HYDROXY ISOBUTRONITRILE-alpha Synonym: ACETONE CYANOHYDRIN	1541	55	21	11				15 -11	17		ACY 27	4	A3.1		9
HYDROXY TRICARBOXYLIC ACID-beta Synonym: CITRIC ACID					916	313					CIT 254				286
HYDROXYACETONITRILE					1909	648	117		166						
HYDROXYANILINE-para	2512	55							166						

HAZARDOUS MATERIALS REFERENCE BOOKS INDEX

CHEMICAL OR MATERIAL NAME	UN/NA Number	Guide Number (DOT)	Firefighter's Hazardous Materials Reference Book	First Aid Manual for Chemical Accidents	Sax's Dangerous Properties of Industrial Materials	Hazardous Chemicals Desk Reference	Rapid Guide to Hazardous Chemicals in the Workplace	Fire Protection Guide on Hazardous Materials (NFPA)	Firefighter's Handbook of Hazardous Materials	Pocket Guide to Chemical Hazards (NIOSH)	Chemical Hazard Response Information System (CHRIS)	Emergency Handling of Hazardous Materials in Surface Transportation (AAR)	Emergency Action Guides (EAG)	Chemical Data Notebook: A User's Manual	Condensed Chemical Dictionary
HYDROXYBENZALDEHYDE-ortho Synonym: SALICYLALDEHYDE					3010			−83	166		SAL1009				620
HYDROXYBENZENE Synonym: PHENOL	1671	55	677	22	2729	928	175	134–78	166	176	PHN 958	743	C2.0	P 1	620
HYDROXYBENZENE Synonym: CARBOLIC ACID	1671	55	185						166			743	C2.0		620
HYDROXYBENZOIC ACID-ortho Synonym: SALICYLIC ACID			740		3011	1027		−83			SLA1047				620
HYDROXYBUTYRALDEHYDE-beta	2839	55							166						620
HYDROXYCYCLOHEXANE Synonym: CYCLOHEXANOL	1993	27	245	14	991	333	59	−30	166	78	CHN 246	280			337
HYDROXYDIMETHYLARSINE OXIDE Synonym: CACODYLIC ACID	1572/2811	53	156		1923	651			166		CDA 235	175			194
HYDROXYETHER Synonym: ETHYLENE GLYCOL MONOETHYL ETHER	1171	26	375					−51	143		EGE 529	432	E2.5		489
HYDROXYETHYL ACRYLATE	1760	60	455						166			534			
HYDROXYETHYL ACRYLATE-beta Synonym: 2-HYDROXYETHYL ACRYLATE								−60	166		HAI 617				621
HYDROXYETHYLAMINE-beta Synonym: MONOETHANOLAMINE	2491	60	594					118–	166		MEA 749		M5.0		795
HYDROXYISOBUTYRONITRILE-alpha Synonym: ACETONE CYANOHYDRIN	1541	55	21	11				15 −11	167		ACY 27	4	A3.1		9
HYDROXYLAMINE			456		1936	653		101–60	167		HDA 629				622
HYDROXYLAMINE SULFATE	2865	60	457								HAS 619				622
HYDROXYPROPENE Synonym: ALLYL ALCOHOL	1098	57	38	11	102	33	7	21 −13	22	34	ALA 39	35	A7.0		38

CHEMICAL OR MATERIAL NAME	UN/NA Number	Guide Number (DOT)	Firefighter's Hazardous Materials Reference Book	First Aid Manual for Chemical Accidents	Sax's Dangerous Properties of Industrial Materials	Hazardous Chemicals Desk Reference	Rapid Guide to Hazardous Chemicals in the Workplace	Fire Protection Guide on Hazardous Materials (NFPA)	Firefighter's Handbook of Hazardous Materials	Pocket Guide to Chemical Hazards (NIOSH)	Chemical Hazard Response Information System (CHRIS)	Emergency Handling of Hazardous Materials in Surface Transportation (AAR)	Emergency Action Guides (EAG)	Chemical Data Notebook: A User's Manual	Condensed Chemical Dictionary
HYDROXYPROPIONIC ACID-alpha Synonym: LACTIC ACID	1760	60	496		2083	703					LTA 711	569			625
HYDROXYPROPYL ACRYLATE	1760	60	458								HPA 640	535			
HYDROXYPROPYL METHACRYLATE											HPM 642				625
HYDROXYTOLUENE Synonym: CRESOL	2076	55	232	14	959	324	55	57 –	167	74	CRS 307	270	C7.0		626
HYDROXYTOLUENE (-ortho) Synonym: CRESOL (-ortho)	2076	55	232	14	960	324	55	57 –29	167	74	CRO 305	270	C7.0		322
HYDROXYTOLUENE-alpha Synonym: BENZYL ALCOHOL	1987	26			395	131		–18	42		BAL 117	118			626
HYDROXYTRICARBALLYLIC ACID-beta Synonym: CITRIC ACID					916	313					CIT 254				286
HYFLUORIC ACID Synonym: HYDROFLUORIC ACID SOLUTION	1790	59	443	18	1897	642	114		164		HFA 635	527	H4.1.1		
HYHDROXYNITROBENZENE-meta Synonym: 3-NITROPHENOL	1663	55			2548	871			215		NIP 833				
HYLENE M50 Synonym: DIPHENYLMETHANE DIISOCYANATE	2489	55	321						127		DPM 467	382			
HYLENE T Synonym: TOLUENE DIISOCYANATE	2078	57	828		3313		215		269		TDI1095	910	T4.1	T 5	1157
HYLENE T Synonym: TOLUENE-2,4-DIISOCYANATE	2078	57	828		3313	1128	215	170–87	269	214	TDI1095	910	T4.1		1157
HYOSCYAMINE					1966	659									
HYPOCHLORITE	1791	60			1966	659			167			536			627
HYPOCHLOROUS ACID					1967				167						627

HAZARDOUS MATERIALS REFERENCE BOOKS INDEX

CHEMICAL OR MATERIAL NAME	UN/NA Number	Guide Number (DOT)	Firefighter's Hazardous Materials Reference Book	First Aid Manual for Chemical Accidents	Sax's Dangerous Properties of Industrial Materials	Hazardous Chemicals Desk Reference	Rapid Guide to Hazardous Chemicals in the Workplace	Fire Protection Guide on Hazardous Materials (NFPA)	Firefighter's Handbook of Hazardous Materials	Pocket Guide to Chemical Hazards (NIOSH)	Chemical Hazard Response Information System (CHRIS)	Emergency Handling of Hazardous Materials in Surface Transportation (AAR)	Emergency Action Guides (EAG)	Chemical Data Notebook: A User's Manual	Condensed Chemical Dictionary
HYPOCHLOROUS ACID CALCIUM SALT	1748	45	172	13	1967	660		46 –	167		CHY 252	182	C1.1	C 2	204
Synonym: CALCIUM HYPOCHLORITE															
HYPONITROUS ACID					1968	660			167						
HYPONITROUS ACID ANHYDRIDE	1070	14	644			660					NTO 858	700	N3.5		833
Synonym: NITROUS OXIDE															
HYPONITROUS ETHER	1194	30							167						
HYTROL O	1915	26	246	14		334	59	-30	88	78	CCH 228	281	C8.2		337
Synonym: CYCLOHEXANONE															
IBN	2284/1993	28	469	18	2023	680	122	-61	172		IBN 663	549			654
Synonym: ISOBUTYRONITRILE															
IGNITER FOR AIRCRAFT THRUST DEVICE	2792	32							168						
ILLUMINATING GAS	1016	18							168						
ILLUMINATING OIL	1223	27	494		2078	701	126		177		KRS 689	566			671
Synonym: KEROSENE															
IMINOBISPROPYLAMINE	2269	60							168						
IMPERIAL GREEN	1585	53	222				55		83		CAA 203	257			631
Synonym: COPPER ACETOARSENITE															
INCIDOL	2085	49	122		392	130	23		42	44			B2.0	B 2	133
Synonym: BENZOYL PEROXIDE															
INDENE					1979	663	118		168						632
INDIUM					1981	663	118		168						633
INEDIBLE TALLOW								-84			TLO1125				1114
Synonym: TALLOW															
INERTON DW DMC	1162	29	305					-42	122		DMD 429	370	D3.1		415
Synonym: DIMETHYLDICHLOROSILANE															
INFECTIOUS SUBSTANCE	2814	24							168						

CHEMICAL OR MATERIAL NAME	UN/NA Number	Guide Number (DOT)	Firefighter's Hazardous Materials Reference Book	First Aid Manual for Chemical Accidents	Sax's Dangerous Properties of Industrial Materials	Hazardous Chemicals Desk Reference	Rapid Guide to Hazardous Chemicals in the Workplace	Fire Protection Guide on Hazardous Materials (NFPA)	Firefighter's Handbook of Hazardous Materials	Pocket Guide to Chemical Hazards (NIOSH)	Chemical Hazard Response Information System (CHRIS)	Emergency Handling of Hazardous Materials in Surface Transportation (AAR)	Emergency Action Guides (EAG)	Chemical Data Notebook: A User's Manual	Condensed Chemical Dictionary
INFECTIOUS SUBSTANCE	2900	24							168						
INHIBINE Synonym: HYDROGEN PEROXIDE	2014	45		18	1901	644	115	99 –	165	126		531	H4.5	H 5	616
INHIBINE Synonym: HYDROGEN PEROXIDE	2015	47	450	18	1901	644	115	100–	165	126	HPO 643	532	H4.5	H 5	616
INK	1210	26							168			538			636
INK MATERIALS NEC	2367/1993								168			538			
INSECTICIDE	1993	27							168			540			637
INSECTICIDE	2902	55							168						637
INSECTICIDE	2588	55							168						637
INSECTICIDE ADHESIVES LIQUID	1133	26										539			
INSECTICIDE ADHESIVES SPREADERS OR STICKERS LIQUID	1760	60													
INSECTICIDE GAS	1967	15							168			539			
INSECTICIDE GAS	1968	12							168			540			
INSULATING OIL Synonym: OILS MISCELLANEOUS TRANSFORMER	1270	27									OTF 914				637
IODATES					1988	665			168						
IODIC ACID					1988				168						640
IODINE			460	18	1988	666	118		169	128					641
IODINE AZIDE					1989	666			169						
IODINE HEPTAFLUORIDE					1989				169						
IODINE MONOCHLORIDE	1792	59			1989	667			169			540			641
IODINE MONOCHLORIDE-alpha	1792	59							169						
IODINE MONOCHLORIDE-beta	1792	59							169						

HAZARDOUS MATERIALS REFERENCE BOOKS INDEX

CHEMICAL OR MATERIAL NAME	UN/NA Number	Guide Number (DOT)	Firefighter's Hazardous Materials Reference Book	First Aid Manual for Chemical Accidents	Sax's Dangerous Properties of Industrial Materials	Hazardous Chemicals Desk Reference	Rapid Guide to Hazardous Chemicals in the Workplace	Fire Protection Guide on Hazardous Materials (NFPA)	Firefighter's Handbook of Hazardous Materials	Pocket Guide to Chemical Hazards (NIOSH)	Chemical Hazard Response Information System (CHRIS)	Emergency Handling of Hazardous Materials in Surface Transportation (AAR)	Emergency Action Guides (EAG)	Chemical Data Notebook: A User's Manual	Condensed Chemical Dictionary
IODINE OXIDE					1990				169						
IODINE PENTAFLUORIDE	2495	44			1990	667			169			540			641
IODOACETIC ACID					1991	667			169						
IODOBUTANE	2390	26							169						
IODOETHANE									170						642
IODOFENPHOS					1993				170						
IODOFORM					1993	668	119		170						642
IODOMETHANE	2644	55							170						642
IODOMETHYLPROPANE	2391/1993	26							170			541			
IODOPROPENE	1723	29							170						
IODOTOLUENE-alpha	2653	53							170						
IONONE-alpha					1998	669		-60							
IONONE-beta					1998	669		-60							
IPA Synonym: ISOPROPANOL	1219	26	480						174			555	14.0		644
IRON (III) CHLORIDE Synonym: FERRIC CHLORIDE SOLUTION	2582	60	386		1698	577	99		147			446	F1.0		
IRON (OUS) SULFATE Synonym: FERROUS SULFATE	9125/3077		397			579	119		147		FRS 593	450			518
IRON AMMONIUM SULFATE Synonym: FERROUS AMMONIUM SULFATE	9122/3077	31	392				119				FAS 579	449			517
IRON CARBONYL	1994	57					119	-60	170						647
IRON CHLORIDE Synonym: FERRIC CHLORIDE SOLUTION	2582	60	386	18	1698	577.	99		147			446	F1.0		
IRON CHLORIDE	1773	60		18			119		170						

CHEMICAL OR MATERIAL NAME	UN/NA Number	Guide Number (DOT)	Firefighter's Hazardous Materials Reference Book	First Aid Manual for Chemical Accidents	Sax's Dangerous Properties of Industrial Materials	Hazardous Chemicals Desk Reference	Rapid Guide to Hazardous Chemicals in the Workplace	Fire Protection Guide on Hazardous Materials (NFPA)	Firefighter's Handbook of Hazardous Materials	Pocket Guide to Chemical Hazards (NIOSH)	Chemical Hazard Response Information System (CHRIS)	Emergency Handling of Hazardous Materials in Surface Transportation (AAR)	Emergency Action Guides (EAG)	Chemical Data Notebook: A User's Manual	Condensed Chemical Dictionary
IRON CHLORIDE SOLUTION	2582	60		18			119								
IRON DICHLORIDE Synonym: FERROUS CHLORIDE	1759	60	394		1701	579	119		147		FEC 582	449			517
IRON FLUORIDE Synonym: FERRIC FLUORIDE	9120/3077		388		1698	577	119		147		FFX 585	447			513
IRON II SULFATE Synonym: FERRIC SULFATE	9121/3082	31	391		1699	577	119		147		FSF 595	448			514
IRON III CHLORIDE Synonym: FERRIC CHLORIDE	1773	60	387		1698	577	99		147		FCL 580	447			512
IRON MASS OR SPONGE	1376	37					119		170			542			647
IRON MURIATE Synonym: FERRIC CHLORIDE SOLUTION	2582	60	386		1698	577	119		147			446	F1.0		
IRON OXIDE	1376	37			2008	671	119		170	130					
IRON PENTACARBONYL	1994/2810	57			2009	672	120		170			542			648
IRON PERCHLORIDE Synonym: FERRIC CHLORIDE	1773	60	387		1698	577	99		147		FCL 580	447			512
IRON PROTOCHLORIDE Synonym: FERROUS CHLORIDE	1759	60	394		1701	579	119		147		FEC 582	449			648
IRON PROTOXALATE Synonym: FERROUS OXALATE			396				119				FOX 592				517
IRON SESQUISULFATE Synonym: FERRIC SULFATE	9 21/3082	31	391		1699	577	119		147		FSF 595	448			514
IRON SULPHATE	9 88/3077				2010		119					542			
IRON SWARF	2793	32					119		170						
IRON TERSULFATE Synonym: FERRIC SULFATE	9121/3082	31	391		1699	577	119		147		FSF 595	448			514

HAZARDOUS MATERIALS REFERENCE BOOKS INDEX

CHEMICAL OR MATERIAL NAME	UN/NA Number	Guide Number (DOT)	Firefighter's Hazardous Materials Reference Book	First Aid Manual for Chemical Accidents	Sax's Dangerous Properties of Industrial Materials	Hazardous Chemicals Desk Reference	Rapid Guide to Hazardous Chemicals in the Workplace	Fire Protection Guide on Hazardous Materials (NFPA)	Firefighter's Handbook of Hazardous Materials	Pocket Guide to Chemical Hazards (NIOSH)	Chemical Hazard Response Information System (CHRIS)	Emergency Handling of Hazardous Materials in Surface Transportation (AAR)	Emergency Action Guides (EAG)	Chemical Data Notebook: A User's Manual	Condensed Chemical Dictionary
IRON TRICHLORIDE	2582	60	386		1698	577	99		147			446	F1.0		
Synonym: FERRIC CHLORIDE SOLUTION															
IRON TRICHLORIDE	1773	60	387		1698		99		147		FCL 580	447			512
Synonym: FERRIC CHLORIDE															
IRON VITRIOL	9125/3077		397		1702	579	119		147		FRS 593	450			518
Synonym: FERROUS SULFATE															
IRRITATING AGENT	1693	58							171						
ISANO OIL								-60	171						
ISCEON 11	3082/9188		838	25	3353	1138	219		273		TCF1083	919			1170
Synonym: TRICHLOROFLUOROMETHANE															
ISCEON 12	1028	12	275	15	1137	380	68		102	86	DCF 360	331	D1.1		379
Synonym: DICHLORODIFLUOROMETHANE															
ISCOBROME	1062	55	559	20	2274		137	110–67	192	146	MTB 805	619	M3.0	M 3	759
Synonym: METHYL BROMIDE															
ISCOBROME D	1605	55	371		1601			85 –	143	110	EDB 518	430	E2.4.1		487
Synonym: ETHYLENE DIBROMIDE															
ISO HEPTYL ALCOHOL	1987	26										543			
ISOAMYHYDRIDE	1265	27	476					-62	173		IPT 680	552	12.0		658
Synonym: ISOPENTANE															
ISOAMYL ACETATE	1104	26			2011	673	120	-60	171	130	IAT 660				649
ISOAMYL ALCOHOL	1987	26	461												649
ISOAMYL ALCOHOL	1105	26	461		2011	673	120	-60	171	130	IAA 655				649
ISOAMYL ALCOHOL	2706	26			2011	674	121								649
ISOAMYL ALCOHOL-tert	1105	26							171						
ISOAMYL BUTYRATE					2012	674		-60	171						650
ISOAMYL CHLORIDE								-61	171						650

CHEMICAL OR MATERIAL NAME	UN/NA Number	Guide Number (DOT)	Firefighter's Hazardous Materials Reference Book	First Aid Manual for Chemical Accidents	Sax's Dangerous Properties of Industrial Materials	Hazardous Chemicals Desk Reference	Rapid Guide to Hazardous Chemicals in the Workplace	Fire Protection Guide on Hazardous Materials (NFPA)	Firefighter's Handbook of Hazardous Materials	Pocket Guide to Chemical Hazards (NIOSH)	Chemical Hazard Response Information System (CHRIS)	Emergency Handling of Hazardous Materials in Surface Transportation (AAR)	Emergency Action Guides (EAG)	Chemical Data Notebook: A User's Manual	Condensed Chemical Dictionary
ISOAMYL ETHANOATE Synonym: ISOAMYL ACETATE	1104	26			2011	673	120	−60	171	130	IAT 660				649
ISOAMYL FORMATE	1109	26			2013	674			171						650
ISOAMYL NITRITE	1113	26							171						650
ISOAMYLENE-alpha	2561	26							171						650
ISOAMYLENE-beta	1108	26							171						650
ISOBORNYL THIOCYANOACETATE									171						651
ISOBUTANAL Synonym: ISOBUTYRALDEHYDE	2045/1993	26	467	18	2022	679		−61	171		BAD 115	548	11.1		653
ISOBUTANE	1969/1075	22	462					−61	171		IBT 665	543	10.5		651
ISOBUTANE MIXTURE	1969/1075	22	462					−61	171		IBT 665	543			
ISOBUTANOL Synonym: ISOBUTYL ALCOHOL	1212/1120	26			2016	676	121	−61	171	132	IAL 658	545			652
ISOBUTENE Synonym: ISOBUTYLENE	1055/1075	22			2015	675			171		IBL 662	547	11.0.5		652
ISOBUTYL 2-PROPENOATE Synonym: BUTYL ACRYLATE-iso	2527	27	145								BAI 116				
ISOBUTYL ACETATE	1213	26	463	18	2015	675	121	−61	171	130	IBA 661	544	11.0		652
ISOBUTYL ACRYLATE	2527/1993	27			2016	676		−61	171			544			652
ISOBUTYL ALCOHOL	1212/1120	26			2016	676	121	−61	171	132	IAL 658	545			652
ISOBUTYL ALDEHYDE	2045	26							171						652
ISOBUTYL BUTYRATE								−61	172						
ISOBUTYL CARBINOL	1105	26							172						652
ISOBUTYL CHLORIDE								−61	172						
ISOBUTYL FORMATE	2393/1993	29			2018	677		−61	172			545			

HAZARDOUS MATERIALS REFERENCE BOOKS INDEX

CHEMICAL OR MATERIAL NAME	UN/NA Number	Guide Number (DOT)	Firefighter's Hazardous Materials Reference Book	First Aid Manual for Chemical Accidents	Sax's Dangerous Properties of Industrial Materials	Hazardous Chemicals Desk Reference	Rapid Guide to Hazardous Chemicals in the Workplace	Fire Protection Guide on Hazardous Materials (NFPA)	Firefighter's Handbook of Hazardous Materials	Pocket Guide to Chemical Hazards (NIOSH)	Chemical Hazard Response Information System (CHRIS)	Emergency Handling of Hazardous Materials in Surface Transportation (AAR)	Emergency Action Guides (EAG)	Chemical Data Notebook: A User's Manual	Condensed Chemical Dictionary
ISOBUTYL HEPTYL KETONE								-61	172						653
ISOBUTYL ISOBUTYRATE	2528/1993	26			2019	677		-61	172			546			653
ISOBUTYL ISOCYANATE	2486	57							172			546			
ISOBUTYL METHACRYLATE	2283/1993	27			2020	678			172		BMI 147	546			653
ISOBUTYL METHYL CARBINOL	2053	26							172						
ISOBUTYL METHYL KETONE Synonym: METHYL ISOBUTYL KETONE	1245/1993	26	571	20				-70	172		MIK 765	630	M3.3		771
ISOBUTYL METHYL KETONE PEROXIDE	2126	51							172						
ISOBUTYL METHYLMETHANOL Synonym: METHYL AMYL ALCOHOL	2053	26	557						191		MAA 718				757
ISOBUTYL PROPIONATE	2394/1993	26							172			546			653
ISOBUTYL VINYL ETHER	1304	26			2022	678			172						653
ISOBUTYLALDEHYDE Synonym: ISOBUTYRALDEHYDE	2045/1993	26	467	18	2022	679		-61			BAD 115	548	11.1		652
ISOBUTYLAMINE	1214	68	465		2017	676	118	-61	172		IAM 659	547			652
ISOBUTYLBENZENE	2709	27			2017	677		-61	172						652
ISOBUTYLCARBINOL Synonym: ISOAMYL ALCOHOL	1105	26	461		2011	673	120	-61	172	130	IAA 655				652
ISOBUTYLENE	1055/1075	22							172		IBL 662	547	11.0.5		653
ISOBUTYLMETHYLCARBINOL Synonym: METHYL AMYL ALCOHOL	2053	26	557			799			172		MAA 718				757
ISOBUTYLMETHYLCARBINOL Synonym: METHYL ISOBUTYL CARBINOL	2053/1993	26	570		2334		145	-70	172	152	MIC 763	630			771
ISOBUTYRALDEHYDE	2045/1993	26	467	18	2022	679		-61			BAD 115	548	11.1		653
ISOBUTYRIC ACID	2529	29	468		2023	679			172		IBR 664	548			654

CHEMICAL OR MATERIAL NAME	UN/NA Number	Guide Number (DOT)	Firefighter's Hazardous Materials Reference Book	First Aid Manual for Chemical Accidents	Sax's Dangerous Properties of Industrial Materials	Hazardous Chemicals Desk Reference	Rapid Guide to Hazardous Chemicals in the Workplace	Fire Protection Guide on Hazardous Materials (NFPA)	Firefighter's Handbook of Hazardous Materials	Pocket Guide to Chemical Hazards (NIOSH)	Chemical Hazard Response Information System (CHRIS)	Emergency Handling of Hazardous Materials in Surface Transportation (AAR)	Emergency Action Guides (EAG)	Chemical Data Notebook: A User's Manual	Condensed Chemical Dictionary
ISOBUTYRIC ACID	2045	26						-61			IBR 664				654
ISOBUTYRIC ALDEHYDE Synonym: ISOBUTYRALDEHYDE	2045/1993	26	467	18	2022	679		-61			BAD 115	548	11.1		653
ISOBUTYRIC ANHYDRIDE	2530	29			2023	679		-61	172			548			654
ISOBUTYRIC ETHER	2385	26							172						
ISOBUTYRONITRILE	2284/1993	28	469	18	2023	680	122	-61	172		IBN 663	549			654
ISOBUTYROYL CHLORIDE	2395	29							172						654
ISOBUTYRYL CHLORIDE	2395	29							172			549			
ISOCTYL TRICHLOROPHENOXYACETATE Synonym: 2,4,5-T ESTERS	2765	55							259		TES1105				
ISOCYANATE	2207	55			2025				172			549			654
ISOCYANATE	2206	55			2025				172						654
ISOCYANATE	3080	28			2025				172						654
ISOCYANATE	2478	28			2025				172			550			654
ISOCYANATOBENZOTRIFLUORIDE	2285	55							172						
ISOCYANATOMETHANE Synonym: METHYL ISOCYANATE	2480	30	572	20	2335	799	146	115-70	195	152	MIS 767	631	M3.4	M 6	771
ISOCYANIC ACID METHYL ESTER Synonym: METHYL ISOCYANATE	2480	30	572	20	2335	799	146	115-70	195	152	MIS 767	631	M3.4	M 6	771
ISOCYANIC ACID METHYL-m-PHENYLENE ESTER Synonym: TOLUENE DIISOCYANATE	2078	57	828		3312		215		269			910	T4.1	T 5	1157
ISOCYANIC ACID METHYLPHENYLENE ESTER Synonym: TOLUENE DIISOCYANATE	2078	57	828		3312		215		269			910	T4.1	T 5	1157
ISODECALDEHYDE	1993	27	470					-61	173		IDA 666				655

HAZARDOUS MATERIALS REFERENCE BOOKS INDEX

CHEMICAL OR MATERIAL NAME	UN/NA Number	Guide Number (DOT)	Firefighter's Hazardous Materials Reference Book	First Aid Manual for Chemical Accidents	Sax's Dangerous Properties of Industrial Materials	Hazardous Chemicals Desk Reference	Rapid Guide to Hazardous Chemicals in the Workplace	Fire Protection Guide on Hazardous Materials (NFPA)	Firefighter's Handbook of Hazardous Materials	Pocket Guide to Chemical Hazards (NIOSH)	Chemical Hazard Response Information System (CHRIS)	Emergency Handling of Hazardous Materials in Surface Transportation (AAR)	Emergency Action Guides (EAG)	Chemical Data Notebook: A User's Manual	Condensed Chemical Dictionary
ISODECALDEHYDE MIXED ISOMERS	1993	27	470					-61	173		IDA 666				655
Synonym: ISODECALDEHYDE															
ISODECANE								-61	173						655
ISODECYL ACRYLATE			471		2026				173		IAI 657				
ISODECYL ALCOHOL			472						173		ISA 682				
ISODIPRENE											CAR 211				
Synonym: CARENE															
ISODODECANE	2286	27							173						
ISODRIN					2027	681			173						655
ISOHEPTANE	2287	27						-62	173						656
ISOHEXANE	2462	26	473		2028	681	122	-62	173						656
ISOHEXANE	1208	27			2028	681	122	-62			IHA 667				656
ISOHEXYL ALCOHOL-tert								-62	173						
ISONITROPROPANE	2608	26			2554	873	163	-74	215	166	NPP 848				831
Synonym: 2-NITROPROPANE															
ISONONANOYL PEROXIDE	2128	52							173						
ISONONANYL PEROXIDE	2128	52							173						
ISOOCTALDEHYDE	1993	27	474								IOC 669				
ISOOCTANE	1262	27						-62	173			551			657
ISOOCTENE	1216	27						-62	173			551			657
ISOOCTYL ALCOHOL	1987	26	475		2032	683	123	-62			IOA 668	552			657
ISOOCTYL NITRATE								-62	173						
ISOOCTYLALDEHYDE	1993	27	474								IOC 669				
Synonym: ISOOCTALDEHYDE															

CHEMICAL OR MATERIAL NAME	UN/NA Number	Guide Number (DOT)	Firefighter's Hazardous Materials Reference Book	First Aid Manual for Chemical Accidents	Sax's Dangerous Properties of Industrial Materials	Hazardous Chemicals Desk Reference	Rapid Guide to Hazardous Chemicals in the Workplace	Fire Protection Guide on Hazardous Materials (NFPA)	Firefighter's Handbook of Hazardous Materials	Pocket Guide to Chemical Hazards (NIOSH)	Chemical Hazard Response Information System (CHRIS)	Emergency Handling of Hazardous Materials in Surface Transportation (AAR)	Emergency Action Guides (EAG)	Chemical Data Notebook: A User's Manual	Condensed Chemical Dictionary
ISOPENTADIENE Synonym: ISOPRENE	1218	27	479		2037	685		102–62	174		IPR 679	554	13.0	1 1	658
ISOPENTALDEHYDE								–62	173						657
ISOPENTANE	1265	27	476					–62	173		IPT 680	552	12.0		658
ISOPENTANOIC ACID	1760	60						–62	173			553			658
ISOPENTENE	2371	27							173			553			
ISOPENTYL ACETATE Synonym: ISOAMYL ACETATE	1104	26			2011	673	120	–60	171	130	IAT 660				649
ISOPENTYL ALCOHOL Synonym: ISOAMYL ALCOHOL	1105	26	461		2011	673	120	–60	173	130	IAA 655				658
ISOPENTYL NITRITE Synonym: AMYL NITRITE-Iso	1113	26	83		2033	684			174		ANI 63				75
ISOPESTOX				18											
ISOPHORONE	1993	27	477		2034	684	123	101–62	174	132	IPH 674	553			658
ISOPHORONE DIISOCYANATE	2290	55			2034	685	123		174		IPD 672	553			658
ISOPHORONEDIAMINE	2289/1760	60							174		IPI 675	554			
ISOPHTHALIC ACID			478		2035						IPL 676				658
ISOPHTHALOYL CHLORIDE					2036			–62	174						658
ISOPRENE	1218	27	479		2037	685		102–62	174		IPR 679	554	13.0	1 1	658
ISOPROPANOL Synonym: ISOPROPYL ALCOHOL	1219	26	480	18	2040	687	124	–63	174	132	IPA 670	555	14.0	1 2	659
ISOPROPANOL AND METHANOL MIXTURE	1987/1986	26										556			
ISOPROPANOLAMINE Synonym: MONOISOPROPANOLAMINE	1760	60	595		2038				174		MPA 782	658			659
ISOPROPANOLAMINE	9188				2038				174						659

HAZARDOUS MATERIALS REFERENCE BOOKS INDEX

CHEMICAL OR MATERIAL NAME	UN/NA Number	Guide Number (DOT)	Firefighter's Hazardous Materials Reference Book	First Aid Manual for Chemical Accidents	Sax's Dangerous Properties of Industrial Materials	Hazardous Chemicals Desk Reference	Rapid Guide to Hazardous Chemicals in the Workplace	Fire Protection Guide on Hazardous Materials (NFPA)	Firefighter's Handbook of Hazardous Materials	Pocket Guide to Chemical Hazards (NIOSH)	Chemical Hazard Response Information System (CHRIS)	Emergency Handling of Hazardous Materials in Surface Transportation (AAR)	Emergency Action Guides (EAG)	Chemical Data Notebook: A User's Manual	Condensed Chemical Dictionary
ISOPROPANOLAMINE DODECYLBENZENESULFONATE	9127/3082								174			559			
ISOPROPENYL ACETATE	2403	26						-62	174			559			659
ISOPROPENYL ACETYLENE								-62	174						659
ISOPROPENYL BENZENE	2303/1993	27							174			559			
ISOPROPENYL METHYL KETONE	1246	26	573		2336	799		-70	174		MPK 787	631			772
Synonym: METHYL ISOPROPENYL KETONE															
ISOPROPENYLBENZENE	2303	27	580		2398	818	149		174	154	MSR 802				780
Synonym: METHYL STYRENE-alpha															
ISOPROPYL 2,4-DICHLOROPHENOXY ACETATE	2765/3077	55	10						90		DES 393	336			
Synonym: 2,4-D ESTERS															
ISOPROPYL ACETATE	1220	26	481	18	2039	686	124	-63	174	132	IAC 656	560			659
ISOPROPYL ACETONE	1245/1993	26	571	20				-70	174		MIK 765	630	M3.3		659
Synonym: METHYL ISOBUTYL KETONE															
ISOPROPYL ACID PHOSPHATE	1793	60							174			560			
ISOPROPYL ALCOHOL	1219/1993	26	482	18	2040	687	124	-63	174	132	IPA 670	560	14.0	I 2	660
Synonym: ISOPROPANOL															
ISOPROPYL BENZOATE					2045			-63	175						
ISOPROPYL BUTYRATE	2405	26							175			561			660
ISOPROPYL CHLORIDE	2356	26						-63	175						660
ISOPROPYL CHLOROACETATE	2947	29							175			561			
ISOPROPYL CHLOROCARBONATE	2407/1993	57			2047	690			175			561	15.0		661
Synonym: ISOPROPYL CHLOROFORMATE															
ISOPROPYL CHLOROFORMATE	2407/1993	57							175			561	15.0		661
ISOPROPYL CHLOROMETHANOATE	2407/1993	57							175			561	15.0		661
Synonym: ISOPROPYL CHLOROFORMATE															

CHEMICAL OR MATERIAL NAME	UN/NA Number	Guide Number (DOT)	Firefighter's Hazardous Materials Reference Book	First Aid Manual for Chemical Accidents	Sax's Dangerous Properties of Industrial Materials	Hazardous Chemicals Desk Reference	Rapid Guide to Hazardous Chemicals in the Workplace	Fire Protection Guide on Hazardous Materials (NFPA)	Firefighter's Handbook of Hazardous Materials	Pocket Guide to Chemical Hazards (NIOSH)	Chemical Hazard Response Information System (CHRIS)	Emergency Handling of Hazardous Materials in Surface Transportation (AAR)	Emergency Action Guides (EAG)	Chemical Data Notebook: A User's Manual	Condensed Chemical Dictionary
ISOPROPYL CHLOROPROPIONATE	2934	29							175						
ISOPROPYL CYANIDE	2284/1993	28	469	18	2023	680	122	-61	175		IBN 663	549			661
Synonym: ISOBUTYRONITRILE															
ISOPROPYL ETHER	1159	26	483		2050	691	125	72 -63	175	134	IPE 673	357	D2.1		661
Synonym: DIISOPROPYL ETHER															
ISOPROPYL FORMALDEHYDE	2045/1993	26	467	18	2022	679		-61			BAD 115	548	I1.1		653
Synonym: ISOBUTYRALDEHYDE															
ISOPROPYL FORMATE	2408	27						103-63	175						
ISOPROPYL GLYCIDYL ETHER					2050	691	126		175	134					661
ISOPROPYL ISOBUTYRATE	2406	28							175			561			
ISOPROPYL ISOCYANATE	2483	28							175			562			
ISOPROPYL LACTATE								-63	175						
ISOPROPYL MERCAPTAN	2703	27	484						175		IPM 677				661
ISOPROPYL MERCAPTAN	2402	27							175		IPM 677	562			661
ISOPROPYL METHANOATE	2408	27							175						
ISOPROPYL NITRATE	1222	27			2057	693			175			562			662
ISOPROPYL PERCARBONATE	2134	52	485						175		IPC 671				662
ISOPROPYL PERCARBONATE	2133	52	486						175		IPC 671				662
ISOPROPYL PEROXYDICARBONATE	2133	52							175						662
ISOPROPYL PEROXYDICARBONATE	2134	52	485						175		IPC 671				662
Synonym: ISOPROPYL PERCARBONATE															
ISOPROPYL PROPIONATE	2409	27							175			563			
ISOPROPYL VINYL ETHER					2060				175						
ISOPROPYLACETONE	1245/1993	26	571	20				-70	174		MIK 765	630	M3.3		659
Synonym: METHYL ISOBUTYL KETONE															

HAZARDOUS MATERIALS REFERENCE BOOKS INDEX

CHEMICAL OR MATERIAL NAME	UN/NA Number	Guide Number (DOT)	Firefighter's Hazardous Materials Reference Book	First Aid Manual for Chemical Accidents	Sax's Dangerous Properties of Industrial Materials	Hazardous Chemicals Desk Reference	Rapid Guide to Hazardous Chemicals in the Workplace	Fire Protection Guide on Hazardous Materials (NFPA)	Firefighter's Handbook of Hazardous Materials	Pocket Guide to Chemical Hazards (NIOSH)	Chemical Hazard Response Information System (CHRIS)	Emergency Handling of Hazardous Materials in Surface Transportation (AAR)	Emergency Action Guides (EAG)	Chemical Data Notebook: A User's Manual	Condensed Chemical Dictionary
ISOPROPYLAMINE	1221	68	487	18	2041	688	124	102–63	176	134	IPP 678	563			660
ISOPROPYLAMINE DODECYLBENZENESULFONATE Synonym: DODECYLBENZENESULFONIC ACID ISOPROPYL-AMINE SALT											DAI 335				
ISOPROPYLAMINOETHANOL	1993	27			2042				174			564			660
ISOPROPYLAMINOETHANOL	1987	26			2042				174						660
ISOPROPYLBENZENE Synonym: CUMENE	1918/1993	28	234			328	56	58 –29	175	76	CUM 325	564	14.5		660
ISOPROPYLBENZENE HYDROPEROXIDE Synonym: CUMENE HYDROPEROXIDE	2116	51	235		2045		125	–29	176		CMH 266				329
ISOPROPYLCARBINOL Synonym: ISOBUTYL ALCOHOL	1212/1120	26			2016	676	121	–61	176	132	IAL 658	545			660
ISOPROPYLCUMYL HYDROPEROXIDE	2171/3109	48							176			565			
ISOPROPYLCYANOHYDRIN Synonym: ACETONE CYANOHYDRIN	1541	55	21	11				15 –11	17		ACY 27	4	A3.1		9
ISOPROPYLCYCLOHEXYLAMINE								–63	176						
ISOPROPYLIDENEACETONE Synonym: MESITYL OXIDE	1229	26	545		2205	757	132	107–64	176	140	MSO 801	606			661
ISOPROPYLTOLUENE	2046/1993	27							176			565			662
ISOPROPYLTOLUENE-para Synonym: CYMENE-para	2046	27	252			340		–31	90		CMP 269	288			
ISOPROPYLTOLUOL Synonym: CYMENE-para	2046	27	252			340		–31	90		CMP 269	288			
ISOQUINOLINE						694			176						662
ISOSAFROLE					2061	694									663
ISOSORBIDE DINITRATE MIXTURE	2907/1325	32							176			565			

CHEMICAL OR MATERIAL NAME	UN/NA Number	Guide Number (DOT)	Firefighter's Hazardous Materials Reference Book	First Aid Manual for Chemical Accidents	Sax's Dangerous Properties of Industrial Materials	Hazardous Chemicals Desk Reference	Rapid Guide to Hazardous Chemicals in the Workplace	Fire Protection Guide on Hazardous Materials (NFPA)	Firefighter's Handbook of Hazardous Materials	Pocket Guide to Chemical Hazards (NIOSH)	Chemical Hazard Response Information System (CHRIS)	Emergency Handling of Hazardous Materials in Surface Transportation (AAR)	Emergency Action Guides (EAG)	Chemical Data Notebook: A User's Manual	Condensed Chemical Dictionary
ISOTRIDECANOL Synonym: TRIDECANOL			842					−88	274		TDN1096				1173
ISOTRIDECYL ALCOHOL Synonym: TRIDECANOL			842					−88	274		TDN1096				1173
ISOTRON 11 Synonym: TRICHLOROFLUOROMETHANE	3082/9188		838	25	3353	1138	219		273		TCF1083	919			1170
ISOTRON 12 Synonym: DICHLORODIFLUOROMETHANE	1028	12	275	15	1137	380	68		102	86	DCF 360	331	D1.1		379
ISOTRON 22 Synonym: CHLORODIFLUOROMETHANE	1018	12		14	791	278	48		74		MCF 737	222			795
ISOTRON 22 Synonym: MONOCHLORODIFLUOROMETHANE	1018	12	593						205				M4.0		265
ISOVALERAL Synonym: ISOVALERALDEHYDE	1989	26	488						176		IVA 683				663
ISOVALERALDEHYDE	1989	26	488						176		IVA 683				664
ISOVALERIC ALDEHYDE Synonym: ISOVALERALDEHYDE	1989	26	488						176		IVA 683				664
ISOVALERONE Synonym: DIISOBUTYL KETONE	1157	26	296		1300	433	75	−40	176	92	DIK 414	355			405
IVALON Synonym: FORMALDEHYDE SOLUTION	1198/2209	29	401	17	1748	591	102	91 −54	150	116	FMS 590	462	F2.0	F 2	
JAPAN LACQUER					2070				176						
JAYSOL S Synonym: ETHYL ALCOHOL	1170	26	344	16	1572	525	90	−48	136		EAL 504	408	E2.2	E 2	478
JET FUEL JP-1 Synonym: KEROSENE	1223	27	489		2071	701	126		176		JPO 685	566			671

HAZARDOUS MATERIALS REFERENCE BOOKS INDEX

CHEMICAL OR MATERIAL NAME	UN/NA Number	Guide Number (DOT)	Firefighter's Hazardous Materials Reference Book	First Aid Manual for Chemical Accidents	Sax's Dangerous Properties of Industrial Materials	Hazardous Chemicals Desk Reference	Rapid Guide to Hazardous Chemicals in the Workplace	Fire Protection Guide on Hazardous Materials (NFPA)	Firefighter's Handbook of Hazardous Materials	Pocket Guide to Chemical Hazards (NIOSH)	Chemical Hazard Response Information System (CHRIS)	Emergency Handling of Hazardous Materials in Surface Transportation (AAR)	Emergency Action Guides (EAG)	Chemical Data Notebook: A User's Manual	Condensed Chemical Dictionary
JET FUELS A AND A1					2071			–63	176						
JET FUELS B					2071			–63	176						
JET FUELS JP-1	1223	27	489		2071				176		JPO 685				
JET FUELS JP-3	1223	27	490								JPT 686				
JET FUELS JP-4	1863	27	491		2071			–63	176		JPF 684				
JET FUELS JP-5	2761	55	492		2071			–63	176		JPV 687				
JET FUELS JP-6					2071			–64	177						
JET THRUST UNIT									177						
JP-1 Synonym: FUEL OIL NO. 1	1993	27	403		2071			–55	176			464	F3.0		
JP-1 Synonym: OILS MISCELLANEOUS RANGE	1223	27			2071				176		ORG 751				
JP-1 Synonym: OILS FUEL NO. 1	1223/1993	27			2071				176		OON 746	566			
JUDEAN PITCH Synonym: ASPHALT	1999	27	105	12	307	97	17	–17	36		ASP 93	99	A12.0		100
JUTE	1372	32													667
JUTE LASHINGS	1372	32													
JUTE REFECTIONS	1372	32													
K STOFF Synonym: METHYL CHLOROFORMATE	1238	28	562	20					193		MHC 760	624	M3.1.1		762
KAOLIN				18	2074	700									668
KAYAFUME Synonym: METHYL BROMIDE	1062	55	559	20	2274	776	137	110–67	192	146	MTB 805	619	M3.0	M 3	759

CHEMICAL OR MATERIAL NAME	UN/NA Number	Guide Number (DOT)	Firefighter's Hazardous Materials Reference Book	First Aid Manual for Chemical Accidents	Sax's Dangerous Properties of Industrial Materials	Hazardous Chemicals Desk Reference	Rapid Guide to Hazardous Chemicals in the Workplace	Fire Protection Guide on Hazardous Materials (NFPA)	Firefighter's Handbook of Hazardous Materials	Pocket Guide to Chemical Hazards (NIOSH)	Chemical Hazard Response Information System (CHRIS)	Emergency Handling of Hazardous Materials in Surface Transportation (AAR)	Emergency Action Guides (EAG)	Chemical Data Notebook: A User's Manual	Condensed Chemical Dictionary
KEL F MONOMER **Synonym: TRIFLUOROCHLOROETHYLENE**	1082	17	850					-89	276		TFC1108	925			1177
KELENE **Synonym: ETHYL CHLORIDE**	1037	27	350		1586	532	92	82 -49	136	108	ECL 514	412	E2.3		483
KELTHANE **Synonym: 4,4-DICHLORO-alpha-TRICHLOROMETHYLBENZHYDROL**	3082	31									DTM 489	566			
KEPONE	2761/9189	55	493		2077	701			177		KPE 688	566			670
KEROSENE	1223	27			2078	701	126		177		KRS 689	566			671
KEROSENE **Synonym: OILS MISCELLANEOUS RANGE**	1223	27			2078	701	126		177		ORG 751	566			671
KEROSENE **Synonym: FUEL OIL NO. 1**	1993	27	403					-55				464	F3.0		671
KEROSENE **Synonym: OILS FUEL NO. 1**	1223	27	494		2078	701	126		177		KRS 689	566			671
KEROSENE HEAVY **Synonym: OILS MISCELLANEOUS SPRAY**	1270	27									OSY 908				
KEROSENE HEAVY **Synonym: JET FUEL JP-5**	2761	55	492		2071			-63	176		JPV 687				
KEROSINE **Synonym: JET FUEL JP-1**	1223	27	489		2071				177		JPO 685				
KEROSINE **Synonym: FUEL OIL NO. 1**	1993	27	403					-55				464	F3.0		
KEROSINE **Synonym: KEROSENE**	1223	27	494		2078	701	126		177		KRS 689	566			671
KEROSINE **Synonym: OILS MISCELLANEOUS RANGE**	1223	27							177		ORG 900				

HAZARDOUS MATERIALS REFERENCE BOOKS INDEX

CHEMICAL OR MATERIAL NAME	UN/NA Number	Guide Number (DOT)	Firefighter's Hazardous Materials Reference Book	First Aid Manual for Chemical Accidents	Sax's Dangerous Properties of Industrial Materials	Hazardous Chemicals Desk Reference	Rapid Guide to Hazardous Chemicals in the Workplace	Fire Protection Guide on Hazardous Materials (NFPA)	Firefighter's Handbook of Hazardous Materials	Pocket Guide to Chemical Hazards (NIOSH)	Chemical Hazard Response Information System (CHRIS)	Emergency Handling of Hazardous Materials in Surface Transportation (AAR)	Emergency Action Guides (EAG)	Chemical Data Notebook: A User's Manual	Condensed Chemical Dictionary
KEROSINE Synonym: OILS FUEL NO. 1	1223/1993	27							177		OON 894	464			
KETENE			495	18	2078	701	126		177	134					671
KETOCYCLOPENTANE	2245	26							177						
KETOHEXAMETHYLENE Synonym: CYCLOHEXANONE	1915	26	246	14	991	334	59	-30	177	78	CCH 228	281	C8.2		672
KETONE	1224	26			2079	702			177			567			676
KETONE OILS	1224	26							177			567			
KETOPROPANE-beta Synonym: ACETONE	1090	26	20	11	22	10	2	-11	17	30	ACT 26	4	A3.0	A 3	9
KETTLE RENDERED LARD Synonym: OILS EDIBLE LARD											OLD 885				
KILLAX Synonym: TETRAETHYL LEAD	1705	15		25	3219	1100					TEP 927				1130
KILLMASTER Synonym: DURSBAN	1615										DUR 495				448
KING'S GOLD Synonym: ARSENIC TRISULFIDE	1557	53	101				16	33 -	35		ART 86	93			98
KING'S GREEN Synonym: COPPER ACETOARSENITE	1585	53	222				55		83		CAA 203	257			673
KING'S YELLOW Synonym: ARSENIC TRISULFIDE	1557	53	101				16	33 -	35		ART 86	93			98
KLOREX Synonym: SODIUM CHLORATE	1495	35	766	24	3070	1048		156-	253		SDC1029	834	S2.1	S 2	1052
KOH Synonym: POTASSIUM HYDROXIDE	1813	60	712	23	2871	984	186	145-	239		‘	781	P6.0	P 5	954

CHEMICAL OR MATERIAL NAME	UN/NA Number	Guide Number (DOT)	Firefighter's Hazardous Materials Reference Book	First Aid Manual for Chemical Accidents	Sax's Dangerous Properties of Industrial Materials	Hazardous Chemicals Desk Reference	Rapid Guide to Hazardous Chemicals in the Workplace	Fire Protection Guide on Hazardous Materials (NFPA)	Firefighter's Handbook of Hazardous Materials	Pocket Guide to Chemical Hazards (NIOSH)	Chemical Hazard Response Information System (CHRIS)	Emergency Handling of Hazardous Materials in Surface Transportation (AAR)	Emergency Action Guides (EAG)	Chemical Data Notebook: A User's Manual	Condensed Chemical Dictionary
KOPFUME Synonym: ETHYLENE DIBROMIDE	1605	55	371					85 –	143	110	EDB 518	430	E2.4.1		487
KRESOL Synonym: CRESOL	2076	55	232	14	959	324	55	57 –	85	74	CRS 307	270	C7.0		322
KRYPTON	1970	21							177			568			676
KRYPTON	1056	12							177			568			676
KUSATOL Synonym: SODIUM CHLORATE	1495	35	766	24	3070	1048		156–	253		SDC1029	834	S2.1	S 2	1052
KWIK KIL Synonym: STRYCHNINE	1692	53	794		3143	1074	195		257	198	STR1063	862			1096
KYANOL Synonym: ANILINE OIL	1547	57	85		254	72	13		32		ANL 64	79	A10.0	A12	79
L.P. GAS	1075	22							177						
LACQUER	1263	26			2083				177						678
LACQUER BASE	2557	32							177						
LACQUER BASE	1263	26							177			568			
LACTIC ACID	1760	60	496		2083	703					LTA 711	569			679
LACTIC ACID ETHYL ESTER Synonym: ETHYL LACTATE	1192	26	357		2084	703		–53	177		ELT 541	1192			495
LACTOL SPIRITS	1115	26							177						
LACTONITRILE					2085	704		–64	177						679
LACTONITRILE 2-METHYL Synonym: ACETONE CYANOHYDRIN	1541	55	21	11				15 –11	17		ACY 27	4	A3.1		9
LAH Synonym: LITHIUM ALUMINUM HYDRIDE	1410	40	520					105–	181		LAH 691	580			680

HAZARDOUS MATERIALS REFERENCE BOOKS INDEX

CHEMICAL OR MATERIAL NAME	UN/NA Number	Guide Number (DOT)	Firefighter's Hazardous Materials Reference Book	First Aid Manual for Chemical Accidents	Sax's Dangerous Properties of Industrial Materials	Hazardous Chemicals Desk Reference	Rapid Guide to Hazardous Chemicals in the Workplace	Fire Protection Guide on Hazardous Materials (NFPA)	Firefighter's Handbook of Hazardous Materials	Pocket Guide to Chemical Hazards (NIOSH)	Chemical Hazard Response Information System (CHRIS)	Emergency Handling of Hazardous Materials in Surface Transportation (AAR)	Emergency Action Guides (EAG)	Chemical Data Notebook: A User's Manual	Condensed Chemical Dictionary
LANARKITE	1794	60			2105	716	128		179		LSF 709	575			692
Synonym: LEAD SULFATE															
LANTHANUM					2087	705			178						681
LANTHANUM SULFATE					2088				178						682
LARD											OLD 885				682
Synonym: OILS EDIBLE LARD															
LARVACIDE	1580	56	210	14	865	296	50	54 –	77	70	CPL 290	229	C4.3		271
Synonym: CHLOROPICRIN															
LATEX LIQUID SYNTHETIC	9188/3082	31	497								LLS 701	569			
LAUGHING GAS	1070	14	644						178		NTO 858	700	N3.5		684
Synonym: NITROUS OXIDE															
LAUREL CAMPHOR	2717	32							178						
LAUROYL PEROXIDE	2893	48	499		2091	706	127		178						684
LAUROYL PEROXIDE	2124	48	498		2091	706	127		178		LPO 705				684
LAURYL ALCOHOL											DDN 376				685
Synonym: DODECANOL															
LAURYL AMMONIUM SULFATE											ALS 45				
Synonym: AMMONIUM LAURYL SULFATE															
LAURYL BROMIDE								–64							685
LAURYL ISOQUINOLINIUM BROMIDE					2092										
LAURYL MAGNESIUM SULFATE											DSM 479				
Synonym: DODECYL SULFATE MAGNESIUM SALT															
LAURYL MERCAPTAN	2124	48	500		2092	706	127				LRM 707				685
LAURYL SODIUM SULFATE											DDS 378				
Synonym: DODECYL SULFATE SODIUM SALT															

CHEMICAL OR MATERIAL NAME	UN/NA Number	Guide Number (DOT)	Firefighter's Hazardous Materials Reference Book	First Aid Manual for Chemical Accidents	Sax's Dangerous Properties of Industrial Materials	Hazardous Chemicals Desk Reference	Rapid Guide to Hazardous Chemicals in the Workplace	Fire Protection Guide on Hazardous Materials (NFPA)	Firefighter's Handbook of Hazardous Materials	Pocket Guide to Chemical Hazards (NIOSH)	Chemical Hazard Response Information System (CHRIS)	Emergency Handling of Hazardous Materials in Surface Transportation (AAR)	Emergency Action Guides (EAG)	Chemical Data Notebook: A User's Manual	Condensed Chemical Dictionary
LAURYL SULFATE DIETHANOLAMINE SALT SOLUTION Synonym: **DODECYL SULFATE DIETHANOLAMINE SALT**											DSD 475				
LAURYL SULFATE MAGNESIUM SALT Synonym: **DODECYL SULFATE MAGNESIUM SALT**											DSM 479				
LAURYL SULFATE SODIUM SALT Synonym: **DODECYL SULFATE SODIUM SALT**											DDS 378				
LAURYL SULFATE TRIETHANOLAMINE SALT Synonym: **DODECYL SULFATE TRIETHANOLAMINE SALT**											DST 482				
LAURYLBENZENE Synonym: **DODECYLBENZENE**											DDB 371				441
LAURYLBENZENESULFONIC ACID Synonym: **DODECYLBENZENESULFONIC ACID**	2584	60	330						130		DSA 474	390	D5.0		441
LEAD	2291	53		19	2093	707	127		178	136					686
LEAD (II) CHLORIDE Synonym: **LEAD CHLORIDE**	2291	53	503		2097	710	128		178		LCL 695	571			687
LEAD ACETATE	1616	53	501	19	2094	708	128		178		LAC 690	570			686
LEAD ACETATE TRIHYDRATE Synonym: **LEAD ACETATE**	1616	53	501	19	2095	709	128		178		LAC 690	570			686
LEAD ARSENATE	1617	53	502	19			128	103–	178		LAR 693	570			687
LEAD ARSENATE ACID Synonym: **LEAD ARSENATE**	1617	53	502	19		709	128	103–	178		LAR 693	570			687
LEAD ARSENITE	1618	53			2095	709	128		178						687
LEAD AZIDE	0129				2095	710	128		178			570			687
LEAD BAGHOUSE FLUE DUST	2811	53					128					571			
LEAD BOTTOMS Synonym: **LEAD SULFATE**	1794	60			2105	716	128		179		LSF 709	575			692

HAZARDOUS MATERIALS REFERENCE BOOKS INDEX

CHEMICAL OR MATERIAL NAME	UN/NA Number	Guide Number (DOT)	Firefighter's Hazardous Materials Reference Book	First Aid Manual for Chemical Accidents	Sax's Dangerous Properties of Industrial Materials	Hazardous Chemicals Desk Reference	Rapid Guide to Hazardous Chemicals in the Workplace	Fire Protection Guide on Hazardous Materials (NFPA)	Firefighter's Handbook of Hazardous Materials	Pocket Guide to Chemical Hazards (NIOSH)	Chemical Hazard Response Information System (CHRIS)	Emergency Handling of Hazardous Materials in Surface Transportation (AAR)	Emergency Action Guides (EAG)	Chemical Data Notebook: A User's Manual	Condensed Chemical Dictionary
LEAD CHLORIDE	2291	53	503		2096	710	128		178		LCL 695	571			687
LEAD CHROMATE				19	2097	710	127								688
LEAD CHROMATE BASIC				19	2097	711	128								
LEAD COMPOUND	2291	53		19	2098	711	127		178	136		571			686
LEAD CYANIDE	1620	53			2098	712	128		178			572			688
LEAD DICHLORIDE Synonym: LEAD CHLORIDE	2291	53	503		2096	710	128		178		LCL 695	571			687
LEAD DIFLUORIDE Synonym: LEAD FLUORIDE	2811/3077	53	504		2101	713	128		178		LFR 698	573			688
LEAD DIOXIDE	1872	42		19	2099	712	128		178			572			688
LEAD DROSS	1794	60			2100	713	128		178						688
LEAD FLUE DUST	2811	53					128					572			
LEAD FLUOBORATE Synonym: LEAD FLUOROBORATE	2291/3077	53	505		2100	713	128		178		LFB 697	573			688
LEAD FLUORIDE	2811/3077	53	504		2101	713	128		178		LFR 698	573			688
LEAD FLUOROBORATE SOLUTION Synonym: LEAD FLUOROBORATE	2291/3077	53	505				128				LFB 697	573			
LEAD FLUOROBORATE	2291/3077	53	505				128				LFB 697	573			
LEAD HYPOSULFITE Synonym: LEAD THIOSULFATE			512				128				LTS 715				689
LEAD IODIDE	2811/3077	53	506				128		178		LID 700	573			689
LEAD IV ACETATE Synonym: LEAD TETRAACETATE			510		2106		128				LTT 716				692
LEAD MONOXIDE Synonym: LITHARGE			518		2102	714	128		179		LTH 713				689

CHEMICAL OR MATERIAL NAME	UN/NA Number	Guide Number (DOT)	Firefighter's Hazardous Materials Reference Book	First Aid Manual for Chemical Accidents	Sax's Dangerous Properties of Industrial Materials	Hazardous Chemicals Desk Reference	Rapid Guide to Hazardous Chemicals in the Workplace	Fire Protection Guide on Hazardous Materials (NFPA)	Firefighter's Handbook of Hazardous Materials	Pocket Guide to Chemical Hazards (NIOSH)	Chemical Hazard Response Information System (CHRIS)	Emergency Handling of Hazardous Materials in Surface Transportation (AAR)	Emergency Action Guides (EAG)	Chemical Data Notebook: A User's Manual	Condensed Chemical Dictionary
LEAD NITRATE	1469	42	507	19	2103	714	128		179		LNT 703	574			689
LEAD OXIDE YELLOW			518				128				LTH 713				690
Synonym: LITHARGE															
LEAD PERCHLORATE	1470	42			2104		128		179			574			690
LEAD PEROXIDE	1872	42					128		179						690
LEAD PHOSPHITE	2989/1325	32					128		179			575			690
LEAD PROTOXIDE			518				128		180		LTH 713				690
Synonym: LITHARGE															
LEAD STEARATE	2311/3077	53	508	19	2105		128		179		LSA 708	575			691
LEAD SULFATE	1794	60			2105	716	128		179		LSF 709	575			692
LEAD SULFIDE	2311/3077	53	509	19	2105		128				LSU 710	576			692
LEAD SULFOCYANATE	2291/3077	53	511		2106		128		179		LTC 712	576			693
Synonym: LEAD THIOCYANATE															
LEAD SULPHATE CRUDE	9188/3077						128					576			
LEAD TETRAACETATE			510		2106		128				LTT 716				692
LEAD TETRACHLORIDE					2106		128		179						
LEAD TETRAETHYL	1649	56	813	25	3219	1100	209	-85	179	208	TEL1102	889			693
Synonym: TETRAETHYL LEAD															
LEAD TETRAMETHYL	1649	56		25	3245	1105	210	-86	179	210	TML1129				1134
Synonym: TETRAMETHYL LEAD															
LEAD THIOCYANATE	2291/3077	53	511		2106		128		179		LTC 712	576			693
LEAD THIOSULFATE			512				128				LTS 715				693
LEAD TRINITRORESORCINATE	0130					716	128		179						693
LEAD TUNGSTATE			513				128				LTU 717				693

HAZARDOUS MATERIALS REFERENCE BOOKS INDEX

CHEMICAL OR MATERIAL NAME	UN/NA Number	Guide Number (DOT)	Firefighter's Hazardous Materials Reference Book	First Aid Manual for Chemical Accidents	Sax's Dangerous Properties of Industrial Materials	Hazardous Chemicals Desk Reference	Rapid Guide to Hazardous Chemicals in the Workplace	Fire Protection Guide on Hazardous Materials (NFPA)	Firefighter's Handbook of Hazardous Materials	Pocket Guide to Chemical Hazards (NIOSH)	Chemical Hazard Response Information System (CHRIS)	Emergency Handling of Hazardous Materials in Surface Transportation (AAR)	Emergency Action Guides (EAG)	Chemical Data Notebook: A User's Manual	Condensed Chemical Dictionary
LEAD WOLFRAMATE Synonym: LEAD TUNGSTATE			513				128				LTU 717				693
LEAF LARD Synonym: OILS EDIBLE LARD											OLD 885				
LEDON 12 Synonym: DICHLORODIFLUOROMETHANE	1028	12	275	15	1137	380	68		102	86	DCF 360	331	D1.1		379
LEUCOL Synonym: QUINOLINE	2656	29	736		2968	1014		−83	246		QNL1005	805			987
LEWISITE	1955	15							179			577			697
LICHENIC ACID Synonym: FUMARIC ACID	9126/9188		407		1760	595			152		FUM 599	465			697
LIFE RAFT	2990	31							179						
LIGHT GASOLINE Synonym: GASOLINE	1203	27	412	18	1779	601	104	−56	153		GAK 602	473	G1.0	G 1	554
LIGHT LIGROIN	1115	26							179						
LIGHT LIGROIN Synonym: PETROLEUM NAPHTHA	1255	27	676		2463		152		227		PTN 999	739	P2.0		890
LIGHT NAPHTHA Synonym: NAPHTHA V,M & P	1255	27	605	21	2463	835	152	−73	207		NVM 863				
LIGHT NAPHTHA Synonym: NAPHTHA SOLVENT	1256	27	603	21	2463	835	152	−73	207		NSV 850	673			
LIGHT OIL Synonym: OILS MISCELLANEOUS COAL TAR	1136	27									OCT 876				698
LIGHTER	1226	26							179						
LIGHTER	1057	17							179			577			
LIGHTER FLUID	1226	26							179						

CHEMICAL OR MATERIAL NAME	UN/NA Number	Guide Number (DOT)	Firefighter's Hazardous Materials Reference Book	First Aid Manual for Chemical Accidents	Sax's Dangerous Properties of Industrial Materials	Hazardous Chemicals Desk Reference	Rapid Guide to Hazardous Chemicals in the Workplace	Fire Protection Guide on Hazardous Materials (NFPA)	Firefighter's Handbook of Hazardous Materials	Pocket Guide to Chemical Hazards (NIOSH)	Chemical Hazard Response Information System (CHRIS)	Emergency Handling of Hazardous Materials in Surface Transportation (AAR)	Emergency Action Guides (EAG)	Chemical Data Notebook: A User's Manual	Condensed Chemical Dictionary
LIGNITE ASH URANIUM BEARING	2912	62										578			
LIME	1910	60		19					179						699
LIME CHLORIDE Synonym: CALCIUM HYPOCHLORITE	1748	45	172	13				46 –	63		CHY 252	182	C1.1	C 2	204
LIME SALTS NEC	1759	60										578			
LIMED WOOD ROSIN Synonym: CALCIUM RESINATE	1313	32	177		671	234			63		CRE 300	184			207
LIMONENE Synonym: DIPENTENE	2052/1993	27	318					-45	179		DPN 468	381			700
LINALOOL (EX BOIS DE ROSE; SYNTHETIC)					2118	721		-64	180						700
LINDANE Synonym: BENZENE HEXACHLORIDE	2761/3082	55	514	19	360	117			180	136	BHC 144	578			701
LINEAR ALCOHOL			515								LAL 692				
LINSEED OIL Synonym: OILS MISCELLANEOUS LINSEED					2121	722					OLS 887				702
LIQUEFIED GAS	1058	12							180			579			
LIQUEFIED HYDROCARBON GAS Synonym: LIQUEFIED PETROLEUM GAS	1075	22	517	19	2123	722	128		180		LPG 704		L1.0	L 1	703
LIQUEFIED NATURAL GAS	1972	22	516												703
LIQUEFIED NATURAL GAS	1058	12									LNG 702				703
LIQUEFIED NONFLAMMABLE GAS CHARGED WITH NITRO-GEN	1058	12							180						
LIQUEFIED PETROLEUM GAS	1075	22	517	19	2123	722	128		180		LPG 704		L1.0		703
LIQUEFIED PHENOL Synonym: CARBOLIC OIL	2821	55	186	22							CBO 220				217

HAZARDOUS MATERIALS REFERENCE BOOKS INDEX

CHEMICAL OR MATERIAL NAME	UN/NA Number	Guide Number (DOT)	Firefighter's Hazardous Materials Reference Book	First Aid Manual for Chemical Accidents	Sax's Dangerous Properties of Industrial Materials	Hazardous Chemicals Desk Reference	Rapid Guide to Hazardous Chemicals in the Workplace	Fire Protection Guide on Hazardous Materials (NFPA)	Firefighter's Handbook of Hazardous Materials	Pocket Guide to Chemical Hazards (NIOSH)	Chemical Hazard Response Information System (CHRIS)	Emergency Handling of Hazardous Materials in Surface Transportation (AAR)	Emergency Action Guides (EAG)	Chemical Data Notebook: A User's Manual	Condensed Chemical Dictionary
LIQUID AMMONIA **Synonym: AMMONIA**	1005	15	48	11	220	57	11	25 -14	180	38	AMA 47	78	A9.0	A 9	62
LIQUID AMMONIA **Synonym: ANHYDROUS AMMONIA**	1005	15	48	11	220	57	11	25 -14	180	38	AMA 47	78	A9.0	A 9	62
LIQUID ARGON	1951	21							180						
LIQUID ASPHALT **Synonym: ASPHALT BLENDING STOCKS ROOFERS FLUX**	1999	27	105	12	307	97	17		36		ARF 84	99	A12.0		
LIQUID ASPHALT **Synonym: OILS MISCELLANEOUS ROAD**	1999	27		12							ORD 899				
LIQUID BLEACH **Synonym: SODIUM HYPOCHLORITE**	1791	60	778	24	3088	1057			254		SHC1043	843			1059
LIQUID CAMPHOR **Synonym: CAMPHOR OIL**	1130	27	179		676	237		-25	180		CPO 292	185			
LIQUID CARBON DIOXIDE	2187	21		13					180						
LIQUID CHLORINE **Synonym: CHLORINE**	1017	20	197	13	756	267	45	49 -	180	62	CLX 261	214	C4.0	C 6	259
LIQUID FLUORINE	9192	25		17					180						
LIQUID GUM CAMPHOR **Synonym: CAMPHOR OIL**	1130	27	179		676	237		-25	64		CPO 292	185			
LIQUID HELIUM	1963	21							180						
LIQUID HYDROGEN **Synonym: HYDROGEN**	1966	22	445		1898	642		97 -59	180		HXX 654	533			614
LIQUID IMPURE CAMPHOR **Synonym: CAMPHOR OIL**	1130	27	179		676	237		-25	64		CPO 292	185			
LIQUID NITROGEN **Synonym: NITROGEN**	1977	21		21	2536	866		127-	180		NXX 864	695	N3.4.3	N 3	827

CHEMICAL OR MATERIAL NAME	UN/NA Number	Guide Number (DOT)	Firefighter's Hazardous Materials Reference Book	First Aid Manual for Chemical Accidents	Sax's Dangerous Properties of Industrial Materials	Hazardous Chemicals Desk Reference	Rapid Guide to Hazardous Chemicals in the Workplace	Fire Protection Guide on Hazardous Materials (NFPA)	Firefighter's Handbook of Hazardous Materials	Pocket Guide to Chemical Hazards (NIOSH)	Chemical Hazard Response Information System (CHRIS)	Emergency Handling of Hazardous Materials in Surface Transportation (AAR)	Emergency Action Guides (EAG)	Chemical Data Notebook: A User's Manual	Condensed Chemical Dictionary
LIQUID NITROGEN DIOXIDE	1067	20		21					180						
LIQUID OXYGEN	1072	14						104–							
LIQUID OXYGEN Synonym: OXYGEN	1073	23	659		2652	901		104–	180		OXY 920	720	O3.0	O 1	862
LIQUID PETROLATUM Synonym: OILS MISCELLANEOUS MINERAL	1270	27									OMN 888				
LIQUID PITCH Synonym: CREOSOTE COAL TAR	1993	27	231	14					84		CCT 233		C6.0		322
LIQUID TAR ACID	2821	55										579			
LITHARGE			518						180		LTH 713				705
LITHIUM	1415	40	519		2123	723			180		LTM 714	584			705
LITHIUM ACETYLIDE ETHYLENEDIAMINE COMPLEX	2813	40			2124	723			180			580			
LITHIUM ALKYL	2445/3049	40							180			580			
LITHIUM ALUMINUM HYDRIDE	1411	40							181			580			706
LITHIUM ALUMINUM HYDRIDE	1410	40	520					105–	181		LAH 691	580			706
LITHIUM AMIDE	1412/2813	40			2124	723			181			581			706
LITHIUM BICHROMATE			521								LBC 694				707
LITHIUM BICHROMATE DIHYDRATE Synonym: LITHIUM BICHROMATE			521								LBC 694				707
LITHIUM BOROHYDRIDE	1413	40							181			582			706
LITHIUM CHROMATE	9134/3077	31	522		2127	725			181		LCR 696	582			707
LITHIUM COMPOUNDS					2127	725			181						705
LITHIUM DICHROMATE Synonym: LITHIUM BICHROMATE			521								LBC 694				707
LITHIUM FERROSILICON	2830	41							181			582			707

HAZARDOUS MATERIALS REFERENCE BOOKS INDEX

CHEMICAL OR MATERIAL NAME	UN/NA Number	Guide Number (DOT)	Firefighter's Hazardous Materials Reference Book	First Aid Manual for Chemical Accidents	Sax's Dangerous Properties of Industrial Materials	Hazardous Chemicals Desk Reference	Rapid Guide to Hazardous Chemicals in the Workplace	Fire Protection Guide on Hazardous Materials (NFPA)	Firefighter's Handbook of Hazardous Materials	Pocket Guide to Chemical Hazards (NIOSH)	Chemical Hazard Response Information System (CHRIS)	Emergency Handling of Hazardous Materials in Surface Transportation (AAR)	Emergency Action Guides (EAG)	Chemical Data Notebook: A User's Manual	Condensed Chemical Dictionary
LITHIUM HYDRIDE	1414	40	523	19	2128	725	129	105–	181	136	LHD 699	583			708
LITHIUM HYDRIDE	2805	40		19	2128	725	129		181	136		583			708
LITHIUM HYDROXIDE MONOHYDRATE	2680	60							181			583			
LITHIUM HYDROXIDE SOLUTION	2679	60							181			583			
LITHIUM HYPOCHLORITE	1471	45			2128	726			181			584			708
LITHIUM METAL	1415	40						104–	181		LTM 714	584			
LITHIUM NITRATE	2722	35							181						709
LITHIUM NITRIDE	2806	37			2129	726			181			584			709
LITHIUM PEROXIDE	1472	47			2129	726			181			585			709
LITHIUM SILICON	1417	40			2130	726			181			585			709
LNG	1972	22							182						710
LNG Synonym: LIQUEFIED NATURAL GAS	1058	12									LNG 702				710
LOBELINE					2133										
LONDON PURPLE	1621	53			2134	727			182			585			711
LONG TIME BURNING OIL Synonym: OILS MISCELLANEOUS MINERAL SEAL	1270	27									OMS 889				
LOROL 20 Synonym: OCTANOL	1987	26	650		2620						OTA 909	706	O1.0		847
LOROL 22 Synonym: DECYL ALCOHOL (-n)	1987	26	258		1032	347			91		DAN 338	291			350
LORSBAN Synonym: DURSBAN	1615										DUR 495				448
LOX	1072	14						104–							711

CHEMICAL OR MATERIAL NAME	UN/NA Number	Guide Number (DOT)	Firefighter's Hazardous Materials Reference Book	First Aid Manual for Chemical Accidents	Sax's Dangerous Properties of Industrial Materials	Hazardous Chemicals Desk Reference	Rapid Guide to Hazardous Chemicals in the Workplace	Fire Protection Guide on Hazardous Materials (NFPA)	Firefighter's Handbook of Hazardous Materials	Pocket Guide to Chemical Hazards (NIOSH)	Chemical Hazard Response Information System (CHRIS)	Emergency Handling of Hazardous Materials in Surface Transportation (AAR)	Emergency Action Guides (EAG)	Chemical Data Notebook: A User's Manual	Condensed Chemical Dictionary
LOX Synonym: OXYGEN	1073	23	659		2652	901		104–	182		OXY 920	720	O3.0	O 1	711
LPG Synonym: LIQUEFIED PETROLEUM GAS	1075	22	517	19	2123	722	128		182	136	LPG 704		L1.0	L 1	711
LUBRICATING OIL Synonym: OILS MISCELLANEOUS MOTOR	1270/1993	27			2136			–64	182		OMT 890	585			712
LUBRICATING OIL SPINDLE								–64	182						
LUCIDOL Synonym: DIBENZOYL PEROXIDE	2085/2087	49	262						182		DPO 469	320			368
LUCIDOL Synonym: BENZOYL PEROXIDE	2085	49	122		392	130	23		182	44			B2.0	B 2	133
LUCITE	1247	26													712
LUMBRICAL Synonym: PIPERAZINE	2579/1760	60	696		2811	959		–80			PPZ 983	766			919
LUNAR CAUSTIC Synonym: SILVER NITRATE	1493	45	749		3049	1040	198		251		SVN1069	828			1039
LUPERCO JDB 50 T Synonym: CYCLOHEXANONE PEROXIDE	2896/3106	51							88		CHP 248	282			337
LUTIDENE	1993	27													
LYE Synonym: CAUSTIC POTASH	1814	60	193						67		CPS 295	781			233
LYE Synonym: CAUSTIC SODA	1824	60	194						182		CSS 314				233
LYE Synonym: SODIUM HYDROXIDE	1823	60	777	24	3087	1056	200	159–	182	198	SHD1044	842	S2.5	S 4	714

CHEMICAL OR MATERIAL NAME	UN/NA Number	Guide Number (DOT)	Firefighter's Hazardous Materials Reference Book	First Aid Manual for Chemical Accidents	Sax's Dangerous Properties of Industrial Materials	Hazardous Chemicals Desk Reference	Rapid Guide to Hazardous Chemicals in the Workplace	Fire Protection Guide on Hazardous Materials (NFPA)	Firefighter's Handbook of Hazardous Materials	Pocket Guide to Chemical Hazards (NIOSH)	Chemical Hazard Response Information System (CHRIS)	Emergency Handling of Hazardous Materials in Surface Transportation (AAR)	Emergency Action Guides (EAG)	Chemical Data Notebook: A User's Manual	Condensed Chemical Dictionary
LYE Synonym: POTASSIUM HYDROXIDE	1814	60	711	23	2871	984	186	145–	239		PTH 995	781	P6.0	P 5	714
LYE Synonym: POTASSIUM HYDROXIDE	1813	60	712	23	2871	984	186	145–	239			781	P6.0	P 5	714
LYE SOLUTION	1824	60							182						
LYNALYL ACETATE (EX BOIS DE ROSE SYNTHETIC)								–64	182						
LYSOFORM Synonym: FORMALDEHYDE SOLUTION	1198/2209	29	401	17	1748	591	102	91 –54	150	116	FMS 590	462	F2.0	F 2	
M B C FUMIGANT Synonym: METHYL BROMIDE	1062	55	559	20	2274	776	137	110–67	192	146	MTB 805	619	M3.0	M 3	759
M B C FUMIGANT MEBR Synonym: METHYL BROMIDE	1062	55	559	20	2274	776	137	110–67	192	146	MTB 805	619	M3.0	M 3	759
MAAC Synonym: METHYL AMYL ACETATE	1233	26	556								MAC 719	642			757
MACQUER'S SALT Synonym: POTASSIUM ARSENATE	1677	53	703						237		PAS 928	776			948
MAGNESIUM	1418	76			2147	731					MGX 758	587			717
MAGNESIUM	1869	76	524		2147	731		106–	182		MGX 758	591			717
MAGNESIUM ALKYL	3053	40							182			587			
MAGNESIUM ALLOY	1869	76				731			183						
MAGNESIUM ALLOY	1418	76				731			183			587			
MAGNESIUM ALUMINUM PHOSPHIDE	1419	41							183						
MAGNESIUM ARSENATE	1622	53							183			587			718
MAGNESIUM BISULFITE SOLUTION	2693	60							183						
MAGNESIUM BROMATE	1473	35							183			588			718

CHEMICAL OR MATERIAL NAME	UN/NA Number	Guide Number (DOT)	Firefighter's Hazardous Materials Reference Book	First Aid Manual for Chemical Accidents	Sax's Dangerous Properties of Industrial Materials	Hazardous Chemicals Desk Reference	Rapid Guide to Hazardous Chemicals in the Workplace	Fire Protection Guide on Hazardous Materials (NFPA)	Firefighter's Handbook of Hazardous Materials	Pocket Guide to Chemical Hazards (NIOSH)	Chemical Hazard Response Information System (CHRIS)	Emergency Handling of Hazardous Materials in Surface Transportation (AAR)	Emergency Action Guides (EAG)	Chemical Data Notebook: A User's Manual	Condensed Chemical Dictionary
MAGNESIUM CHLORATE	2723/1479	35			2148	732			183			588			719
MAGNESIUM DIAMIDE	2004	37							183			588			
MAGNESIUM DIETHYL	1367	40							183						
MAGNESIUM DIOXIDE	1476	35							183						719
MAGNESIUM DIPHENYL	2005	40							183			588			
MAGNESIUM DODECYL SULFATE Synonym: DODECYL SULFATE MAGNESIUM SALT											DSM 479				
MAGNESIUM FLUOROSILICATE	2853	53							183			589			719
MAGNESIUM FLUOSILICATE	2853	53							183						
MAGNESIUM GRANULES	2950	40							183			589			
MAGNESIUM HYDRIDE	2010/2813	40			2150	733			183			589			
MAGNESIUM LAURYL SULFATE Synonym: DODECYL SULFATE MAGNESIUM SALT											DSM 479				720
MAGNESIUM NITRATE	1474	35			2150	733			183			589			720
MAGNESIUM OXIDE	1693	58			2151	733	129								720
MAGNESIUM PERCHLORATE	1475	35	525						183		MPC 783	590			721
MAGNESIUM PERCHLORATE ANHYDROUS Synonym: MAGNESIUM PERCHLORATE	1475	35	525						183		MPC 783	590			721
MAGNESIUM PERCHLORATE HEXAHYDRATE Synonym: MAGNESIUM PERCHLORATE	1475	35	525						183		MPC 783	590			721
MAGNESIUM PEROXIDE	1476	35			2151	734			183			590			721
MAGNESIUM PHOSPHIDE	2011/2813	41			2151	734			183			590			
MAGNESIUM POWDER	1418	76			2147	731			183		MGX 758	587			
MAGNESIUM SILICIDE	2624	40							183			591			722
MAGNESIUM SILICOFLUORIDE	2853	53							183						722

HAZARDOUS MATERIALS REFERENCE BOOKS INDEX

CHEMICAL OR MATERIAL NAME	UN/NA Number	Guide Number (DOT)	Firefighter's Hazardous Materials Reference Book	First Aid Manual for Chemical Accidents	Sax's Dangerous Properties of Industrial Materials	Hazardous Chemicals Desk Reference	Rapid Guide to Hazardous Chemicals in the Workplace	Fire Protection Guide on Hazardous Materials (NFPA)	Firefighter's Handbook of Hazardous Materials	Pocket Guide to Chemical Hazards (NIOSH)	Chemical Hazard Response Information System (CHRIS)	Emergency Handling of Hazardous Materials in Surface Transportation (AAR)	Emergency Action Guides (EAG)	Chemical Data Notebook: A User's Manual	Condensed Chemical Dictionary
MAGNESIUM-o-ARSENATE	1622	53							183						
MAGNETS NEC	2807														
MAGNICIDE H Synonym: ACROLEIN	1092	30	31	11	63	25	5	17 −12	20	32	ARL 85	12	A4.0	A 5	18
MALATHION	2783/3082	55	526	19	2153	734	126		183	138	MLT 773	591			724
MALAZIDE Synonym: MALEIC HYDRAZIDE			529								MLH 770				724
MALEIC ACID	2215	60	527		2154	736			183		MLI 771	591			724
MALEIC ACID ANHYDRIDE Synonym: MALEIC ANHYDRIDE	2215	60	528	19	2155	736	130	106–64	184	138	MLA 769	592	M1.0		724
MALEIC ACID HYDRAZIDE Synonym: MALEIC HYDRAZIDE			529								MLH 770				724
MALEIC ANHYDRIDE	2215	60	528	19	2155	736	130	106–64	184	138	MLA 769	592	M1.0		724
MALEIC HYDRAZIDE			529								MLH 770				724
MALEINIC ACID Synonym: MALEIC ACID	2215	60	527		2154	736			184		MLI 771	591			724
MALENIC ACID Synonym: MALEIC ACID	2215	60	527		2154	736			183		MLI 771	591			724
MALIX Synonym: ENDOSULFAN	2761	55	332		1519	505	86		131		ESF 559	398			463
MALONIC ACID									184						725
MALONIC DINITRILE	2647	53							184						725
MALONIC ETHYL ESTER NITRILE	2666	55							184						725
MALONIC MONONITRILE Synonym: CYANOACETIC ACID	1935	55	240		975			59 −	184		CYA 327				725
MALONONITRILE	2647/2811	53		19	2157	737	130		184			592			725

CHEMICAL OR MATERIAL NAME	UN/NA Number	Guide Number (DOT)	Firefighter's Hazardous Materials Reference Book	First Aid Manual for Chemical Accidents	Sax's Dangerous Properties of Industrial Materials	Hazardous Chemicals Desk Reference	Rapid Guide to Hazardous Chemicals in the Workplace	Fire Protection Guide on Hazardous Materials (NFPA)	Firefighter's Handbook of Hazardous Materials	Pocket Guide to Chemical Hazards (NIOSH)	Chemical Hazard Response Information System (CHRIS)	Emergency Handling of Hazardous Materials in Surface Transportation (AAR)	Emergency Action Guides (EAG)	Chemical Data Notebook: A User's Manual	Condensed Chemical Dictionary
MANEB	2210	37							184						726
MANEB	2968/2813	37							184			592			726
MANGANESE					2159	738	130		184	138					726
MANGANESE COMPOUNDS					2161	738	131		184	138					
MANGANESE DIOXIDE	1479	35			2162	739	131		184						727
MANGANESE NITRATE	2724	35					131		184			593			728
MANGANESE OXIDE					2163	740	131		184						
MANGANESE RESINATE	1330/1325	32					131		184			593			728
MANGANESE RESINATE	1325	32					131								728
MANGANESE TRICARBONYL METHYLCYCLOPENTADIENYL						740	131		184						
MANGANOUS ETHYLENE BIS DITHIO CARBAMATE	2210	37													
MANNITOL HEXANITRATE	0133				2165	741			185			593			731
MAOH Synonym: METHYL AMYL ALCOHOL	2053	26	557						191		MAA 718				757
MAOH Synonym: METHYL ISOBUTYL CARBINOL	2053/1993	26	570		2334	799	145	-70	195	152	MIC 763	630			771
MAPO	2501	55							185						
MAPP	1060	17							185						
MAPP GAS Synonym: METHYL ACETYLENE PROPADIENE MIXTURE	1060	17	552		2248	769	134		185	142	MAP 725	640			755
MARLATE 50 Synonym: METHOXYCHLOR	2761/3077	55	550		2224	763	133		191	140	MOC 780	615			753
MARSH GAS Synonym: METHANE	1971	17	547	20	2213	760		-65	185		MTH 809			M 1	731

CHEMICAL OR MATERIAL NAME	UN/NA Number	Guide Number (DOT)	Firefighter's Hazardous Materials Reference Book	First Aid Manual for Chemical Accidents	Sax's Dangerous Properties of Industrial Materials	Hazardous Chemicals Desk Reference	Rapid Guide to Hazardous Chemicals in the Workplace	Fire Protection Guide on Hazardous Materials (NFPA)	Firefighter's Handbook of Hazardous Materials	Pocket Guide to Chemical Hazards (NIOSH)	Chemical Hazard Response Information System (CHRIS)	Emergency Handling of Hazardous Materials in Surface Transportation (AAR)	Emergency Action Guides (EAG)	Chemical Data Notebook: A User's Manual	Condensed Chemical Dictionary
MARSHITE Synonym: **COPPER IODIDE**							55				CID 253				312
MASSICOT Synonym: **LITHARGE**			518						180		LTH 713				732
MATCHES	2254	32							185						
MATCHES	1945	32							185			594			
MATCHES	1944	32							185			594			
MATCHES	1331	32							185			594			
MB Synonym: **METHYL BROMIDE**	1062	55	559	20	2274	776	137	110–67	192	146	MTB 805	619	M3.0	M 3	759
MBX Synonym: **METHYL BROMIDE**	1062	55	559	20	2274	776	137	110–67	192	146	MTB 805	619	M3.0	M 3	759
MCA Synonym: **CHLOROACETIC ACID**	1751	60	199	13	761	270		50 –26	72		MCA 734	216	C3.9		734
MCB Synonym: **CHLOROBENZENE**	1134	27	203	13	768	273	47	51 –26	73	64	CRB 298	219	C4.1		263
MCP MONOCALCIUM PHOSPHATE MONOHYDRATE Synonym: **CALCIUM PHOSPHATE**					671	234					CAL 207				206
MDI Synonym: **DIPHENYLMETHANE DIISOCYANATE**	2489	55	321						127		DPM 467	382			734
MEA Synonym: **MONOETHANOLAMINE**	2491	60	594					118–	206		MEA 749		M5.0		734
MEADOW GREEN Synonym: **COPPER ACETOARSENITE**	1585	53	222				55		83		CAA 203	257			310
MECURIALIN Synonym: **METHYLAMINE**	1235	68		20	2253	771	136	109–66	200	144	MTA 804				756

CHEMICAL OR MATERIAL NAME	UN/NA Number	Guide Number (DOT)	Firefighter's Hazardous Materials Reference Book	First Aid Manual for Chemical Accidents	Sax's Dangerous Properties of Industrial Materials	Hazardous Chemicals Desk Reference	Rapid Guide to Hazardous Chemicals in the Workplace	Fire Protection Guide on Hazardous Materials (NFPA)	Firefighter's Handbook of Hazardous Materials	Pocket Guide to Chemical Hazards (NIOSH)	Chemical Hazard Response Information System (CHRIS)	Emergency Handling of Hazardous Materials in Surface Transportation (AAR)	Emergency Action Guides (EAG)	Chemical Data Notebook: A User's Manual	Condensed Chemical Dictionary
MEDICINES	1851	11							185			595			
MEK Synonym: METHYL ETHYL KETONE	1193	26	566	20	2319	793	143	-69	185		MEK 751	627	M3.2	M 5	736
MEMTETRAHYDROPHTHALIC ANHYDRIDE	1760	60										596			737
MENDRIN Synonym: ENDRIN	2761/2996	55	334		1520	507	86		131	102	EDR 520	399			463
MENITE Synonym: PHOSDRIN	2783	55	681	22					231	180	PHD 955				906
MENTHA-1,8-DIENE-*para* Synonym: DIPENTENE	2052/1993	27	318		2176	743		-45	185		DPN 468	381			427
MENTHOL					2179	744									738
MEP Synonym: METHYLETHYLPYRIDINE	2300	60	567						194		MEP 753	648			739
MERCAPTAN	3071	28			2182	746			185			598			739
MERCAPTAN MIXTURE	1228	28			2182	746			185			597			
MERCAPTOACETIC ACID	1940	60							185						739
MERCAPTODIMETHUR	2757	55	530						185		MCD 736				
MERCAPTOETHANE Synonym: ETHYL MERCAPTAN	2363	27	358		1631	550	97	-53	138	112		418	E2.9		496
MERCAPTOETHANE Synonym: ETHYL MERCAPTAN	1228	28			1631						EMC 543				496
MERCAPTOMETHANE Synonym: METHYL MERCAPTAN	1064	13	574		2338	801	147	-70	195	154	MMC 774	633	M3.5		496
MERCURCURIC POTASSIUM CYANIDE	1626	53													
MERCURIALIN Synonym: MONOMETHYLAMINE	1061	19	596						206				M6.0		796

CHEMICAL OR MATERIAL NAME	UN/NA Number	Guide Number (DOT)	Firefighter's Hazardous Materials Reference Book	First Aid Manual for Chemical Accidents	Sax's Dangerous Properties of Industrial Materials	Hazardous Chemicals Desk Reference	Rapid Guide to Hazardous Chemicals in the Workplace	Fire Protection Guide on Hazardous Materials (NFPA)	Firefighter's Handbook of Hazardous Materials	Pocket Guide to Chemical Hazards (NIOSH)	Chemical Hazard Response Information System (CHRIS)	Emergency Handling of Hazardous Materials in Surface Transportation (AAR)	Emergency Action Guides (EAG)	Chemical Data Notebook: A User's Manual	Condensed Chemical Dictionary
MERCURIC ACETATE	1629	53	531		2188	747			186		MAT 726				740
MERCURIC ACETYLIDE									186						
MERCURIC AMMONIUM CHLORIDE	1630	53							186		MCC 735	598			
MERCURIC ARSENATE	1623/2811	53							186			598			740
MERCURIC BENZOATE	1631	35										598			740
MERCURIC BROMIDE	1634	53							186			599			740
MERCURIC CHLORIDE	1624	53	532	19					186		MRC 792	599			740
MERCURIC CHLORIDE AMMONIATED Synonym: MERCURIC AMMONIUM CHLORIDE	1630	53							186		MCC 735	598			741
MERCURIC CYANIDE	1636	53	533					107–	186		MCN 741	599			741
MERCURIC IODIDE	1638	53	534						186		MID 764	600			741
MERCURIC IODIDE RED Synonym: MERCURIC IODIDE	1638	53	534						186		MID 764	600			741
MERCURIC NITRATE	1625	42	536						186		MNT 778	600			741
MERCURIC OLEATE	1640	53							186			600			741
MERCURIC OXIDE	1641	53	537		2189	747			186		MOX 781	601			
MERCURIC OXIDE RED Synonym: MERCURIC OXIDE	1641	53	537		2189	747			186		MOX 781	601			742
MERCURIC OXIDE YELLOW Synonym: MERCURIC OXIDE	1641	53	537		2189	747			186		MOX 781	601			742
MERCURIC OXYCYANIDE	1642	53							186			601			742
MERCURIC POTASSIUM CYANIDE	1626	53							186			601			742
MERCURIC SALICYLATE	1644	53			2190	748			186			601			742
MERCURIC SUBSULFATE	2025	53							186			602			
MERCURIC SULFATE	1645	53	538						186		MRS 796	602			743

CHEMICAL OR MATERIAL NAME	UN/NA Number	Guide Number (DOT)	Firefighter's Hazardous Materials Reference Book	First Aid Manual for Chemical Accidents	Sax's Dangerous Properties of Industrial Materials	Hazardous Chemicals Desk Reference	Rapid Guide to Hazardous Chemicals in the Workplace	Fire Protection Guide on Hazardous Materials (NFPA)	Firefighter's Handbook of Hazardous Materials	Pocket Guide to Chemical Hazards (NIOSH)	Chemical Hazard Response Information System (CHRIS)	Emergency Handling of Hazardous Materials in Surface Transportation (AAR)	Emergency Action Guides (EAG)	Chemical Data Notebook: A User's Manual	Condensed Chemical Dictionary
MERCURIC SULFIDE	2025	53	539						186		MSF 800				
MERCURIC SULFIDE BLACK Synonym: MERCURIC SULFIDE	2025	53	539						186		MSF 800				743
MERCURIC SULFIDE RED Synonym: MERCURIC SULFIDE	2025	53	539						186		MSF 800				743
MERCURIC SULFOCYANATE Synonym: MERCURIC THIOCYANATE	1646	53	540		2190	748			186		MRT 797	603			743
MERCURIC SULFOCYANIDE Synonym: MERCURIC THIOCYANATE	1645	53	540								MRT 797				743
MERCURIC THIOCYANATE	1646	53	540						186		MRT 797				743
MERCURICSULFOCYANATE Synonym: MERCURIC THIOCYANATE	1646	53				748			186		MRT 797	603			743
MERCUROL	1639	53			2191	749			186			603			
MERCUROUS ACETATE	1629	53							186						743
MERCUROUS BROMIDE	1634	53							186			599			743
MERCUROUS CHLORIDE	2025	53	541	19	2192	749			187		MRR 795				744
MERCUROUS GLUCONATE	1637	53							187			603			
MERCUROUS IODIDE	1638	53		19					187			600			744
MERCUROUS NITRATE	1627	42	542						187		MRN 794	603			744
MERCUROUS NITRATE MONOHYDRATE Synonym: MERCUROUS NITRATE	1627	42	542						187		MRN 794	603			744
MERCUROUS OXIDE BLACK	1641	53							187						
MERCUROUS SULFATE	1628	53							187			604			744
MERCURY	2809	60	543	19	2192	750	131		187		MCR 744				744

CHEMICAL OR MATERIAL NAME	UN/NA Number	Guide Number (DOT)	Firefighter's Hazardous Materials Reference Book	First Aid Manual for Chemical Accidents	Sax's Dangerous Properties of Industrial Materials	Hazardous Chemicals Desk Reference	Rapid Guide to Hazardous Chemicals in the Workplace	Fire Protection Guide on Hazardous Materials (NFPA)	Firefighter's Handbook of Hazardous Materials	Pocket Guide to Chemical Hazards (NIOSH)	Chemical Hazard Response Information System (CHRIS)	Emergency Handling of Hazardous Materials in Surface Transportation (AAR)	Emergency Action Guides (EAG)	Chemical Data Notebook: A User's Manual	Condensed Chemical Dictionary
MERCURY (II) CHLORIDE Synonym: MERCURIC CHLORIDE	1624	53	532	19	2195	751	132		186		MRC 792	599			740
MERCURY (II) CYANIDE Synonym: MERCURIC CYANIDE	1636	53	533		2196	752	132	107–	186		MCN 741	599			741
MERCURY (II) NITRATE Synonym: MERCURIC NITRATE	1625	42	536		2201	755	132		186		MNT 778	600			741
MERCURY (II) SULFATE (1:1) Synonym: MERCURIC SULFATE	1645	53	538		2202	756	132		186		MRS 796	602			743
MERCURY ACETATE	1629	53		20			132		187			604			
MERCURY AMIDE CHLORIDE Synonym: MERCURIC AMMONIUM CHLORIDE	1630	53			2193	750	132		187		MCC 735	598			
MERCURY AMMONIUM CHLORIDE Synonym: MERCURIC AMMONIUM CHLORIDE	1630	53	544				132		187		MCC 735	598			
MERCURY BASED PESTICIDE	3011	28							188			605			
MERCURY BASED PESTICIDE	3012/2777	55							188			605			
MERCURY BASED PESTICIDE	2777	55							188			605			
MERCURY BASED PESTICIDE	2778	28							188			604			
MERCURY BENZOATE	1631	53			2193	750	132		187						
MERCURY BICHLORIDE Synonym: MERCURIC CHLORIDE	1624	53	532	19			132		186		MRC 792	599			745
MERCURY BINIODIDE Synonym: MERCURIC IODIDE	1638	53	534	19			132		186		MID 764	600			741
MERCURY BISULFATE Synonym: MERCURIC SULFATE	1645	53	538				132		187		MRS 796	602			743
MERCURY BROMIDE	1634	53			2194	751	132		187						
MERCURY COMPOUND	2025	53		20	2196	752	132		187			606			745

CHEMICAL OR MATERIAL NAME	UN/NA Number	Guide Number (DOT)	Firefighter's Hazardous Materials Reference Book	First Aid Manual for Chemical Accidents	Sax's Dangerous Properties of Industrial Materials	Hazardous Chemicals Desk Reference	Rapid Guide to Hazardous Chemicals in the Workplace	Fire Protection Guide on Hazardous Materials (NFPA)	Firefighter's Handbook of Hazardous Materials	Pocket Guide to Chemical Hazards (NIOSH)	Chemical Hazard Response Information System (CHRIS)	Emergency Handling of Hazardous Materials in Surface Transportation (AAR)	Emergency Action Guides (EAG)	Chemical Data Notebook: A User's Manual	Condensed Chemical Dictionary
MERCURY COMPOUND	2024	53		20	2196	752	132		187			606			745
MERCURY CYANIDE Synonym: MERCURIC CYANIDE	1636	53	533		2196	752	132	107–	187		MCN 741	599			741
MERCURY FULMINATE	0135			20	2198	753	132		187						745
MERCURY GLUCONATE	1637	53			2198	753	132		187						
MERCURY IODIDE	1638	53			2198	754	132		187						
MERCURY METAL	2809	60		19	2192	750	131		187		MCR 744				
MERCURY MONOCHLORIDE Synonym: MERCUROUS CHLORIDE	2025	53	541	19	2192	749	132		187		MRR 795				744
MERCURY NITRATE MONOHYDRATE Synonym: MERCURIC NITRATE	1625	42	536			755	132		186		MNT 778	600			741
MERCURY NUCLEATE	1639	53					132		188						
MERCURY OLEATE	1640	53			2201	755	132		188						
MERCURY OXIDE Synonym: MERCURIC OXIDE	1641	53	537		2202	756	132		188		MOX 781	601			
MERCURY OXYCYANIDE	1642	53		20			132		188						
MERCURY PERCHLORIDE Synonym: MERCURIC CHLORIDE	1624	53	532	19			132		186		MRC 792	599			740
MERCURY PERNITRATE Synonym: MERCURIC NITRATE	1625	42	536				132		186		MNT 778	600			741
MERCURY PERSULFATE Synonym: MERCURIC SULFATE	1645	53	538				132		188		MRS 796	602			743
MERCURY POTASSIUM IODIDE	1643	53					132		188						
MERCURY PROTOCHLORIDE Synonym: MERCUROUS CHLORIDE	2025	53	541	19	2192		132		187		MRR 795				744

HAZARDOUS MATERIALS REFERENCE BOOKS INDEX

CHEMICAL OR MATERIAL NAME	UN/NA Number	Guide Number (DOT)	Firefighter's Hazardous Materials Reference Book	First Aid Manual for Chemical Accidents	Sax's Dangerous Properties of Industrial Materials	Hazardous Chemicals Desk Reference	Rapid Guide to Hazardous Chemicals in the Workplace	Fire Protection Guide on Hazardous Materials (NFPA)	Firefighter's Handbook of Hazardous Materials	Pocket Guide to Chemical Hazards (NIOSH)	Chemical Hazard Response Information System (CHRIS)	Emergency Handling of Hazardous Materials in Surface Transportation (AAR)	Emergency Action Guides (EAG)	Chemical Data Notebook: A User's Manual	Condensed Chemical Dictionary
MERCURY PROTONITRATE Synonym: MERCUROUS NITRATE	1627	42	542				132		187		MRN 794	603			744
MERCURY RHODANIDE Synonym: MERCURIC THIOCYANATE	1646	53	540				132		186		MRT 797				743
MERCURY SALICYLATE	1644	53					132		188						
MERCURY SUBCHLORIDE Synonym: MERCUROUS CHLORIDE	2025	53	541	19	2192	749	132		187		MRR 795				744
MERCURY SULFATE	1645	53			2202	756	132								
MERCURY THIOCYANATE	1646	53					132		188						
MERCURY VAPOR	2809	60			2192		131			140	MCR 744				
MEREX Synonym: KEPONE	2761/9189	55	493		2077	701	123		177		KPE 688	566			670
MESITYL OXIDE	1229	26	545		2205	757	132	107–64	188	140	MSO 801	606			746
MESITYLENE	2325	26							188						746
MESUROL Synonym: MERCAPTODIMETHUR	2757	55	530						185		MCD 736				968
METACETONE	1156	26							188						747
METACETONIC ACID Synonym: PROPIONIC ACID	1848	60	721		2909	997	188	148–81	243		PNA 968	791	P7.1		
METAFUME Synonym: METHYL BROMIDE	1062	55	559	20	2274	776	137	110–67	192	146	MTB 805	619	M3.0	M 3	759
METAL ALKYL	2003	40							188			609			
METAL ALKYL HALIDE	3049	40							188			608			
METAL ALKYL HYDRIDE	3050	40							188			608			
METAL ALKYL SOLUTION	9195	40							188			608			

CHEMICAL OR MATERIAL NAME	UN/NA Number	Guide Number (DOT)	Firefighter's Hazardous Materials Reference Book	First Aid Manual for Chemical Accidents	Sax's Dangerous Properties of Industrial Materials	Hazardous Chemicals Desk Reference	Rapid Guide to Hazardous Chemicals in the Workplace	Fire Protection Guide on Hazardous Materials (NFPA)	Firefighter's Handbook of Hazardous Materials	Pocket Guide to Chemical Hazards (NIOSH)	Chemical Hazard Response Information System (CHRIS)	Emergency Handling of Hazardous Materials in Surface Transportation (AAR)	Emergency Action Guides (EAG)	Chemical Data Notebook: A User's Manual	Condensed Chemical Dictionary
METALDEHYDE	1332/1325	32						-65	188			609			747
METALLIC RESINATE Synonym: CALCIUM RESINATE	1313	32	177		671	234			63		CRE 300	184			207
METALLIC SODIUM Synonym: SODIUM	1428	40	752	24	3056	1042		155-	251		SDU1036	844	S2.0	S 1	1047
METEPA	2501	55							188						749
METHACIDE Synonym: TOLUENE	1294	27	826	25	3309	1125	215	170-87	269	214	TOL1135	909	T4.0	T 4	1157
METHACRYLALDEHYDE	2396	28							188						749
METHACRYLATE MONOMER Synonym: METHYL METHACRYLATE	1247	26	575	20	2342	803	147	115-70	195	154	MMM 775	633	M3.6		773
METHACRYLIC ACID	2531/1760	60			2209	758	133	108-65	189		MAD 720	610			749
METHACRYLIC ACID 2,3-EPOXYPROPYL ESTER Synonym: GLYCIDYL METHACRYLATE	1993	27	420								GCM 605				
METHACRYLIC ACID BUTYL ESTER Synonym: BUTYL METHACRYLATE-n	2227/1993	26							58		BMN 148	156			187
METHACRYLIC ACID ETHYL ESTER Synonym: ETHYL METHACRYLATE	2277/1993	26	359		1633	551		-53	138		ETM 569	419	E3.0		496
METHACRYLIC ACID GLACIAL	1760	60										610			
METHACRYLIC ACID METHYL ESTER Synonym: METHYL METHACRYLATE	1247	26	575	20	2342	803	147	115-70	195	154	MMM 775	633	M3.6		773
METHACRYLIC ACID-alpha	2531	60							189						750
METHACRYLIC ACID-beta	2823	60							189						750
METHACRYLONITRILE	3079	28		20				-65	189		MET 755	611			750
METHALLYL ALCOHOL	2614/1987	26	546					-65	189			611			750
METHALLYL CHLORIDE	2554/1993	26						-65	189		MCL 739	611			

CHEMICAL OR MATERIAL NAME	UN/NA Number	Guide Number (DOT)	Firefighter's Hazardous Materials Reference Book	First Aid Manual for Chemical Accidents	Sax's Dangerous Properties of Industrial Materials	Hazardous Chemicals Desk Reference	Rapid Guide to Hazardous Chemicals in the Workplace	Fire Protection Guide on Hazardous Materials (NFPA)	Firefighter's Handbook of Hazardous Materials	Pocket Guide to Chemical Hazards (NIOSH)	Chemical Hazard Response Information System (CHRIS)	Emergency Handling of Hazardous Materials in Surface Transportation (AAR)	Emergency Action Guides (EAG)	Chemical Data Notebook: A User's Manual	Condensed Chemical Dictionary
METHAMYL CARBINOL	1993	27										612			
METHANAL Synonym: FORMALDEHYDE SOLUTION	1198/2209	29	401	17	1748	591	102	91 −54	150	116	FMS 590	462	F2.0	F 2	750
METHANAL SOLUTION Synonym: FORMALDEHYDE SOLUTION	1198/2209	29	401	17	1748	591	102	91 −54	150	116	FMS 590	462	F2.0	F 2	750
METHANAMINE Synonym: MONOMETHYLAMINE	1061	19	596						206				M6.0		796
METHANAMINE-n,n-DIMETHYL Synonym: TRIMETHYLAMINE	1083	19	853	26	3399	1146	221	178–	278		TMA1126	930	T6.0	T 7	1181
METHANE	1972	22		20	2213	760		−65	189			612		M 1	750
METHANE	1971	17	547	20	2213	760		−65	189		MTH 809			M 1	750
METHANE CARBOXYLIC ACID Synonym: ACETIC ACID	2789	29		11	15	6	1		189	30	AAC 1	3	A2.0	A 2	751
METHANE SULFONIC ACID					2213	760			189						751
METHANE TETRACHLORIDE Synonym: CARBON TETRACHLORIDE	1846	55	192	13	701	246	43	48 –	66	60	CBT 223	193	C2.3	C 5	221
METHANEARSONIC ACID SODIUM SALT	1557	53	548								MSA 799				
METHANETHIOL Synonym: METHYL MERCAPTAN	1064	13	574	20	2338	801	147	−70	189	154	MMC 774	633	M3.5		751
METHANETHIOMETHANE Synonym: DIMETHYL SULFIDE	1164	27	303					75 −44	189		DSL 478	363			751
METHANO INDANE	2762	28							190						
METHANOIC ACID Synonym: FORMIC ACID	1779	60	402	17	1750	592	102	92 −55	190	118	FMA 589	463			751
METHANOL	1987	26													751

CHEMICAL OR MATERIAL NAME	UN/NA Number	Guide Number (DOT)	Firefighter's Hazardous Materials Reference Book	First Aid Manual for Chemical Accidents	Sax's Dangerous Properties of Industrial Materials	Hazardous Chemicals Desk Reference	Rapid Guide to Hazardous Chemicals in the Workplace	Fire Protection Guide on Hazardous Materials (NFPA)	Firefighter's Handbook of Hazardous Materials	Pocket Guide to Chemical Hazards (NIOSH)	Chemical Hazard Response Information System (CHRIS)	Emergency Handling of Hazardous Materials in Surface Transportation (AAR)	Emergency Action Guides (EAG)	Chemical Data Notebook: A User's Manual	Condensed Chemical Dictionary
METHANOL Synonym: METHYL ALCOHOL	1230	28	549	20	2251	770	136	–66	190	144	MAL 722	613	M2.0	M 2	751
METHENAMINE	1328	32							190						751
METHENEAMINE Synonym: HEXAMETHYLENETETRAMINE	1328	32	434		1856	625			160		HMT 639				598
METHENYL TRICHLORIDE Synonym: CHLOROFORM	1888	55	205	14	815	282	49	52 –	76	68	CRF 301	223	C4.2	C 7	266
METHIOCARB Synonym: MERCAPTODIMETHUR	2757	55	530						190		MCD 736				751
METHMERCAPTURON Synonym: MERCAPTODIMETHUR	2757	55	530						185		MCD 736				
METHOGAS Synonym: METHYL BROMIDE	1062	55	559	20	2274	776	137	110–67	192	146	MTB 805	619	M3.0	M 3	759
METHOMYL					2216	761	133		190						752
METHOXY ANILINE-ortho	2431	55							190						752
METHOXY ANILINE-para	2431	55							190						752
METHOXY BUTYL ACETATE	2708	2							190						
METHOXY DDT Synonym: METHOXYCHLOR	2761/3077	55	550		2224	763	133		191	140	MOC 780	615			753
METHOXY ETHYL PHTHALATE					2230			–65	190						
METHOXYBENZALDEHYDE-ortho								–65	190						
METHOXYBENZENE	2222	26							190						752
METHOXYCARBONYL CHLORIDE Synonym: METHYL CHLOROFORMATE	1238	28	562	20					193		MHC 760	624	M3.1.1		762
METHOXYCHLOR	2761/3077	55	550		2224	763	133		191	140	MOC 780	615			753

CHEMICAL OR MATERIAL NAME	UN/NA Number	Guide Number (DOT)	Firefighter's Hazardous Materials Reference Book	First Aid Manual for Chemical Accidents	Sax's Dangerous Properties of Industrial Materials	Hazardous Chemicals Desk Reference	Rapid Guide to Hazardous Chemicals in the Workplace	Fire Protection Guide on Hazardous Materials (NFPA)	Firefighter's Handbook of Hazardous Materials	Pocket Guide to Chemical Hazards (NIOSH)	Chemical Hazard Response Information System (CHRIS)	Emergency Handling of Hazardous Materials in Surface Transportation (AAR)	Emergency Action Guides (EAG)	Chemical Data Notebook: A User's Manual	Condensed Chemical Dictionary
METHOXYCHLOROMETHANE Synonym: METHYL CHLOROMETHYL ETHER	1239	57						111–	193			643	M3.1.2		762
METHOXYDIHYDROPYRAN	1993	27										615			
METHOXYETHANE	1039	26							191						
METHOXYETHYLENE Synonym: VINYL METHYL ETHER	1087	17	875					–93	191		VME 177	951			753
METHOXYMETHYL ISOCYANATE	2605	57							191			616			
METHOXYMETHYLCHLORIDE Synonym: METHYL CHLOROMETHYL ETHER	1239	57						111–	193			643	M3.1.2		762
METHOXYMETHYLPENTANONE	2293	27							191						
METHYL 2-METHYL-2-PROPENOATE Synonym: METHYL METHACRYLATE	1247	26	575	20	2342	803	147	115–70	195	154	MMM 775	633	M3.6		773
METHYL 2-PROPENOATE Synonym: METHYL ACRYLATE	1919	26	553	20	2250	769	135	–66	191	142	MAM 723	618			755
METHYL ABIETATE					2246	767		–65							754
METHYL ACETALDEHYDE Synonym: PROPIONALDEHYDE	1275	26	720	23	2908	997		147–	243		PAD 923	791	P7.0		968
METHYL ACETATE	1231	26	551	20	2247	767	134	–65	191	142	MTT 813	617			754
METHYL ACETIC ESTER	1231	26							191						
METHYL ACETOACETATE	1993	27						–65	191		MAE 721	617			754
METHYL ACETONE	1232	26													754
METHYL ACETONE Synonym: METHYL ETHYL KETONE	1193	26	566	20	2319	793	143	–69	191		MEK 751	627	M3.2	M 5	754
METHYL ACETYLENE	1060	17			2248	768	134		191	142					755
METHYL ACETYLENE PROPADIENE MIXTURE	1060	17	552		2248	769	134		200	142	MAP 725	640			755

CHEMICAL OR MATERIAL NAME	UN/NA Number	Guide Number (DOT)	Firefighter's Hazardous Materials Reference Book	First Aid Manual for Chemical Accidents	Sax's Dangerous Properties of Industrial Materials	Hazardous Chemicals Desk Reference	Rapid Guide to Hazardous Chemicals in the Workplace	Fire Protection Guide on Hazardous Materials (NFPA)	Firefighter's Handbook of Hazardous Materials	Pocket Guide to Chemical Hazards (NIOSH)	Chemical Hazard Response Information System (CHRIS)	Emergency Handling of Hazardous Materials in Surface Transportation (AAR)	Emergency Action Guides (EAG)	Chemical Data Notebook: A User's Manual	Condensed Chemical Dictionary
METHYL ACROLEIN-beta Synonym: CROTONALDEHYDE	1143	28	233	14	964	325	56	57 -29	191	76	CTA 317	271	C7.1		755
METHYL ACRYLATE	1919	26	553	20	2250	769	135	-66	191	142	MAM 723	618			755
METHYL ACRYLIC ACID	2531	60							191						
METHYL ACRYLIC ACID-alpha Synonym: METHYL METHACRYLATE	1247	26	575	20	2342	803	147	115-70	195	154	MMM 775	633	M3.6		773
METHYL ALCOHOL Synonym: METHANOL	1230	28	554	20	2251	770	136	-66	191	144	MAL 722	613	M2.0	M 2	755
METHYL ALLYL CHLORIDE	2554/1993	26	555						191			618			
METHYL alpha-METHYLACRYLATE Synonym: METHYL METHACRYLATE	1247	26	575	20	2342	803	147	115-70	195	154	MMM 775	633	M3.6		773
METHYL ALUMINUM SESQUIBROMIDE	1926	40			2252				191						756
METHYL ALUMINUM SESQUICHLORIDE	1927	40			2253			-66	191						756
METHYL AMYL ACETATE	1233	26	556								MAC 719	642			757
METHYL AMYL ALCOHOL	2053	26	557						191		MAA 718				757
METHYL AMYL KETONE Synonym: AMYL METHYL KETONE-n	1110	26	558		2257	772	137	-66	192		AMK 53	76			757
METHYL BENZOATE	2938	31						-66	192		MBZ 733	619			758
METHYL BORATE	2416	26						-67	192						759
METHYL BROMIDE	1062	55	559	20	2274	776	137	110-67	192	146	MTB 805	619	M3.0	M 3	759
METHYL BROMIDE AND CHLOROPICRIN MIXTURE	1581	55							192			620			
METHYL BROMIDE AND ETHYLENE DIBROMIDE MIXTURE	1647	55							192			619			
METHYL BROMIDE AND NONFLAMMABLE COMPRESSED GAS MIXTURE	1955	15							192			620			
METHYL BROMOACETATE	2643/2810	58			2275	777			192			621			759

HAZARDOUS MATERIALS REFERENCE BOOKS INDEX

CHEMICAL OR MATERIAL NAME	UN/NA Number	Guide Number (DOT)	Firefighter's Hazardous Materials Reference Book	First Aid Manual for Chemical Accidents	Sax's Dangerous Properties of Industrial Materials	Hazardous Chemicals Desk Reference	Rapid Guide to Hazardous Chemicals in the Workplace	Fire Protection Guide on Hazardous Materials (NFPA)	Firefighter's Handbook of Hazardous Materials	Pocket Guide to Chemical Hazards (NIOSH)	Chemical Hazard Response Information System (CHRIS)	Emergency Handling of Hazardous Materials in Surface Transportation (AAR)	Emergency Action Guides (EAG)	Chemical Data Notebook: A User's Manual	Condensed Chemical Dictionary
METHYL BUTANONE	2397	26							192						
METHYL BUTENE	2460	26	560									621			
METHYL BUTYL ETHER	2398	26							192						
METHYL BUTYL KETONE Synonym: METHYL n-BUTYL KETONE	1993/1224	27	576	20				-67	200		MBK 729	622			760
METHYL BUTYLAMINE-n	2945	29			2277				192						760
METHYL BUTYRATE	1237	26			2279	779			192		MBU 732	522			761
METHYL CARBINOL	1170	26							192						
METHYL CARBITOL Synonym: DIETHYLENE GLYCOL MONOMETHYL ETHER					1221			-38	192		DGM 402				761
METHYL CARBITOL ACETATE					2283				192						761
METHYL CARBONATE	1161	26			2283	780		-67	192						
METHYL CARBONIMIDE Synonym: METHYL ISOCYANATE	2480	30	572	20	2335	799	141	115–70	195	152	MIS 767	631	M3.4	M 6	771
METHYL CELLOSOLVE Synonym: ETHYLENE GLYCOL MONOMETHYL ETHER	1188	26						-51	193	146	EME 544	433			762
METHYL CELLOSOLVE ACETATE	1189	26						-67	193	146					762
METHYL CHLORIDE	1063	18	561	20	2284	781	138	111–67	193	146	MTC 806	623	M3.1	M 4	762
METHYL CHLORIDE AND CHLOROPICRIN MIXTURE	1582	18							193						
METHYL CHLORIDE AND METHYLENE CHLORIDE MIXTURE	1912	22							193			623			
METHYL CHLOROACETATE	2295/1993	57			2285	781		-67	193		MED 750	623			762
METHYL CHLOROCARBONATE Synonym: METHYL CHLOROFORMATE	1238	28	562	20	2285	782			193		MHC 760	624	M3.1.1		762
METHYL CHLOROETHANOATE	2295	57							193						
METHYL CHLOROFORM	2831	74			2287	782	138		193	148					762

CHEMICAL OR MATERIAL NAME	UN/NA Number	Guide Number (DOT)	Firefighter's Hazardous Materials Reference Book	First Aid Manual for Chemical Accidents	Sax's Dangerous Properties of Industrial Materials	Hazardous Chemicals Desk Reference	Rapid Guide to Hazardous Chemicals in the Workplace	Fire Protection Guide on Hazardous Materials (NFPA)	Firefighter's Handbook of Hazardous Materials	Pocket Guide to Chemical Hazards (NIOSH)	Chemical Hazard Response Information System (CHRIS)	Emergency Handling of Hazardous Materials in Surface Transportation (AAR)	Emergency Action Guides (EAG)	Chemical Data Notebook: A User's Manual	Condensed Chemical Dictionary
METHYL CHLOROFORMATE	1238	28	562	20					193		MHC 760	624	M3.1.1		762
METHYL CHLOROMETHYL ETHER	1239	57						111–	193				M3.1.2		762
METHYL CHLOROMETHYL ETHER ANHYDROUS Synonym: CHLOROMETHYL METHYL ETHER	1239	57	207		832	286	49	111–	193	68	CME 265	643	M3.1.2		
METHYL CHLOROPROPIONATE	2933	29							193						
METHYL CHLOROSULFONATE					2289				193						763
METHYL CYANIDE Synonym: ACETONITRILE	1648	28	563	11	24	12	3	15 –12	193	32	ATN 104	625			763
METHYL CYCLOHEXANE	2296	27			2294	785	139		201	148	MCY 747	644			763
METHYL CYCLOHEXANOL	2617/1987	26			2295	785	139		193	148		625			764
METHYL CYCLOHEXANONE PEROXIDE	3046	52							193						
METHYL CYCLOHEXANONE-ortho	2297	26								148					764
METHYL CYCLOHEXYLAMINE	1760	60			2296				201			626			764
METHYL CYCLOHEXYLAMINE	2924/1993	29			2296				201			626			764
METHYL CYCLOPENTADIENE								–68	193		MCK 738				
METHYL CYCLOPENTANE	2298	26	564		2297	786		112–	193		MCP 743	626			764
METHYL DECAHYDRONAPHTHALENE	1993	27										626			
METHYL DEMETON					2298	786	140		193						765
METHYL DIBORANE					2301				193						
METHYL DICHLOROACETATE	2299	60							194		MDC 748	627			765
METHYL DICHLOROETHANOATE	2299	60							194						
METHYL DICHLOROSILANE	1242	29	565					113–	194		MCS 745	646	M3.1.3		765
METHYL ETHER Synonym: DIMETHYL ETHER	1033	22	301		2317	792		74 –68	194		DIM 416	361			767

HAZARDOUS MATERIALS REFERENCE BOOKS INDEX

CHEMICAL OR MATERIAL NAME	UN/NA Number	Guide Number (DOT)	Firefighter's Hazardous Materials Reference Book	First Aid Manual for Chemical Accidents	Sax's Dangerous Properties of Industrial Materials	Hazardous Chemicals Desk Reference	Rapid Guide to Hazardous Chemicals in the Workplace	Fire Protection Guide on Hazardous Materials (NFPA)	Firefighter's Handbook of Hazardous Materials	Pocket Guide to Chemical Hazards (NIOSH)	Chemical Hazard Response Information System (CHRIS)	Emergency Handling of Hazardous Materials in Surface Transportation (AAR)	Emergency Action Guides (EAG)	Chemical Data Notebook: A User's Manual	Condensed Chemical Dictionary
METHYL ETHYL ETHER	1039	26						-69	194						
METHYL ETHYL KETONE	1193	26	566	20	2319	793	143	-69	194		MEK 751	627	M3.2	M 5	768
METHYL ETHYL KETONE PEROXIDE	2550/9188	51			2319		143		194			627			768
METHYL ETHYL KETONE PEROXIDE	2563/9188	51							194			627			768
METHYL ETHYL KETONE PEROXIDE	3068/9188	48							194			627			768
METHYL ETHYL KETONE PEROXIDE	2127/9188	49			2319	794	143		194			627			768
METHYL ETHYL KETOXIME								-69	194						
METHYL ETHYL METHANE	1011	22							194						
METHYL ETHYL PYRIDINE	2300	60	567						194		MEP 753	648			
METHYL FLUORIDE	2454	22							194			628	M3.4.1		768
METHYL FORMAL	1234	26	568								MTF 808				
METHYL FORMAMIDE-n					2323	796									
METHYL FORMATE	1243	26	569		2323	796	144	114-69	194	150	MFM 757	628	M3.4.4		769
METHYL GLYCOL ACETATE	1189	26						-69	194						
METHYL GUTHION	2783	55							194						
METHYL HEPTINE CARBONATE								-69	194						
METHYL HEPTYL KETONE								-69	194		MHK 761				
METHYL HEXYL KETONE								-69	194						770
METHYL HYDRAZINE	1244	57			2327	797	144	114-	195	152	MHZ 762	648	M3.4.6		770
METHYL HYDRIDE	1971	17							195						770
METHYL HYDROXIDE Synonym: METHANOL	1230	28	549									613	M2.0		751
METHYL IODIDE	2644	55			2333	798	145		195	152	MIO 766				771
METHYL ISOAMYL KETONE	2302/1224	26		20	2334	798	145	-70	195			630			771

CHEMICAL OR MATERIAL NAME	UN/NA Number	Guide Number (DOT)	Firefighter's Hazardous Materials Reference Book	First Aid Manual for Chemical Accidents	Sax's Dangerous Properties of Industrial Materials	Hazardous Chemicals Desk Reference	Rapid Guide to Hazardous Chemicals in the Workplace	Fire Protection Guide on Hazardous Materials (NFPA)	Firefighter's Handbook of Hazardous Materials	Pocket Guide to Chemical Hazards (NIOSH)	Chemical Hazard Response Information System (CHRIS)	Emergency Handling of Hazardous Materials in Surface Transportation (AAR)	Emergency Action Guides (EAG)	Chemical Data Notebook: A User's Manual	Condensed Chemical Dictionary
METHYL ISOBUTENYL KETONE	1229	26	545		2205	757	132	107–64	188	140	MSO 801	606			771
Synonym: MESITYL OXIDE															
METHYL ISOBUTYL CARBINOL	2053/1993	26	570		2334	799	145	–70	195	152	MIC 763	630			771
METHYL ISOBUTYL KETONE	1245/1993	26	571	20				–70	195		MIK 765	630	M3.3		771
METHYL ISOBUTYL KETONE PEROXIDE	2126/3107	51							195			631			
METHYL ISOCYANATE	2480	30	572	20	2335	799	146	115–70	195	152	MIS 767	631	M3.4	M 6	771
METHYL ISOCYANIDE					2335				195						
METHYL ISOPROPENYL KETONE	1246	26	573		2336	799		–70	195		MPK 787	631			772
METHYL ISOPROPYL KETONE	2397/1993	26		20	2336	800	146		195			632			772
METHYL ISOTHIOCYANATE	2477	28							195		MIT 768	632			772
METHYL ISOVALERATE	2400/1993	27							195			649			
METHYL KETONE	1090	26	20	11	22	10	2	–11	17	30	ACT 26	4	A3.0	A 3	9
Synonym: ACETONE															
METHYL LACTATE								–70	195						772
METHYL MAGNESIUM BROMIDE IN ETHYL ETHER	1928	37			2338	800			195			632			
METHYL MERCAPTAN	1064	13	574	20	2338	801	147	–70	195	154	MMC 774	633	M3.5		773
METHYL MERCAPTOPROPIONALDEHYDE	2922/1993	59							195			649			
METHYL METHACRYLATE	1247	26	575	20	2342	803	147	115–70	195	154	MMM 775	633	M3.6		773
METHYL METHANOATE	1243	26	569		2323	796	144	114–69	195	150	MFM 757	628	M3.4.4		769
Synonym: METHYL FORMATE															
METHYL METHOXY PENTANONE	1993	27										634			
METHYL MUSTARD OIL	2477	28							195						
METHYL n-AMYL KETONE	1110	26			2257	772	137		195	144					757
METHYL n-BUTYL KETONE	1993	27	576	20					200		MBK 729				760

HAZARDOUS MATERIALS REFERENCE BOOKS INDEX

CHEMICAL OR MATERIAL NAME	UN/NA Number	Guide Number (DOT)	Firefighter's Hazardous Materials Reference Book	First Aid Manual for Chemical Accidents	Sax's Dangerous Properties of Industrial Materials	Hazardous Chemicals Desk Reference	Rapid Guide to Hazardous Chemicals in the Workplace	Fire Protection Guide on Hazardous Materials (NFPA)	Firefighter's Handbook of Hazardous Materials	Pocket Guide to Chemical Hazards (NIOSH)	Chemical Hazard Response Information System (CHRIS)	Emergency Handling of Hazardous Materials in Surface Transportation (AAR)	Emergency Action Guides (EAG)	Chemical Data Notebook: A User's Manual	Condensed Chemical Dictionary
METHYL NITRATE				20	2352	805			195						774
METHYL NITRITE	2455	17			2352	805			195						774
METHYL NITROBENZENE Synonym: NITROTOLUENE (-para)	1664	55		21	2591	882	166	130–75	216		NTT 861				774
METHYL NONYL KETONE								–70	195						775
METHYL NORBORNENE DICARBOXYLIC ANHYDRIDE	1760	60													775
METHYL o-SILICATE Synonym: METHYL ORTHOSILICATE	2606/1993	57							196			634	M3.6.1		
METHYL ORTHOSILICATE	2606/1993	57							196			634	M3.6.1		
METHYL OXIDE	1033	22						–70	196						776
METHYL OXIRANE Synonym: PROPYLENE OXIDE	1280	26	729	23	2922	1003	190	150–82	203	190	POX 971	798	P8.1		971
METHYL PARATHION	2783	55	577	20	2369	809	148		196		MPT 790	635			776
METHYL PARATHION MIXTURE	2783	55	577	20	2369	809			196		MPT 790	635			
METHYL PENTYL KETONE Synonym: AMYL METHYL KETONE (-n)	1110	26	81						31		AMK 53	76			
METHYL PHENOL Synonym: CRESOL	2076	55	232	14	959	324	55	57 –	196	74	CRS 307	270	C7.0		322
METHYL PHENYL CARBINYL ACETATE					2377			–71	196						
METHYL PHENYL KETONE Synonym: ACETOPHENONE	9207				26	13		–12	196		ACP 24				10
METHYL PHENYLACETATE								–71	203						777
METHYL PHOSPHINE					2381				196						
METHYL PHOSPHONIC DICHLORIDE	9206	39			2382	812			196			637			
METHYL PHOSPHONOTHIOIC DICHLORIDE	1760	60	578		2382						MPD 784				

CHEMICAL OR MATERIAL NAME	UN/NA Number	Guide Number (DOT)	Firefighter's Hazardous Materials Reference Book	First Aid Manual for Chemical Accidents	Sax's Dangerous Properties of Industrial Materials	Hazardous Chemicals Desk Reference	Rapid Guide to Hazardous Chemicals in the Workplace	Fire Protection Guide on Hazardous Materials (NFPA)	Firefighter's Handbook of Hazardous Materials	Pocket Guide to Chemical Hazards (NIOSH)	Chemical Hazard Response Information System (CHRIS)	Emergency Handling of Hazardous Materials in Surface Transportation (AAR)	Emergency Action Guides (EAG)	Chemical Data Notebook: A User's Manual	Condensed Chemical Dictionary
METHYL PHOSPHONOUS DICHLORIDE	2845	40			2383	812			196			638			
METHYL PHTHALYL ETHYL GLYCOLATE					2383			-71							
METHYL PIPERAZINE-n					2383	812			196						778
METHYL PROPENYL KETONE	1246	26							197						
METHYL PROPIONATE	1248	26			2389	814		-72	197			638			779
METHYL PROPYL ACETYLENE								-72	197						
METHYL PROPYL BENZENE	2046	27							197						779
METHYL PROPYL CARBINOL	1105	26						-72	197						779
METHYL PROPYL ETHER	2612	26							197						
METHYL PROPYL KETONE	1249	26						-72	197			638			779
METHYL SALICYLATE					2396	817		-72	197		MES 754				780
METHYL SILICATE Synonym: METHYL ORTHOSILICATE	2606/1993	57			2397	817	149		197			634	M3.6.1		780
METHYL STEARATE								-72							780
METHYL STYRENE	2618	27	579		2398	818	144		197						
METHYL STYRENE-alpha	2303	27	580		2398	818	149		197	154	MSR 802				780
METHYL SULFATE Synonym: DIMETHYL SULFATE	1595	57	302			472	80	75 -44	197	98	DSF 477	362	D3.1.1		781
METHYL SULFHYDRATE Synonym: METHYL MERCAPTAN	1064	13	574	20	2338	801	147	-70	195	154	MMC 774	633	M3.5		773
METHYL SULFIDE Synonym: DIMETHYL SULFIDE	1164	27	303					75 -44	197		DSL 478	363			781
METHYL SULFOXIDE Synonym: DIMETHYL SULFOXIDE					1417	472		-44	197		DMS 438				781
METHYL TETRAHYDROFURAN	2536/1993	26			2402				204			651			

HAZARDOUS MATERIALS REFERENCE BOOKS INDEX

CHEMICAL OR MATERIAL NAME	UN/NA Number	Guide Number (DOT)	Firefighter's Hazardous Materials Reference Book	First Aid Manual for Chemical Accidents	Sax's Dangerous Properties of Industrial Materials	Hazardous Chemicals Desk Reference	Rapid Guide to Hazardous Chemicals in the Workplace	Fire Protection Guide on Hazardous Materials (NFPA)	Firefighter's Handbook of Hazardous Materials	Pocket Guide to Chemical Hazards (NIOSH)	Chemical Hazard Response Information System (CHRIS)	Emergency Handling of Hazardous Materials in Surface Transportation (AAR)	Emergency Action Guides (EAG)	Chemical Data Notebook: A User's Manual	Condensed Chemical Dictionary
METHYL THIRAM Synonym: THIRAM	2771	55	822		3285	1118	213		268	212	THR1119	904			1146
METHYL TOLUENE SULFONATE								-72	197						
METHYL TRICHLOROACETATE	2533	53			2408				197			651			
METHYL TRICHLOROSILANE	1250	29	581		2408	819		117–72	197		MTS 812	651	M3.6.2		782
METHYL TRIETHOXYSILANE					2409				197						
METHYL TRINITROBENZENE									197						
METHYL TRITHION					2411	820			197						782
METHYL TUADS Synonym: THIRAM	2771	55	822		3285	1118	213		268	212	THR1119	904			1146
METHYL VALERALDEHYDE	2367	27			2412							652			
METHYL VINYL ETHER Synonym: VINYL METHYL ETHER	1087	17	875		2413	820		-93	197		VME1177	951			782
METHYL VINYL KETONE	1251	28	582		2414			117–73	198		MVK 814	639			782
METHYL-alpha-METHYLACRYLATE Synonym: METHYL METHACRYLATE	1247	26	575	20	2342	803	147	115–70	195	154	MMM 775	633	M3.6		773
METHYL-alpha-PYRROLIDONE-n Synonym: 1-METHYLPYRROLIDONE											MPY 791				779
METHYL-n-NITRO-n-NITROSOGUANIDINE-n	1325	32			2356	806			200			639			
METHYL-n-NITROSOANILINE-n					2360	807	143								
METHYL-tert-BUTYL ETHER	2398/1993	26			2278	779			200		MBE 728	640	M3.5		760
METHYLACETIC ACID Synonym: PROPIONIC ACID	1848	60	721		2909	997	188	148–81	191		PNA 968	791	P7.1		754
METHYLACETIC ANHYDRIDE Synonym: PROPIONIC ANHYDRIDE	2496	29	722		2909	998		148–81	243		PAH 924	792			968

CHEMICAL OR MATERIAL NAME	UN/NA Number	Guide Number (DOT)	Firefighter's Hazardous Materials Reference Book	First Aid Manual for Chemical Accidents	Sax's Dangerous Properties of Industrial Materials	Hazardous Chemicals Desk Reference	Rapid Guide to Hazardous Chemicals in the Workplace	Fire Protection Guide on Hazardous Materials (NFPA)	Firefighter's Handbook of Hazardous Materials	Pocket Guide to Chemical Hazards (NIOSH)	Chemical Hazard Response Information System (CHRIS)	Emergency Handling of Hazardous Materials in Surface Transportation (AAR)	Emergency Action Guides (EAG)	Chemical Data Notebook: A User's Manual	Condensed Chemical Dictionary
METHYLACETYLENE ALLENE MIXTURE Synonym: **METHYL ACETYLENE PROPADIENE MIXTURE**	1060	17	552		2248	769	134		200	142	MAP 725	640			755
METHYLACROLEIN-beta Synonym: **CROTONALDEHYDE**	1143	28	233	14	964	325	56	57 −29	191	76	CTA 317	271	C7.1		755
METHYLACRYLATE	1919	26		20	2250	769	135	109−	191	142	MAM 723	618			755
METHYLACRYLONITRILE					2251	770	135		200						
METHYLAL Synonym: **METHYL FORMAL**	1234	26	568		2251	770	135	−66	200	144	MTF 808	641			755
METHYLALDEHYDE Synonym: **FORMALDEHYDE SOLUTION**	1198/2209	29	401	17	1748	591	102	91 −54	150	116	FMS 590	462	F2.0	F 2	
METHYLALLENE	1010	17							191						
METHYLAMINE	1235	68		20	2253	771	136	−66	200	144	MSZ 803				756
METHYLAMINE	1061	19	583	20	2253	771	136	109−66	200	144	MTA 804	641			756
METHYLAMINOBENZENE-n Synonym: **METHYLANILINE (-n)**	2294	57	584		2257	772	137		201		MAN 724				757
METHYLAMYL ALCOHOL Synonym: **METHYL ISOBUTYL CARBINOL**	2053/1993	26	570		2334	799	145	−70	191	152	MAA 718	630			757
METHYLANILINE	2294	57	584		2257	772	137		201		MAN 724				757
METHYLANILINE (mono) Synonym: **METHYLANILINE (-n)**	2294	57	584		2257	772	137		201		MAN 724				
METHYLANILINE-n Synonym: **METHYLANILINE**	2294	57	584		2257	772	137		201		MAN 724				757
METHYLANILINE-ortho Synonym: **TOLUIDINE (-ortho)**	1708	55	830	25	3317	1129	216	171−87	201	216	TLI1124				1159
METHYLBENZENE Synonym: **TOLUENE**	1294	27	826	25	3309	1125	215	170−87	192	214	TOL1135	909	T4.0	T 4	758

HAZARDOUS MATERIALS REFERENCE BOOKS INDEX

CHEMICAL OR MATERIAL NAME	UN/NA Number	Guide Number (DOT)	Firefighter's Hazardous Materials Reference Book	First Aid Manual for Chemical Accidents	Sax's Dangerous Properties of Industrial Materials	Hazardous Chemicals Desk Reference	Rapid Guide to Hazardous Chemicals in the Workplace	Fire Protection Guide on Hazardous Materials (NFPA)	Firefighter's Handbook of Hazardous Materials	Pocket Guide to Chemical Hazards (NIOSH)	Chemical Hazard Response Information System (CHRIS)	Emergency Handling of Hazardous Materials in Surface Transportation (AAR)	Emergency Action Guides (EAG)	Chemical Data Notebook: A User's Manual	Condensed Chemical Dictionary
METHYLBENZENESULFONIC ACID Synonym: TOLUENESULFONIC ACID-para	2583	60			3314			–87			TAP1075				1158
METHYLBENZOL Synonym: TOLUENE	1294	27	826	25	3309	1125	215	170–87	269	214	TOL1135	909	T4.0	T 4	1157
METHYLBENZYL ALCOHOL	2937	55							201			642			
METHYLBIVINYL-beta Synonym: ISOPRENE	1218	27	479		2037	685		102–62	174		IPR 679	554	13.0	I 1	658
METHYLBUTENE	2460	26										621			
METHYLBUTYLAMINE (-n)	2945	29			2277			–67	192						760
METHYLCARBAMATE Synonym: CARBOFURAN	2757	55	184		2280	779	40		65		CBF 218	188			216
METHYLCARBINOL Synonym: ETHYL ALCOHOL	1170	26	344	16	1572	525	90	–48	192		EAL 504	408	E2.2	E 2	478
METHYLCARBYLAMINE Synonym: METHYL ISOCYANATE	2480	30	572	20	2335	799	146	115–70	195	152	MIS 767	631	M3.4	M 6	771
METHYLCHLOROFORM Synonym: TRICHLOROETHANE	2831	74		20	2287	782			193		TCE1082		T4.2		762
METHYLCHLOROSILANE	2534	29							193			644			763
METHYLCYCLOHEXANONE	2297	26			2295	786		–68	193			625			
METHYLCYCLOHEXYL ACETATE								–68	193						
METHYLCYCLOPENTADIENYLMANGANESE TRICARBONYL	2810	55							193		MCT 746	645			764
METHYLCYCLOPENTANE	2298	26			2297	786		–68	193		MCP 743	645			764
METHYLDICHLOROARSINE	1556/1955	55							194			645			765
METHYLDICHLOROSILANE	1242	29	565					113–	194		MCS 745	646	M3.1.3		765
METHYLDIETHANOLAMINE (-n)	1693	58						–68				646			765

CHEMICAL OR MATERIAL NAME	UN/NA Number	Guide Number (DOT)	Firefighter's Hazardous Materials Reference Book	First Aid Manual for Chemical Accidents	Sax's Dangerous Properties of Industrial Materials	Hazardous Chemicals Desk Reference	Rapid Guide to Hazardous Chemicals in the Workplace	Fire Protection Guide on Hazardous Materials (NFPA)	Firefighter's Handbook of Hazardous Materials	Pocket Guide to Chemical Hazards (NIOSH)	Chemical Hazard Response Information System (CHRIS)	Emergency Handling of Hazardous Materials in Surface Transportation (AAR)	Emergency Action Guides (EAG)	Chemical Data Notebook: A User's Manual	Condensed Chemical Dictionary
METHYLENE BIS(4-CYCLOHEXYLISOCYANATE)					2309	788	141		201						766
METHYLENE BIS(4-PHENYL ISOCYANATE)	2489	55							201						
METHYLENE BISPHENYL ISOCYANATE	2489	53			2310	789	141			150					
METHYLENE BROMIDE	2364/3082	74							201			646			766
METHYLENE CHLORIDE Synonym: DICHLOROMETHANE	1593	74	585	20	2311	790	142	113–68	201	150	DCM 364	335	D1.1.15		767
METHYLENE CHLOROBROMIDE	1887	58							201						767
METHYLENE DIANILINE-p-p'	2651	53							202						
METHYLENE DICHLORIDE Synonym: DICHLOROMETHANE	1593	74	277						104		DCM 364	335	D1.1.15		767
METHYLENE DIISOCYANATE								–68	202						
METHYLENE DIMETHYL ETHER Synonym: METHYL FORMAL	1234	26	568						202		MTF 808				
METHYLENE DIPHENYLENE DIISOCYANATE	2489	55							202						
METHYLENE OXIDE Synonym: FORMALDEHYDE SOLUTION	1198/2209	29	401	17	1748	591	102	91 –54	150	116	FMS 590	462	F2.0	F 2	
METHYLENEDIANILINE	2651	53			2312			–68	202						
METHYLEPHEDRINE-n (1R,2S)					2316										
METHYLETHANOLAMINE (-n)	1993	27						–68				647			767
METHYLETHYLCARBINOL Synonym: BUTYL ALCOHOL (-sec)	1120	26			593	197	31	–21	194	52	BAS 120		B7.1		767
METHYLETHYLENE Synonym: PROPYLENE	1077	22	727	23				149–82	202		PPL 977		P8.0.2	P 7	970
METHYLETHYLENE GLYCOL Synonym: PROPYLENE GLYCOL			728	23				–82	244		PPG 975				

HAZARDOUS MATERIALS REFERENCE BOOKS INDEX

CHEMICAL OR MATERIAL NAME	UN/NA Number	Guide Number (DOT)	Firefighter's Hazardous Materials Reference Book	First Aid Manual for Chemical Accidents	Sax's Dangerous Properties of Industrial Materials	Hazardous Chemicals Desk Reference	Rapid Guide to Hazardous Chemicals in the Workplace	Fire Protection Guide on Hazardous Materials (NFPA)	Firefighter's Handbook of Hazardous Materials	Pocket Guide to Chemical Hazards (NIOSH)	Chemical Hazard Response Information System (CHRIS)	Emergency Handling of Hazardous Materials in Surface Transportation (AAR)	Emergency Action Guides (EAG)	Chemical Data Notebook: A User's Manual	Condensed Chemical Dictionary
METHYLETHYLENE OXIDE Synonym: PROPYLENE OXIDE	1280	26	729	23	2922	1002	190	150–82	202	190	POX 971	798	P8.1		971
METHYLETHYLENEIMINE Synonym: PROPYLENEIMINE	1921	30	731		2921	1002	190		244	188	PII 959	799	P8.0.5		971
METHYLETHYLNITROSAMINE-n,n					2320		144								
METHYLETHYLPYRIDINE	2300	60							194		MEP 753	648			
METHYLFURAN	2301	26			2324	796			202			648			
METHYLHEPTENONE	2271	26			2326			–69	202						770
METHYLHEXANONE	2302	26							202						
METHYLHYDRAZINE	1244	57	586		2327	797	144	–69	195	152	MHZ 762	648			770
METHYLHYDRAZINE HYDROCHLORIDE					2328	797	145								
METHYLHYDROXYBENZENE-para Synonym: CRESOL (-para)	2076	55	232	14	961	325	55	57 –29	85	74	CSO 313	270	C7.0		322
METHYLISOBUTYLCARBINOL Synonym: METHYL AMYL ALCOHOL	2053	26	557		2334	799	145		195		MAA 718				771
METHYLISOBUTYLCARBINOL ACETATE	1233	26						–70	195						771
METHYLISOBUTYLCARBINYL ACETATE Synonym: METHYL AMYL ACETATE	1233	26	556								MAC 719	642			757
METHYLMERCURY				20	2340	802	147								
METHYLMETHANAMINE-n Synonym: DIMETHYLAMINE	1032	19	304	16	1324	442	77	73 –41	120	94	DMA 427	367	D3.0		411
METHYLMETHANE Synonym: ETHANE	1035	22	337	16	1553	517		–46	203		ETH 566	402			773
METHYLMORPHOLINE Synonym: METHYL MORPHOLINE-n	2535/2924	29			2349	804			203			650			773

CHEMICAL OR MATERIAL NAME	UN/NA Number	Guide Number (DOT)	Firefighter's Hazardous Materials Reference Book	First Aid Manual for Chemical Accidents	Sax's Dangerous Properties of Industrial Materials	Hazardous Chemicals Desk Reference	Rapid Guide to Hazardous Chemicals in the Workplace	Fire Protection Guide on Hazardous Materials (NFPA)	Firefighter's Handbook of Hazardous Materials	Pocket Guide to Chemical Hazards (NIOSH)	Chemical Hazard Response Information System (CHRIS)	Emergency Handling of Hazardous Materials in Surface Transportation (AAR)	Emergency Action Guides (EAG)	Chemical Data Notebook: A User's Manual	Condensed Chemical Dictionary
METHYLOL Synonym: METHANOL	1230	28	549						190			613	M2.0		751
METHYLOXIRANE Synonym: PROPYLENE OXIDE	1280	26	729	23	2922	1003	190	150–82	203	190	POX 971	798	P8.1		971
METHYLPENTADIENE	2461	26				810			196			636			776
METHYLPENTANE	2462	26							203			637			
METHYLPHENOL Synonym: CRESOL	2076	55	232	14	959	324	55	57 –	196	74	CRS 307	270	C7.0		322
METHYLPHENOL-meta Synonym: CRESOL (-meta)	2076	55	232	14	960	324	55	57 –29	203	74	CRL 303	270	C7.0		322
METHYLPHENOL-para Synonym: CRESOL (-para)	2076	55	232	14	961	325	55	57 –29	85	74	CSO 313	270	C7.0		322
METHYLPHENYL CARBINOL	2937	55						–71	196						777
METHYLPHENYLAMINE Synonym: METHYLANILINE (-n)	2294	57	584		2257	772	137		196		MAN 724				757
METHYLPHENYLDICHLOROSILANE	2437	29							203						778
METHYLPHENYLENE ISOCYANATE Synonym: TOLUENE DIISOCYANATE	2078	57	828		3312		215		269		TDI1095	910	T4.1	T 5	1157
METHYLPIPERIDINE (-n)	2399	26			2384	813			196						
METHYLPROPIONALDEHYDE-alpha Synonym: ISOBUTYRALDEHYDE	2045/1993	26	467	18	2022	679		–61	172		BAD 115	548	11.1		653
METHYLPROPYL ETHANDATE-beta Synonym: ISOBUTYL ACETATE	1213	26	463	18	2015	675	121	–61	171	130	IBA 661	544	11.0		652
METHYLPROPYL ETHANOATE-beta Synonym: ISOBUTYL ACETATE	1213	26	463	18	2015	675	121	–61	171	130	IBA 661	544	11.0		652

CHEMICAL OR MATERIAL NAME	UN/NA Number	Guide Number (DOT)	Firefighter's Hazardous Materials Reference Book	First Aid Manual for Chemical Accidents	Sax's Dangerous Properties of Industrial Materials	Hazardous Chemicals Desk Reference	Rapid Guide to Hazardous Chemicals in the Workplace	Fire Protection Guide on Hazardous Materials (NFPA)	Firefighter's Handbook of Hazardous Materials	Pocket Guide to Chemical Hazards (NIOSH)	Chemical Hazard Response Information System (CHRIS)	Emergency Handling of Hazardous Materials in Surface Transportation (AAR)	Emergency Action Guides (EAG)	Chemical Data Notebook: A User's Manual	Condensed Chemical Dictionary
METHYLPROPYLBENZENE Synonym: CYMENE-para	2046/1993	27	252		1011	340		−31	197		CMP 269	288			779
METHYLPROPYLCARBINYLAMINE								−72	197						
METHYLPYRIDINE-alpha Synonym: 2-METHYLPYRIDINE	2313	27			2391	815			204		MPR 789				
METHYLPYRROLE	1993	27						−72	197			420			779
METHYLPYRROLIDINE								−72							779
METHYLPYRROLIDINONE-n Synonym: 1-METHYLPYRROLIDONE							148				MPY 791				779
METHYLPYRROLIDONE (-n) Synonym: 1-METHYLPYRROLIDONE			587		2394	816	148				MPY 791				779
METHYLSTYRENE	2618	27	579			818		−72	197						
METHYLSTYRENE-para	1993	27													
METHYLSTYRENE-para Synonym: VINYL TOLUENE	2618/1993	27	876		2395	1176	229	182–94	286	224	VNT1179	952			1218
METHYLTHIOALCOHOL Synonym: METHYL MERCAPTAN	1064	13	574	20	2338	801	147	−70	195	154	MMC 774	633	M3.5		773
METHYLTHIOMETHANE	1164	27							204						
METHYLZINC Synonym: DIMETHYLZINC	1370	40	309		1425				123		DMZ 441	373			
METRAMINE	1328	32							204						
METRON Synonym: METHYL PARATHION	2783	55	577	20	2369	809	148		196		MPT 790	635			776
MEVINPHOS Synonym: PHOSDRIN	2783/3018	55	681	22	2419	822	149		204	180	PHD 955	652			783

CHEMICAL OR MATERIAL NAME	UN/NA Number	Guide Number (DOT)	Firefighter's Hazardous Materials Reference Book	First Aid Manual for Chemical Accidents	Sax's Dangerous Properties of Industrial Materials	Hazardous Chemicals Desk Reference	Rapid Guide to Hazardous Chemicals in the Workplace	Fire Protection Guide on Hazardous Materials (NFPA)	Firefighter's Handbook of Hazardous Materials	Pocket Guide to Chemical Hazards (NIOSH)	Chemical Hazard Response Information System (CHRIS)	Emergency Handling of Hazardous Materials in Surface Transportation (AAR)	Emergency Action Guides (EAG)	Chemical Data Notebook: A User's Manual	Condensed Chemical Dictionary
MEXACARBATE **Synonym: ZECTRAN**	1615										ZEC1205				783
MEXACARBATE	2757/2992	55							204			652			783
MH **Synonym: MALEIC HYDRAZIDE**			529								MLH 770				783
MIBC **Synonym: METHYL ISOBUTYL CARBINOL**	2053/1993	26	570		2334	799	145	−70	195	152	MIC 763	630			783
MIBK **Synonym: METHYL ISOBUTYL KETONE**	1245/1993	26	571	20				−70	195		MIK 765	630	M3.3		783
MIC **Synonym: METHYL ISOBUTYL CARBINOL**	2053	26	570			545	145	−70	195	152	MIC 763	6300			757
MIC **Synonym: METHYL ISOCYANATE**	2480	30	572	20	2335	799	146	115–70	195	152	MIS 767	631	M3.4	M 6	771
MIC **Synonym: METHYL AMYL ALCOHOL**	2053	26	557		2334				191		MAA 718				771
MICA					2431	822	150			154					783
MICHLER'S KETONE					2421	822	150								784
MICROLYSIN **Synonym: CHLOROPICRIN**	1580	56	210	14	865	296	50	54 –	77	70	CPL 290	229	C4.3		271
MIDDLE OIL **Synonym: CARBOLIC OIL**	2821	55	186								CBO 220				785
MIK **Synonym: METHYL ISOBUTYL KETONE**	1245/1993	26	571	20				−70	195		MIK 765	630	M3.3		771
MILD MERCURY CHLORIDE **Synonym: MERCUROUS CHLORIDE**	2025	53	541	19	2192	749			187		MRR 795				744

HAZARDOUS MATERIALS REFERENCE BOOKS INDEX

CHEMICAL OR MATERIAL NAME	UN/NA Number	Guide Number (DOT)	Firefighter's Hazardous Materials Reference Book	First Aid Manual for Chemical Accidents	Sax's Dangerous Properties of Industrial Materials	Hazardous Chemicals Desk Reference	Rapid Guide to Hazardous Chemicals in the Workplace	Fire Protection Guide on Hazardous Materials (NFPA)	Firefighter's Handbook of Hazardous Materials	Pocket Guide to Chemical Hazards (NIOSH)	Chemical Hazard Response Information System (CHRIS)	Emergency Handling of Hazardous Materials in Surface Transportation (AAR)	Emergency Action Guides (EAG)	Chemical Data Notebook: A User's Manual	Condensed Chemical Dictionary
MILK ACID	1760	60	496		2083	703					LTA 711	569			679
Synonym: LACTIC ACID															
MILK WHITE	1794	60			2105	716	128		179		LSF 709	575			692
Synonym: LEAD SULFATE															
MILLERS FUMIGRAIN	1093	30	34	11	68	27	6	19 -12	20	34	ACN 23	13	A5.1	A 7	19
Synonym: ACRYLONITRILE															
MINERAL CARBON	1361	32	195		739	258			68		CHC 244	203			248
Synonym: CHARCOAL															
MINERAL COLZA OIL	1270	27									OMS 889				
Synonym: OILS MISCELLANEOUS MINERAL SEAL															
MINERAL OIL					2426	824	150	-73							787
MINERAL PITCH	1999	27	105	12	307	97	17	-17	36		ASP 93	99	A12.0		787
Synonym: ASPHALT															
MINERAL SEAL OIL TYPICAL								-73							
MINERAL SPIRITS	1300/1993	27	588					-73	204		MNS 777	653			787
MINING REAGENT	2022	55			2431				204						
MIPAFOX	2783	55										654			787
MIRBANE OIL	1662	55	631	21	2518	861	158	124–74	205	162	NTB 852		N3.1		787
Synonym: NITROBENZENE															
MIREX	1615		589		1431	828			205		MRX 798				
MISCHMETAL	1333	32													787
MITIS GREEN	1585	53	222				55		83		CAA 203	257			310
Synonym: COPPER ACETOARSENITE															
MIXED ACID	1796	73	626						205			683	N1.0		788
Synonym: NITRATING ACID															
MIXED ACID	1826	60							205						788

CHEMICAL OR MATERIAL NAME	UN/NA Number	Guide Number (DOT)	Firefighter's Hazardous Materials Reference Book	First Aid Manual for Chemical Accidents	Sax's Dangerous Properties of Industrial Materials	Hazardous Chemicals Desk Reference	Rapid Guide to Hazardous Chemicals in the Workplace	Fire Protection Guide on Hazardous Materials (NFPA)	Firefighter's Handbook of Hazardous Materials	Pocket Guide to Chemical Hazards (NIOSH)	Chemical Hazard Response Information System (CHRIS)	Emergency Handling of Hazardous Materials in Surface Transportation (AAR)	Emergency Action Guides (EAG)	Chemical Data Notebook: A User's Manual	Condensed Chemical Dictionary
MIXED CHLOROSILANES	2924	29										655			
MIXED PRIMARY AMYL NITRATES Synonym: AMYL NITRATE (-n)	1112/1993	26	82		246	71		29 –15	31		ANT 67	76			75
MIXTURE OF BENZENE TOLUENE XYLENES Synonym: NAPHTHA COAL TAR	2553	27	602		2463		152	–73		158	NCT 826				
MMH Synonym: METHYL HYDRAZINE	1244	57	586		2327	797	144	114–69	195	152	MHZ 762	648	M3.4.6		789
MOHR'S SALT Synonym: FERROUS AMMONIUM SULFATE	9122/3077	31	392						147		FAS 579	449			790
MOLASSES ALCOHOL Synonym: ETHYL ALCOHOL	1170	26	344	16	1572	525	90	–48	136		EAL 504	408	E2.2	E 2	478
MOLECULAR CHLORINE Synonym: CHLORINE	1017	20	197	13	756	267	45	49 –	69	62	CLX 261	214	C4.0	C 6	259
MOLTEN ADIPIC ACID Synonym: ADIPIC ACID	9077/3077		35		79	28		–12	20		ADA 28	15	A6.0	A 8	24
MOLYBDENUM					2435	829	150			156					792
MOLYBDENUM PENTACHLORIDE	2508	60			2437	829	151		205			656			793
MOLYBDENUM TRIOXIDE Synonym: MOLYBDIC TRIOXIDE			590		2437	830	151		205		MTO 811				794
MOLYBDIC ACID (50%) Synonym: AMMONIUM MOLYBDATE					229	63					AMB 48				67
MOLYBDIC ANHYDRIDE Synonym: MOLYBDIC TRIOXIDE			590								MTO 811				
MOLYBDIC TRIOXIDE			590								MTO 811				
MONAZITE SAND Synonym: THORIUM ORE	2912	62	824									906	T2.0		

HAZARDOUS MATERIALS REFERENCE BOOKS INDEX

CHEMICAL OR MATERIAL NAME	UN/NA Number	Guide Number (DOT)	Firefighter's Hazardous Materials Reference Book	First Aid Manual for Chemical Accidents	Sax's Dangerous Properties of Industrial Materials	Hazardous Chemicals Desk Reference	Rapid Guide to Hazardous Chemicals in the Workplace	Fire Protection Guide on Hazardous Materials (NFPA)	Firefighter's Handbook of Hazardous Materials	Pocket Guide to Chemical Hazards (NIOSH)	Chemical Hazard Response Information System (CHRIS)	Emergency Handling of Hazardous Materials in Surface Transportation (AAR)	Emergency Action Guides (EAG)	Chemical Data Notebook: A User's Manual	Condensed Chemical Dictionary
MONDUR TDS Synonym: **TOLUENE-2,4-DIISOCYANATE**	2078	57	828		3313	1128	215	170–87	269	214	TDI1095	910	T4.1		1157
MONDUR TDS Synonym: **TOLUENE DIISOCYANATE**	2078	57	828		3312		215		269		TDI1095	910	T4.1	T 5	1157
MONO PE Synonym: **PENTAERYTHRITOL**			667		2682	912	172				PET 949				880
MONO-(TRICHLORO)-TETRAMONOPOTASSIUM DICHLORO-PENTA-S-	2468	45										656			
MONO-n-BUTYLAMINE Synonym: **BUTYLAMINE (-n)**	1125	68		13	594	197	32	41 –	205	52	BAM 118	169			182
MONO-n-PROPYLAMINE	1277	68							205						
MONO-sec-HEXYLAMINE	2379	27							205						
MONO-tert-BUTYL-META-CRESOL	1993	27										657			
MONOALLYLAMINE Synonym: **ALLYLAMINE**	2334	28			103	34	8	21 –13	22			40	A7.1		38
MONOAMMONIUM ORTHOPHOSPHATE Synonym: **AMMONIUM PHOSPHATE**					232	66					APP 78				68
MONOAMMONIUM SALT Synonym: **AMMONIUM BICARBONATE**	9081/3077		51		221	58			27		ABC 7	56			63
MONOBROMOBENZENE Synonym: **BROMOBENZENE**	2514	26	137					−19	51		BBZ 125	142			170
MONOBROMOMETHANE Synonym: **METHYL BROMIDE**	1062	55	559	20	2274	776	137	110–67	192	146	MTB 805	619	M3.0	M 3	759
MONOCHLORACETIC ACID Synonym: **CHLOROACETIC ACID**	1751	60	199	13	761	270		50 –26	72		MCA 734	216	C3.9		261
MONOCHLORATED ACETONE	1695	59							205						

CHEMICAL OR MATERIAL NAME	UN/NA Number	Guide Number (DOT)	Firefighter's Hazardous Materials Reference Book	First Aid Manual for Chemical Accidents	Sax's Dangerous Properties of Industrial Materials	Hazardous Chemicals Desk Reference	Rapid Guide to Hazardous Chemicals in the Workplace	Fire Protection Guide on Hazardous Materials (NFPA)	Firefighter's Handbook of Hazardous Materials	Pocket Guide to Chemical Hazards (NIOSH)	Chemical Hazard Response Information System (CHRIS)	Emergency Handling of Hazardous Materials in Surface Transportation (AAR)	Emergency Action Guides (EAG)	Chemical Data Notebook: A User's Manual	Condensed Chemical Dictionary
MONOCHLORETHANE Synonym: ETHYL CHLORIDE	1037	27	350		1586	532	92	82 -49	136	108	ECL 514	412	E2.3		483
MONOCHLORETHANOIC ACID ETHYL ESTER Synonym: ETHYL CHLOROACETATE	1181/1993	55	351					-49	136		ECA 510	412			483
MONOCHLORIDE Synonym: CHLOROSULFONIC ACID	1754	39	212					55 -			CSA 309	233	C4.5	C 8	273
MONOCHLOROACETIC ACID	1750	59	590						205						795
MONOCHLOROACETIC ACID Synonym: CHLOROACETIC ACID	1751	60	592	13	761	270		50 -26	72		MCA 734	216	C3.9		795
MONOCHLOROACETONE	1695	59							205						795
MONOCHLOROBENZENE Synonym: CHLOROBENZENE	1134	27	203	13	768	273	47	51 -26	205	64	CRB 298	219	C4.1		795
MONOCHLOROBROMOMETHANE	1887	58							205						
MONOCHLORODIFLUOROMETHANE Synonym: CHLORODIFLUOROMETHANE	1018	12	593	14	791	278	48		205		MCF 737	222	M4.0		795
MONOCHLORODIMETHYL ETHER Synonym: METHYL CHLOROMETHYL ETHER	1239	57						111-	193			643	M3.1.2		762
MONOCHLOROETHANE Synonym: ETHYL CHLORIDE	1037	27	350		1586	532	92	82 -49	136	108	ECL 514	412	E2.3		795
MONOCHLOROETHANOIC ACID Synonym: CHLOROACETIC ACID	1751	60	199	13	761	270		50 -26	72		MCA 734	216	C3.9		261
MONOCHLOROETHYLENE Synonym: VINYL CHLORIDE	1086	17	872	26	3495	1174	227	180-93	205	224	VCM1174	949	V1.1	V 1	1215
MONOCHLOROMETHANE Synonym: METHYL CHLORIDE	1063	18	561	20	2284	781	138	111-67	206	146	MTC 806	623	M3.1	M 4	762

HAZARDOUS MATERIALS REFERENCE BOOKS INDEX

CHEMICAL OR MATERIAL NAME	UN/NA Number	Guide Number (DOT)	Firefighter's Hazardous Materials Reference Book	First Aid Manual for Chemical Accidents	Sax's Dangerous Properties of Industrial Materials	Hazardous Chemicals Desk Reference	Rapid Guide to Hazardous Chemicals in the Workplace	Fire Protection Guide on Hazardous Materials (NFPA)	Firefighter's Handbook of Hazardous Materials	Pocket Guide to Chemical Hazards (NIOSH)	Chemical Hazard Response Information System (CHRIS)	Emergency Handling of Hazardous Materials in Surface Transportation (AAR)	Emergency Action Guides (EAG)	Chemical Data Notebook: A User's Manual	Condensed Chemical Dictionary
MONOCHLOROMETHYL ETHER Synonym: CHLOROMETHYL METHYL ETHER	1239	57	207		832	286	49		76	68	CME 265				
MONOCHLOROMONOBROMO METHANE	1887	58							206						
MONOCHLOROPENTAFLUOROETHANE	1020	12							206						795
MONOCHLOROTETRAFLUOROETHANE	1021	12							206		MTE 807				795
MONOCHLOROTRIFLUOROMETHANE	1022	12							206		MCM 740				795
MONOCROTOPHOS	2783	55			2440	830	151		206						795
MONOETHANOLAMINE	2491	60	594					118–	206		MEA 749		M5.0		795
MONOETHYLAMINE Synonym: ETHYLAMINE	1036	68	366	16	1573	526	90	119–48	206	106	EAM 505	427	E2.2.1		795
MONOETHYLENE GLYCOL Synonym: ETHYLENE GLYCOL			373	17	1604	541	94	–50	143		EGL 531		E2.4.7		487
MONOFLUOPHOSPHORIC ACID	1776	59							206						
MONOFLUOROETHYLENE Synonym: VINYL FLUORIDE	1860	17	874	26	3498	1175	228	–93	285		VFI1176	950			1217
MONOFLUOROPHOSPHORIC ACID	1776	59							206						
MONOGLYME Synonym: ETHYLENE GLYCOL DIMETHYL ETHER	2252	27						–51	143		EGD 528				796
MONOHYDROXYBENZENE Synonym: CARBOLIC ACID	1671	55	185						65			743	C2.0		217
MONOHYDROXYMETHANE Synonym: METHANOL	1230	28	549						190			613	M2.0		751
MONOISOBUTYLAMINE Synonym: ISOBUTYLAMINE	1214	68	465		2017	676	122	–61	172		IAM 659	547			652
MONOISOPROPANOLAMINE	1760	60	595						206		MPA 782	658			

CHEMICAL OR MATERIAL NAME	UN/NA Number	Guide Number (DOT)	Firefighter's Hazardous Materials Reference Book	First Aid Manual for Chemical Accidents	Sax's Dangerous Properties of Industrial Materials	Hazardous Chemicals Desk Reference	Rapid Guide to Hazardous Chemicals in the Workplace	Fire Protection Guide on Hazardous Materials (NFPA)	Firefighter's Handbook of Hazardous Materials	Pocket Guide to Chemical Hazards (NIOSH)	Chemical Hazard Response Information System (CHRIS)	Emergency Handling of Hazardous Materials in Surface Transportation (AAR)	Emergency Action Guides (EAG)	Chemical Data Notebook: A User's Manual	Condensed Chemical Dictionary
MONOISOPROPYLAMINE Synonym: ISOPROPYLAMINE	1221	68	487	18	2014	688	124	102–63	206	134	IPP 678	563			660
MONOMETHYL ANILINE	2294	57							206	156					
MONOMETHYLAMINE	1061	19	596						206				M6.0		796
MONOMETHYLAMINE Synonym: METHYLAMINE	1235	68		20	2253	771	136	109–66	206	144	MTA 804	658			796
MONOMETHYLAMINE ANHYDROUS Synonym: MONOMETHYLAMINE	1061	19	596						206				M6.0		796
MONOMETHYLHYDRAZINE Synonym: METHYL HYDRAZINE	1244	57	586	21	2327	797	144	114–69	206	152	MHZ 762	648	M3.4.6		770
MONONITROGEN MONOXIDE Synonym: NITRIC OXIDE	1660	20	628	21	2509	857	157		210	162	NTX 862	685	N2.5	N 2	823
MONOSODIUM METHANE ARSONATE Synonym: METHANEARSONIC ACID SODIUM SALT	1557	53	548								MSA 799				
MONOSODIUM METHYL ARSONATE Synonym: METHANEARSONIC ACID SODIUM SALT	1557	53	548		2443	832					MSA 799				
MONOVINYL CHLORIDE Synonym: VINYL CHLORIDE	1086	17	872	26	3495	1174	227	180–93	285	224	VCM1174	949	V1.1	V 1	1215
MONOXIDE Synonym: CARBON MONOXIDE	1016	18	191	13	700	245	42	48 –25	66	60	CMO 268	192	C2.2.1	C 4	221
MORICID Synonym: FORMALDEHYDE SOLUTION	1198/2209	29	401	17	1748	591	102	91 –54	150	116	FMS 590	462	F2.0	F 2	
MORPHINE					2446	832									797
MORPHINE SULFATE					2447	833									
MORPHOLINE	2054	29	597		2447	833	151	120–73	206	156	MPL 788	659			798
MORPHOLINE	1760	60			2447	833	151	–73	207			660			798

HAZARDOUS MATERIALS REFERENCE BOOKS INDEX

CHEMICAL OR MATERIAL NAME	UN/NA Number	Guide Number (DOT)	Firefighter's Hazardous Materials Reference Book	First Aid Manual for Chemical Accidents	Sax's Dangerous Properties of Industrial Materials	Hazardous Chemicals Desk Reference	Rapid Guide to Hazardous Chemicals in the Workplace	Fire Protection Guide on Hazardous Materials (NFPA)	Firefighter's Handbook of Hazardous Materials	Pocket Guide to Chemical Hazards (NIOSH)	Chemical Hazard Response Information System (CHRIS)	Emergency Handling of Hazardous Materials in Surface Transportation (AAR)	Emergency Action Guides (EAG)	Chemical Data Notebook: A User's Manual	Condensed Chemical Dictionary
MORPHOTHION					2450	833			207						798
MORTOPAL Synonym: TETRAETHYL LEAD	1705	15		25	3219	1100					TEP 927				1130
MOSS GREEN Synonym: COPPER ACETOARSENITE	1585	53	222				55		83		CAA 203	257			310
MOTOR FUEL Synonym: GASOLINE	1203	27	412	18	1779	601	104	−56	207		GAK 602	473	G1.0	G 1	554
MOTOR FUEL ANTI KNOCK COMPOUND	1649	56	598					121–	207		MFA 756	660	M7.0		
MOTOR FUEL ANTI KNOCK MIXTURE	1649	56									MFA 756		M7.0		
MOTOR FUEL CONSISTING OF ALCOHOL CASTER OIL NI-TROMETHANE	1993	27													
MOTOR OIL Synonym: OILS MISCELLANEOUS LUBRICATING	1270	27									OLB 884				
MOTOR SPIRIT Synonym: GASOLINES AUTOMOTIVE	1203	27		18	1779	601	104		207		GAT 603		G1.0	G 1	554
MOTOR SPIRIT Synonym: GASOLINE	1203	27	412	18	1779	601	104	−56	207		GAK 602	473	G1.0	G 1	
MOUSE TOX Synonym: STRYCHNINE	1692	53	794		3143	1074	202		257	198	STR1063	862			1096
MPT Synonym: METHYL PARATHION	2783	55	577	20	2369	809	148		196		MPT 790	635			776
MPTD Synonym: METHYL PHOSPHONOTHIOIC DICHLORIDE	1760	60	578		2382						MPD 784				
MSMA Synonym: METHANEARSONIC ACID SODIUM SALT	1557	53	548								MSA 799				

CHEMICAL OR MATERIAL NAME	UN/NA Number	Guide Number (DOT)	Firefighter's Hazardous Materials Reference Book	First Aid Manual for Chemical Accidents	Sax's Dangerous Properties of Industrial Materials	Hazardous Chemicals Desk Reference	Rapid Guide to Hazardous Chemicals in the Workplace	Fire Protection Guide on Hazardous Materials (NFPA)	Firefighter's Handbook of Hazardous Materials	Pocket Guide to Chemical Hazards (NIOSH)	Chemical Hazard Response Information System (CHRIS)	Emergency Handling of Hazardous Materials in Surface Transportation (AAR)	Emergency Action Guides (EAG)	Chemical Data Notebook: A User's Manual	Condensed Chemical Dictionary
MULTRATHANE M **Synonym: DIPHENYLMETHANE DIISOCYANATE**	2489	55	321						127		DPM 467				
MURIATIC ACID **Synonym: HYDROCHLORIC ACID**	1789	60	441	18	1892	639	113		163		HCL 625	524	H4.0		800
MURIATIC ACID **Synonym: HYDROCHLORIC ACID SOLUTION**	1789	60	441	18	1892	639	113		163		HCL 625	524	H4.0		800
MURIATIC ETHER **Synonym: ETHYL CHLORIDE**	1037	27	350		1586	532	92	82 −73	207	108	ECL 514	412	E2.3		483
MUSK XYLENE	2956	32							207						
MUSTARD OIL	1955	15						−73							801
MVC **Synonym: VINYL CHLORIDE**	1086	17	872	26	3495	1174	227	180–93	285	224	VCM1174	949	V1.1	V 1	1215
MYRISTIC ALCOHOL **Synonym: TETRADECANOL**			809		3215	1100		−85			TTN1156				1128
MYRISTYL ALCOHOL **Synonym: TETRADECANOL**			809		3215	1100		−85			TTN1156				802
n-(2-AMINOETHYL) ETHANOLAMINE (-n) **Synonym: AMINOETHYLETHANOLAMINE**	1760	60	47		172				26		AEA 30	50			57
n-b-HYDROXYETHYLETHYLENEDIAMINE (-n) **Synonym: AMINOETHYLETHANOLAMINE**	1760	60	47		172				26		AEA 30	50			57
NABAM	1609		599						207		NAB 816				803
NACCONATE 100 **Synonym: TOLUENE DIISOCYANATE**	2078	57	828		3312		215		269		TDI1095	910	T4.1	T 5	1157
NACCONATE 100 **Synonym: TOLUENE-2,4-DIISOCYANATE**	2078	57	828		3313	1128	215	170–87	269	214	TDI1095	910	T4.1		1157

CHEMICAL OR MATERIAL NAME	UN/NA Number	Guide Number (DOT)	Firefighter's Hazardous Materials Reference Book	First Aid Manual for Chemical Accidents	Sax's Dangerous Properties of Industrial Materials	Hazardous Chemicals Desk Reference	Rapid Guide to Hazardous Chemicals in the Workplace	Fire Protection Guide on Hazardous Materials (NFPA)	Firefighter's Handbook of Hazardous Materials	Pocket Guide to Chemical Hazards (NIOSH)	Chemical Hazard Response Information System (CHRIS)	Emergency Handling of Hazardous Materials in Surface Transportation (AAR)	Emergency Action Guides (EAG)	Chemical Data Notebook: A User's Manual	Condensed Chemical Dictionary
NACCONATE 300	2489	55	321						127		DPM 467				
Synonym: DIPHENYLMETHANE DIISOCYANATE															
NACCONOL 988 A	2584	60	330						130		DSA 474	390	D5.0		441
Synonym: DODECYLBENZENESULFONIC ACID															
NADONE	1915	26	246	14	991	334	59	-30	88	78	CCH 228	281	C8.2		337
Synonym: CYCLOHEXANONE															
NAK	1422	40							207						803
NALED	2783/3082	55	600		2461	835	152		207		NLD 838	673			804
NAPHTHA	2553	27	602		2463	835	152		207	158	NCT 826	673			804
NAPHTHA	1300/1993	27	588					-73	204		MNS 777	653			804
Synonym: MINERAL SPIRITS															
NAPHTHA	1256	27	603		2463	835	152		207		NSV 850				804
NAPHTHA	1255	27	676		2463	835	152		207		PTN 999	739	P2.0		804
Synonym: PETROLEUM NAPHTHA															
NAPHTHA COAL TAR	2553	27	602		2463		152	-73		158	NCT 826				804
NAPHTHA DISTILLATE	1268	27	604		2463				207						
NAPHTHA PETROLEUM	1255	27			2463										
NAPHTHA SOLVENT	1256	27	603		2463	835	152	-73	207		NSV 850	673			
NAPHTHA STODDARD SOLVENT	1268	27	604								NSS 849				
NAPHTHA V,M & P	1255	27	605		2463	835	152	-73	207		NVM 863				
NAPHTHACENE									207						805
NAPHTHALENE	2304	32		21	2464	835	152	-73	207		NTM 857	675	N0.5.5		805
NAPHTHALENE	1334	32	606	21	2464	835	152	121-	207	158			N0.5.5		805
NAPHTHALIN	2304	32		21	2464	835	152	-73	207		NTM 857	675	N0.5.5		805
Synonym: NAPHTHALENE															

CHEMICAL OR MATERIAL NAME	UN/NA Number	Guide Number (DOT)	Firefighter's Hazardous Materials Reference Book	First Aid Manual for Chemical Accidents	Sax's Dangerous Properties of Industrial Materials	Hazardous Chemicals Desk Reference	Rapid Guide to Hazardous Chemicals in the Workplace	Fire Protection Guide on Hazardous Materials (NFPA)	Firefighter's Handbook of Hazardous Materials	Pocket Guide to Chemical Hazards (NIOSH)	Chemical Hazard Response Information System (CHRIS)	Emergency Handling of Hazardous Materials in Surface Transportation (AAR)	Emergency Action Guides (EAG)	Chemical Data Notebook: A User's Manual	Condensed Chemical Dictionary
NAPHTHANE Synonym: DECAHYDRONAPHTHALENE	1147	27	257		1027	345		-31	207		DHN 405	289			349
NAPHTHENE	1145	26													806
NAPHTHENIC ACID	9137/3082		607		2466	837					NTI 855	673			806
NAPHTHOL-beta								-73	208						806
NAPHTHYL METHYL CARBAMATE	2811	53										674			
NAPHTHYLAMINE	2077	55	608	21	2471				208						
NAPHTHYLAMINE	1550/2811	55		21	2471				208			674			
NAPHTHYLAMINE-alpha Synonym: 1-NAPHTHYLAMINE	2077	55			2471	838	154	-73	208	158	NAO 820				807
NAPHTHYLAMINE-beta	1650	55			2471	839	153		208	158					807
NAPHTHYLTHIOUREA	1651	53							208			674			
NAPHTHYLUREA	1652	53							208						
NAPTHALANE Synonym: DECAHYDRONAPHTHALENE	1147	27	257		1027	345		-31	90		DHN 405	289			349
NARCOTILE Synonym: ETHYL CHLORIDE	1037	27	350		1586	532	92	82 -49	136	108	ECL 514	412	E2.3		483
NATRIUM Synonym: SODIUM	1428	40	752	24	3056	1042		155-	208		SDU1036	844	S2.0	S 1	809
NATURAL GAS	1972	22	609					122-55	208			612		M 1	810
NATURAL GAS Synonym: METHANE	1971	17	547	20	2213	760		-73	208		MTH 809			M 1	810
NATURAL GASOLINE Synonym: GASOLINE CASINGHEAD	1257	27	414	18	1779	601		-56	153		GCS 607	675		G 1	810
NATURAL GASOLINE Synonym: GASOLINE	1203	27	412	18	1779	601	104	-56	208		GAK 602	473	G1.0	G 1	810

CHEMICAL OR MATERIAL NAME	UN/NA Number	Guide Number (DOT)	Firefighter's Hazardous Materials Reference Book	First Aid Manual for Chemical Accidents	Sax's Dangerous Properties of Industrial Materials	Hazardous Chemicals Desk Reference	Rapid Guide to Hazardous Chemicals in the Workplace	Fire Protection Guide on Hazardous Materials (NFPA)	Firefighter's Handbook of Hazardous Materials	Pocket Guide to Chemical Hazards (NIOSH)	Chemical Hazard Response Information System (CHRIS)	Emergency Handling of Hazardous Materials in Surface Transportation (AAR)	Emergency Action Guides (EAG)	Chemical Data Notebook: A User's Manual	Condensed Chemical Dictionary
NAUGATUCK DO 14 Synonym: PROPARGITE	2765/3082	55	718						242		PRG 986	790			967
NAVY SPECIAL FUEL OIL Synonym: FUEL OIL NO. 5	1993	27	406					-55	151			464	F3.3		
NECATORINA Synonym: CARBON TETRACHLORIDE	1846	55	192	13	701	246	43	48 –	66	60	CBT 223	193	C2.3	C 5	221
NECATORINE Synonym: CARBON TETRACHLORIDE	1846	55	192	13	701	246	43	48 –	66	60	CBT 223	193	C2.3	C 5	221
NEO PENTANOIC ACID	1993	27										675			813
NEOHEXANE	1208	27	610						209		NHX 830	676			812
NEON	1913	21			2481	842			209			676			813
NEON	1065	12			2481	842			209			676			813
NEOPENTANE	2044	22			2481	842	154		209						813
NEOPENTANE	1265	27			2481	842	154								813
NEOPRENE	1991	30			2482	843	154		209						813
NEOSOL Synonym: ETHYL ALCOHOL	1170	26	344	16	1572	525	90	-48	136		EAL 504	408	E2.2	E 2	813
NEPHIS Synonym: ETHYLENE DIBROMIDE	1605	55	371					85 –	143	110	EDB 518	430	E2.4.1		487
NEUTRAL AMMONIUM CHROMATE Synonym: AMMONIUM CHROMATE	9086/3077		57		224	61			28		ACH 19	58			65
NEUTRAL AMMONIUM FLUORIDE Synonym: AMMONIUM FLUORIDE	2505	54	61		226	61		26 –	28		AFR 35	59			65
NEUTRAL ANHYDROUS CALCIUM HYPOCHLORITE Synonym: CALCIUM HYPOCHLORITE	1748	45	172	13				46 –	63		CHY 252	182	C1.1	C 2	204

CHEMICAL OR MATERIAL NAME	UN/NA Number	Guide Number (DOT)	Firefighter's Hazardous Materials Reference Book	First Aid Manual for Chemical Accidents	Sax's Dangerous Properties of Industrial Materials	Hazardous Chemicals Desk Reference	Rapid Guide to Hazardous Chemicals in the Workplace	Fire Protection Guide on Hazardous Materials (NFPA)	Firefighter's Handbook of Hazardous Materials	Pocket Guide to Chemical Hazards (NIOSH)	Chemical Hazard Response Information System (CHRIS)	Emergency Handling of Hazardous Materials in Surface Transportation (AAR)	Emergency Action Guides (EAG)	Chemical Data Notebook: A User's Manual	Condensed Chemical Dictionary
NEUTRAL LEAD ACETATE Synonym: LEAD ACETATE	1616	53	501	19	2094	708	128		178		LAC 690	570			686
NEUTRAL LEAD STEARATE Synonym: LEAD STEARATE	2811/3077	53	508	19	2105		128		179		LSA 708	575			691
NEUTRAL NICOTINE SULFATE Synonym: NICOTINE SULFATE	1658	55	623		2501	854			210		NCS 825	681			
NEUTRAL POTASSIUM CHROMATE Synonym: POTASSIUM CHROMATE	9 42/3077		706	23	2863	980			238		PCH 936	778			951
NEUTRAL VERDIGRIS Synonym: COPPER ACETATE	9 06/3077		221		944	319	55				COP 279	257			310
NIA 12 40 Synonym: ETHION	2783	55	339		1558		88		134		ETO 571	404			476
NIAGARA 10,242 Synonym: CARBOFURAN	2757	55	184		695	242	40		65		CBF 218	188			216
NIALATE Synonym: ETHION	2783	55	339		1558	520	88		134		ETO 571	404			476
NIAX CATALYST ESN					2488	845			209						
NICKEL (II) FLUOBORATE Synonym: NICKEL FLUOROBORATE			617		2493	848	156		209		NFB 828				
NICKEL ACETATE			611		2489	846	156				NKA 834				817
NICKEL ACETATE TETRAHYDRATE Synonym: NICKEL ACETATE			611		2490	846	156				NKA 834				817
NICKEL AMMONIUM SULFATE	9138/3077	31	612		2490		156		209		NAS 821	676			818
NICKEL AMMONIUM SULFATE HEXAHYDRATE Synonym: NICKEL AMMONIUM SULFATE	9138/3077		612		2490		156		209		NAS 821	676			818
NICKEL AND COMPOUNDS	1325	32		21	2492	845	155		209	160		679			817

CHEMICAL OR MATERIAL NAME	UN/NA Number	Guide Number (DOT)	Firefighter's Hazardous Materials Reference Book	First Aid Manual for Chemical Accidents	Sax's Dangerous Properties of Industrial Materials	Hazardous Chemicals Desk Reference	Rapid Guide to Hazardous Chemicals in the Workplace	Fire Protection Guide on Hazardous Materials (NFPA)	Firefighter's Handbook of Hazardous Materials	Pocket Guide to Chemical Hazards (NIOSH)	Chemical Hazard Response Information System (CHRIS)	Emergency Handling of Hazardous Materials in Surface Transportation (AAR)	Emergency Action Guides (EAG)	Chemical Data Notebook: A User's Manual	Condensed Chemical Dictionary
NICKEL BROMIDE			613				156				NBR 822				818
NICKEL BROMIDE TRIHYDRATE Synonym: NICKEL BROMIDE			613				156				NBR 822				818
NICKEL CARBONYL	1259	57	614	21	2491	846	155	122-74	209	160	NKC 835	677	N0.5		818
NICKEL CATALYST	1378/1325	32					156		209			677			
NICKEL CATALYST	2881/1325	37					156	123-	209			677			
NICKEL CHLORIDE	1378/9188	32					156				NCL 823	678			818
NICKEL CHLORIDE HEXAHYDRATE Synonym: NICKEL CHLORIDE	1378/9188	32			2491	847	156				NCL 823	678			818
NICKEL CYANIDE	1653	53	616		2492	848	156		209		NCN 824	678			818
NICKEL DIHYDROXIDE Synonym: NICKEL HYDROXIDE	9140/3077	31	619		2494	849	156		209		NKH 836	680			818
NICKEL FLUOROBORATE			617		2493		156		209		NFB 828				
NICKEL FLUOROBORATE SOLUTION Synonym: NICKEL FLUOROBORATE			617		2493		156		209		NFB 828				
NICKEL FORMATE			618				156				NFM 829				818
NICKEL FORMATE DIHYDRATE Synonym: NICKEL FORMATE			618				156				NFM 829				818
NICKEL HYDROXIDE	9140/3077	31	619		2494	849	156		209		NKH 836	680			818
NICKEL IRON SCRAP	2793	32													
NICKEL NITRATE	2725	35	620		2495	850	156		209		NNT 843				819
NICKEL NITRATE HEXAHYDRATE Synonym: NICKEL NITRATE	2725	35	620		2495	850	156		209		NNT 843				819
NICKEL NITRITE	2726	35			2495		156		209			678			
NICKEL PLATING SOLUTION	1760	60										678			

CHEMICAL OR MATERIAL NAME	UN/NA Number	Guide Number (DOT)	Firefighter's Hazardous Materials Reference Book	First Aid Manual for Chemical Accidents	Sax's Dangerous Properties of Industrial Materials	Hazardous Chemicals Desk Reference	Rapid Guide to Hazardous Chemicals in the Workplace	Fire Protection Guide on Hazardous Materials (NFPA)	Firefighter's Handbook of Hazardous Materials	Pocket Guide to Chemical Hazards (NIOSH)	Chemical Hazard Response Information System (CHRIS)	Emergency Handling of Hazardous Materials in Surface Transportation (AAR)	Emergency Action Guides (EAG)	Chemical Data Notebook: A User's Manual	Condensed Chemical Dictionary
NICKEL PLATING SOLUTION	2810	55										679			
NICKEL SULFATE	9141/9188	58	621		2498	852	156		209		NKS 837	679			820
NICKEL SULFATE COPPER REFINERY CRUDE	9188						156								
NICKEL TETRACARBONYL Synonym: NICKEL CARBONYL	1259	57	614	21	2491	846	155	122–74	209	160	NKC 835	677	N0.5		820
NICKELOUS ACETATE Synonym: NICKEL ACETATE			611		2489	846	156				NKA 834				817
NICKELOUS HYDROXIDE Synonym: NICKEL HYDROXIDE	9140/3077	31	619		2494	849	156		209		NKH 836	680			819
NICKELOUS SULFATE Synonym: NICKEL SULFATE	9141/9188	58	621		2498	852	156		209		NKS 837	679			820
NICOTINE	1654	55	622		2500	853	157	-74	209	160	NIC 831	682			820
NICOTINE COMPOUND	1555/3144	55							209			680			
NICOTINE HYDROCHLORIDE	1656	55			2501	854			209			680			
NICOTINE SALICYLATE	1657	53							209			680			
NICOTINE SULFATE	1658	55	623		2501	854			210		NCS 825	681			
NICOTINE TARTRATE	1659	53			2501	854			210			681			
NIOBE OIL	2938	31						-74	210						
NITAL Synonym: NITRIC ACID FUMING	2032	44		21	2509	856		123-	210	160		684	N2.0	N 1	822
NITER Synonym: POTASSIUM NITRATE	1486	35	713		2874	985			210			782	P6.1	P 6	822
NITRALIN	1609		625								NTL 856				822
NITRAM Synonym: AMMONIUM NITRATE	1942	43	63		229	64		26 -	29		AMN 56	62	A8.1	A11	67

CHEMICAL OR MATERIAL NAME	UN/NA Number	Guide Number (DOT)	Firefighter's Hazardous Materials Reference Book	First Aid Manual for Chemical Accidents	Sax's Dangerous Properties of Industrial Materials	Hazardous Chemicals Desk Reference	Rapid Guide to Hazardous Chemicals In the Workplace	Fire Protection Guide on Hazardous Materials (NFPA)	Firefighter's Handbook of Hazardous Materials	Pocket Guide to Chemical Hazards (NIOSH)	Chemical Hazard Response Information System (CHRIS)	Emergency Handling of Hazardous Materials In Surface Transportation (AAR)	Emergency Action Guides (EAG)	Chemical Data Notebook: A User's Manual	Condensed Chemical Dictionary
NITRAN **Synonym: METHYL PARATHION**	2783	55	577	20	2369	809	148		196		MPT 790	635			776
NITRANILINE-ortho **Synonym: 2-NITROANILINE**	1661	55							210		NTA 851	687			824
NITRATE	1477	35			2507	855			210			682			
NITRATE OF SODIUM AND POTASH	1478	35							210						
NITRATINE **Synonym: SODIUM NITRATE**	1498	35	780		3097	1060			210		SDN1033	845	S2.6	S 5	1063
NITRATING ACID	1796	73	626						210			683	N1.0		822
NITRATING ACID MIXTURE	1826	60							210			682			
NITRE **Synonym: POTASSIUM NITRATE**	1486	35	713		2874	985			210			782	P6.1	P 6	955
NITREX NITROGEN SOLUTIONS **Synonym: AMMONIUM NITRATE UREA SOLUTION**											ANU 68				
NITRIC ACID **Synonym: AMMONIUM NITRATE**	1942	43	63	21	229	64		26 –	29		AMN 56	62	A8.1	A11	823
NITRIC ACID	1760	60		21	2507		157	123–	210	160				N 1	823
NITRIC ACID	2031	44	627	21	2507	856	157	123–	210	160	NAC 817	684		N 1	823
NITRIC ACID **Synonym: ALUMINUM NITRATE**	1438	35	44	21	130	42	9		24		ALN 44	45			823
NITRIC ACID FUMING	2032	44		21	2509	856		123–	210	160		684	N2.0	N 1	823
NITRIC ACID IRON II SALT **Synonym: FERRIC NITRATE**	1466	35	390						147		FNT 591	447			513
NITRIC ACID POTASSIUM SALT **Synonym: POTASSIUM NITRATE**	1486	35	713		2874	985			239			782	P6.1	P 6	955
NITRIC ETHER	1993	27							210						

CHEMICAL OR MATERIAL NAME	UN/NA Number	Guide Number (DOT)	Firefighter's Hazardous Materials Reference Book	First Aid Manual for Chemical Accidents	Sax's Dangerous Properties of Industrial Materials	Hazardous Chemicals Desk Reference	Rapid Guide to Hazardous Chemicals in the Workplace	Fire Protection Guide on Hazardous Materials (NFPA)	Firefighter's Handbook of Hazardous Materials	Pocket Guide to Chemical Hazards (NIOSH)	Chemical Hazard Response Information System (CHRIS)	Emergency Handling of Hazardous Materials in Surface Transportation (AAR)	Emergency Action Guides (EAG)	Chemical Data Notebook: A User's Manual	Condensed Chemical Dictionary
NITRIC OXIDE	1660	20	628	21	2509	857	157		210	162	NTX 862	685	N2.5	N 2	823
NITRIC OXIDE AND NITROGEN TETROXIDE MIXTURE	1975	20							210						
NITRIC SULFURIC ACID MIXTURES Synonym: NITRATING ACID	1796	73	626						210			683	N1.0		822
NITRILOACETONITRILE Synonym: CYANOGEN	1026	18	241		979	331	57	−29	87		CYG 330	277	C8.0		333
NITRILOTRIACETIC ACID AND SALTS			629								NAA 815				
NITRITE	2627	35			2511	858			210						
NITRO SIL Synonym: ANHYDROUS AMMONIA	1005	15	48	11	220	57	11	25 –	32		AMA 47	78	A9.0	A 9	62
NITROANILINE	1661	55		21							NTA 851	687			
NITROANILINE-meta	1661	55		21	2514	859			211						824
NITROANILINE-ortho Synonym: 2-NITROANILINE	1661	55		21	2514	859			211		NTA 851	687	N3.0		824
NITROANILINE-ortho Synonym: NITROANILINE	1661	55		21	2514	859			211		NTA 851	687	N3.0		824
NITROANILINE-para Synonym: 4-NITROANILINE	1661	55	630	21	2514	860	158	124–74	211	162	NAL 818				824
NITROANISOLE	2730	55							211			687			
NITROBENZENE	1662	55	631	21	2518	861	158	124–74	212	162	NTB 852	687	N3.1		825
NITROBENZOL Synonym: NITROBENZENE	1662	55	631	21	2518	861	158	124–74	212	162	NTB 852	687	N3.1		825
NITROBENZOTRIFLUORIDE	2306/2810	54							212			688			
NITROBIPHENYL								−74							
NITROBROMOBENZENE	2732	55							212			688			

HAZARDOUS MATERIALS REFERENCE BOOKS INDEX

CHEMICAL OR MATERIAL NAME	UN/NA Number	Guide Number (DOT)	Firefighter's Hazardous Materials Reference Book	First Aid Manual for Chemical Accidents	Sax's Dangerous Properties of Industrial Materials	Hazardous Chemicals Desk Reference	Rapid Guide to Hazardous Chemicals in the Workplace	Fire Protection Guide on Hazardous Materials (NFPA)	Firefighter's Handbook of Hazardous Materials	Pocket Guide to Chemical Hazards (NIOSH)	Chemical Hazard Response Information System (CHRIS)	Emergency Handling of Hazardous Materials in Surface Transportation (AAR)	Emergency Action Guides (EAG)	Chemical Data Notebook: A User's Manual	Condensed Chemical Dictionary
NITROCARBOL	1261	26	639	21	2543	870	162	128–74	214	166	NMT 839	696			829
Synonym: NITROMETHANE															
NITROCELLULOSE	2060	26							212						826
NITROCELLULOSE	2555	33						125–	212			689			826
NITROCELLULOSE	2557	32							212			689			826
NITROCELLULOSE	2059	26						125–	212			689			826
NITROCELLULOSE	2556	33							212			690			826
NITROCELLULOSE GUM	2059	56	220		940	319		–28	82		CLD 257	249			300
Synonym: COLLODION															
NITROCELLULOSE SOLUTION	2059/2095	26	220		940	319		–28	212		CLD 257	689			300
Synonym: COLLODION															
NITROCHLOROBENZENE	1578	55						126–74	212						826
NITROCHLOROBENZENE-meta	1578	55	632						212			690	N3.2		
NITROCHLOROBENZENE-ortho	1578	55	633		841				212		CNO 273	690	N3.3		269
Synonym: CHLORONITROBENZENE-ortho															
NITROCHLOROBENZENE-para	1578	55	634	21	2525	863	159	–74	212	164		691	N3.4		
NITROCHLOROBENZOL-para	1578	55							212						
NITROCHLOROBENZOTRIFLUORIDE	2307	54							212						
NITROCHLOROFORM	1580	56	210	14	865	296	50	54 –	212	70	CPL 290	229	C4.3		826
Synonym: CHLOROPICRIN															
NITROCHLOROMETHANE	1580	56							212						
NITROCRESOL	2446	55							213			691			
NITROCYCLOHEXANE					2527			–74	213						
NITROETHANE	2842	26	635		2530	864	159	127–74	213	164	NTE 854	692			827
NITROGEN	1977	21		21	2536	866		127–	213		NXX 864	695	N3.4.3	N 3	827

CHEMICAL OR MATERIAL NAME	UN/NA Number	Guide Number (DOT)	Firefighter's Hazardous Materials Reference Book	First Aid Manual for Chemical Accidents	Sax's Dangerous Properties of Industrial Materials	Hazardous Chemicals Desk Reference	Rapid Guide to Hazardous Chemicals in the Workplace	Fire Protection Guide on Hazardous Materials (NFPA)	Firefighter's Handbook of Hazardous Materials	Pocket Guide to Chemical Hazards (NIOSH)	Chemical Hazard Response Information System (CHRIS)	Emergency Handling of Hazardous Materials in Surface Transportation (AAR)	Emergency Action Guides (EAG)	Chemical Data Notebook: A User's Manual	Condensed Chemical Dictionary
NITROGEN	1066	12	636	21	2536	866		127–	213			695	N3.4.3	N 3	827
NITROGEN CHLORIDE					2536	866			213						827
NITROGEN DIOXIDE Synonym: NITROGEN TETROXIDE	1067	20	637	21	2536	867	160	128–	213	164	NOX 845	692			828
NITROGEN FERTILIZER SOLUTION CONTAINING 2% FORMALDEHYDE	1043/9188	16							213			692			
NITROGEN FLUORIDE OXIDE	9271	20			2537	867									
NITROGEN HYDROGEN GAS MIXTURE	1954	22										694			
NITROGEN HYDROGEN GAS MIXTURE	1956	12													
NITROGEN MONOXIDE Synonym: NITRIC OXIDE	1660	20	628	21	2507	857	157		213	162	NTX 862	685	N2.5	N 2	828
NITROGEN OXIDE	1070	14			2537	867	160		213						828
NITROGEN OXIDE	1955	15													828
NITROGEN OXIDE	2201	23			2537	867	160								828
NITROGEN OXIDES							160		213						828
NITROGEN OXYCHLORIDE	1069	16							213						
NITROGEN PEROXIDE Synonym: NITROGEN TETROXIDE	1067	20	637		2537	867		128–	213		NOX 845	693	N3.4.6		828
NITROGEN SALTS AGRICULTURAL CRUDE	1479	35										693			
NITROGEN TETROXIDE	1067	20	637		2537	867		128–	213		NOX 845		N3.4.6		828
NITROGEN TRICHLORIDE									214						828
NITROGEN TRIFLUORIDE	2451	15		21	2538	868	161		214	164		694			828
NITROGEN TRIIODIDE					2538	868			214						828
NITROGEN TRIOXIDE	2421/1955	20						128–	214			694	N3.4.8		828
NITROBENZENESULFONIC ACID	2305	60													

HAZARDOUS MATERIALS REFERENCE BOOKS INDEX

CHEMICAL OR MATERIAL NAME	UN/NA Number	Guide Number (DOT)	Firefighter's Hazardous Materials Reference Book	First Aid Manual for Chemical Accidents	Sax's Dangerous Properties of Industrial Materials	Hazardous Chemicals Desk Reference	Rapid Guide to Hazardous Chemicals in the Workplace	Fire Protection Guide on Hazardous Materials (NFPA)	Firefighter's Handbook of Hazardous Materials	Pocket Guide to Chemical Hazards (NIOSH)	Chemical Hazard Response Information System (CHRIS)	Emergency Handling of Hazardous Materials in Surface Transportation (AAR)	Emergency Action Guides (EAG)	Chemical Data Notebook: A User's Manual	Condensed Chemical Dictionary
NITROGLYCERIN	3064	26		21				-74	214	166					828
NITROGLYCERIN	0143/0144			21	2539	868	161	-74	214			695			828
NITROGLYCERIN MIXED ETHYLENE GLYCOL DINITRATE(1:1)					2539	869	161		214						
NITROGLYCERIN SOLUTION	1204	26		21	2539	868	161	-74							828
NITROGUANIDINE	1336	33							214			695			829
NITROHYDROCHLORIC ACID	1798	60							214			696			829
NITROIMINODIETHYLENEDIISOCYANIC ACID					2542	870	162								
NITROMETHANE	1261	26	639	21	2543	870	162	128-74	214	166	NMT 839	696			829
NITROMURIATIC ACID	1798	60							214						
NITROPHENOL	1663/3077	55	640	21								697			
NITROPHENOL-meta Synonym: 3-NITROPHENOL	1663	55		21	2548	871			215		NIP 833				830
NITROPHENOL-ortho Synonym: 2-NITROPHENOL	1663	55		21	2548	871			215		NTP 859				830
NITROPHENOL-para Synonym: 4-NITROPHENOL	1663	55		21	2548	871		129-	215		NPH 846				830
NITROPROPANE	2608	26	641					129-	215			697			
NITROPROPANE-sec Synonym: 2-NITROPROPANE	2608	26			2554	872	163	-74	215	166	NPP 848				831
NITROSO-n-ETHYL ANILINE-n					2566	876	164								
NITROSODIETHYLAMINE-n					2562	875	164								
NITROSODIISOPROPYLAMINE-n					2563	875	164								
NITROSODIMETHYLAMINE-n					2564	875	164		215	168					831
NITROSODIMETHYLANILINE	1369	32							215						
NITROSODIMETHYLANILINE-para	1369	32							215						

CHEMICAL OR MATERIAL NAME	UN/NA Number	Guide Number (DOT)	Firefighter's Hazardous Materials Reference Book	First Aid Manual for Chemical Accidents	Sax's Dangerous Properties of Industrial Materials	Hazardous Chemicals Desk Reference	Rapid Guide to Hazardous Chemicals in the Workplace	Fire Protection Guide on Hazardous Materials (NFPA)	Firefighter's Handbook of Hazardous Materials	Pocket Guide to Chemical Hazards (NIOSH)	Chemical Hazard Response Information System (CHRIS)	Emergency Handling of Hazardous Materials in Surface Transportation (AAR)	Emergency Action Guides (EAG)	Chemical Data Notebook: A User's Manual	Condensed Chemical Dictionary
NITROSOGUANIDINE	0473/9052				2568	877			215			697			831
NITROSOIMINO DIETHANOL					2570	878	165								
NITROSOPHENOL-para Synonym: NITROSOPHENOL					2580				215						832
NITROSOPIPERIDINE-n					2581	880	165								
NITROSOPYRROLIDINE-n					2584	880	165								
NITROSTARCH	1337	33			2588	881			215			698			832
NITROSYL CHLORIDE	1069	16	642		2589	881			216		NTC 853	699			832
NITROSYL FLUORIDE					2589	881			216						832
NITROSYLSULFURIC ACID	2308/1760	60			2590	881			216			699			832
NITROTOLUENE	1664	55		21				130–		168		699			
NITROTOLUENE-meta	1664	55	643		2591	882	165	130–74	216		NTR 860				832
NITROTOLUENE-mixo					2592	883	165		216						
NITROTOLUENE-ortho	1664	55			2591	882	166	130–75	216		NIE 832	699			832
NITROTOLUENE-para	1664	55			2591	882	166	130–75	216		NTT 861				833
NITROTOLUOL	1664	55							216						
NITROTRICHLOROMETHANE Synonym: CHLOROPICRIN	1580	56	210	14	865	296	50	54 –	216	70	CPL 290	229	C4.3		833
NITROUREA	0 47/0151				2592	883			216			687			833
NITROUS ACID					2593	884			216						833
NITROUS ETHER Synonym: ETHYL NITRITE	1194	30	360		1643	554		89 –53	216		ETN 570	420			497
NITROUS OXIDE	1070	14	644						216		NTO 858	700	N3.5		833
NITROUS OXIDE	2201	23							216			700	N3.5		833

HAZARDOUS MATERIALS REFERENCE BOOKS INDEX

CHEMICAL OR MATERIAL NAME	UN/NA Number	Guide Number (DOT)	Firefighter's Hazardous Materials Reference Book	First Aid Manual for Chemical Accidents	Sax's Dangerous Properties of Industrial Materials	Hazardous Chemicals Desk Reference	Rapid Guide to Hazardous Chemicals in the Workplace	Fire Protection Guide on Hazardous Materials (NFPA)	Firefighter's Handbook of Hazardous Materials	Pocket Guide to Chemical Hazards (NIOSH)	Chemical Hazard Response Information System (CHRIS)	Emergency Handling of Hazardous Materials in Surface Transportation (AAR)	Emergency Action Guides (EAG)	Chemical Data Notebook: A User's Manual	Condensed Chemical Dictionary
NITROXYLENE	1665	55			2593	884			216			700			833
NITROXYLOL	1665	55							216						
NITROXYLOL-meta	1665	55							216						
NITROXYLOL-ortho	1665	55							216						
NITROXYLOL-para	1665	55							216						
NITRYL CHLORIDE					2593	884			216						833
NITRYL FLUORIDE					2593	884			216						834
NITRYL HYDROXIDE	2032	44		21	2509	856		123–	210	160		684	N2.0	N 1	823
Synonym: NITRIC ACID FUMING															
NONANE	1920/1993	27	645	21	2596	885	167	–75	216		NAN 819	701			835
NONANE-n	1920/1993	27	645	21	2596	885	167	–75	216		NAN 819	701			835
Synonym: NONANE															
NONANOL	1987	26	646								NNN 841				835
NONENE								–75	217		NON 844				835
NONFLAMMABLE GAS	1956	12													
NONYL ACETATE					2598	885		–75	217						836
NONYL ALCOHOL			646		2598	885			217		NNN 841				836
Synonym: NONANOL															
NONYL MERCAPTAN-tert								–75	217						
NONYL TRICHLOROSILANE	1799	60			2599	886			217			701			837
NONYLCARBINOL	1987	26	258		1032	347			217		DAN 338	291			350
Synonym: DECYL ALCOHOL (-n)															
NONYLETHYLENE-n											UDC1164				
Synonym: 1-UNDECENE															
NONYLNAPHTHALENE								–75	217						

CHEMICAL OR MATERIAL NAME	UN/NA Number	Guide Number (DOT)	Firefighter's Hazardous Materials Reference Book	First Aid Manual for Chemical Accidents	Sax's Dangerous Properties of Industrial Materials	Hazardous Chemicals Desk Reference	Rapid Guide to Hazardous Chemicals in the Workplace	Fire Protection Guide on Hazardous Materials (NFPA)	Firefighter's Handbook of Hazardous Materials	Pocket Guide to Chemical Hazards (NIOSH)	Chemical Hazard Response Information System (CHRIS)	Emergency Handling of Hazardous Materials in Surface Transportation (AAR)	Emergency Action Guides (EAG)	Chemical Data Notebook: A User's Manual	Condensed Chemical Dictionary
NONYLPHENOL	1760	60	647		2598			−75	217		NNP 842	701			837
NORANONE-S **Synonym: DI-n-BUTYL KETONE**	1224	26									DBK 346				
NORBORNADIENE	2251	26			2600				217						
NORMAL HEPTANE **Synonym: HEPTANE (-n)**	1206	27	425	18	1830	616	106	−57		120	HPT 644	505	H1.0.3		592
NORMAL HEXANE **Synonym: HEXANE (-n)**	1208	27	435	18	1859	627	109	−58			HXA 649	517	H2.0		599
NORMAL IS BUTANOL **Synonym: BUTYL ALCOHOL (-n)**	1120	26			593	196	31	−21		52	BAN 119	150	B7.1		181
NORMAL LEAD ACETATE **Synonym: LEAD ACETATE**	1616	53	501	19	2094	708	128				LAC 690	570			686
NORMAL PENTANE **Synonym: PENTANE**	1265	27	668	22	2687		172	−77			PTA 990	730	P1.0		881
NORVALAMINE **Synonym: BUTYLAMINE (-n)**	1125	68		13	594	197	32	41 –	60	52	BAM 118	169	B7.3		182
NSC 8,819 **Synonym: ACROLEIN**	1092	30	31	11	63	25	5	17 −12	20	32	ARL 85	12	A4.0	A 5	18
NSFO **Synonym: FUEL OIL NO. 5**	1993	27	406					−55	151			464	F3.3		
NTA **Synonym: NITRILOTRIACETIC ACID AND SALTS**			629								NAA 815				839
NUCLEAR REACTOR FUEL	2918	63	648									702			
NUX VOMICA **Synonym: STRYCHNINE**	1692	53	794		2611	1074	202		257	198	STR1063	862			1096

HAZARDOUS MATERIALS REFERENCE BOOKS INDEX

CHEMICAL OR MATERIAL NAME	UN/NA Number	Guide Number (DOT)	Firefighter's Hazardous Materials Reference Book	First Aid Manual for Chemical Accidents	Sax's Dangerous Properties of Industrial Materials	Hazardous Chemicals Desk Reference	Rapid Guide to Hazardous Chemicals in the Workplace	Fire Protection Guide on Hazardous Materials (NFPA)	Firefighter's Handbook of Hazardous Materials	Pocket Guide to Chemical Hazards (NIOSH)	Chemical Hazard Response Information System (CHRIS)	Emergency Handling of Hazardous Materials in Surface Transportation (AAR)	Emergency Action Guides (EAG)	Chemical Data Notebook: A User's Manual	Condensed Chemical Dictionary
O,O-DIETHYL O-(p-NITROPHENYL) PHOSPHOROTHIOATE Synonym: PARATHION	2784	28		22	2667	907			222		PTO1000				872
OCTA KLOR Synonym: CHLORDANE	2762	28	196		749	263	44		217	60	CDN 237	211			258
OCTACHLOROCAMPHENE Synonym: TOXAPHENE	2761	55	832						270		TXP1159	911			1161
OCTACHLORONAPHTHALENE					2614	889	167		217	168					845
OCTACHLOROTETRAHYDROINDANE	2762	28							217						
OCTADECANOIC ACID Synonym: STEARIC ACID			791		3131	1070		−84			SRA1055				846
OCTADECYL TRICHLOROSILANE	1800	39			2616	890		−75	217			703			846
OCTADECYLIC ACID-n Synonym: STEARIC ACID			791		3131	1070		−84			SRA1055				1088
OCTADIENE	2309/1993	27							217			703			
OCTAFLUOROCYCLOBUTANE	1976/1956	12							217			704			846
OCTAFLUOROPROPANE	2424	12							217			704			846
OCTAMETHYL PYROPHOSPHORAMIDE					2618	891			218						847
OCTAN-1-OL-n Synonym: OCTANOL	1987	26	650		2620						OTA 909	706	O1.0		847
OCTANAL	1191/1993	26							218			705			
OCTANE	1262	27	649	21	2619	892	168		218	168	OAN 867	705			847
OCTANE-n Synonym: OCTANE	1262	27	649	21	2619	892	168		218	168	OAN 867	705			847
OCTANOL	1987	26	650		2620						OTA 909	706	O1.0		847
OCTANOL-n Synonym: OCTANOL	1987	26	650		2620						OTA 909	706	O1.0		847

CHEMICAL OR MATERIAL NAME	UN/NA Number	Guide Number (DOT)	Firefighter's Hazardous Materials Reference Book	First Aid Manual for Chemical Accidents	Sax's Dangerous Properties of Industrial Materials	Hazardous Chemicals Desk Reference	Rapid Guide to Hazardous Chemicals In the Workplace	Fire Protection Guide on Hazardous Materials (NFPA)	Firefighter's Handbook of Hazardous Materials	Pocket Guide to Chemical Hazards (NIOSH)	Chemical Hazard Response Information System (CHRIS)	Emergency Handling of Hazardous Materials In Surface Transportation (AAR)	Emergency Action Guides (EAG)	Chemical Data Notebook: A User's Manual	Condensed Chemical Dictionary
OCTANOYL PEROXIDE (-n)	2129	52							218						
OCTILIN Synonym: OCTANOL	1987	26	650		2620						OTA 909	706	O1.0		847
OCTOIL Synonym: DIOCTYL PHTHALATE (-n)					1445	481		-45			DOP 455				425
OCTYCARBINOL Synonym: NONANOL			646								NNN 841				835
OCTYL ACETATE					2621				218						848
OCTYL ACETATE-n	1993	27			2621										848
OCTYL ALCOHOL Synonym: OCTANOL	1987	26	650		2622	893		-76			OTA 909	706	O1.0		848
OCTYL ALCOHOL-n Synonym: OCTANOL	1987	26	650		2622				218		OTA 909	706	O1.0		848
OCTYL ALDEHYDE Synonym: ETHYLHEXALDEHYDE	1191	26	356						218		EHA 535	438			848
OCTYL CHLORIDE								-76	218						848
OCTYL EPOXY TALLATE			652								OET 878				
OCTYL HYDROPEROXIDE-tert	2160	48							218						
OCTYL MERCAPTAN	3023	57							218						849
OCTYL MERCAPTAN-tert	3023	57							218						849
OCTYL PEROXY-2-ETHYLHEXANOATE-tert	2161	52							218						
OCTYL TRICHLOROSILANE	1801	60			2625	894			218			707			850
OCTYLAMINE					2622			-76	218						848
OCTYLENE-alpha Synonym: 1-OCTENE	1993	27						-75	218		OTE 913				847

HAZARDOUS MATERIALS REFERENCE BOOKS INDEX

CHEMICAL OR MATERIAL NAME	UN/NA Number	Guide Number (DOT)	Firefighter's Hazardous Materials Reference Book	First Aid Manual for Chemical Accidents	Sax's Dangerous Properties of Industrial Materials	Hazardous Chemicals Desk Reference	Rapid Guide to Hazardous Chemicals in the Workplace	Fire Protection Guide on Hazardous Materials (NFPA)	Firefighter's Handbook of Hazardous Materials	Pocket Guide to Chemical Hazards (NIOSH)	Chemical Hazard Response Information System (CHRIS)	Emergency Handling of Hazardous Materials in Surface Transportation (AAR)	Emergency Action Guides (EAG)	Chemical Data Notebook: A User's Manual	Condensed Chemical Dictionary
OIL	1270/1993	27	653						218			708			850
OIL GAS	1071	22			2626	894			218			708			851
OIL MIST										170					
OIL OF BITTER ALMONDS Synonym: NITROBENZENE	1662	55	631	21	2518	861	158	124–74	212	162	NTB 852		N3.1		851
OIL OF MIRBANE Synonym: NITROBENZENE	1662	55	631	21	2518	861	158	124–74	218	162	NTB 852		N3.1		851
OIL OF MYRBANE Synonym: NITROBENZENE	1662	55	631	21	2518	861	158	124–74	219	162	NTB 852		N3.1		825
OIL OF PINE Synonym: PINE OIL	1272	26	695		2810	959		–80	235		OPI 895	764	P5.0		918
OIL OF VITRIOL Synonym: SULFURIC ACID	1830	39	802	24	3163	1082	204	164–	219	200	SFA1037	871	S4.2	S 8	851
OILS CLARIFIED											OCF 872				
OILS CRUDE	1267	27									OIL 882	273			
OILS DIESEL	1270	27									ODS 877				
OILS EDIBLE CASTOR											OCA 870				
OILS EDIBLE COCONUT											OCC 871				
OILS EDIBLE COTTONSEED											OCS 875				
OILS EDIBLE FISH											OFS 880				
OILS EDIBLE LARD											OLD 885				
OILS EDIBLE OLIVE								–76			OOL 893				
OILS EDIBLE PALM								–76			OPM 896				
OILS EDIBLE PEANUT								–76			OPN 897				
OILS EDIBLE SAFFLOWER											OSF 905				

CHEMICAL OR MATERIAL NAME	UN/NA Number	Guide Number (DOT)	Firefighter's Hazardous Materials Reference Book	First Aid Manual for Chemical Accidents	Sax's Dangerous Properties of Industrial Materials	Hazardous Chemicals Desk Reference	Rapid Guide to Hazardous Chemicals in the Workplace	Fire Protection Guide on Hazardous Materials (NFPA)	Firefighter's Handbook of Hazardous Materials	Pocket Guide to Chemical Hazards (NIOSH)	Chemical Hazard Response Information System (CHRIS)	Emergency Handling of Hazardous Materials in Surface Transportation (AAR)	Emergency Action Guides (EAG)	Chemical Data Notebook: A User's Manual	Condensed Chemical Dictionary
OILS EDIBLE SOYA BEAN											OSB 903				
OILS EDIBLE TUCUM											OTC 911				
OILS EDIBLE VEGETABLE											OVG 918				
OILS FUEL 1-D	1270	27									OOD 892				
OILS FUEL 2	1223/1993	27									OTW 917	464			
OILS FUEL 2-D	1270	27									OTD 912				
OILS FUEL 4	1223/1993	27									OFR 879	464			
OILS FUEL 5	1223/1993	27									OFV 881	464			
OILS FUEL 6	1223	27									OSX 907				
OILS FUEL NO. 1	1223/1993	27									OON 894	464			
OILS MISCELLANEOUS ABSORPTION	1270	27									OAS 869				
OILS MISCELLANEOUS COAL TAR	1136	27									OCT 876				
OILS MISCELLANEOUS CROTON											OCR 874				
OILS MISCELLANEOUS LINSEED											OLS 887				
OILS MISCELLANEOUS LUBRICATING	1270	27									OLB 884				
OILS MISCELLANEOUS MINERAL	1270	27									OMN 888				
OILS MISCELLANEOUS MINERAL SEAL	1270	27									OMS 889				
OILS MISCELLANEOUS MOTOR	1270	27									OMT 890				
OILS MISCELLANEOUS NEATSFOOT											ONF 891				
OILS MISCELLANEOUS PENETRATING	1270	27									OPT 898				
OILS MISCELLANEOUS RANGE	1223	27									ORG 900				
OILS MISCELLANEOUS RESIN	1286	26									ORS 902				
OILS MISCELLANEOUS ROAD	1999	27									ORD 899				
OILS MISCELLANEOUS ROSIN	1286	26						-83			ORN 901				

HAZARDOUS MATERIALS REFERENCE BOOKS INDEX

CHEMICAL OR MATERIAL NAME	UN/NA Number	Guide Number (DOT)	Firefighter's Hazardous Materials Reference Book	First Aid Manual for Chemical Accidents	Sax's Dangerous Properties of Industrial Materials	Hazardous Chemicals Desk Reference	Rapid Guide to Hazardous Chemicals in the Workplace	Fire Protection Guide on Hazardous Materials (NFPA)	Firefighter's Handbook of Hazardous Materials	Pocket Guide to Chemical Hazards (NIOSH)	Chemical Hazard Response Information System (CHRIS)	Emergency Handling of Hazardous Materials in Surface Transportation (AAR)	Emergency Action Guides (EAG)	Chemical Data Notebook: A User's Manual	Condensed Chemical Dictionary
OILS MISCELLANEOUS SPERM								–83			OSP 906				
OILS MISCELLANEOUS SPINDLE	1270	27						–83			OSD 904				
OILS MISCELLANEOUS SPRAY	1270	27									OSY 908				
OILS MISCELLANEOUS TALL											OTL 915				
OILS MISCELLANEOUS TANNER'S											OTN 916				
OILS MISCELLANEOUS TRANSFORMER	1270	27									OTF 914				
OILS MISCELLANEOUS TURBINE											OTB 910				
OLAMINE Synonym: MONOETHANOLAMINE	2491	60	594		2629			118–	206		MEA 749		M5.0		795
OLEFIANT GAS Synonym: ETHYLENE	1962	21	368	16	1597	536		83 –50	142		ETL 568	436	E2.4	E 4	485
OLEFIANT GAS Synonym: ETHYLENE	1038	22		16	1597	536		83 –50	142		ETL 568	436	E2.4	E 4	485
OLEIC ACID			654		2630	896		–76			OLA 883				852
OLEIC ACID AMMONIUM SALT Synonym: AMMONIUM OLEATE											AOL 69				67
OLEIC ACID POTASSIUM SALT			655		2630	896					OAP 868				
OLEIC ACID SODIUM SALT			656		2630	896					OAC 866				
OLEUM	1831	39	657						219		OLM 886		O2.0		852
OMAL Synonym: TRICHLOROPHENOL	2020	53	839								TPH1139				
OMITE Synonym: PROPARGITE	2765/3082	55	718						242		PRG 986	790			967
ONA Synonym: 2-NITROANILINE	1661	55									NTA 851	687			824

CHEMICAL OR MATERIAL NAME	UN/NA Number	Guide Number (DOT)	Firefighter's Hazardous Materials Reference Book	First Aid Manual for Chemical Accidents	Sax's Dangerous Properties of Industrial Materials	Hazardous Chemicals Desk Reference	Rapid Guide to Hazardous Chemicals in the Workplace	Fire Protection Guide on Hazardous Materials (NFPA)	Firefighter's Handbook of Hazardous Materials	Pocket Guide to Chemical Hazards (NIOSH)	Chemical Hazard Response Information System (CHRIS)	Emergency Handling of Hazardous Materials in Surface Transportation (AAR)	Emergency Action Guides (EAG)	Chemical Data Notebook: A User's Manual	Condensed Chemical Dictionary
ONCB Synonym: NITROCHLOROBENZENE-ortho	1578	55	633						212			690	N3.3		269
ONP Synonym: 2-NITROPHENOL	1663	55			2548						NTP 859				
ORGANIC COMPOUND OF ARSENIC	1556	55							219						
ORGANIC PEROXIDE	9187	52							219						
ORGANIC PEROXIDE	1993	52							219						
ORGANIC PEROXIDE	9183	52							219						
ORGANIC PEROXIDE	2899	52							219						
ORGANIC PEROXIDE	2255	48							219						
ORGANIC PEROXIDE MIXTURE	2756	52							219						
ORGANIC PHOSPHATE COMPOUND	2783/3018	55							219			712			
ORGANIC PHOSPHORUS COMPOUND	1955/1967	15							219			711			
ORGANOCHLORINE PESTICIDE	2996/2761	55		21					219			712			
ORGANOCHLORINE PESTICIDE	2761	55		21					219			712			
ORGANOCHLORINE PESTICIDE	2995	28		21					219						
ORGANOCHLORINE PESTICIDE	2762	28		21					219			711			
ORGANOPHOSPHORUS PESTICIDE	2783	55										713			
ORGANOPHOSPHORUS PESTICIDE	2784	28							219			713			
ORGANOPHOSPHORUS PESTICIDE	3017	28							219						
ORGANOPHOSPHORUS PESTICIDE	3018/2783	55							219			712			
ORGANOTIN PESTICIDE	2787	28							219			714			
ORGANOTIN PESTICIDE	3019	28							219						
ORGANOTIN PESTICIDE	3020/2786	55							219			714			

HAZARDOUS MATERIALS REFERENCE BOOKS INDEX

CHEMICAL OR MATERIAL NAME	UN/NA Number	Guide Number (DOT)	Firefighter's Hazardous Materials Reference Book	First Aid Manual for Chemical Accidents	Sax's Dangerous Properties of Industrial Materials	Hazardous Chemicals Desk Reference	Rapid Guide to Hazardous Chemicals in the Workplace	Fire Protection Guide on Hazardous Materials (NFPA)	Firefighter's Handbook of Hazardous Materials	Pocket Guide to Chemical Hazards (NIOSH)	Chemical Hazard Response Information System (CHRIS)	Emergency Handling of Hazardous Materials in Surface Transportation (AAR)	Emergency Action Guides (EAG)	Chemical Data Notebook: A User's Manual	Condensed Chemical Dictionary
ORGANOTIN PESTICIDE	2786	55							219			715			
ORM A	1693	58							219						
ORM B	1760	60							219						
ORPIMENT Synonym: ARSENIC TRISULFIDE	1557	53	101				16	33 –	35		ART 86	93			858
ORTHO C 11 Synonym: SODIUM CHLORATE	1495	35	766	24	3070	1048		156–	253		SDC1029	834	S2.1	S 2	1052
ORTHOARSENIC ACID Synonym: ARSENIC ACID	1554	53	96			86	16		35		ASA 88	91			858
ORTHOBORIC ACID Synonym: BORIC ACID				12	529	174					BAC 114				858
ORTHOCIDE Synonym: CAPTAN	9099/3082		180		682	238	40		64		CPT 296	186			214
ORTHODICHLOROBENZENE Synonym: DICHLOROBENZENE-ortho	1591	58	272	15	1127	376	66	64 –34	101	84	DBO 350	329	D1.0.1		377
ORTHOPHOSPHORIC ACID Synonym: PHOSPHORIC ACID	1805	60	684	22	2786	946	180	137–	232	182	PAC 922	753	P4.0		809
ORTHOTITANIC ACID TETRABUTYL ESTER Synonym: TETRABUTYL TITANATE			806						261		TBT1078				1126
ORTHOXYLENE	1307	27							220						858
OSMIC ACID	2471	55		21											
OSMIC ACID ANHYDRIDE	2471	55		21											
OSMIUM TETROXIDE	2471/2811	55			2637	898	168		220	170		716			859
OXACYCLOPENTADIENE Synonym: FURAN	2389	26			1763	596		-55	152		FUR 600	467	F3.3.1		545

CHEMICAL OR MATERIAL NAME	UN/NA Number	Guide Number (DOT)	Firefighter's Hazardous Materials Reference Book	First Aid Manual for Chemical Accidents	Sax's Dangerous Properties of Industrial Materials	Hazardous Chemicals Desk Reference	Rapid Guide to Hazardous Chemicals in the Workplace	Fire Protection Guide on Hazardous Materials (NFPA)	Firefighter's Handbook of Hazardous Materials	Pocket Guide to Chemical Hazards (NIOSH)	Chemical Hazard Response Information System (CHRIS)	Emergency Handling of Hazardous Materials in Surface Transportation (AAR)	Emergency Action Guides (EAG)	Chemical Data Notebook: A User's Manual	Condensed Chemical Dictionary
OXACYCLOPENTANE Synonym: TETRAHYDROFURAN	2056	26	818		3227	1103	210	167–86	263	210	THF1116	894	T1.0	T 2	1131
OXAL Synonym: GLYOXAL			421		1802	609			220		GOS 610				570
OXALATES	2449	54			2639	899			220			717			
OXALDEHYDE Synonym: GLYOXAL			421		1802	609			155		GOS 610				570
OXALIC ACID	2449	54	658	21	2639	899	169	131–	220	170	OXA 919				860
OXALIC ACID DIAMMONIUM SALT Synonym: AMMONIUM OXALATE	2449/3077	54	65		230	64			29		AOX 70	65			67
OXALIC ACID DINITRILE Synonym: CYANOGEN	1026	18	241		979	331	57	–29	87		CYG 330	277	C8.0		333
OXALIC ACID FERROUS SALT Synonym: FERROUS OXALATE			396								FOX 592				517
OXALIC ETHER	2525	54						–76	220						
OXALONITRILE Synonym: CYANOGEN	1026	18	241		979	331	57	–29	220		CYG 330	277	C8.0		860
OXALYL CHLORIDE					2640				220						860
OXALYL CYANIDE Synonym: CYANOGEN	1026	18	241		979	331	57	–29	220		CYG 330	277	C8.0		333
OXAMMONIUM Synonym: HYDROXYLAMINE			456		1936	653		101–60	220		HDA 629				860
OXAMMONIUM SULFATE Synonym: HYDROXYLAMINE SULFATE	2865	60	457		2641	900					HAS 619				622
OXANE Synonym: ETHYLENE OXIDE	1040	69	377	17	1611	544	96	87 –52	144	112	EOX 550	434	E2.7	E 6	491

HAZARDOUS MATERIALS REFERENCE BOOKS INDEX

CHEMICAL OR MATERIAL NAME	UN/NA Number	Guide Number (DOT)	Firefighter's Hazardous Materials Reference Book	First Aid Manual for Chemical Accidents	Sax's Dangerous Properties of Industrial Materials	Hazardous Chemicals Desk Reference	Rapid Guide to Hazardous Chemicals in the Workplace	Fire Protection Guide on Hazardous Materials (NFPA)	Firefighter's Handbook of Hazardous Materials	Pocket Guide to Chemical Hazards (NIOSH)	Chemical Hazard Response Information System (CHRIS)	Emergency Handling of Hazardous Materials in Surface Transportation (AAR)	Emergency Action Guides (EAG)	Chemical Data Notebook: A User's Manual	Condensed Chemical Dictionary
OXIDES OF NITROGEN Synonym: NITROGEN TETROXIDE	1067	20	637		2537	867		128–	213		NOX 845		N3.4.6		858
OXIDIZER	9194	45							220						
OXIDIZER	1479	35							220						
OXIDIZER	9200	42							220						
OXIDIZER	9199	44							220						
OXIDIZER	9193	45							220						
OXIDIZING MATERIAL	1479	35							220						
OXIDIZING SUBSTANCE	1479/3139	35							220			718			
OXIDOETHANE Synonym: ETHYLENE OXIDE	1040	69	377	17	1611	544	96	87 –52	144	112	EOX 550	434	E2.7	E 6	491
OXIRANE Synonym: ETHYLENE OXIDE	1040	69	377	17	1611	544	96	87 –52	221	112	EOX 550	434	E2.7	E 6	861
OXITOL Synonym: ETHYLENE GLYCOL MONOETHYL ETHER	1171	26	375					–51	143		EGE 529	432	E2.5		489
OXO OCTALDEHYDE Synonym: ISOOCTALDEHYDE	1993	27	474								IOC 669				
OXO OCTYL ALCOHOL Synonym: ISOOCTYL ALCOHOL	1987	26	475		2032	683	123	–62			IOA 668	552			657
OXODIPHENYLMETHANE-alpha Synonym: BENZOPHENONE					379	125					BZP 200				132
OXODITANE-alpha Synonym: BENZOPHENONE					379	125					BZP 200				132
OXOLANE Synonym: TETRAHYDROFURAN	2056	26	818		3227	1103	210	167–86	263	210	THF1116	894	T1.0	T 2	1131

CHEMICAL OR MATERIAL NAME	UN/NA Number	Guide Number (DOT)	Firefighter's Hazardous Materials Reference Book	First Aid Manual for Chemical Accidents	Sax's Dangerous Properties of Industrial Materials	Hazardous Chemicals Desk Reference	Rapid Guide to Hazardous Chemicals in the Workplace	Fire Protection Guide on Hazardous Materials (NFPA)	Firefighter's Handbook of Hazardous Materials	Pocket Guide to Chemical Hazards (NIOSH)	Chemical Hazard Response Information System (CHRIS)	Emergency Handling of Hazardous Materials in Surface Transportation (AAR)	Emergency Action Guides (EAG)	Chemical Data Notebook: A User's Manual	Condensed Chemical Dictionary
OXOLE **Synonym: FURAN**	2389	26			1763	596		−55	221		FUR 600	467	F3.3.1		545
OXOMETHANE **Synonym: FORMALDEHYDE SOLUTION**	1198/2209	29	401	17	1748	591	102	91 −54	150	116	FMS 590	462	F2.0	F 2	
OXOTRIDECYL ALCOHOL **Synonym: TRIDECANOL**			842		3365			−88	274		TDN1096				1173
OXYACYCLOPROPANE **Synonym: ETHYLENE OXIDE**	1040	69	377	17	1611	544	96	87 −52	144	112	EOX 550	434	E2.7	E 6	491
OXYBENZENE **Synonym: CARBOLIC ACID**	1671	55	185						65			743	C2.0		217
OXYFUME 12 **Synonym: ETHYLENE OXIDE**	1040	69	377	17	1611	544	96	87 −52	144	112	EOX 550	434	E2.7	E 6	491
OXYGEN	1072	14			2652	901			221				O3.0	O 1	862
OXYGEN	1073	23	659		2652	901			221		OXY 920	720	O3.0	O 1	862
OXYGEN DIFLUORIDE	2190/1955	20	660	21	2653	902	169		221	170		720			
OXYGEN FLUORIDE	2190	20							221						863
OXYGEN NITROGEN GAS MIXTURE	1956	12										720			
OXYISOBUTYRIC NITRILE	1541	55							221						
OXYLITE **Synonym: BENZOYL PEROXIDE**	2085	49	122			130	23		42	44			B2.0	B 2	133
OXYLITE **Synonym: DIBENZOYL PEROXIDE**	2085/2087	49	262		392				97		DPO 469	320			368
OXYMETHYLENE **Synonym: FORMALDEHYDE SOLUTION**	1198/2209	29	401	17	1748	591	102	91 −54	150	116	FMS 590	462	F2.0	F 2	863
OXYPHENIC ACID **Synonym: CATECHOL**					718	252	44		67		CTC 318				232

HAZARDOUS MATERIALS REFERENCE BOOKS INDEX

CHEMICAL OR MATERIAL NAME	UN/NA Number	Guide Number (DOT)	Firefighter's Hazardous Materials Reference Book	First Aid Manual for Chemical Accidents	Sax's Dangerous Properties of Industrial Materials	Hazardous Chemicals Desk Reference	Rapid Guide to Hazardous Chemicals in the Workplace	Fire Protection Guide on Hazardous Materials (NFPA)	Firefighter's Handbook of Hazardous Materials	Pocket Guide to Chemical Hazards (NIOSH)	Chemical Hazard Response Information System (CHRIS)	Emergency Handling of Hazardous Materials in Surface Transportation (AAR)	Emergency Action Guides (EAG)	Chemical Data Notebook: A User's Manual	Condensed Chemical Dictionary
OXYTOL ACETATE Synonym: ETHYLENE GLYCOL MONOETHYL ETHER ACETATE	1172	26	376					-51	143		EGA 526	432	E2.6		489
OXYTOLUENE Synonym: CRESOL	2076	55	232	14	959	324	55	57 –	221	74	CRS 307	270	C7.0		322
OZONE				21	2655	903	169		221	172					864
PADISCOL Synonym: ETHYL ALCOHOL	1170	26	344	16	1572	525	90	-48	136		EAL 504	408	E2.2	E 2	478
PAINT	1760/3066	60							221			722			866
PAINT	1263	26							221						866
PAINT DRIER Synonym: COPPER NAPHTHENATE	1168/1993	26	226				55		83		CNN 272	260			313
PAINT RELATED MATERIAL	1760/3066	60							221			722			
PAINT RELATED MATERIAL	1263	26							221						
PAINTER'S NAPHTHA Synonym: NAPHTHA V,M & P	1255	27	605		2463	835	152	-73	207		NVM 863				
PALM BUTTER Synonym: OILS EDIBLE PALM								-76			OPM 896				868
PALM FRUIT OIL Synonym: OILS EDIBLE PALM								-76			OPM 896				
PALM OIL Synonym: OILS EDIBLE PALM					2659			-76			OPM 896				868
PALM SEED OIL Synonym: OILS EDIBLE TUCUM											OTC 911				
PAN Synonym: PHTHALIC ANHYDRIDE	2214/1759	60	694	22	2798	954	182	141–80	234	184	PAN 927	761	P4.4		913

CHEMICAL OR MATERIAL NAME	UN/NA Number	Guide Number (DOT)	Firefighter's Hazardous Materials Reference Book	First Aid Manual for Chemical Accidents	Sax's Dangerous Properties of Industrial Materials	Hazardous Chemicals Desk Reference	Rapid Guide to Hazardous Chemicals in the Workplace	Fire Protection Guide on Hazardous Materials (NFPA)	Firefighter's Handbook of Hazardous Materials	Pocket Guide to Chemical Hazards (NIOSH)	Chemical Hazard Response Information System (CHRIS)	Emergency Handling of Hazardous Materials in Surface Transportation (AAR)	Emergency Action Guides (EAG)	Chemical Data Notebook: A User's Manual	Condensed Chemical Dictionary
PANOXYL Synonym: **BENZOYL PEROXIDE**	2085	49	122		392	130	23		42	44			B2.0	B 2	133
PAPER	1379	32							221			723			870
PAPI Synonym: **POLYMETHYLENE POLYPHENYL ISOCYANATE**			699								PPI 976				870
PARACETALDEHYDE Synonym: **PARALDEHYDE**	1264	26	662		2665	906		132–76	222		PDH 944	726			870
PARADI Synonym: **DICHLOROBENZENE-para**	1592	58	273	15	1128	377	67	–34	101	84	DBP 351	330			378
PARADIAMINOBENZENE	1673	53							222						
PARADICHLOROBENZENE Synonym: **DICHLOROBENZENE-para**	1592	58	273	15	1128	377	67	–34	222	84	DBP 351	330			378
PARADOW Synonym: **DICHLOROBENZENE-para**	1592	58	273	15	1128	377	67	–34	101	84	DBP 351	330			378
PARAFFIN				21	2664	905	170								871
PARAFFIN OIL				21				–76	222						871
PARAFORM	2213	32							222						
PARAFORM Synonym: **FORMALDEHYDE SOLUTION**	1198/2209	29	401	17	1748	591	102	91 –54	150	116	FMS 590	462	F2.0	F 2	
PARAFORMALDEHYDE	2213	32	661		2664	906		131–76	222		PFA 950	725			871
PARALDEHYDE	1264	26	662		2665	906		132–76	222		PDH 944	726			871
PARAMENTHANE HYDROPEROXIDE	2125	51							222						
PARAMOTH Synonym: **DICHLOROBENZENE-para**	1592	58	273	15	1128	337	67	–34	101	84	DBP 351	330			378
PARANAPHTHALENE Synonym: **ANTHRACENE**			87		262	76	14	–17	33		ATH 101				83

HAZARDOUS MATERIALS REFERENCE BOOKS INDEX

CHEMICAL OR MATERIAL NAME	UN/NA Number	Guide Number (DOT)	Firefighter's Hazardous Materials Reference Book	First Aid Manual for Chemical Accidents	Sax's Dangerous Properties of Industrial Materials	Hazardous Chemicals Desk Reference	Rapid Guide to Hazardous Chemicals in the Workplace	Fire Protection Guide on Hazardous Materials (NFPA)	Firefighter's Handbook of Hazardous Materials	Pocket Guide to Chemical Hazards (NIOSH)	Chemical Hazard Response Information System (CHRIS)	Emergency Handling of Hazardous Materials in Surface Transportation (AAR)	Emergency Action Guides (EAG)	Chemical Data Notebook: A User's Manual	Condensed Chemical Dictionary
PARANITROANILINE	1661	55							222			726			
PARAQUAT	2588	53		22	2665	906	170		222	172					872
PARAQUAT DICHLORIDE					2666	907	163		222						
PARAQUAT DIMETHYL SULPHATE					2666		170		222						
PARATHION	2784	28		22	2667	907			222		PTO1000				872
PARATHION AND COMPRESSED GAS MIXTURE	1967	15			2668				222			726			
PARATHION METHYL	2783	55	577	20	2369	809	148		196		MPT 790	635			776
Synonym: **METHYL PARATHION**															
PARATHION MIXTURE	2783/3018	55	663	22	2668	908	170		222	172		727			
PARAWET	2783	55							222						
PARAXYLENE	1307	27													
PARIDOL	2783	55	577	20	2369	809	148		196		MPT 790	635			776
Synonym: **METHYL PARATHION**															
PARIS GREEN	1585	53	222				55		83		CAA 203	257			873
Synonym: **COPPER ACETOARSENITE**															
PARROT GREEN	1585	53	222				55		83		CAA 203	257			310
Synonym: **COPPER ACETOARSENITE**															
PATENT ALUMINUM	9078/3077		46		132	43			24		ALM 43	47	A7.1.01		48
Synonym: **ALUMINUM SULFATE**															
PCB	2315/3151	31	698	23	2831	964	185	142–	222		PCB 934	773			875
Synonym: **POLYCHLORINATED BIPHENYL**															
PCBs	2315/3151	31	698	23	2831	964	185	142–	222		PCB 934	773			875
Synonym: **POLYCHLORINATED BIPHENYL**															
PCNB	1578	55	634	21	2525	863	159	–74	212	164			N3.4		875
Synonym: **NITROCHLOROBENZENE-para**															
PCP	2020	53							222						875

CHEMICAL OR MATERIAL NAME	UN/NA Number	Guide Number (DOT)	Firefighter's Hazardous Materials Reference Book	First Aid Manual for Chemical Accidents	Sax's Dangerous Properties of Industrial Materials	Hazardous Chemicals Desk Reference	Rapid Guide to Hazardous Chemicals in the Workplace	Fire Protection Guide on Hazardous Materials (NFPA)	Firefighter's Handbook of Hazardous Materials	Pocket Guide to Chemical Hazards (NIOSH)	Chemical Hazard Response Information System (CHRIS)	Emergency Handling of Hazardous Materials in Surface Transportation (AAR)	Emergency Action Guides (EAG)	Chemical Data Notebook: A User's Manual	Condensed Chemical Dictionary
PE Synonym: PENTAERYTHRITOL			667		2682	912	172				PET 949				876
PEAR OIL Synonym: ISOAMYLACETATE	1104	26			2011	673	120	-60	222	130	IAT 660				876
PEAR OIL Synonym: AMYL ACETATE-sec	1104	26		12	241	68	13	-14	222	38	AAS 5				876
PEARL WHITE Synonym: BISMUTH OXYCHLORIDE						167					BOC 153				876
PELARGONIC ALCOHOL Synonym: NONANOL			646								NNN 841				877
PELARGONYL PEROXIDE	2130	52							222						877
PENT ACETATE	1104	26						-77	222						
PENTA Synonym: PENTACHLOROPHENOL	2020/3082	53		22	2679	912	172	133-	222	174	PCP 940	729			879
PENTA BENZENESULFONIC ACID Synonym: ALKYLBENZENESULFONIC ACID											ABS 11				
PENTABORANE	1380	75	664	22	2676	910	171	132-76	222	172	PTB 991	728			879
PENTABORON ENNEAHYDRIDE	1380	75							223						
PENTACHLORO DIPHENYL OXIDE					2678	911	171								
PENTACHLOROETHANE	1669/2810	55		22	2678	911	171		223		PCE 935	729			879
PENTACHLORONAPHTHALENE					2679	911	171		223	174					879
PENTACHLORONITROBENZENE					2679	911			223						879
PENTACHLOROPHENOL	2020/3082	53		22	2679	912	172	133-	223	174	PCP 940	729			879
PENTADECANOL			666								PDC 942				
PENTADECANOL Synonym: LINEAR ALCOHOLS			515								LAL 692				

HAZARDOUS MATERIALS REFERENCE BOOKS INDEX

CHEMICAL OR MATERIAL NAME	UN/NA Number	Guide Number (DOT)	Firefighter's Hazardous Materials Reference Book	First Aid Manual for Chemical Accidents	Sax's Dangerous Properties of Industrial Materials	Hazardous Chemicals Desk Reference	Rapid Guide to Hazardous Chemicals in the Workplace	Fire Protection Guide on Hazardous Materials (NFPA)	Firefighter's Handbook of Hazardous Materials	Pocket Guide to Chemical Hazards (NIOSH)	Chemical Hazard Response Information System (CHRIS)	Emergency Handling of Hazardous Materials in Surface Transportation (AAR)	Emergency Action Guides (EAG)	Chemical Data Notebook: A User's Manual	Condensed Chemical Dictionary
PENTADECYL ALCOHOL Synonym: PENTADECANOL			666								PDC 942				
PENTAERYTHRITE Synonym: PENTAERYTHRITOL			667		2682	912	172				PET 949				880
PENTAERYTHRITOL			667		2682	912	172				PET 949				880
PENTAERYTHRITOL TETRANITRATE	0411				2682	912			223						880
PENTALIN	1669	55							223						880
PENTAMETHYL HEPTANE	2286	27							223			730			
PENTAMETHYLENE Synonym: CYCLOPENTANE	1146	27	250		1003	338	61	−31	223		CYP 331	286	C8.3		881
PENTAMETHYLENE DICHLORIDE	1152	27							223						
PENTAMETHYLENE OXIDE								−77	223						
PENTANAL Synonym: VALERALDEHYDE (-n)	2058/1993	26	867		3475	1167	226	−92	223		VAL1171	945			881
PENTANE	1265	27	668	22	2687	913	172	−77	223		PTA 990	730	P1.0		881
PENTANE-2,4-DIONE	2310/1993	26							223			730			
PENTANE-n Synonym: PENTANE	1265	27	668	22	2687	913	172	−77	223	174	PTA 990	730	P1.0		881
PENTANOIC ACID	1760	60						−77	224						881
PENTANONE-3	1156	26							224						881
PENTAPHEN								−77	224						
PENTEK Synonym: PENTAERYTHRITOL			667		2682	912	172				PET 949				880
PENTENENITRILE	1992	28													
PENTOL	2705	60							224						

CHEMICAL OR MATERIAL NAME	UN/NA Number	Guide Number (DOT)	Firefighter's Hazardous Materials Reference Book	First Aid Manual for Chemical Accidents	Sax's Dangerous Properties of Industrial Materials	Hazardous Chemicals Desk Reference	Rapid Guide to Hazardous Chemicals in the Workplace	Fire Protection Guide on Hazardous Materials (NFPA)	Firefighter's Handbook of Hazardous Materials	Pocket Guide to Chemical Hazards (NIOSH)	Chemical Hazard Response Information System (CHRIS)	Emergency Handling of Hazardous Materials in Surface Transportation (AAR)	Emergency Action Guides (EAG)	Chemical Data Notebook: A User's Manual	Condensed Chemical Dictionary
PENTYL ACETATE-tert **Synonym: AMYL ACETATE-tert**	1104	26		12	241						AYA 111				
PENTYL ACETATES **Synonym: AMYL ACETATE (-n)**	1104	26	77	12	241	68	12	-14	224	38	AML 54	73			883
PENTYL ALCOHOL **Synonym: AMYL ALCOHOL (-n)**	1105/1987	26	78		242	69			224		AAN 4	74			73
PENTYL METHYL KETONE **Synonym: AMYL METHYL KETONE (-n)**	1110	26	81						31		AMK 53	76			
PENTYLAMINE	1106	68			2695	915			225						883
PENTYLCARBINOL-sec **Synonym: ETHYLBUTANOL**	2275	26	347						141		EBT 509				
PENTYLSILICON TRICHLORIDE **Synonym: AMYLTRICHLOROSILANE (-n)**	1728	29	84						32		ATS 107	78			76
PERACETIC ACID	2131	51	670	22				133–	225		PAA 921	732			883
PERBORATE	1480			22					225						
PERCARBAMIDE **Synonym: UREA PEROXIDE**	1511	35	866						283		UPO1166	945			884
PERCHLORATE	1481	35			2699	917			225			732			
PERCHLORETHYLENE	1897	74							225						
PERCHLORIC ACID	1873	47	672	22	2699			134–	225		PCL 937	732			884
PERCHLORIC ACID	1802	45	671	22	2699	917		134–	225		PCL 937	733			884
PERCHLORIC ACID SOLUTION **Synonym: PERCHLORIC ACID**	1873	47	672	22	2699	917			225		PCL 937	732			884
PERCHLOROCYCLOPENTADIENE **Synonym: HEXACHLOROCYCLOPENTADIENE**	2646	55	428		1841	619	108		158		HCC 623	507	H1.0.4		884

CHEMICAL OR MATERIAL NAME	UN/NA Number	Guide Number (DOT)	Firefighter's Hazardous Materials Reference Book	First Aid Manual for Chemical Accidents	Sax's Dangerous Properties of Industrial Materials	Hazardous Chemicals Desk Reference	Rapid Guide to Hazardous Chemicals in the Workplace	Fire Protection Guide on Hazardous Materials (NFPA)	Firefighter's Handbook of Hazardous Materials	Pocket Guide to Chemical Hazards (NIOSH)	Chemical Hazard Response Information System (CHRIS)	Emergency Handling of Hazardous Materials in Surface Transportation (AAR)	Emergency Action Guides (EAG)	Chemical Data Notebook: A User's Manual	Condensed Chemical Dictionary
PERCHLORODIHOMOCUBANE Synonym: MIREX	1615		589		2431	828			205		MRX 798				
PERCHLOROETHYLENE Synonym: TETRACHLOROETHYLENE	1897/1993	74	673	25	2701	919	173	166–	225	208	TTE1153	733	T0.5	T 1	884
PERCHLOROETHYLENE AND TRICHLOROETHANE BLEND	2810	55										733			
PERCHLOROMETHANE Synonym: CARBON TETRACHLORIDE	1846	55	192	13		246	43	48 –	66	60	CBT 223	193	C2.3	C 5	884
PERCHLOROMETHYL MERCAPTAN	1670	55	674	22	2702	919	174		225	176	PCM 938	734			884
PERCHLORYL FLUORIDE	3083	20		22	2702	920	174			176			P1.5		884
PERCHLORYL FLUORIDE	1955	15		22	2702	920	174		225						884
PERCLENE Synonym: TETRACHLOROETHYLENE	1897	74	808	25				166–	262	208	TTE1153	888	T0.5	T 1	1127
PERFLUORO-2-BUTENE	2422	12										734			
PERFLUOROETHYLENE	1081	17							225						885
PERFLUOROPROPANE	2424	12							226						885
PERFUMERY PRODUCTS	1266/1993	26							226			734			
PERHYDROL Synonym: HYDROGEN PEROXIDE	2015	47	450	18	1901	644	115	100–	165	126	HPO 643	532	H4.5	H 5	616
PERHYDROL Synonym: HYDROGEN PEROXIDE	2014	45		18	1901	644	115	99 –	165	126		531	H4.5	H 5	616
PERHYDRONAPTHALENE Synonym: DECAHYDRONAPHTHALENE	1147	27	257			345		−31	90		DHN 405	289			349
PERIODIC ACID									226						885
PERK Synonym: TETRACHLOROETHYLENE	1897	74	808	25				166–	262	208	TTE1153	888	T0.5	T 1	1127
PERLITE					2707	920									887

CHEMICAL OR MATERIAL NAME	UN/NA Number	Guide Number (DOT)	Firefighter's Hazardous Materials Reference Book	First Aid Manual for Chemical Accidents	Sax's Dangerous Properties of Industrial Materials	Hazardous Chemicals Desk Reference	Rapid Guide to Hazardous Chemicals in the Workplace	Fire Protection Guide on Hazardous Materials (NFPA)	Firefighter's Handbook of Hazardous Materials	Pocket Guide to Chemical Hazards (NIOSH)	Chemical Hazard Response Information System (CHRIS)	Emergency Handling of Hazardous Materials in Surface Transportation (AAR)	Emergency Action Guides (EAG)	Chemical Data Notebook: A User's Manual	Condensed Chemical Dictionary
PERMANGANATE	1482	35			2707	921			226			735			
PEROXAN Synonym: HYDROGEN PEROXIDE	2014	45		18	1901	644	115	99 –	165	126		531	H4.5	H 5	616
PEROXAN Synonym: HYDROGEN PEROXIDE	2015	47	450	18		644	115	100–	165	126	HPO 643	532	H4.5	H 5	616
PEROXIDE	1483	35			2708	921			226			735			888
PEROXIDE Synonym: HYDROGEN PEROXIDE	2015	47	450	18	1901	921	115	100–	165	126	HPO 643	532	H4.5	H 5	888
PEROXIDE Synonym: HYDROGEN PEROXIDE	2014	45		18	1901	921	115	99 –	165	126		531	H4.5	H 5	888
PEROXYACETIC ACID	3045	51			2709				226						888
PEROXYACETIC ACID Synonym: PERACETIC ACID	2131	51	670	22		922	174	133–	226		PAA 921	732			888
PEROXYDICARBONIC ACID BIS(1-METHYLETHYL) ESTER Synonym: ISOPROPYL PERCARBONATE	2134	52	485						175		IPC 671				662
PEROXYDISULFURIC ACID DIAMMONIUM SALT Synonym: AMMONIUM PERSULFATE	1444	35	67		2362	65	12		30		APE 73	66			68
PERSADOX Synonym: BENZOYL PEROXIDE	2085	49	122		392	130	23		42	44			B2.0	B 2	133
PERSIAN INSECT POWDER Synonym: PYRETHRINS	9 84/3082	31	732	23	2944	1007	191		245		PRR 989	800			
PESTICIDE	2903	28							226						
PESTICIDE	2210	37													
PESTICIDE	2902	55							226			736			
PESTICIDE	3021	28							226						
PESTICIDE	2588	55							226			736			

HAZARDOUS MATERIALS REFERENCE BOOKS INDEX

CHEMICAL OR MATERIAL NAME	UN/NA Number	Guide Number (DOT)	Firefighter's Hazardous Materials Reference Book	First Aid Manual for Chemical Accidents	Sax's Dangerous Properties of Industrial Materials	Hazardous Chemicals Desk Reference	Rapid Guide to Hazardous Chemicals in the Workplace	Fire Protection Guide on Hazardous Materials (NFPA)	Firefighter's Handbook of Hazardous Materials	Pocket Guide to Chemical Hazards (NIOSH)	Chemical Hazard Response Information System (CHRIS)	Emergency Handling of Hazardous Materials in Surface Transportation (AAR)	Emergency Action Guides (EAG)	Chemical Data Notebook: A User's Manual	Condensed Chemical Dictionary
PESTMASTER Synonym: ETHYLENE DIBROMIDE	1605	55	371		1601			85 –	143	110	EDB 518	430	E2.4.1		487
PESTMASTER Synonym: METHYL BROMIDE	1062	55	559	20	2274	776	137	110–67	192	146	MTB 805	619	M3.0	M 3	759
PETROHOL Synonym: ISOPROPYL ALCOHOL	1219/1993	26	482	18	2041	687	124	–63	227	132	IPA 670	560		I 2	660
PETROL Synonym: GASOLINE	1203	27	412	18	1779	601	104	–56	227		GAK 602	473	G1.0	G 1	
PETROL Synonym: GASOLINE AUTOMOTIVE	1203	27		18	1779				227		GAT 603	473	G1.0	G 1	554
PETROLATUM			675								PTL 997				889
PETROLATUM JELLY Synonym: PETROLATUM			675								PTL 997				889
PETROLEUM Synonym: OILS CRUDE	1267	27			2713	923					OIL 882	737			889
PETROLEUM ASPHALT Synonym: OILS MISCELLANEOUS ROAD	1999	27			2713	924			36		ORD 899				100
PETROLEUM ASPHALT Synonym: ASPHALT	1999	27	105	12	2713	924	17	–17			ASP 93	99	A12.0		
PETROLEUM CRUDE OIL	1267	27						–78				737			
PETROLEUM DISTILLATE Synonym: DISTILLATES FLASHED FEED STOCKS	1268	27			2714	924	175			176	DFF 398	737			
PETROLEUM DISTILLATE Synonym: DISTILLATES STRAIGHT RUN	1268	27				924	175			176	DSR 480	737			
PETROLEUM DISTILLATE Synonym: PETROLEUM NAPHTHA	1255	27	676		2714		152				PTN 999	739	P2.0		890

CHEMICAL OR MATERIAL NAME	UN/NA Number	Guide Number (DOT)	Firefighter's Hazardous Materials Reference Book	First Aid Manual for Chemical Accidents	Sax's Dangerous Properties of Industrial Materials	Hazardous Chemicals Desk Reference	Rapid Guide to Hazardous Chemicals in the Workplace	Fire Protection Guide on Hazardous Materials (NFPA)	Firefighter's Handbook of Hazardous Materials	Pocket Guide to Chemical Hazards (NIOSH)	Chemical Hazard Response Information System (CHRIS)	Emergency Handling of Hazardous Materials in Surface Transportation (AAR)	Emergency Action Guides (EAG)	Chemical Data Notebook: A User's Manual	Condensed Chemical Dictionary
PETROLEUM ETHER Synonym: PETROLEUM NAPHTHA	1255	27	676	22			152				PTN 999	739	P2.0		889
PETROLEUM ETHER	1271	26		22				−78				738			889
PETROLEUM GAS	1075	22							227			738			890
PETROLEUM INSULATING OIL Synonym: OILS MISCELLANEOUS TRANSFORMER	1270	27									OTF 914				
PETROLEUM JELLY Synonym: PETROLATUM			675								PTL 997				890
PETROLEUM NAPHTHA	1255	27	676						227		PTN 999	739	P2.0		890
PETROLEUM NEUTRALIZING AGENT Synonym: CAUSTIC SODA SOLUTION SPENT	1824	60	194						67		CSS 314		C3.0		
PETROLEUM OIL Synonym: PETROLEUM NAPHTHA	1255	27	676			835	152				PTN 999	739	P2.0		890
PETROLEUM OIL	1270	27							227			740			
PETROLEUM PITCH Synonym: ASPHALT BLENDING STOCKS STRAIGHT RUN RESIDUE	1999	27	105	12	307	97	17		36		ASR 94	99	A12.0		100
PETROLEUM PITCH Synonym: ASPHALT	1999	27	105	12	307	97	17	−17			ASP 93	99	A12.0		
PETROLEUM PROPYLENE TETRAMER	1993	27										741			
PETROLEUM REFRINERY SULFIDE WASTE	1760	60													
PETROLEUM RESIDUE Synonym: ASPHALT BLENDING STOCKS STRAIGHT RUN RESIDUE	1999	27	105	12	307	97	17				ASR 94	99	A12.0		
PETROLEUM SOLVENT Synonym: NAPHTHA V,M & P	1255	27	605	12	2463	835	152	−73	207		NVM 863				890

CHEMICAL OR MATERIAL NAME	UN/NA Number	Guide Number (DOT)	Firefighter's Hazardous Materials Reference Book	First Aid Manual for Chemical Accidents	Sax's Dangerous Properties of Industrial Materials	Hazardous Chemicals Desk Reference	Rapid Guide to Hazardous Chemicals in the Workplace	Fire Protection Guide on Hazardous Materials (NFPA)	Firefighter's Handbook of Hazardous Materials	Pocket Guide to Chemical Hazards (NIOSH)	Chemical Hazard Response Information System (CHRIS)	Emergency Handling of Hazardous Materials in Surface Transportation (AAR)	Emergency Action Guides (EAG)	Chemical Data Notebook: A User's Manual	Condensed Chemical Dictionary
PETROLEUM SOLVENT Synonym: NAPHTHA SOLVENT	1256	27	603	12		835	152	-73	207		NSV 850	673			
PETROLEUM SOLVENT Synonym: PETROLEUM NAPHTHA	1255	27	676	12	2463	835	152				PTN 999	739	P2.0		
PETROLEUM SOLVENT Synonym: NAPHTHA STODDARD SOLVENT	1268	27	604	12	2463						NSS 849	739			
PETROLEUM SPIRIT	1271	26			2714	924	175								890
PETROLEUM SPIRITS Synonym: MINERAL SPIRITS	1300/1993	27	588					-73	204		MNS 777	653			890
PETROLEUM SPIRITS	1115	26							227						890
PETROLEUM TAILINGS Synonym: ASPHALT BLENDING STOCKS ROOFERS FLUX	1999	27	105	12	307	97	17				ARF 84	99	A12.0		
PETROLEUM WAX Synonym: WAXES PARAFFIN	1993	27	879					-94			WPF1185				890
PHELLANDRENE Synonym: DIPENTENE	2052/1993	27	318					-45	126		DPN 468	381			427
PHELLANDRENE-beta								-78	227						890
PHENACHLOR Synonym: TRICHLOROPHENOL	2020	53	839								TPH1139				
PHENACYL BROMIDE	2645/2810	55							227			742			
PHENACYL CHLORIDE Synonym: CHLOROACETOPHENONE (-alpha)	1697	55	200		762	271	47	-26	227	64	CRA 297	217			892
PHENANTHRENE					2717	924	175	-78							892
PHENARSAZINE CHLORIDE	1698	55			2719	925			227						892
PHENCAPTON	2783	55			2721	926			227						

CHEMICAL OR MATERIAL NAME	UN/NA Number	Guide Number (DOT)	Firefighter's Hazardous Materials Reference Book	First Aid Manual for Chemical Accidents	Sax's Dangerous Properties of Industrial Materials	Hazardous Chemicals Desk Reference	Rapid Guide to Hazardous Chemicals in the Workplace	Fire Protection Guide on Hazardous Materials (NFPA)	Firefighter's Handbook of Hazardous Materials	Pocket Guide to Chemical Hazards (NIOSH)	Chemical Hazard Response Information System (CHRIS)	Emergency Handling of Hazardous Materials in Surface Transportation (AAR)	Emergency Action Guides (EAG)	Chemical Data Notebook: A User's Manual	Condensed Chemical Dictionary
PHENE Synonym: BENZENE	1114	27	115	12	356	115	21	35 –17	40	44	BNZ 152	112	B1.0	B 1	128
PHENETHYL ALCOHOL					2723	926		–78	227						892
PHENETHYLENE Synonym: STYRENE MONOMER	2055	27	795		3145	1074	202		257		STY1064	862	S3.0	S 6	1097
PHENETIDINE-ortho	2311	55							227			742			893
PHENETIDINE-para	2311	55							227			742			893
PHENETOLE					2727				227						893
PHENIC ACID Synonym: CARBOLIC ACID	1671	55	185						65			743	C2.0	P 1	893
PHENIC ACID Synonym: PHENOL	1671	55	677	22	2729	928	175	134–78	227	176	PHN 958	743	C2.0	P 1	893
PHENOL	2312	55		22	2729	928	175		227	176		744		P 1	894
PHENOL	1671	55	677	22	2729	928	175	134–78	227	176	PHN 958	743		P 1	894
PHENOL ACETATE	1993	27							227			743			
PHENOL OCTYL	1693	58													
PHENOL SOLUTION	2821	55		22	2729	928	175		227	176				P 1	
PHENOLSULFONIC ACID	18031760	60							227			744			895
PHENOLSULPHONIC ACID	1760/1803	60										744			
PHENOTHIAZINE					2730	929	176								895
PHENOXY PESTICIDE	2765	55							228			745			
PHENOXY PESTICIDE	2999	28							227						
PHENOXY PESTICIDE	3000/2765	55							228			745			
PHENOXY PESTICIDE	2766	28							228			744			

HAZARDOUS MATERIALS REFERENCE BOOKS INDEX

CHEMICAL OR MATERIAL NAME	UN/NA Number	Guide Number (DOT)	Firefighter's Hazardous Materials Reference Book	First Aid Manual for Chemical Accidents	Sax's Dangerous Properties of Industrial Materials	Hazardous Chemicals Desk Reference	Rapid Guide to Hazardous Chemicals in the Workplace	Fire Protection Guide on Hazardous Materials (NFPA)	Firefighter's Handbook of Hazardous Materials	Pocket Guide to Chemical Hazards (NIOSH)	Chemical Hazard Response Information System (CHRIS)	Emergency Handling of Hazardous Materials in Surface Transportation (AAR)	Emergency Action Guides (EAG)	Chemical Data Notebook: A User's Manual	Condensed Chemical Dictionary
PHENOXYBENZENE Synonym: DIPHENYL ETHER	2489	55							127		DPE 462				430
PHENYL ACETATE					2736			−78	228						896
PHENYL ALCOHOL Synonym: CARBOLIC ACID	1671	55	185						65			743	C2.0	P 1	217
PHENYL BROMIDE Synonym: BROMOBENZENE	2514	26	137		2744	934		−19	228		BBZ 125	142			897
PHENYL CHLORIDE Synonym: CHLOROBENZENE	1134	27	203	13			47	51 −26	228	64	CRB 298	219	C4.1		898
PHENYL CHLORIDE	1134	27			768	273			228						898
PHENYL CHLOROFORMATE	2746/2810	55							229			747			
PHENYL CHLOROMETHYLKETONE Synonym: CHLOROACETOPHENONE (-alpha)	1697	55	200		762	271	47	−26	229	64	CRA 297	217			898
PHENYL CYANIDE	2224	55							228						898
PHENYL ETHER					2754	937	177		228	178					899
PHENYL ETHER Synonym: DIPHENYL ETHER	2489	55			2754	937	177		228		DPE 462				899
PHENYL ETHER BIPHENYL MIXTURE					2754	937	177			178					
PHENYL GLYCIDYL ETHER					2759	938	177		228	178					900
PHENYL GLYCYDYL ETHER					2759		177								
PHENYL HYDRATE Synonym: CARBOLIC ACID	1671	55	185						65			743	C2.0	P 1	217
PHENYL HYDRIDE Synonym: BENZENE	1114	27	115	12	356	115	21	35 −17	230	44	BNZ 152	112	B1.0	B 1	128
PHENYL HYDROXIDE Synonym: CARBOLIC ACID	1671	55	185		2729	928			65			743	C2.0	P 1	894

CHEMICAL OR MATERIAL NAME	UN/NA Number	Guide Number (DOT)	Firefighter's Hazardous Materials Reference Book	First Aid Manual for Chemical Accidents	Sax's Dangerous Properties of Industrial Materials	Hazardous Chemicals Desk Reference	Rapid Guide to Hazardous Chemicals in the Workplace	Fire Protection Guide on Hazardous Materials (NFPA)	Firefighter's Handbook of Hazardous Materials	Pocket Guide to Chemical Hazards (NIOSH)	Chemical Hazard Response Information System (CHRIS)	Emergency Handling of Hazardous Materials in Surface Transportation (AAR)	Emergency Action Guides (EAG)	Chemical Data Notebook: A User's Manual	Condensed Chemical Dictionary
PHENYL HYDROXIDE Synonym: PHENOL	1671	55	677	22			175	134–78	227	176	PHN 958	743	C2.0	P 1	217
PHENYL ISOCYANATE	2487	55			2761	939			228						901
PHENYL MERCAPTAN	2337	57			2762	939	178		228			745	P3.0.5		901
PHENYL METHYL ETHER	2222	26							228						
PHENYL METHYL KETONE	1993	27										746			902
PHENYL PHOSPHORUS DICHLORIDE	2798	39							228			746			
PHENYL PHOSPHORUS THIODICHLORIDE	2799	39							229			751			
PHENYL UREA PESTICIDE	3002/2767	55							229			750			
PHENYL UREA PESTICIDE	2768	28							229			750			
PHENYL UREA PESTICIDE	2767	55							229			751			
PHENYL UREA PESTICIDE	3001	28							229						
PHENYLACETALDEHYDE DIMETHYL ACETAL						930			229						896
PHENYLACETAMIDE					2735										896
PHENYLACETIC ACID					2736	930		−78							896
PHENYLACETONITRILE	2470	55			2737	931			229			746			897
PHENYLACETYL CHLORIDE	2577	60							229						897
PHENYLAMINE Synonym: ANILINE OIL	1547	57	85		254	72	13		229		ANL 64	79	A10.0	A12	897
PHENYLAMINE Synonym: ANILINE	1547	57	85	12	254	72	13	29 −16	229	40	ANL 64	79		A12	897
PHENYLANILINE (-n) Synonym: DIPHENYLAMINE			319	16	1457	485	84	−45			DAM 337				897
PHENYLARSENIC DICHLORIDE Synonym: PHENYLDICHLOROARSINE	1556	55	678						228		PDL 945	747			898

HAZARDOUS MATERIALS REFERENCE BOOKS INDEX

CHEMICAL OR MATERIAL NAME	UN/NA Number	Guide Number (DOT)	Firefighter's Hazardous Materials Reference Book	First Aid Manual for Chemical Accidents	Sax's Dangerous Properties of Industrial Materials	Hazardous Chemicals Desk Reference	Rapid Guide to Hazardous Chemicals in the Workplace	Fire Protection Guide on Hazardous Materials (NFPA)	Firefighter's Handbook of Hazardous Materials	Pocket Guide to Chemical Hazards (NIOSH)	Chemical Hazard Response Information System (CHRIS)	Emergency Handling of Hazardous Materials in Surface Transportation (AAR)	Emergency Action Guides (EAG)	Chemical Data Notebook: A User's Manual	Condensed Chemical Dictionary
PHENYLCARBINOL					395	131		−18	228		BAL 117				898
Synonym: BENZYL ALCOHOL															
PHENYLCARBYLAMINE CHLORIDE	1672/2810	55							228			747			898
PHENYLCHLORIDE	1134	27	203	13	768	273	47	51 −26	228	64	CRB 298	219	C4.1		898
Synonym: CHLOROBENZENE															
PHENYLCHLOROFORM	2226	60							229						898
PHENYLCHLOROMETHYLKETONE	1697	55							229						898
PHENYLCYANIDE	2224	55	120		377	125			228		BZN 198	117			898
Synonym: BENZONITRILE															
PHENYLCYCLOHEXANE					2746				229						898
PHENYLDICHLOROARSINE	1556	55	678						228		PDL 945	747			898
PHENYLENEDIAMINE	1673	53							230			748			
PHENYLENEDIAMINE-meta	1673	53			2751	935	176		229			748			899
PHENYLENEDIAMINE-ortho	1673	53			2752	936	176		229						899
PHENYLENEDIAMINE-para	1673	53		22		936	177		229	178					899
PHENYLENEDIAMINE-para	1993	27		22	2752										899
PHENYLETHANE	1175	26	367	16	1579	528	91	81 −48	230	106	ETB 561	409	E2.2.3		899
Synonym: ETHYL BENZENE															
PHENYLETHANOLAMINE								135−	230						899
PHENYLETHENE	2055	27	795		3145	1074	202		257		STY1064	862	S3.0	S 6	1097
Synonym: STYRENE MONOMER															
PHENYLETHYLENE	2055	27	795		3145	1074	202		230		STY1064	862	S3.0	S 6	900
Synonym: STYRENE MONOMER															
PHENYLETHYLENE	2055	27	795	24	3145	1074	202	162–84	230	200	STY1064	862		S 6	900
Synonym: STYRENE															

CHEMICAL OR MATERIAL NAME	UN/NA Number	Guide Number (DOT)	Firefighter's Hazardous Materials Reference Book	First Aid Manual for Chemical Accidents	Sax's Dangerous Properties of Industrial Materials	Hazardous Chemicals Desk Reference	Rapid Guide to Hazardous Chemicals in the Workplace	Fire Protection Guide on Hazardous Materials (NFPA)	Firefighter's Handbook of Hazardous Materials	Pocket Guide to Chemical Hazards (NIOSH)	Chemical Hazard Response Information System (CHRIS)	Emergency Handling of Hazardous Materials in Surface Transportation (AAR)	Emergency Action Guides (EAG)	Chemical Data Notebook: A User's Manual	Condensed Chemical Dictionary
PHENYLHYDRAZINE	2572/2810	53	679	22	2759	938	178	−79	230	180		748			900
PHENYLHYDRAZINE HYDROCHLORIDE	2572	53			2760	939	178		230		PHH 957				
PHENYLHYDRAZINIUM CHLORIDE Synonym: PHENYLHYDRAZINE HYDROCHLORIDE	2572	53			2760	939	178		230		PHH 957				
PHENYLHYDRIDE	1114	27							230						
PHENYLIC ACID Synonym: CARBOLIC ACID	1671	55	185						65			743	C2.0	P 1	901
PHENYLIC ACID	2821	55							230						901
PHENYLIC ALCOHOL Synonym: CARBOLIC ACID	1671	55	185						65			743	C2.0	P 1	217
PHENYLIMINOPHOSGENE	1672	55							230						
PHENYLMAGNESIUM BROMIDE					2762				230						901
PHENYLMERCURIC ACETATE	1574/2811	55		22				135–	228		PMA 965	749			901
PHENYLMERCURIC COMPOUND	2026	53							230			749			
PHENYLMERCURIC HYDROXIDE	1894	53			2763				230						901
PHENYLMERCURIC NITRATE	1395/2811	53							230			749			902
PHENYLMETHANE Synonym: TOLUENE	1294	27	826	25	3309	1125	215	170–87	230	214	TOL1135	909	T4.0	T 4	902
PHENYLMETHANOL Synonym: BENZYL ALCOHOL	1987				395	131		−18	42		BAL 117	118			902
PHENYLMETHYL ALCOHOL Synonym: BENZYL ALCOHOL	1987				395	131		−18	42		BAL 117	118			134
PHENYLMETHYL AMINE Synonym: BENZYLAMINE	2294	57			396				231		BZM 197				134
PHENYLMETHYL ETHANOL AMINE								−79	230						902

HAZARDOUS MATERIALS REFERENCE BOOKS INDEX

CHEMICAL OR MATERIAL NAME	UN/NA Number	Guide Number (DOT)	Firefighter's Hazardous Materials Reference Book	First Aid Manual for Chemical Accidents	Sax's Dangerous Properties of Industrial Materials	Hazardous Chemicals Desk Reference	Rapid Guide to Hazardous Chemicals in the Workplace	Fire Protection Guide on Hazardous Materials (NFPA)	Firefighter's Handbook of Hazardous Materials	Pocket Guide to Chemical Hazards (NIOSH)	Chemical Hazard Response Information System (CHRIS)	Emergency Handling of Hazardous Materials in Surface Transportation (AAR)	Emergency Action Guides (EAG)	Chemical Data Notebook: A User's Manual	Condensed Chemical Dictionary
PHENYLPHOSPHINE					2768	941	179		228						903
PHENYLPHOSPHINE DICHLORIDE Synonym: BENZENE PHOSPHORUS DICHLORIDE	2798	39	117						40		BPD 156	746			129
PHENYLPHOSPHINE THIODICHLORIDE Synonym: BENZENE PHOSPHORUS THIODICHLORIDE	2799	39	118						40		BPT 163				
PHENYLPHOSPHONOTHIOIC DICHLORIDE Synonym: BENZENE PHOSPHORUS THIODICHLORIDE	2799	39	118						40		BPT 163				
PHENYLPHOSPHONOUS DICHLORIDE Synonym: BENZENE PHOSPHORUS DICHLORIDE	2798	39	117						40		BPD 156	746			129
PHENYLPHOSPHORUS DICHLORIDE Synonym: BENZENE PHOSPHORUS DICHLORIDE	2798	39	117						228		BPD 156	746			129
PHENYLPROPYLENE Synonym: METHYL STYRENE-alpha	2303	27	580		2398	818	149		197	154	MSR 802				780
PHENYLTRICHLOROSILANE	1804	39							231			750			905
PHORATE				22	2779	943	179		231						905
PHORONE	1993	27			2782	944		-80	231			751			905
PHOSDRIN	2783	55	681	22					231	180	PHD 955				906
PHOSFENE Synonym: PHOSDRIN	2783	55	681	22					231	180	PHD 955				906
PHOSGENE	1076	15	682	22	2782	944	179	136–	231	180	PHG 956	752	P3.0	P 2	906
PHOSPHABICYCLONONANE	2940	37							231						
PHOSPHATES					2783	945			231						906
PHOSPHATIC FERTILIZER SOLUTION	1805	60													
PHOSPHIDES					2783	945			231						
PHOSPHINE	2199	18	683	22	2783	945	180	136–80	231	180		752	P3.1		907

CHEMICAL OR MATERIAL NAME	UN/NA Number	Guide Number (DOT)	Firefighter's Hazardous Materials Reference Book	First Aid Manual for Chemical Accidents	Sax's Dangerous Properties of Industrial Materials	Hazardous Chemicals Desk Reference	Rapid Guide to Hazardous Chemicals in the Workplace	Fire Protection Guide on Hazardous Materials (NFPA)	Firefighter's Handbook of Hazardous Materials	Pocket Guide to Chemical Hazards (NIOSH)	Chemical Hazard Response Information System (CHRIS)	Emergency Handling of Hazardous Materials in Surface Transportation (AAR)	Emergency Action Guides (EAG)	Chemical Data Notebook: A User's Manual	Condensed Chemical Dictionary
PHOSPHINIC ACID AMMONIUM SALT Synonym: AMMONIUM HYPOPHOSPHITE					228				29		AHP 37				66
PHOSPHONIUM IODIDE					2785				232						907
PHOSPHOPHOSPHORUS CHLORIDE	1809	39													
PHOSPHORATED HYDROGEN Synonym: PHOSPHINE	2199	18	683	22	2783	945	180	136–80	231	180		752	P3.1		907
PHOSPHORETTED HYDROGEN	2199	18							232						
PHOSPHORIC ACID	1805	60	684	22	2786	946	180	137–	232	182	PAC 922	753	P4.0		908
PHOSPHORIC ACID ANHYDRIDE	1807	39		22					232						908
PHOSPHORIC ACID TRIETHYLENEIMINE	2501	55	685						232						
PHOSPHORIC ACID-ortho Synonym: PHOSPHORIC ACID	1805	60	684	22	2786	946	180	137–	232	182	PAC 922	753	P4.0		908
PHOSPHORIC ANHYDRIDE	1807	39							232			754			908
PHOSPHORIC SULFIDE Synonym: PHOSPHORUS PENTASULFIDE	1340	41	688		2794	951	181	139–	232	182	PPP 979	757	P4.2		908
PHOSPHORODICHLORIDIC ACID ETHYL ESTER Synonym: ETHYL PHOSPHORODICHLORIDATE	1760/2927	60	363		1654	558			139		EPP 556	422			
PHOSPHOROTHIOIC ACID O,O-DIETHYL O-p-NITROPHENYL ESTER Synonym: PARATHION	2784	28		22	2667	907			222		PTO1000				872
PHOSPHOROUS (III) HYDRIDE HYDROGEN PHOSPHIDE Synonym: PHOSPHINE	2199	18	683	22	2783	945	180	136–80	231	180		752	P3.1		907
PHOSPHOROUS ACID (ortho)	2834	60							232						908
PHOSPHOROUS BROMIDE	1808	39							232						
PHOSPHORUS Synonym: PHOSPHORUS YELLOW & WHITE	1381	38	693	22	2791	949	180	137–	232	182	PPW 982	760	P4.1		908

CHEMICAL OR MATERIAL NAME	UN/NA Number	Guide Number (DOT)	Firefighter's Hazardous Materials Reference Book	First Aid Manual for Chemical Accidents	Sax's Dangerous Properties of Industrial Materials	Hazardous Chemicals Desk Reference	Rapid Guide to Hazardous Chemicals in the Workplace	Fire Protection Guide on Hazardous Materials (NFPA)	Firefighter's Handbook of Hazardous Materials	Pocket Guide to Chemical Hazards (NIOSH)	Chemical Hazard Response Information System (CHRIS)	Emergency Handling of Hazardous Materials in Surface Transportation (AAR)	Emergency Action Guides (EAG)	Chemical Data Notebook: A User's Manual	Condensed Chemical Dictionary
PHOSPHORUS	2447	38		22	2791	949	180		232			760			908
PHOSPHORUS (III) CHLORIDE Synonym: PHOSPHORUS TRICHLORIDE	1809	39	690	22	2796	953	182	140–	233	184	PPT 981	758	P4.3	P 3	909
PHOSPHORUS ACID-o	2834	60							232						
PHOSPHORUS BLACK			691								PPB 973				
PHOSPHORUS BROMIDE Synonym: PHOSPHORUS TRIBROMIDE	1808	39	689		2795	952		140–	233		PBR 932	758			910
PHOSPHORUS CHLORIDE Synonym: PHOSPHORUS OXYCHLORIDE	1810	39	686	22	2793	950	181	138–	233		PPO 978	755	P4.1.1		909
PHOSPHORUS CHLORIDE Synonym: PHOSPHORUS TRICHLORIDE	1809	39	690	22	2796	953	182	140–	233	184	PPT 981	758	P4.3	P 3	909
PHOSPHORUS CHLORIDE OXIDE Synonym: PHOSPHORUS OXYCHLORIDE	1810	39	686		2793	950	181	138–	233		PPO 978	755	P4.1.1		909
PHOSPHORUS HEPTASULFIDE	1339	32			2793	950			233			754			909
PHOSPHORUS OXIDE TRICHLORIDE Synonym: PHOSPHORUS OXYCHLORIDE	1810	39	686		2793	950	181	138–	233		PPO 978	755	P4.1.1		909
PHOSPHORUS OXYBROMIDE	1939	39							233			755			909
PHOSPHORUS OXYBROMIDE	2576	39							233			755			909
PHOSPHORUS OXYCHLORIDE	1810	39	686		2793	950	181	138–	233		PPO 978	755	P4.1.1		909
PHOSPHORUS OXYTRICHLORIDE Synonym: PHOSPHORUS OXYCHLORIDE	1810	39	686		2793	950	181	138–	233		PPO 978	755	P4.1.1		909
PHOSPHORUS PENTABROMIDE	2691/1759	39			2794	951			233			756			910
PHOSPHORUS PENTACHLORIDE	1806	39	687	22	2794		181	139–	233	182		756			910
PHOSPHORUS PENTAFLUORIDE	2198/1955	15			2794	951			233			756			910
PHOSPHORUS PENTASULFIDE	1340	41	688	22	2794	951	181	139–	233	182	PPP 979	757	P4.2		910

CHEMICAL OR MATERIAL NAME	UN/NA Number	Guide Number (DOT)	Firefighter's Hazardous Materials Reference Book	First Aid Manual for Chemical Accidents	Sax's Dangerous Properties of Industrial Materials	Hazardous Chemicals Desk Reference	Rapid Guide to Hazardous Chemicals in the Workplace	Fire Protection Guide on Hazardous Materials (NFPA)	Firefighter's Handbook of Hazardous Materials	Pocket Guide to Chemical Hazards (NIOSH)	Chemical Hazard Response Information System (CHRIS)	Emergency Handling of Hazardous Materials in Surface Transportation (AAR)	Emergency Action Guides (EAG)	Chemical Data Notebook: A User's Manual	Condensed Chemical Dictionary
PHOSPHORUS PENTASULPHIDE Synonym: **PHOSPHORUS PENTASULFIDE**	1340	41	688	22	2794	951	181	139-	233	182	PPP 979	757	P4.2		910
PHOSPHORUS PENTOXIDE	1807	39			2795	952	182		233			754			910
PHOSPHORUS PERSULFIDE Synonym: **PHOSPHORUS PENTASULFIDE**	1340	41	688	22	2794	951	181	139-	233	182	PPP 979	757	P4.2		910
PHOSPHORUS RED	1338	32	692		2791	948		138-	232		PPR 980	759			
PHOSPHORUS SESQUISULFIDE	1341	41			2795	952			233			757			910
PHOSPHORUS SULFIDE Synonym: **PHOSPHORUS PENTASULFIDE**	1340	41	688	22	2794	951	181	139-	233	182	PPP 979	757	P4.2		910
PHOSPHORUS TRIBROMIDE	1808	39	689		2795	952		140-	233		PBR 932	758			910
PHOSPHORUS TRICHLORIDE	1809	39	690	22	2796	953	182	140-	233	184	PPT 981	758	P4.3	P 3	910
PHOSPHORUS TRICHLORIDE OXIDE Synonym: **PHOSPHORUS OXYCHLORIDE**	1810	39	686		2793	950	181	138-	233		PPO 978	755	P4.1.1		909
PHOSPHORUS TRIFLUORIDE	9273	15							233						
PHOSPHORUS TRIHYDRIDE Synonym: **PHOSPHINE**	2199	18	683	22	2783	945	180	136-80	233	180		752	P3.1		907
PHOSPHORUS TRIOXIDE	2578/1759	60			2796	953			233			758			
PHOSPHORUS TRISULFIDE	1343	41			2796	953			233			759			911
PHOSPHORUS YELLOW & WHITE	1381	38	693	22	2791	949	180	137-	232	182	PPW 982	760	P4.1		
PHOSPHORYL CHLORIDE Synonym: **PHOSPHORUS OXYCHLORIDE**	1810	39	686		2793	950	181	138-	233		PPO 978	755	P4.1.1		911
PHOSPHOTUNGSTIC ACID					2797				234						911
PHOTOGRAPHIC FILM NEC UNEXPOSED	1324	32										761			
PHOTOPHOR Synonym: **CALCIUM PHOSPHIDE**	1360	41	176		671	234			63		CPP 293	184			912

HAZARDOUS MATERIALS REFERENCE BOOKS INDEX

CHEMICAL OR MATERIAL NAME	UN/NA Number	Guide Number (DOT)	Firefighter's Hazardous Materials Reference Book	First Aid Manual for Chemical Accidents	Sax's Dangerous Properties of Industrial Materials	Hazardous Chemicals Desk Reference	Rapid Guide to Hazardous Chemicals in the Workplace	Fire Protection Guide on Hazardous Materials (NFPA)	Firefighter's Handbook of Hazardous Materials	Pocket Guide to Chemical Hazards (NIOSH)	Chemical Hazard Response Information System (CHRIS)	Emergency Handling of Hazardous Materials in Surface Transportation (AAR)	Emergency Action Guides (EAG)	Chemical Data Notebook: A User's Manual	Condensed Chemical Dictionary
PHTHALANDIONE Synonym: PHTHALIC ANHYDRIDE	2214/1759	60	694	22	2798	954	182	141–80	234	184	PAN 927	761	P4.4		913
PHTHALIC ACID					2798	954		-80	234						913
PHTHALIC ACID ANHYDRIDE Synonym: PHTHALIC ANHYDRIDE	2214/1759	60	694	22	2798	954	182	141–80	234	184	PAN 927	761	P4.4		913
PHTHALIC ACID BENZYL BUTYL ETHER Synonym: BUTYL BENZYL PHTHALATE	3082/9188										BPH 159	151			183
PHTHALIC ACID BIS (2-ETHYLHEXYL ESTER) Synonym: DIOCTYL PHTHALATE (-n)					1445	481		-45			DOP 455				425
PHTHALIC ACID BIS (8-METHYLNONYL) ESTER Synonym: DIISODECYL PHTHALATE								-40			DID 410				406
PHTHALIC ACID DIAMYL ESTER Synonym: DI-n-AMYL PHTHALATE											DAP 339				364
PHTHALIC ACID DIBUTYL ESTER Synonym: DIBUTYL PHTHALATE	9095		265		1119	375	66	-34		84	DPA 458				374
PHTHALIC ACID DIETHYL ESTER Synonym: DIETHYL PHTHALATE	9188/3082		288		1236	415	74	-39	108		DPH 465	348			396
PHTHALIC ACID DIHEPTYL ESTER Synonym: DIHEPTYL PHTHALATE											DHP 406				
PHTHALIC ACID DIISODECYL ESTER Synonym: DIISODECYL PHTHALATE								-40			DID 410				406
PHTHALIC ACID DIPENTYL ESTER Synonym: DI-n-AMYL PHTHALATE											DAP 339				364
PHTHALIC ACID-meta Synonym: ISOPHTHALIC ACID			478		2035						IPL 676				658
PHTHALIC ANHYDRIDE	2214/1759	60	694	22	2798	954	182	141–80	234	184	PAN 927	761	P4.4		913

CHEMICAL OR MATERIAL NAME	UN/NA Number	Guide Number (DOT)	Firefighter's Hazardous Materials Reference Book	First Aid Manual for Chemical Accidents	Sax's Dangerous Properties of Industrial Materials	Hazardous Chemicals Desk Reference	Rapid Guide to Hazardous Chemicals in the Workplace	Fire Protection Guide on Hazardous Materials (NFPA)	Firefighter's Handbook of Hazardous Materials	Pocket Guide to Chemical Hazards (NIOSH)	Chemical Hazard Response Information System (CHRIS)	Emergency Handling of Hazardous Materials in Surface Transportation (AAR)	Emergency Action Guides (EAG)	Chemical Data Notebook: A User's Manual	Condensed Chemical Dictionary
PHTHALIMIDE DERIVATIVE PESTICIDE	3008/2767	55							234			762			
PHTHALIMIDE DERIVATIVE PESTICIDE	3007	28							234						
PHTHALIMIDE DERIVATIVE PESTICIDE	2773	55							234			762			
PHTHALIMIDE DERIVATIVE PESTICIDE	2774	28							234			761			
PHTHALONITRILE					2800	955			234						913
PHYTAR Synonym: SODIUM CACODYLATE	1688/2811	53	765						252		SCD1020	834			1051
PIC CLOR PS Synonym: CHLOROPICRIN	1580	56	210	14	865	296	50	54 –		70	CPL 290	229	C4.3		271
PICFUME Synonym: CHLOROPICRIN	1580	56	210	14	865	296	50	54 –	77	70	CPL 290	229	C4.3		271
PICOLINE Synonym: 2-METHYLPYRIDINE	2313	27			2391	815			234		MPR 789	762			
PICOLINE-alpha Synonym: 2-METHYLPYRIDINE	2313/1993	27			2391	815			234		MPR 789	762			915
PICOLINE-beta	2313/1993	27			2803				234			762			915
PICOLINE-y	2313/1993	27							234			762			915
PICRAMIC ACID									234						916
PICRATES	0081				2805				234			763			
PICRATOL	1347	33							234						
PICRIC ACID	1344	33		22	2805	956	183	141–	234	184		763		P 4	916
PICRIC ACID	0154			22	2805	956	183		234			763		P 4	916
PICRIDE Synonym: CHLOROPICRIN	1580	56	210	14	865	296	50	54 –	77	70	CPL 290	229	C4.3		271
PICRITE	1336	33							234						

HAZARDOUS MATERIALS REFERENCE BOOKS INDEX

CHEMICAL OR MATERIAL NAME	UN/NA Number	Guide Number (DOT)	Firefighter's Hazardous Materials Reference Book	First Aid Manual for Chemical Accidents	Sax's Dangerous Properties of Industrial Materials	Hazardous Chemicals Desk Reference	Rapid Guide to Hazardous Chemicals in the Workplace	Fire Protection Guide on Hazardous Materials (NFPA)	Firefighter's Handbook of Hazardous Materials	Pocket Guide to Chemical Hazards (NIOSH)	Chemical Hazard Response Information System (CHRIS)	Emergency Handling of Hazardous Materials in Surface Transportation (AAR)	Emergency Action Guides (EAG)	Chemical Data Notebook: A User's Manual	Condensed Chemical Dictionary
PIGMENT WHITE 3	1794	60			2105		128		179		LSF 709	575			692
Synonym: **LEAD SULFATE**															
PILOCARPINE					2806	957									
PIMELIC KETONE	1915	26	246	14	991	334	59	-30	235	78	CCH 228	281	C8.2		918
Synonym: **CYCLOHEXANONE**															
PINANE HYDROPEROXIDE	2162	51							235						
PINANYL HYDROPEROXIDE	2162	51							235			764			
PINDONE	2472	53			2809	958	184		235	184		764			918
PINE MILL	1299	27													
PINE OIL	1272	26	695		2810	959		-80	235			764	P5.0		918
PINE TAR	1993	27							235			765			919
PINE TAR	1272	26							235						919
PINE TAR OIL	1272	26							235						919
PINENE	2368	26							235		PIN 960	765			
PINENE-beta	2368	26							235						
PINTSCH GAS	1075	22										765			919
PIPERAZIDINE	2579/1760	60	696		2811	959		-80			PPZ 983	766			919
Synonym: **PIPERAZINE**															
PIPERAZINE	2579/1760	60	696		2811	959		-80			PPZ 983	766			919
PIPERAZINE DIHYDROCHLORIDE					2812	959	184								919
PIPERIDINE	2401/1993	29			2814	960		-80	235			766			919
PIPERYLENE	1993	27							235			766			920
PIVALOYL CHLORIDE	2438	29							235						
PLANAVIN	1609		625								NTL 856				922
Synonym: **NITRALIN**															

CHEMICAL OR MATERIAL NAME	UN/NA Number	Guide Number (DOT)	Firefighter's Hazardous Materials Reference Book	First Aid Manual for Chemical Accidents	Sax's Dangerous Properties of Industrial Materials	Hazardous Chemicals Desk Reference	Rapid Guide to Hazardous Chemicals in the Workplace	Fire Protection Guide on Hazardous Materials (NFPA)	Firefighter's Handbook of Hazardous Materials	Pocket Guide to Chemical Hazards (NIOSH)	Chemical Hazard Response Information System (CHRIS)	Emergency Handling of Hazardous Materials in Surface Transportation (AAR)	Emergency Action Guides (EAG)	Chemical Data Notebook: A User's Manual	Condensed Chemical Dictionary
PLANT SPRAY OIL Synonym: OILS MISCELLANEOUS SPRAY	1270	27									OSY 908				
PLASTIAZAN 60 Synonym: ETHYLENE GLYCOL MONOETHYL ETHER	1171	26	375					-51	143		EGE 529	432	E2.5		489
PLASTIC LATEX Synonym: LATEX LIQUID SYNTHETIC	3082/9188		497								LLS 701	569			
PLASTIC MOULDING MATERIAL	2211	32							235						
PLASTIC SOLVENT	1263	26										767			
PLASTIC SOLVENT	1993	27							235						
PLASTICIZED DDP Synonym: DIISODECYL PHTHALATE								-40			DID 410				406
PLASTICIZER ALCOHOL	1987	26										767			
PLASTICS	2006	37							235			767			922
PLATINUM				22	2825	963	184			186					924
PLATINUM FULMINATE				22	2826				236						
PLOSOPHORIC CHEMICALS	1479/3139	35										768			
PLUMBOUS ARSENATE Synonym: LEAD ARSENATE	1617	53	502	19	2095	709	128	103-	178		LAR 693	570			687
PLUMBOUS CHLORIDE Synonym: LEAD CHLORIDE	2291	53	503		2096	710	128		178		LCL 695	571			687
PLUMBOUS FLUORIDE Synonym: LEAD FLUORIDE	2811/3077	53	504		2101	713	128		178		LFR 698	573			688
PLUMBOUS OXIDE Synonym: LITHARGE			518						180		LTH 713				926
PLUMBOUS SULFIDE Synonym: LEAD SULFIDE	2811/3077	53	509	19	2105	716	128				LSU 710	576			926

HAZARDOUS MATERIALS REFERENCE BOOKS INDEX

CHEMICAL OR MATERIAL NAME	UN/NA Number	Guide Number (DOT)	Firefighter's Hazardous Materials Reference Book	First Aid Manual for Chemical Accidents	Sax's Dangerous Properties of Industrial Materials	Hazardous Chemicals Desk Reference	Rapid Guide to Hazardous Chemicals in the Workplace	Fire Protection Guide on Hazardous Materials (NFPA)	Firefighter's Handbook of Hazardous Materials	Pocket Guide to Chemical Hazards (NIOSH)	Chemical Hazard Response Information System (CHRIS)	Emergency Handling of Hazardous Materials in Surface Transportation (AAR)	Emergency Action Guides (EAG)	Chemical Data Notebook: A User's Manual	Condensed Chemical Dictionary
PLURACOL POLYOL Synonym: POLYPROPYLENE GLYCOL					2847	971		-80			PGC 952				940
PLUTONIUM AND COMPOUNDS					2827	963			236						926
PNA Synonym: 4-NITROANILINE	1661	55			2514		158		211		NAL 818				
PNCB Synonym: NITROCHLOROBENZENE-para	1578	55	634	21	2525	863	159	−74	212	164		691	N3.4		
PNP Synonym: 4-NITROPHENOL	1663	55			2548	871			215		NPH 846				
POISON	2928	59							236						927
POISON	2930	34							236						927
POISON	2927	59							236						927
POISON	2929	57							236						927
POISON B LIQUID	2810	55							236						
POISONOUS GAS	1955	15							236						
POISONOUS GAS	1953	18							236			769			
POISONOUS LIQUID	1953	18							236						
POISONOUS LIQUID	2929	57							236			770			
POISONOUS LIQUID	2810	55							236			769			
POISONOUS SOLID	2928	59							236			770			
POISONOUS SOLID	2930	34							236			771			
POISONOUS SOLID	2811	53							236			771			
POLISH	1263	26							236						928
POLISH	1142	26							236						928
POLONIUM AND COMPOUNDS					2830	964			236						929

CHEMICAL OR MATERIAL NAME	UN/NA Number	Guide Number (DOT)	Firefighter's Hazardous Materials Reference Book	First Aid Manual for Chemical Accidents	Sax's Dangerous Properties of Industrial Materials	Hazardous Chemicals Desk Reference	Rapid Guide to Hazardous Chemicals in the Workplace	Fire Protection Guide on Hazardous Materials (NFPA)	Firefighter's Handbook of Hazardous Materials	Pocket Guide to Chemical Hazards (NIOSH)	Chemical Hazard Response Information System (CHRIS)	Emergency Handling of Hazardous Materials in Surface Transportation (AAR)	Emergency Action Guides (EAG)	Chemical Data Notebook: A User's Manual	Condensed Chemical Dictionary
POLONIUM CARBONYL					2830	964			236						
POLY (OXYETHYL) DODECYL ETHER Synonym: ETHOXYLATED DODECANOL											EOD 547				
POLY (OXYETHYL) LAURYL ETHER Synonym: ETHOXYLATED DODECANOL											EOD 547				
POLY (OXYETHYL) MYRISTYL ETHER Synonym: ETHOXYLATED TETRADECANOL											EOT 549				
POLY (OXYETHYL) PENTADECYL ETHER Synonym: ETHOXYLATED PENTADECANOL											EOP 548				
POLY (OXYETHYL) TETRADECYL ETHER Synonym: ETHOXYLATED TETRADECANOL											EOT 549				
POLY (OXYETHYL) TRIDECYL ETHER Synonym: ETHOXYLATED TRIDECANOL											ETD 563				
POLY (PROPYLENE GLYCOL) METHYL ETHER Synonym: POLYPROPYLENE GLYCOL METHYL ETHER									244		PGM 953				
POLY DIMETHYLSILOXANE Synonym: DIMETHYLPOLYSILOXANE					2836	967					DMP 437				931
POLY SOLV Synonym: DIETHYLENE GLYCOL DIMETHYL ETHER								−38	111		DGD 400				941
POLY SOLV DB Synonym: DIETHYLENE GLYCOL MONOBUTYL ETHER					1221			−38	111		DME 430				392
POLY SOLV DE Synonym: DIETHYLENE GLYCOL MONOETHYL ETHER	1993	27						−38	111		DGE 401	351			392
POLY SOLV DM Synonym: DIETHYLENE GLYCOL MONOMETHYL ETHER					1221			−38	111		DGM 402				393
POLY SOLV EB Synonym: ETHYLENE GLYCOL MONOBUTYL ETHER	2369	26						−51	143		EGM 532	432			489

CHEMICAL OR MATERIAL NAME	UN/NA Number	Guide Number (DOT)	Firefighter's Hazardous Materials Reference Book	First Aid Manual for Chemical Accidents	Sax's Dangerous Properties of Industrial Materials	Hazardous Chemicals Desk Reference	Rapid Guide to Hazardous Chemicals in the Workplace	Fire Protection Guide on Hazardous Materials (NFPA)	Firefighter's Handbook of Hazardous Materials	Pocket Guide to Chemical Hazards (NIOSH)	Chemical Hazard Response Information System (CHRIS)	Emergency Handling of Hazardous Materials in Surface Transportation (AAR)	Emergency Action Guides (EAG)	Chemical Data Notebook: A User's Manual	Condensed Chemical Dictionary
POLY SOLV EE Synonym: ETHYLENE GLYCOL MONOETHYL ETHER	1171	26	375					-51	143		EGE 529	432	E2.5		489
POLY SOLV EE ACETATE Synonym: ETHYLENE GLYCOL MONOETHYL ETHER ACE-TATE	1172	26	376					-51	143		EGA 526	432	E2.6		489
POLY SOLV EM Synonym: ETHYLENE GLYCOL MONOMETHYL ETHER	1188	26						-51	143		EME 544	433			489
POLYALKYLAMINE	2733	29							236						
POLYALKYLAMINE	2735	60							236						
POLYALKYLAMINE	2734	29							236						
POLYALKYLAMINES	2733	29							236						
POLYALKYLAMINES	2734	29							236						
POLYALKYLAMINES	2735	60							236						
POLYBUTENE			697								PLB 962				930
POLYBUTYLENE TEREPHTHALATE THERMOPLASTIC RESIN	1133	26										773			
POLYCHLOR AGRICULTURAL FUNGICIDES	2811	53										773			
POLYCHLORINATED BIPHENYL	2315/3151	31	698	23	2831	964	185	142-	236		PCB 934	773			931
POLYCHLORINATED BIPHENYL (AROCLOR 1242)				23	2832	965	185								
POLYCHLORINATED BIPHENYL (AROCLOR 1254)				23	2832	966	185								
POLYCHLOROBIPHENYLS	2315	31							236						
POLYCHLOROPOLYPHENYLS Synonym: POLYCHLORINATED BIPHENYL	2315/9188	31	698	23	2831	964	185	142-	236		PCB 934	773			931
POLYETHYLBENZENE	1993	27										774			
POLYETHYLLIGROIN	1271	26													
POLYETHYLLIGROIN	1993	27										774			

CHEMICAL OR MATERIAL NAME	UN/NA Number	Guide Number (DOT)	Firefighter's Hazardous Materials Reference Book	First Aid Manual for Chemical Accidents	Sax's Dangerous Properties of Industrial Materials	Hazardous Chemicals Desk Reference	Rapid Guide to Hazardous Chemicals in the Workplace	Fire Protection Guide on Hazardous Materials (NFPA)	Firefighter's Handbook of Hazardous Materials	Pocket Guide to Chemical Hazards (NIOSH)	Chemical Hazard Response Information System (CHRIS)	Emergency Handling of Hazardous Materials in Surface Transportation (AAR)	Emergency Action Guides (EAG)	Chemical Data Notebook: A User's Manual	Condensed Chemical Dictionary
POLYFORMALDEHYDE Synonym: PARAFORMALDEHYDE	2213	32	661		2664	906		131–76	237		PFA 950	725			934
POLYISOBUTYLENE PLASTICS Synonym: POLYBUTENE			697								PLB 962				930
POLYISOBUTYLENE RESINS Synonym: POLYBUTENE			697								PLB 962				930
POLYISOBUTYLENE WAXES Synonym: POLYBUTENE			697								PLB 962				930
POLYMETHYLENE POLYPHENYL ISOCYANATE			699								PPI 976				938
POLYOXPROPYLENE GLYCOL Synonym: POLYPROPYLENE GLYCOL					2847	971		-80			PGC 952				940
POLYOXYMETHYLENE Synonym: PARAFORMALDEHYDE	2213	32	661		2664	906		131–76	222		PFA 950	725			939
POLYOXYMETHYLENE GLYCOL Synonym: PARAFORMALDEHYDE	2213	32	661		2664	906		131–76	237		PFA 950	725			871
POLYOXYPROPYLENE GLYCOL Synonym: POLYPROPYLENE GLYCOL					2847	971		-80			PGC 952				940
POLYOXYPROPYLENE GLYCOL METHYL ETHER Synonym: POLYPROPYLENE GLYCOL METHYL ETHER											PGM 953				
POLYPHOSPHORIC ACID	1760	60	700						237		PPA 972				940
POLYPROPYLENE			701		2847						PLP 963				940
POLYPROPYLENE GLYCOL					2847	971		-80			PGC 952				940
POLYPROPYLENE GLYCOL METHYL ETHER											PGM 953				
POLYPROPYLENE GLYCOLS P400 TO P4000 Synonym: POLYPROPYLENE GLYCOL					2847	971		-80			PGC 952				940
POLYSTYRENE BEADS	2211	32							237			775			

HAZARDOUS MATERIALS REFERENCE BOOKS INDEX

CHEMICAL OR MATERIAL NAME	UN/NA Number	Guide Number (DOT)	Firefighter's Hazardous Materials Reference Book	First Aid Manual for Chemical Accidents	Sax's Dangerous Properties of Industrial Materials	Hazardous Chemicals Desk Reference	Rapid Guide to Hazardous Chemicals in the Workplace	Fire Protection Guide on Hazardous Materials (NFPA)	Firefighter's Handbook of Hazardous Materials	Pocket Guide to Chemical Hazards (NIOSH)	Chemical Hazard Response Information System (CHRIS)	Emergency Handling of Hazardous Materials in Surface Transportation (AAR)	Emergency Action Guides (EAG)	Chemical Data Notebook: A User's Manual	Condensed Chemical Dictionary
POLYVINYL ALCOHOL	1693	58			2851	974			237						943
POLYVINYL CHLORIDE				23	2852	974	185		237						943
PORTLAND CEMENT					2855	976	185			186					947
POTASAN					2855				237						947
POTASH LIQUOR	1814	60							237						
POTASH NITRATES	1486	35	713		2874	985			239			782	P6.1		955
Synonym: POTASSIUM NITRATE															
POTASSA	1813	60	712		2871	984	186	145–	239			781	P6.0	P 5	954
Synonym: POTASSIUM HYDROXIDE															
POTASSIUM	2257	40	702	23	2855	976			237		PTM 998	785			947
POTASSIUM	1420	40		23	2855	976			237			786			947
POTASSIUM ACID ARSENATE	1677	53	703						237		PAS 928	776			948
Synonym: POTASSIUM ARSENATE															
POTASSIUM ACID OXALATE											PBO 930				948
Synonym: POTASSIUM BINOXALATE															
POTASSIUM ALKALI SALTS CRUDE DRY	1479	35										776			
POTASSIUM AMIDE					2857				237						
POTASSIUM ANTIMONYL TARTRATE	1551	53	90		274	80			33		APT 81	83			948
Synonym: ANTIMONY POTASSIUM TARTRATE															
POTASSIUM ARSENATE	1677	53	703						237		PAS 928	776			948
POTASSIUM ARSENITE	1678	54	704		2857	977			237		POA 969	776			949
POTASSIUM BICHROMATE	1479	35	710	23	2859	978			237		PTD 993	779			949
Synonym: POTASSIUM DICHROMATE															
POTASSIUM BIFLUORIDE	1811	60							237			776			949
POTASSIUM BINOXALATE											PBO 930				949

CHEMICAL OR MATERIAL NAME	UN/NA Number	Guide Number (DOT)	Firefighter's Hazardous Materials Reference Book	First Aid Manual for Chemical Accidents	Sax's Dangerous Properties of Industrial Materials	Hazardous Chemicals Desk Reference	Rapid Guide to Hazardous Chemicals in the Workplace	Fire Protection Guide on Hazardous Materials (NFPA)	Firefighter's Handbook of Hazardous Materials	Pocket Guide to Chemical Hazards (NIOSH)	Chemical Hazard Response Information System (CHRIS)	Emergency Handling of Hazardous Materials in Surface Transportation (AAR)	Emergency Action Guides (EAG)	Chemical Data Notebook: A User's Manual	Condensed Chemical Dictionary
POTASSIUM BISULFATE	2509	60			2859	978			237						949
POTASSIUM BOROHYDRIDE	1870	40							237			777			950
POTASSIUM BROMATE	1484	35			2860	979			238			777			950
POTASSIUM CHLORATE	1485	35	705	23	2861	979		143–	238		PCR 941	777			950
POTASSIUM CHLORATE	2427	31		23	2861	979			238			778			950
POTASSIUM CHROMATE	9142/3077		706	23	2863	980			238		PCH 936	778			951
POTASSIUM CHROMATE (VI) Synonym: POTASSIUM CHROMATE	9142/3077		706	23	2863	980			238		PCH 936	778			951
POTASSIUM COPPER CYANIDE	1679	53							238						951
POTASSIUM CUPROCYANIDE	1679	53			2864				238						951
POTASSIUM CYANIDE SOLID	1680	55	708	23	2865	982	185	144–	238		PTC 992	779			951
POTASSIUM CYANIDE SOLUTION	1680	55	707	23	2864	981	185		238		PTC 992	778			
POTASSIUM DICHLORO-s-TRIAZINETRIONE	2465	45						144–	238		PDT 947				952
POTASSIUM DICHLOROISOCYANURATE Synonym: POTASSIUM DICHLORO-s-TRIAZINETRIONE	2465	45	709		2865	982		144–	238		PDT 947				952
POTASSIUM DICHROMATE	1479	35	710	23					238		PTD 993	779			952
POTASSIUM DIHYDROGEN ARSENATE Synonym: POTASSIUM ARSENATE	1677	53	703						237		PAS 928	776			948
POTASSIUM FLUORIDE	1812	54		23	2868	982			238			779			953
POTASSIUM FLUOROACETATE	2628/2811	53			2868	983			238			780			
POTASSIUM FLUOROSILICATE	2655	53							238			780			
POTASSIUM FLUOSILICATE	2655	53		23					238						953
POTASSIUM FLUOZIRCONATE Synonym: ZIRCONIUM POTASSIUM FLUORIDE	9162/3077	31	911		2868		235		290		ZPF1214	972			953

HAZARDOUS MATERIALS REFERENCE BOOKS INDEX

CHEMICAL OR MATERIAL NAME	UN/NA Number	Guide Number (DOT)	Firefighter's Hazardous Materials Reference Book	First Aid Manual for Chemical Accidents	Sax's Dangerous Properties of Industrial Materials	Hazardous Chemicals Desk Reference	Rapid Guide to Hazardous Chemicals in the Workplace	Fire Protection Guide on Hazardous Materials (NFPA)	Firefighter's Handbook of Hazardous Materials	Pocket Guide to Chemical Hazards (NIOSH)	Chemical Hazard Response Information System (CHRIS)	Emergency Handling of Hazardous Materials in Surface Transportation (AAR)	Emergency Action Guides (EAG)	Chemical Data Notebook: A User's Manual	Condensed Chemical Dictionary
POTASSIUM HEXAFLUOROZIRCONATE Synonym: ZIRCONIUM POTASSIUM FLUORIDE	9162/3077	31	911				235		290		ZPF1214	972			1252
POTASSIUM HYDRATE Synonym: POTASSIUM HYDROXIDE	1813	60	712	23	2871	984	186	145–	238			781	P6.0	P 5	954
POTASSIUM HYDRIDE					2870	983			238						954
POTASSIUM HYDROGEN FLUORIDE	1811	60							238						954
POTASSIUM HYDROGEN SULFATE	2509	60							239			780			
POTASSIUM HYDROSULFITE	1929	32							239			780			
POTASSIUM HYDROXIDE	1814	60	711	23	2871	984	186	145–	239		PTH 995	781	P6.0	P 5	954
POTASSIUM HYDROXIDE	1813	60	712	23	2871	984	186	145–	239			781	P6.0	P 5	954
POTASSIUM HYDROXIDE SOLUTION Synonym: CAUSTIC POTASH	1814	60	711	23	2871	984			239		CPS 295	781		P 5	233
POTASSIUM HYPOCHLORITE	1791	60			2872				239						
POTASSIUM IODIDE					2872	984					PTI 996				954
POTASSIUM METAARSENITE Synonym: POTASSIUM ARSENITE	1678	54	704		2857	977			237		POA 969	776			955
POTASSIUM METABISULFITE	2693	60										782			955
POTASSIUM METAL	2257	40						143–	239		PYM 998	785			947
POTASSIUM METAL	1420	40							239			786			947
POTASSIUM METAVANADATE	2864	53			2873				239						
POTASSIUM MONOXIDE	2033	60							239			782			
POTASSIUM NITRATE	1486	35	713		2874	985			239			782	P6.1	P 6	955
POTASSIUM NITRATE AND SODIUM NITRITE MIXTURE	1487	35			2874				239			782			
POTASSIUM NITRITE	1488	35			2874	985			239			783			955

CHEMICAL OR MATERIAL NAME	UN/NA Number	Guide Number (DOT)	Firefighter's Hazardous Materials Reference Book	First Aid Manual for Chemical Accidents	Sax's Dangerous Properties of Industrial Materials	Hazardous Chemicals Desk Reference	Rapid Guide to Hazardous Chemicals in the Workplace	Fire Protection Guide on Hazardous Materials (NFPA)	Firefighter's Handbook of Hazardous Materials	Pocket Guide to Chemical Hazards (NIOSH)	Chemical Hazard Response Information System (CHRIS)	Emergency Handling of Hazardous Materials in Surface Transportation (AAR)	Emergency Action Guides (EAG)	Chemical Data Notebook: A User's Manual	Condensed Chemical Dictionary
POTASSIUM OLEATE Synonym: OLEIC ACID POTASSIUM SALT			655		2630	896					OAP 868				955
POTASSIUM OXALATE			714								PTS1003				956
POTASSIUM OXALATE MONOHYDRATE Synonym: POTASSIUM OXALATE			714								PTS1003				956
POTASSIUM OXIDE	2033	60		23					239			782			956
POTASSIUM PERCHLORATE	1489	35		23	2877	986		145–	239			783			956
POTASSIUM PERIODATE					2877				239						956
POTASSIUM PERMANGANATE	1490	35	715		2877	986	186		239		PTP1001	783			956
POTASSIUM PEROXIDE	1491	47	716		2878	987		146–	239		POP 970	784			956
POTASSIUM PEROXYDISULFATE	1492	35							240						957
POTASSIUM PERSULFATE	1492	35							240			784			957
POTASSIUM PHOSPHIDE	2012	41			2879	987			240			784			
POTASSIUM PICRATE					2879				240						
POTASSIUM SALTS NEC	1479	35													
POTASSIUM SELENATE	2630	53			2880	987			240						
POTASSIUM SELENITE	2630	53							240						
POTASSIUM SILICOFLUORIDE	2655	53							240						958
POTASSIUM SODIUM ALLOY	1422	40			2881	987									
POTASSIUM SULFIDE	1382	32			2882	988		146–	240			784			959
POTASSIUM SULFIDE	1847	60			2882	988			240						959
POTASSIUM SUPEROXIDE	2466	47							240			785			
POTASSIUM SUPEROXIDE Synonym: POTASSIUM PEROXIDE	1491	47	716		2878	987		146–	239		POP 970	784			956

HAZARDOUS MATERIALS REFERENCE BOOKS INDEX

CHEMICAL OR MATERIAL NAME	UN/NA Number	Guide Number (DOT)	Firefighter's Hazardous Materials Reference Book	First Aid Manual for Chemical Accidents	Sax's Dangerous Properties of Industrial Materials	Hazardous Chemicals Desk Reference	Rapid Guide to Hazardous Chemicals in the Workplace	Fire Protection Guide on Hazardous Materials (NFPA)	Firefighter's Handbook of Hazardous Materials	Pocket Guide to Chemical Hazards (NIOSH)	Chemical Hazard Response Information System (CHRIS)	Emergency Handling of Hazardous Materials in Surface Transportation (AAR)	Emergency Action Guides (EAG)	Chemical Data Notebook: A User's Manual	Condensed Chemical Dictionary
POTASSIUM XANTHATE								-81	240						960
POTASSIUM ZINC CHROMATE Synonym: ZINC POTASSIUM CHROMATE			905				234				ZPC1213				1247
POTASSIUM ZIRCONIUM FLUORIDE Synonym: ZIRCONIUM POTASSIUM FLUORIDE	9162/3077	31	911				235		290		ZPF1214	972			1252
POTASSIUM-m-VANADATE	2864	53							241						
POTATO ALCOHOL Synonym: ETHYL ALCOHOL	1170	26	344	16	1572	525	90	-48	136		EAL 504	408	E2.2	E 2	478
POTATO SPIRIT OIL Synonym: ISOAMYL ALCOHOL	1105	26	461		2011	673	120	-60	171	130	IAA 655				649
POTCRATE Synonym: POTASSIUM CHLORATE	1485	35	705	23	2861	979		143-	238		PCR 941	777			950
PRESERVATIVE OIL Synonym: OILS MISCELLANEOUS PENETRATING	1270	27									OPT 898				
PRIMARY n-AMYL ALCOHOL	1987	26													74
PRIME STEAM LARD Synonym: OILS EDIBLE LARD											OLD 885				
PRODUCER GAS					2895	992			241						964
PROFUME Synonym: METHYL BROMIDE	1062	55	559	20	2274	776	137	110-67	192	146	MTB 805	619	M3.0	M 3	759
PROFUME A Synonym: CHLOROPICRIN	1580	56	210	14	865	296	50	54 -	77	70	CPL 290	229	C4.3		271
PROMETHIUM					2898	993			241						965
PROP-2-EN-1-AL Synonym: ACROLEIN	1092	30	31	11	63	25	5	17 -12	20	32	ARL 85	12	A4.0	A 5	18
PROPADIENE	2200	22							241			789			966

CHEMICAL OR MATERIAL NAME	UN/NA Number	Guide Number (DOT)	Firefighter's Hazardous Materials Reference Book	First Aid Manual for Chemical Accidents	Sax's Dangerous Properties of Industrial Materials	Hazardous Chemicals Desk Reference	Rapid Guide to Hazardous Chemicals in the Workplace	Fire Protection Guide on Hazardous Materials (NFPA)	Firefighter's Handbook of Hazardous Materials	Pocket Guide to Chemical Hazards (NIOSH)	Chemical Hazard Response Information System (CHRIS)	Emergency Handling of Hazardous Materials in Surface Transportation (AAR)	Emergency Action Guides (EAG)	Chemical Data Notebook: A User's Manual	Condensed Chemical Dictionary
PROPADIENE METHYL ACETYLENE MIXTURE Synonym: **METHYL ACETYLENE PROPADIENE MIXTURE**	1060	17	552		2248	769	134			142	MAP 725	640			755
PROPANAL Synonym: **PROPIONALDEHYDE**	1275	26	720	23	2908	997		147–81	241		PAD 923	791	P7.0		966
PROPANE	1075	22	717	23	2899	993	186					789			966
PROPANE	1978	22		23	2899	993	186	–81	241	186	PRP 988				966
PROPANE BUTANE (PROPYLENE) Synonym: **LIQUEFIED PETROLEUM GAS**	1075	22	517		2123	722	128		180		LPG 704	789	L1.0	L 1	703
PROPANE-1,1-DICHLORO Synonym: **1,1-DICHLOROPROPANE**	1279	27			1163						DPB 459				
PROPANE-1-THIOL Synonym: **PROPYL MERCAPTAN-n**	2704				2902						PMN 967				972
PROPANE-2-THIOL Synonym: **ISOPROPYL MERCAPTAN**	2402	27							241		IPM 677	562			661
PROPANE-2-THIOL Synonym: **ISOPROPYL MERCAPTAN**	2703	27	484						175		IPM 677				661
PROPANECARBOXYLIC ACID Synonym: **BUTYRIC ACID (-n)**	2820	60	154		643	216		44 –25	61		BRA 164	173			192
PROPANETHIOL	2704				2902	994									
PROPANETHIOL	2402	27				994	187		242			790			
PROPANOIC ACID Synonym: **PROPIONIC ACID**	1848	60	721		2909	997	188	148–81	242		PNA 968	791	P7.1		966
PROPANOIC ACID METHYLACETIC ACID Synonym: **PROPIONIC ACID**	1848	60	721		2909	997	188	148–81	243		PNA 968	791	P7.1		968
PROPANOIC ANHYDRIDE Synonym: **PROPIONIC ANHYDRIDE**	2496	29	722		2909	998		148–81	242		PAH 924	792			968

CHEMICAL OR MATERIAL NAME	UN/NA Number	Guide Number (DOT)	Firefighter's Hazardous Materials Reference Book	First Aid Manual for Chemical Accidents	Sax's Dangerous Properties of Industrial Materials	Hazardous Chemicals Desk Reference	Rapid Guide to Hazardous Chemicals in the Workplace	Fire Protection Guide on Hazardous Materials (NFPA)	Firefighter's Handbook of Hazardous Materials	Pocket Guide to Chemical Hazards (NIOSH)	Chemical Hazard Response Information System (CHRIS)	Emergency Handling of Hazardous Materials in Surface Transportation (AAR)	Emergency Action Guides (EAG)	Chemical Data Notebook: A User's Manual	Condensed Chemical Dictionary
PROPANOL	1274	26							242				P6.5		
PROPANOLIDE Synonym: PROPIOLACTONE-beta			719		2907	997	187		242	186	PLT 964				
PROPANONE Synonym: ACETONE	1090	26	20	11	22	10	2	−11	242	30	ACT 26	4	A3.0	A 3	9
PROPARGIL Synonym: PROPARGITE	2765/3082	55	718						242		PRG 986	790			967
PROPARGITE	2765/3082	55	718						242		PRG 986	790			967
PROPARGYL ALCOHOL	1986	28			2902	994	187	146–81	242		PRO 987	791			967
PROPARGYL BROMIDE	2345	29			2905	995		−81	242						967
PROPELLANT 12 Synonym: DICHLORODIFLUOROMETHANE	1028	12	275	15	1137	380	68		102	86	DCF 360	331	D1.1		379
PROPEN-1-OL-3 Synonym: ALLYL ALCOHOL	1098	57	38	11	102	33	7	21 −13	242	34	ALA 39	35	A7.0		38
PROPENAL Synonym: ACROLEIN	1092	30	31	11	63	25	5	17 −12	20	32	ARL 85	12	A4.0	A 5	18
PROPENAMIDE (50%) Synonym: ACRYLAMIDE	2074	55	32	11	64	26	5	18 −	242	34	AAM 3	12			18
PROPENE Synonym: PROPYLENE	1077/1075	22	727	23	2904	995		149–82	242		PPL 977		P8.0.2	P 7	967
PROPENE ACID Synonym: ACRYLIC ACID	2218	29	33		65	26	6	18 −12	242		ACR 25	13	A5.0	A 6	18
PROPENE NITRILE Synonym: ACRYLONITRILE	1093	30	34	11	68	27	6	19 −12	242	34	ACN 23	13	A5.1	A 7	967
PROPENE OXIDE Synonym: PROPYLENE OXIDE	1280	26	729	23	2922	1003	190	150–82	242	190	POX 971	798	P8.1		971

CHEMICAL OR MATERIAL NAME	UN/NA Number	Guide Number (DOT)	Firefighter's Hazardous Materials Reference Book	First Aid Manual for Chemical Accidents	Sax's Dangerous Properties of Industrial Materials	Hazardous Chemicals Desk Reference	Rapid Guide to Hazardous Chemicals in the Workplace	Fire Protection Guide on Hazardous Materials (NFPA)	Firefighter's Handbook of Hazardous Materials	Pocket Guide to Chemical Hazards (NIOSH)	Chemical Hazard Response Information System (CHRIS)	Emergency Handling of Hazardous Materials in Surface Transportation (AAR)	Emergency Action Guides (EAG)	Chemical Data Notebook: A User's Manual	Condensed Chemical Dictionary
PROPENE POLYMER Synonym: POLYPROPYLENE			701		2904	996					PLP 963				940
PROPENEOXIDE Synonym: PROPYLENE OXIDE	1280	26	729	23	2922	1003	190	150–82	242	190	POX 971	798	P8.1		971
PROPENOIC ACID Synonym: ACRYLIC ACID	2218	29	33		65	26	6	18 –12	20		ACR 25	13	A5.0	A 6	967
PROPENOL Synonym: ALLYL ALCOHOL	1098	57	38	11	102	33	7	21 –13	242	34	ALA 39	35	A7.0		38
PROPENYL ACETATE	2403	26							242						
PROPENYL ALCOHOL Synonym: ALLYL ALCOHOL	1098	57	38	11	102	33	7	21 –13	242	34	ALA 39	35	A7.0		967
PROPENYL ETHYL ETHER								-81	242						
PROPIOLACTONE-beta			719		2907	997	187		242	186	PLT 964				
PROPIONAL Synonym: PROPIONALDEHYDE	1275	26	720	23	2908	997		147–	243		PAD 923	791	P7.0		968
PROPIONALDEHYDE	1275	26	720	23	2908	997		147–	243		PAD 923	791	P7.0		968
PROPIONE	1156	26							243						968
PROPIONIC ACID	1848	60	721		2909	997	188	148–81	243		PNA 968	791	P7.1		968
PROPIONIC ALDEHYDE Synonym: PROPIONALDEHYDE	1275	26	720	23	2908	997		147–	243		PAD 923		P7.0		968
PROPIONIC ANHYDRIDE	2496	29	722		2909	998		148–81	243		PAH 924	792			968
PROPIONIC ETHER	1195	26							243						
PROPIONIC NITRILE	2404	28						-81	243						
PROPIONITRITE	1935	55													
PROPIONOLACTONE-beta Synonym: PROPIOLACTONE-beta			719		2907	997	187		242	186	PLT 964				

CHEMICAL OR MATERIAL NAME	UN/NA Number	Guide Number (DOT)	Firefighter's Hazardous Materials Reference Book	First Aid Manual for Chemical Accidents	Sax's Dangerous Properties of Industrial Materials	Hazardous Chemicals Desk Reference	Rapid Guide to Hazardous Chemicals in the Workplace	Fire Protection Guide on Hazardous Materials (NFPA)	Firefighter's Handbook of Hazardous Materials	Pocket Guide to Chemical Hazards (NIOSH)	Chemical Hazard Response Information System (CHRIS)	Emergency Handling of Hazardous Materials in Surface Transportation (AAR)	Emergency Action Guides (EAG)	Chemical Data Notebook: A User's Manual	Condensed Chemical Dictionary
PROPIONONITRILE	2404	28			2910	998	188								
PROPIONYL CHLORIDE	1815/1993	29			2910	999		-81	243			793			968
PROPIONYL OXIDE Synonym: PROPIONIC ANHYDRIDE	2496	29	722		2909	998		148-81	243		PAH 924	792			968
PROPIONYL PEROXIDE	2132	52							243						968
PROPRIETARY ANTI FREEZE OR ENGINE COOLANT PREPARATIONS	1993	27										793			
PROPYL ACETATE Synonym: PROPYL ACETATE (-n)	1276	26	723	23	2914	999	189	-81	243		PAT 929	794	P8.0		969
PROPYL ACETATE-n Synonym: PROPYL ACETATE	1276	26	723	23	2914	999	189	-81	243	188	PAT 929	794	P8.0		969
PROPYL ALCOHOL Synonym: PROPYL ALCOHOL-n	1274	26	724	23	2915	1000	189	-81	243	188	PAL 925	709			969
PROPYL ALCOHOL-n	1274	26	724	23	2915	1000	189		243	188	PAL 925	709			969
PROPYL ALCOHOL-sec Synonym: ISOPROPANOL	1219	26	480						243			555	14.0		659
PROPYL ALCOHOL-sec Synonym: ISOPROPYL ALCOHOL	1219/1993	26	482		2040	687	124	-63	243	132	IPA 670	560		1 2	969
PROPYL ALDEHYDE Synonym: PROPIONALDEHYDE	1275	26	720	23	2908	997		147-	243		PAD 923	791	P7.0		969
PROPYL BENZENE	2364/1993	26						-81	243		PBZ 933	794			969
PROPYL BENZENE-n Synonym: PROPYL BENZENE	2364/1993	26							243		PBZ 933	794			969
PROPYL CARBINOL	1120	26						-81	243						
PROPYL CHLORIDE	1278	26						-82	243			795			970
PROPYL CHLOROTHIOLFORMATE								-82	243						

CHEMICAL OR MATERIAL NAME	UN/NA Number	Guide Number (DOT)	Firefighter's Hazardous Materials Reference Book	First Aid Manual for Chemical Accidents	Sax's Dangerous Properties of Industrial Materials	Hazardous Chemicals Desk Reference	Rapid Guide to Hazardous Chemicals in the Workplace	Fire Protection Guide on Hazardous Materials (NFPA)	Firefighter's Handbook of Hazardous Materials	Pocket Guide to Chemical Hazards (NIOSH)	Chemical Hazard Response Information System (CHRIS)	Emergency Handling of Hazardous Materials in Surface Transportation (AAR)	Emergency Action Guides (EAG)	Chemical Data Notebook: A User's Manual	Condensed Chemical Dictionary
PROPYL CYANIDE	2411	28							243						970
PROPYL ETHER	2384	26			2923	1003			243						
PROPYL FORMATE (-n)	1281	26			2923	1004		-82	243			795			972
PROPYL ISOCYANATE	2482	28			2925	1004			243						
PROPYL MERCAPTAN	2402	27	725						244			795			972
PROPYL MERCAPTAN-n	2704										PMN 967				972
PROPYL METHANOATE	1281	26							244						
PROPYL METHANOL	1120	26							244						
PROPYL NITRATE Synonym: PROPYL NITRATE-n	1865	30		23	2926	1004	191	151-82	244	190					972
PROPYL NITRITE					2926				244						
PROPYL PEROXYDICARBONATE	2133	52							244						
PROPYL PROPIONATE					2928			-82	244						972
PROPYL TRICHLOROSILANE (-n)	1816	29			2930	1005		152-82	244			796			973
PROPYL-1-PROPANAMINE-n Synonym: DI-n-PROPYLAMINE	2383/1993	68			1470	488			93		DNA 442	314	D4.0		432
PROPYLAMINE	1277	68	726	23	2915	1000		-81	244		PRA 984	796			969
PROPYLAMINE-iso Synonym: ISOPROPYLAMINE	1221	68	487	23	2014	688	124	102-63	176	134	IPP 678	563			660
PROPYLAMINE-MONO-n Synonym: PROPYLAMINE-n	1277	68	726	23	2915	1000		149-	244		PRA 984	796			969
PROPYLAMINE-n Synonym: PROPYLAMINE	1277	68	726	23		1000		149-	244		PRA 984	796			969
PROPYLAMINE-n Synonym: DI-n-PROPYLAMINE	2383/1993	68			2915				93		DNA 442	314	D4.0		969

CHEMICAL OR MATERIAL NAME	UN/NA Number	Guide Number (DOT)	Firefighter's Hazardous Materials Reference Book	First Aid Manual for Chemical Accidents	Sax's Dangerous Properties of Industrial Materials	Hazardous Chemicals Desk Reference	Rapid Guide to Hazardous Chemicals in the Workplace	Fire Protection Guide on Hazardous Materials (NFPA)	Firefighter's Handbook of Hazardous Materials	Pocket Guide to Chemical Hazards (NIOSH)	Chemical Hazard Response Information System (CHRIS)	Emergency Handling of Hazardous Materials in Surface Transportation (AAR)	Emergency Action Guides (EAG)	Chemical Data Notebook: A User's Manual	Condensed Chemical Dictionary
PROPYLCARBINOL-n	1120	26				196	31	–21		52	BAN 119	150	B7.1		181
Synonym: BUTYL ALCOHOL (-n)															
PROPYLCARBINOL-para	1120	26				196	31	–21	56	52	BAN 119	150	B7.1		181
Synonym: BUTYL ALCOHOL (-n)															
PROPYLCHLOROFORMATE	2740	57							244						
PROPYLENE	1077/1075	22	727	23				149–82	244		PPL 977	796	P8.0.2	P 7	970
PROPYLENE ALDEHYDE	1143	28	233			325	56	57 –29	244	76	CTA 317	271	C7.1		325
Synonym: CROTONALDEHYDE															
PROPYLENE BUTYLENE POLYMER											PBP 931				
PROPYLENE CHLOROHYDRIN	2611/2810	57						–82	244			797			970
PROPYLENE DICHLORIDE	1279	27	4		2919	1001	189	150–82	244	188	DPP 470	798	P8.0.3		970
Synonym: 1,2-DICHLOROPROPANE															
PROPYLENE GLYCOL			728	23				–82	244		PPG 975				
PROPYLENE GLYCOL ACETATE	1189	26							244						
PROPYLENE GLYCOL DINITRATE					2920	1002	190								971
PROPYLENE GLYCOL METHYL ETHER								–82	244		PME 966				
PROPYLENE GLYCOL MONOACRYLATE	1760	60	458								HPA 640	535			
Synonym: HYDROXYPROPYL ACRYLATE															
PROPYLENE GLYCOL MONOMETHACRYLATE			459								HPM 642				625
Synonym: HYDROXYPROPYL METHACRYLATE															
PROPYLENE GLYCOL MONOMETHYL ETHER				23	2921	1002	190		244						971
PROPYLENE IMINE	1921	30	731		2921	1002	190		244	188	PII 959	799	P8.0.5		971
Synonym: PROPYLENEIMINE															
PROPYLENE OXIDE	1280	26	729	23	2922	1003	190	150–82	244	190	POX 971	798	P8.1		971
PROPYLENE SULFIDE					2922				244						

CHEMICAL OR MATERIAL NAME	UN/NA Number	Guide Number (DOT)	Firefighter's Hazardous Materials Reference Book	First Aid Manual for Chemical Accidents	Sax's Dangerous Properties of Industrial Materials	Hazardous Chemicals Desk Reference	Rapid Guide to Hazardous Chemicals in the Workplace	Fire Protection Guide on Hazardous Materials (NFPA)	Firefighter's Handbook of Hazardous Materials	Pocket Guide to Chemical Hazards (NIOSH)	Chemical Hazard Response Information System (CHRIS)	Emergency Handling of Hazardous Materials in Surface Transportation (AAR)	Emergency Action Guides (EAG)	Chemical Data Notebook: A User's Manual	Condensed Chemical Dictionary
PROPYLENE TETRAMER	2850/1993	27	730						244		PTT1004	799			
PROPYLENE TETRAMER Synonym: DODECENE	1993	27									DOD 454	390			441
PROPYLENE TRIMER Synonym: NONENE								-75	217		NON 844				835
PROPYLENEDIAMINE	2258	29						-82	244						970
PROPYLENEIMINE	1921	30	731			1002	190		244	188	PII 959	799	P8.0.5		971
PROPYLENIMINE Synonym: PROPYLENEIMINE	1921	30	731		2921	1002	190		244	188	PII 959	799	P8.0.5		971
PROPYLETHYLENE Synonym: 1-PENTENE	1108	26			2691				244		PTE 994				882
PROPYLIC ALDEHYDE Synonym: PROPIONALDEHYDE	1275	26	720	23	2908	997		147–	244		PAD 923	791	P7.0		968
PROPYLIDENE CHLORIDE Synonym: 1,1-DICHLOROPROPANE	1279	27			1163						DPB 459				
PROPYNE								-82	244						973
PROZOIN Synonym: PROPIONIC ACID	1848	60	721		2909	997	188	148–81			PNA 968	791	P7.1		968
PRUSSIC ACID Synonym: HYDROCYANIC ACID	1051	13	442			641	114	-59	245			529	H4.1	H 2	975
PRUSSIC ACID Synonym: HYDROGEN CYANIDE	1051	13	448	18	1896			98 –	245	126	HCN 626	529			975
PSEUDOACETIC ACID Synonym: PROPIONIC ACID	1848	60	721		2909	997	188	148–81			PNA 968	791	P7.1		968
PSEUDOACONITINE					2937	1006			245						
PSEUDOCUMENE	1993	27							245			799			975

HAZARDOUS MATERIALS REFERENCE BOOKS INDEX

CHEMICAL OR MATERIAL NAME	UN/NA Number	Guide Number (DOT)	Firefighter's Hazardous Materials Reference Book	First Aid Manual for Chemical Accidents	Sax's Dangerous Properties of Industrial Materials	Hazardous Chemicals Desk Reference	Rapid Guide to Hazardous Chemicals in the Workplace	Fire Protection Guide on Hazardous Materials (NFPA)	Firefighter's Handbook of Hazardous Materials	Pocket Guide to Chemical Hazards (NIOSH)	Chemical Hazard Response Information System (CHRIS)	Emergency Handling of Hazardous Materials in Surface Transportation (AAR)	Emergency Action Guides (EAG)	Chemical Data Notebook: A User's Manual	Condensed Chemical Dictionary
PSEUDOHEXYL ALCOHOL	2275	26	347						141		EBT 509				975
Synonym: ETHYL BUTANOL															
PURSSITE	1026	18	241		979	331	57	−29			CYG 330	277	C8.0		333
Synonym: CYANOGEN															
PYRAZINE HEXAHYDRIDE	2579/1760	60	696		2811	959		−80			PPZ 983	766			978
Synonym: PIPERAZINE															
PYRAZOLE					2943				245						978
PYRENE					2944	1007	191		245						979
PYRETHRIN I	9184				2944	1007	191				PRR 989				979
PYRETHRIN II	9184				2945	1007	191				PRR 989				979
PYRETHRINS ACIDS	9184	31	732		2945	1007	191				PRR 989				
PYRETHRUM	9184	31			2945				245	190					979
PYRETHRUM FLOWERS	9184/3082	31	732	23	2945	1007	191		245		PRR 989	800			
Synonym: PYRETHRINS															
PYRIDINE	1282	26	734		2946	1008	192	152–83	245	190	PRD 985	800	P9.0		979
PYROACETIC ETHER DIMETHYL KETAL	1090	26	20	11	22	10	2	11	17	30	ACT 26	4	A3.0	A 3	9
Synonym: ACETONE															
PYROBENZOL	1114	27	115	12	356	115	21	35 −17	40	44	BNZ 152	112	B1.0	B 1	128
Synonym: BENZENE															
PYROBENZOLE	1114	27	115	12	356	115	21	35 −17	40	44	BNZ 152	112	B1.0	B 1	128
Synonym: BENZENE															
PYROCATECHIN					718	252	44		245		CTC 318				232
Synonym: CATECHOL															
PYROCATECHINIC ACID					718	252	44		67		CTC 318				232
Synonym: CATECHOL															
PYROCATECHOL									245						981

CHEMICAL OR MATERIAL NAME	UN/NA Number	Guide Number (DOT)	Firefighter's Hazardous Materials Reference Book	First Aid Manual for Chemical Accidents	Sax's Dangerous Properties of Industrial Materials	Hazardous Chemicals Desk Reference	Rapid Guide to Hazardous Chemicals in the Workplace	Fire Protection Guide on Hazardous Materials (NFPA)	Firefighter's Handbook of Hazardous Materials	Pocket Guide to Chemical Hazards (NIOSH)	Chemical Hazard Response Information System (CHRIS)	Emergency Handling of Hazardous Materials in Surface Transportation (AAR)	Emergency Action Guides (EAG)	Chemical Data Notebook: A User's Manual	Condensed Chemical Dictionary
PYROFAX Synonym: LIQUEFIED PETROLEUM GAS	1075	22	517		2123	722	128		180		LPG 704		L1.0	L 1	703
PYROGALLIC ACID			735						245		PGA 951				
PYROGALLOL Synonym: PYROGALLIC ACID			735			1010			245		PGA 951				981
PYROGENTISIC ACID Synonym: HYDROQUINONE	2662	53	453		1906	647	116	-59	165	128	HDQ 631	534			618
PYROLAN					2958				245						981
PYROLIGNEOUS ACID	1993	27													981
PYROMUCIC ALDEHYDE Synonym: FURFURAL	1199	29	408	17	1765	597	103	92 -55	245	118	FFA 583	467			982
PYROPHORIC FUEL	1375	37							245						
PYROPHORIC LIQUID	2845	40							245						
PYROPHORIC METAL	1383	37							246						
PYROPHORIC SOLID	2846	37							246						
PYROSULFURIC ACID	1831	39							246						983
PYROSULFURYL CHLORIDE	1817	39			2958	1010			246			803			983
PYROXYLIC SPIRIT Synonym: METHANOL	1230	28	549		2251				190			613	M2.0		755
PYROXYLIC SPIRIT Synonym: METHYL ALCOHOL	1230	28	554	20		770	136	-66	191	144	MAL 722	613		M 2	751
PYROXYLIN	1325	32							246						983
PYROXYLIN PLASTIC	1325/2006	32						153-	246			803			
PYROXYLIN SOLUTION Synonym: COLLODION	2059/1993	26	220		940	319		-83	246		CLD 257	249			300

HAZARDOUS MATERIALS REFERENCE BOOKS INDEX

CHEMICAL OR MATERIAL NAME	UN/NA Number	Guide Number (DOT)	Firefighter's Hazardous Materials Reference Book	First Aid Manual for Chemical Accidents	Sax's Dangerous Properties of Industrial Materials	Hazardous Chemicals Desk Reference	Rapid Guide to Hazardous Chemicals in the Workplace	Fire Protection Guide on Hazardous Materials (NFPA)	Firefighter's Handbook of Hazardous Materials	Pocket Guide to Chemical Hazards (NIOSH)	Chemical Hazard Response Information System (CHRIS)	Emergency Handling of Hazardous Materials in Surface Transportation (AAR)	Emergency Action Guides (EAG)	Chemical Data Notebook: A User's Manual	Condensed Chemical Dictionary
PYROXYLIN SOLVENT	2059	26							246						
PYRROLE					2958	1010		-83	246						
PYRROLIDINE	1922	29			2959	1010		-83	246						
PYRROLYLENE	1010	17	139	12				40 –	53	48	BDI 134	148	B4.0	B 4	
Synonym: BUTADIENE															
QUAKERAL	1199	29	408	17	1765	597	103	92 –55	152	118	FFA 583	467			545
Synonym: FURFURAL															
QUICKLIME	1910	60	174	13	669	233	38	47 –	246	56	CAO 209	183			987
Synonym: CALCIUM OXIDE															
QUICKSILVER	2809	60	543	20	2192	750	131		246		MCR 744				987
Synonym: MERCURY															
QUINALDINE	1993	27			2965				246			804			987
QUINOL	2662	53	453		1906	647	116	-59	165	128	HDQ 631	534			987
Synonym: HYDROQUINONE															
QUINOLINE	2656	29	736		2968	1014		-83	246		QNL1005	805			987
QUINONE	2587	55		23	2970	1015	192		246	192					988
R 20	1888	55	205	14	815	282	49	52 –	76	68	CRF 301	223	C4.2		266
Synonym: CHLOROFORM															
R 22	1018	12	593						205				M4.0		795
Synonym: MONOCHLORODIFLUOROMETHANE															
RACEMIC LACTIC ACID	1760	60	496		2083	703					LTA 711	569			679
Synonym: LACTIC ACID															
RADIOACTIVE MATERIAL	2982	63							246			807			
RADIOACTIVE MATERIAL	2908	61							246						
RADIOACTIVE MATERIAL	2918	63							246			806			
RADIOACTIVE MATERIAL	2910/2911	61							246			807			

CHEMICAL OR MATERIAL NAME	UN/NA Number	Guide Number (DOT)	Firefighter's Hazardous Materials Reference Book	First Aid Manual for Chemical Accidents	Sax's Dangerous Properties of Industrial Materials	Hazardous Chemicals Desk Reference	Rapid Guide to Hazardous Chemicals in the Workplace	Fire Protection Guide on Hazardous Materials (NFPA)	Firefighter's Handbook of Hazardous Materials	Pocket Guide to Chemical Hazards (NIOSH)	Chemical Hazard Response Information System (CHRIS)	Emergency Handling of Hazardous Materials in Surface Transportation (AAR)	Emergency Action Guides (EAG)	Chemical Data Notebook: A User's Manual	Condensed Chemical Dictionary
RADIOACTIVE MATERIAL	2912	62							246			806			
RADIOACTIVE MATERIAL	2911/2910	61							246			807			
RADIOACTIVE MATERIAL	2909/2910	61							246			808			
RADIOACTIVE MATERIAL	2974	63	737						246			807			
RADIUM AND COMPOUNDS					2975	1017			247						991
RADON					2975	1017			247						992
RAGS	1856								247						
RANGE OIL Synonym: KEROSENE	1223	27	494				126		247		KRS 689	566			671
RANGE OIL Synonym: OILS FUEL NO. 1	1223/1993	27							247		OON 894	464			
RANGE OIL Synonym: FUEL OIL NO. 1	1993/1202	27	403		2071	701		−55				464	F3.0		
RANGE OIL Synonym: JET FUEL JP-1	1223	27	489		2078				247		JPO 685				
RARE GAS MIXTURE	1979	12							247			809			
RARE GAS NITROGEN MIXTURE	1981	12							247			809			
RARE GAS OXYGEN MIXTURE	1980	14							247			809			
RASPITE Synonym: LEAD TUNGSTATE			513				128				LTU 717				693
RATOX Synonym: THALLIUM SULFATE	1707	53	820		3263	1110	212		266		TSU1150	901			1138
RAW LINSEED OIL Synonym: OILS MISCELLANEOUS LINSEED											OLS 887				
RC PLASTICIZER DBP Synonym: DIBUTYL PHTHALATE	9095		265	15	1119	375	66	−34		84	DPA 458				374

CHEMICAL OR MATERIAL NAME	UN/NA Number	Guide Number (DOT)	Firefighter's Hazardous Materials Reference Book	First Aid Manual for Chemical Accidents	Sax's Dangerous Properties of Industrial Materials	Hazardous Chemicals Desk Reference	Rapid Guide to Hazardous Chemicals in the Workplace	Fire Protection Guide on Hazardous Materials (NFPA)	Firefighter's Handbook of Hazardous Materials	Pocket Guide to Chemical Hazards (NIOSH)	Chemical Hazard Response Information System (CHRIS)	Emergency Handling of Hazardous Materials in Surface Transportation (AAR)	Emergency Action Guides (EAG)	Chemical Data Notebook: A User's Manual	Condensed Chemical Dictionary
REALGAR Synonym: ARSENIC DISULFIDE	1557	53	97				16		35		ARD 83				97
RECEPTACLES	2037	17							247			810			
RED ARSENIC GLASS Synonym: ARSENIC DISULFIDE	1557	53	97				16		35		ARD 83				97
RED ARSENIC SULFIDE Synonym: ARSENIC DISULFIDE	1557	53	97				16		35		ARD 83				97
RED FUMING ACID Synonym: NITRIC ACID FUMING	2032	44		21	2509	856		123–	210	160		684	N2.0	N 1	823
RED FUMING NITRIC ACID	2032	44							247						
RED OIL Synonym: OLEIC ACID			654		2630	896		–76			OLA 883				996
RED ORPIMENT Synonym: ARSENIC DISULFIDE	1557	53	97				16		35		ARD 83				97
RED OXIDE OF NITROGEN Synonym: NITROGEN TETROXIDE	1067	20	637		2537	867		128–	213		NOX 845		N3.4.6		828
RED TR BASE Synonym: 4-CHLORO-o-TOLUIDINE	2239	55			879	299	51				CTD 319				274
REDUCING LIQUID	1142	26							247						
REFRIGERANT 12 Synonym: DICHLORODIFLUOROMETHANE	1028	12	275	15	2237	380	68		102	86	DCF 360	331	D1.1		379
REFRIGERANT 152a Synonym: 1,1-DIFLUOROETHANE	1030	22	2						112		DFE 397				399
REFRIGERANT GAS	1954	22							247			810			
REFRIGERANT GAS	1078	12							247			811			
REFRIGERATING MACHINE	2857	21							247			811			

CHEMICAL OR MATERIAL NAME	UN/NA Number	Guide Number (DOT)	Firefighter's Hazardous Materials Reference Book	First Aid Manual for Chemical Accidents	Sax's Dangerous Properties of Industrial Materials	Hazardous Chemicals Desk Reference	Rapid Guide to Hazardous Chemicals in the Workplace	Fire Protection Guide on Hazardous Materials (NFPA)	Firefighter's Handbook of Hazardous Materials	Pocket Guide to Chemical Hazards (NIOSH)	Chemical Hazard Response Information System (CHRIS)	Emergency Handling of Hazardous Materials in Surface Transportation (AAR)	Emergency Action Guides (EAG)	Chemical Data Notebook: A User's Manual	Condensed Chemical Dictionary
REGULOX Synonym: **MALEIC HYDRAZIDE**			529								MLH 770				724
REMOVING LIQUID	1142	26							247						
RESIDUAL ASPHALT Synonym: **ASPHALT BLENDING STOCKS STRAIGHT RUN RESIDUE**	1999	27	105	12	307	97	17				ASR 94	99	A12.0		
RESIDUAL FUEL NO. 4 Synonym: **FUEL OIL NO. 4**	1993/1202	27	405					–55	151			464	F3.2		
RESIDUAL FUEL NO. 5 Synonym: **FUEL OIL NO. 5**	1993/1202	27	406					–55	151			464	F3.3		
RESIDUAL FUEL OIL NO. 4 Synonym: **OILS FUEL NO. 4**	1223/1993	27									OFR 879	464			
RESIDUAL FUEL OIL NO. 5 Synonym: **OILS FUEL NO. 5**	1223/1993	27									OFV 881	464			
RESIDUAL FUEL OIL NO. 6 Synonym: **OILS FUEL NO. 6**	1223	27									OSX 907				
RESIDUAL OIL Synonym: **ASPHALT BLENDING STOCKS ROOFERS FLUX**	1999	27	105	12	307	97	16				ARF 84	99	A12.0		1001
RESIN COMPOUND	1866	26			2983				247			812			1001
RESIN OIL Synonym: **OILS MISCELLANEOUS ROSIN**	1286	26						–83			ORN 901				
RESIN SOLUTION	1866	26			2983	1019			247			812			
RESIN SOLUTION	1896	55				1019			247						
RESIN SOLUTION	2868					1019									
RESINOUS PETROLEUM RESIDUE	1993	27										812			

CHEMICAL OR MATERIAL NAME	UN/NA Number	Guide Number (DOT)	Firefighter's Hazardous Materials Reference Book	First Aid Manual for Chemical Accidents	Sax's Dangerous Properties of Industrial Materials	Hazardous Chemicals Desk Reference	Rapid Guide to Hazardous Chemicals in the Workplace	Fire Protection Guide on Hazardous Materials (NFPA)	Firefighter's Handbook of Hazardous Materials	Pocket Guide to Chemical Hazards (NIOSH)	Chemical Hazard Response Information System (CHRIS)	Emergency Handling of Hazardous Materials in Surface Transportation (AAR)	Emergency Action Guides (EAG)	Chemical Data Notebook: A User's Manual	Condensed Chemical Dictionary
RESORCIN Synonym: RESORCINOL	2876	55	738	23	2984	1019	192	-83	247		RSC1006	813			1003
RESORCINOL	2876	55	738	23	2984	1019	192	-83	247		RSC1006	813			1003
RETARDER W Synonym: SALICYLIC ACID			740		3011	1027		-83			SLA1047				1019
RETINOL Synonym: OILS MISCELLANEOUS ROSIN	1286	26						-83			ORN 901				1004
RETINOL Synonym: OILS MISCELLANEOUS RESIN	1286	26									ORS 902				1004
RHODANATE Synonym: SODIUM THIOCYANATE			788	24							SCY1026				1071
RHODIUM					2989	1021	193			192					1006
RICIN					2992				247						
ROAD ASPHALT Synonym: ASPHALT	1999	27					17		247		ASP 93	99			100
ROAD BINDER Synonym: ASPHALT BLENDING STOCKS STRAIGHT RUN RESIDUE	1999	27	105	12	307	97	17				ASR 94	99	A12.0		
ROAD OIL Synonym: ASPHALT BLENDING STOCKS ROOFERS FLUX	1999	27	105	12	307	97	17				ARF 84	99	A12.0		
RODENTICIDES	1681	53							247			816			1011
RONNEL				23	2997	1022	193		248	192					
ROSIN OIL Synonym: OILS MISCELLANEOUS RESIN	1286	26						-83	248		ORS 902	817			
ROSIN SOLUTION	1993	27	739									818	R1.0		1012

CHEMICAL OR MATERIAL NAME	UN/NA Number	Guide Number (DOT)	Firefighter's Hazardous Materials Reference Book	First Aid Manual for Chemical Accidents	Sax's Dangerous Properties of Industrial Materials	Hazardous Chemicals Desk Reference	Rapid Guide to Hazardous Chemicals in the Workplace	Fire Protection Guide on Hazardous Materials (NFPA)	Firefighter's Handbook of Hazardous Materials	Pocket Guide to Chemical Hazards (NIOSH)	Chemical Hazard Response Information System (CHRIS)	Emergency Handling of Hazardous Materials in Surface Transportation (AAR)	Emergency Action Guides (EAG)	Chemical Data Notebook: A User's Manual	Condensed Chemical Dictionary
ROSINOL Synonym: OILS MISCELLANEOUS ROSIN	1286	26						-83			ORN 901				
ROSINOL Synonym: OILS MISCELLANEOUS RESIN	1286	26									ORS 902				
ROTENONE					3000	1023	193		248	194					1012
ROTEX Synonym: METHYL BROMIDE	1062	55	559	20	2274	776	137	110-67	192	146	MTB 805	619	M3.0	M 3	759
RUBBER REGENERATED	1345	32							248						
RUBBER SCRAP	1345/1325	32			3001				248			819			
RUBBER SOLUTION	1287	26							248			819			
RUBBER SOLVENT	1287	26			3001	1023			248						
RUBBER SOLVENT Synonym: PETROLEUM NAPHTHA	1255	27	676		3001	1023			227		PTN 999	739	P2.0		890
RUBBING ALCOHOL	1987	26										819			
RUBBING ALCOHOL Synonym: ISOPROPYL ALCOHOL	1219/1993	26	482	18	2040	687	124	-63	174	132	IPA 670	560		I 2	660
RUBBING ALCOHOL Synonym: ISOPROPANOL	1219	26	480						174			555	I4.0		659
RUBIDIUM HYDRATE	2677	60													
RUBIDIUM HYDROXIDE	2678/1759	60			3003	1024			248			820			1015
RUBIDIUM HYDROXIDE SOLUTION	2677/1760	60			3003	1024			248			820			1015
RUBIDIUM METAL	1423	40			3002	1023			248			820			1014
RUBY ARSENIC Synonym: ARSENIC DISULFIDE	1557	53	97				16		35		ARD 83				97
RUM DENATURED									248						

CHEMICAL OR MATERIAL NAME	UN/NA Number	Guide Number (DOT)	Firefighter's Hazardous Materials Reference Book	First Aid Manual for Chemical Accidents	Sax's Dangerous Properties of Industrial Materials	Hazardous Chemicals Desk Reference	Rapid Guide to Hazardous Chemicals in the Workplace	Fire Protection Guide on Hazardous Materials (NFPA)	Firefighter's Handbook of Hazardous Materials	Pocket Guide to Chemical Hazards (NIOSH)	Chemical Hazard Response Information System (CHRIS)	Emergency Handling of Hazardous Materials in Surface Transportation (AAR)	Emergency Action Guides (EAG)	Chemical Data Notebook: A User's Manual	Condensed Chemical Dictionary
SACCHAROSE Synonym: SUCROSE			796		3151	1077					SRS1056				1017
SACCHARUM Synonym: SUCROSE			796		3151	1077					SRS1056				1099
SAFETY SOLVENT	1256	27							248						
SAFFLOWER OIL Synonym: OILS EDIBLE SAFFLOWER					3008	1026					OSF 905				1018
SAFFLOWER SEED OIL Synonym: OILS EDIBLE SAFFLOWER											OSF 905				1018
SAFROLE						1026		-83							1018
SAL ACETOSELLA Synonym: POTASSIUM BINOXALATE											PBO 930				949
SAL AMMONIAC Synonym: AMMONIUM CHLORIDE	9085/9188		56		223	60	11		28		AMC 49	57			1018
SAL VOLATILE Synonym: AMMONIUM CARBONATE	9084/3077	31	55	11	223	60			28		ACB 14	57			64
SALICYLAL Synonym: SALICYLALDEHYDE					3010			-83	248		SAL1009				1019
SALICYLALDEHYDE					3010			-83	248		SAL1009				1019
SALICYLIC ACID			740		3011	1027		-83			SLA1047				1019
SALICYLIC ALDEHYDE Synonym: SALICYLALDEHYDE					3010			-83	248		SAL1009				1019
SALITHION									248						
SALMIAC Synonym: AMMONIUM CHLORIDE	9085/9188		56		223	60	11		28		AMC 49	57			64

CHEMICAL OR MATERIAL NAME	UN/NA Number	Guide Number (DOT)	Firefighter's Hazardous Materials Reference Book	First Aid Manual for Chemical Accidents	Sax's Dangerous Properties of Industrial Materials	Hazardous Chemicals Desk Reference	Rapid Guide to Hazardous Chemicals in the Workplace	Fire Protection Guide on Hazardous Materials (NFPA)	Firefighter's Handbook of Hazardous Materials	Pocket Guide to Chemical Hazards (NIOSH)	Chemical Hazard Response Information System (CHRIS)	Emergency Handling of Hazardous Materials in Surface Transportation (AAR)	Emergency Action Guides (EAG)	Chemical Data Notebook: A User's Manual	Condensed Chemical Dictionary
SALT OF SATURN Synonym: LEAD ACETATE	1616	53	501	19	2094	708	128		178		LAC 690	570			686
SALT OF SORREL Synonym: POTASSIUM BINOXALATE											PBO 930				949
SALTPETER Synonym: POTASSIUM NITRATE	1486	35	713		2874	985			248			782	P6.1	P 6	1020
SALUFER Synonym: SODIUM SILICOFLUORIDE	2674	53	786						255		SFR1042				1069
SAMARIUM					3014	1027			248						1020
SAND ACID Synonym: HYDROFLUOROSILICIC ACID	1778	60							164				H4.2		614
SAND ACID Synonym: FLUOSILICIC ACID	1778	60	399						150		FSL 596				532
SANTOCHLOR Synonym: DICHLOROBENZENE-para	1592	58	273	15	1128	377	67	−34	101	84	DBP 351	330			1021
SANTOPHEN 20 Synonym: PENTACHLOROPHENOL	2020/3082	53		22	2679	912	172	133–	223	174	PCP 940	729			879
SARALEX Synonym: DIAZINON	2783	55	261	15	1083	363	64		96		DZN 415				366
SCHEELE'S GREEN Synonym: COPPER ARSENITE	1586	53	223				55		83		CPA 283	257			1024
SCHEELITE Synonym: LEAD TUNGSTATE			513				128				LTU 717				1024
SCHWEINFURT GREEN	1585	53							248						
SCHWEINFURTH GREEN Synonym: COPPER ACETOARSENITE	1585	53	222				55		83		CAA 203	257			310

CHEMICAL OR MATERIAL NAME	UN/NA Number	Guide Number (DOT)	Firefighter's Hazardous Materials Reference Book	First Aid Manual for Chemical Accidents	Sax's Dangerous Properties of Industrial Materials	Hazardous Chemicals Desk Reference	Rapid Guide to Hazardous Chemicals in the Workplace	Fire Protection Guide on Hazardous Materials (NFPA)	Firefighter's Handbook of Hazardous Materials	Pocket Guide to Chemical Hazards (NIOSH)	Chemical Hazard Response Information System (CHRIS)	Emergency Handling of Hazardous Materials in Surface Transportation (AAR)	Emergency Action Guides (EAG)	Chemical Data Notebook: A User's Manual	Condensed Chemical Dictionary
SCONATEX Synonym: VINYLIDENE CHLORIDE	1303	26	878	26	3498	1175	229	181–93	286		VCI1173	952	V1.2		1217
SEAL COATING MATERIAL Synonym: ASPHALT BLENDING STOCKS STRAIGHT RUN RESIDUE	1999	27	105	12	307	97	17				ASR 94	99	A12.0		
SECONDARY AMMONIUM PHOSPHATE Synonym: AMMONIUM PHOSPHATE					232	66					APP 78				68
SECONDARY BUTYL ACETATE	1123	26													
SEED CAKE	1386	32							249			821			
SEED CAKE	2217	31							248			822			
SELANE Synonym: HYDROGEN SELENIDE	2202	13	451	18	1903	645	116		165	128		532	H4.6		617
SELENATES	2630	53							249						
SELENIC ACID	1905	59			3026	1030			249			822			1028
SELENIC ANHYDRIDE Synonym: SELENIUM TRIOXIDE			743								STO1062				
SELENIOUS ACID DISODIUM SALT Synonym: SODIUM SELENITE	2630	53	784		3112	1064			255		SSE1058	850			1068
SELENIOUS ANHYDRIDE Synonym: SELENIUM DIOXIDE	2811	53	741		3029				249		SLD1048				1028
SELENITES	2630	53													
SELENIUM	2658	53			3027	1030	194		249	194					1028
SELENIUM ANHYDRIDE Synonym: HYDROGEN SELENIDE	2202	13	451	18	1903	645	116		165	128		532	H4.6		617
SELENIUM DIHYDRIDE Synonym: HYDROGEN SELENIDE	2202	13	451	18	1903	645	116		165	128		532	H4.6		617

CHEMICAL OR MATERIAL NAME	UN/NA Number	Guide Number (DOT)	Firefighter's Hazardous Materials Reference Book	First Aid Manual for Chemical Accidents	Sax's Dangerous Properties of Industrial Materials	Hazardous Chemicals Desk Reference	Rapid Guide to Hazardous Chemicals in the Workplace	Fire Protection Guide on Hazardous Materials (NFPA)	Firefighter's Handbook of Hazardous Materials	Pocket Guide to Chemical Hazards (NIOSH)	Chemical Hazard Response Information System (CHRIS)	Emergency Handling of Hazardous Materials in Surface Transportation (AAR)	Emergency Action Guides (EAG)	Chemical Data Notebook: A User's Manual	Condensed Chemical Dictionary
SELENIUM DIOXIDE	2811	53	741		3029				249		SLD1048				1028
SELENIUM DISULFIDE	2657/2811	55			3029	1031			249	194		822			
SELENIUM HEXAFLUORIDE	2194/1955	15	742	24	3030	1031	194		249	194		822			
SELENIUM HYDRIDE Synonym: HYDROGEN SELENIDE	2202	13	451	18	1903	645	116		249	128			H4.6		617
SELENIUM METAL	2658	53							249	194					1028
SELENIUM OXIDE Synonym: SELENIUM DIOXIDE	2811	53	741		3029				249		SLD1048				1028
SELENIUM OXYCHLORIDE	2879	59			3030	1032			249			823			
SELENIUM SULFIDE	2657	55							249	194					1028
SELENIUM TRIOXIDE			743								STO1062				
SELF REACTIVE SUBSTANCES	3031	71							249						
SELF REACTIVE SUBSTANCES	3032	71							249						
SENARMONTITE Synonym: ANTIMONY TRIOXIDE	1549/3077	60	94								ATX 109	87			90
SENTRY Synonym: CALCIUM HYPOCHLORITE	1748	45	172	13				46 –	63		CHY 252	182	C1.1	C 2	1029
SENTRY GRAIN PRESERVER Synonym: PROPIONIC ACID	1848	60	721		2909	997	188	148–81	243		PNA 968	791	P7.1		968
SEVIN Synonym: CARBARYL	2757/3077	55	182		688	241	40		249	58	CBY 224	187			1031
SEWER GAS Synonym: HYDROGEN SULFIDE	1053	13	452	18	1903	646	116	100–59	165	128	HDS 632	532	H4.7	H 6	617
SEXTONE Synonym: CYCLOHEXANONE	1915	26	246	14	991	334	59	-30	88	78	CCH 228	281	C8.2		337
SHALE OIL	1288	27							249			824			1031

CHEMICAL OR MATERIAL NAME	UN/NA Number	Guide Number (DOT)	Firefighter's Hazardous Materials Reference Book	First Aid Manual for Chemical Accidents	Sax's Dangerous Properties of Industrial Materials	Hazardous Chemicals Desk Reference	Rapid Guide to Hazardous Chemicals in the Workplace	Fire Protection Guide on Hazardous Materials (NFPA)	Firefighter's Handbook of Hazardous Materials	Pocket Guide to Chemical Hazards (NIOSH)	Chemical Hazard Response Information System (CHRIS)	Emergency Handling of Hazardous Materials in Surface Transportation (AAR)	Emergency Action Guides (EAG)	Chemical Data Notebook: A User's Manual	Condensed Chemical Dictionary
SHED A LEAF Synonym: SODIUM CHLORATE	1495	35	766	24	3070			156–	253		SDC1029	834	S2.1	S 2	1052
SHELL CHARCOAL Synonym: CHARCOAL	1361	32	195		739	258			68		CHC 244	203			248
SHELLAC	1263	26							249			825			1032
SHOE BLEACH	1993	27													
SIGNAL OIL Synonym: OILS MISCELLANEOUS MINERAL SEAL	1270	27							249		OMS 889				
SILANE	2203	17		24				153–83	249						1034
SILIBOND Synonym: ETHYL SILICATE	1292	29	365		1659	560	98	–54	139	114	ESC 558				500
SILICA AMORPHOUS FUMED					3038	1033	194			194					
SILICA AMORPHOUS HYDRATED					3038	1033	195			194					
SILICA CRYSTALLINE				24	3038	1033	195								
SILICA CRYSTALLINE CRISTOBALITE Synonym: SILICA CRYSTALLINE					3039	1035	195			196					
SILICA CRYSTALLINE QUARTZ Synonym: SILICA CRYSTALLINE					3040	1035	195			196					
SILICA CRYSTALLINE TRIDYMITE Synonym: SILICA CRYSTALLINE					3040	1035	196			196					
SILICA FUSED					3040	1035	196								1034
SILICA GEL AND AMORPHOUS PRECIPITATED					3041	1036	196								
SILICANE	2203	17							249						
SILICATE SOAPSTONE					3041	1036									
SILICIC ACID Synonym: METHYL ORTHOSILICATE	2606/1993	57							196			634	M3.6.1		1035

CHEMICAL OR MATERIAL NAME	UN/NA Number	Guide Number (DOT)	Firefighter's Hazardous Materials Reference Book	First Aid Manual for Chemical Accidents	Sax's Dangerous Properties of Industrial Materials	Hazardous Chemicals Desk Reference	Rapid Guide to Hazardous Chemicals in the Workplace	Fire Protection Guide on Hazardous Materials (NFPA)	Firefighter's Handbook of Hazardous Materials	Pocket Guide to Chemical Hazards (NIOSH)	Chemical Hazard Response Information System (CHRIS)	Emergency Handling of Hazardous Materials in Surface Transportation (AAR)	Emergency Action Guides (EAG)	Chemical Data Notebook: A User's Manual	Condensed Chemical Dictionary
SILICOCHLOROFORM Synonym: TRICHLOROSILANE	1295	38	840		3361	1141		174–88	249		TCS1090	922			1035
SILICOFLUORIC ACID Synonym: HYDROFLUOROSILICIC ACID	1778	60			3041	1036			249				H4.2		532
SILICOFLUORIC ACID Synonym: FLUOSILICIC ACID	1778	60	399	17	3041	1036			249		FSL 596				614
SILICOFLUORIDE	2856	53							249						
SILICON	1346/1325	31			3042	1037	196					827			1035
SILICON BROMIDE					3042				249						
SILICON CARBIDE					3042	1037	197								1036
SILICON CHLORIDE Synonym: SILICON TETRACHLORIDE	1818	39	744		3042	1037		154–	250		STC1060	826	S1.0		
SILICON DIBROMIDE SULFIDE					3043				250						
SILICON FLUORIDE	1859	16			3043	1038			250						
SILICON POWDER	1346/1325	32			3042	1037	196		250			827			1035
SILICON TETRACHLORIDE Synonym: SILICON CHLORIDE	1818	39	744		3042	1037		154–	250		STC1060	826	S1.0		
SILICON TETRAFLUORIDE	1859	16						155–	250			827			
SILICON TETRAHYDRIDE	2203	17			3043	1038	197		250						
SILICONE FLUIDS Synonym: DIMETHYLPOLYSILOXANE					3043	470					DMP 437				1036
SILVER					3044	1038	197			196					1038
SILVER ACETATE			745								SVA1065				1038
SILVER ACETYLIDE					3045	1038			250						1038
SILVER AMMONIUM COMPOUNDS					3045				250						

HAZARDOUS MATERIALS REFERENCE BOOKS INDEX

CHEMICAL OR MATERIAL NAME	UN/NA Number	Guide Number (DOT)	Firefighter's Hazardous Materials Reference Book	First Aid Manual for Chemical Accidents	Sax's Dangerous Properties of Industrial Materials	Hazardous Chemicals Desk Reference	Rapid Guide to Hazardous Chemicals in the Workplace	Fire Protection Guide on Hazardous Materials (NFPA)	Firefighter's Handbook of Hazardous Materials	Pocket Guide to Chemical Hazards (NIOSH)	Chemical Hazard Response Information System (CHRIS)	Emergency Handling of Hazardous Materials in Surface Transportation (AAR)	Emergency Action Guides (EAG)	Chemical Data Notebook: A User's Manual	Condensed Chemical Dictionary
SILVER ARSENITE	1683	53			3046				250						
SILVER AZIDE					3046	1038			250						
SILVER CARBONATE			746								SVC1066				1039
SILVER CYANIDE	1684	53			3047	1039			250			827			1039
SILVER FLUORIDE			747		3048	1039			250		SVF1067				1039
SILVER FULMINATE					3048				251						
SILVER IODATE			748								SVI1068				
SILVER MONOFLUORIDE Synonym: SILVER FLUORIDE			747		3048	1039			250		SVF1067				1039
SILVER NITRATE	1493	45	749		3049	1040	198		251		SVN1069	828			1039
SILVER NITRIDE					3050				251						1040
SILVER NITROPRUSSIDE					3050				251						
SILVER OXALATE					3050				251						
SILVER OXIDE			750		3051				251		SVO1070				1040
SILVER PICRATE	1347	33							251						1040
SILVER SULFATE			751								SVS1071				1041
SILVER TETRAZOL	1347	33													
SILVEX Synonym: 2-(2,4,5-TRICHLOROPHENOXY) PROPANOIC ACID					3358						TPA1136				1041
SILVISAR 510 Synonym: CACODYLIC ACID	1572/2811	53	156						62		CDA 235	175			194
SIPOL L8 Synonym: OCTANOL	1987	26	650		3620						OTA 909	706	O1.0		847
SKELLYSOLVE B Synonym: HEXANE (-n)	1208	27	435	18	1859	627	109	-58	160		HXA 649	517	H2.0		599

CHEMICAL OR MATERIAL NAME	UN/NA Number	Guide Number (DOT)	Firefighter's Hazardous Materials Reference Book	First Aid Manual for Chemical Accidents	Sax's Dangerous Properties of Industrial Materials	Hazardous Chemicals Desk Reference	Rapid Guide to Hazardous Chemicals in the Workplace	Fire Protection Guide on Hazardous Materials (NFPA)	Firefighter's Handbook of Hazardous Materials	Pocket Guide to Chemical Hazards (NIOSH)	Chemical Hazard Response Information System (CHRIS)	Emergency Handling of Hazardous Materials in Surface Transportation (AAR)	Emergency Action Guides (EAG)	Chemical Data Notebook: A User's Manual	Condensed Chemical Dictionary
SKELLYSOLVE C Synonym: HEPTANE (-n)	1206	27	425	18	1830	616	106	-57	157	120	HPT 644	505	H1.0.3		592
SLAKED LIME Synonym: CALCIUM HYDROXIDE			171	13		232	37		63		CAH 206				1043
SLIMICIDE Synonym: ACROLEIN	1092	30	31	11	63	25	5	17 -12	20	32	ARL 85	12	A4.0	A 5	1043
SLOW CURING ASPHALT Synonym: OILS MISCELLANEOUS ROAD	1999	27									ORD 899				
SLUDGE ACID	1906	60			3054	1040			251			828			
SMITHSONITE Synonym: ZINC CARBONATE	9157/3077		893		3538		234		288		ZCB1197	965			1243
SMOKELESS POWDER	1325	32			3054				251			829			1044
SODA CHLORATE Synonym: SODIUM CHLORATE	1495	35	766	24	3070	1048		156–	251		SDC1029	834	S2.1	S 2	1052
SODA LIME	1907	60			3056				251			829			1046
SODA LYE Synonym: SODIUM HYDROXIDE	1823	60	777	24	3087	1056	200	159–	254	198	SHD1044	842	S2.5	S 4	1058
SODA NITER Synonym: SODIUM NITRATE	1498	35	780		3097	1060			251		SDN1033	845	S2.6	S 5	1046
SODA SALTPETER Synonym: SODIUM NITRATE	1498	35	780		3097	1060			254		SDN1033	845	S2.6	S 5	1063
SODAMIDE Synonym: SODIUM AMIDE	1425	76	755		3059	1044			251		SAM1010	830			1046
SODIUM	1428	40	752	24	3057	1042		155–	251		SDU1036	844	S2.0	S 1	1047
SODIUM	1429	40		24	3056	1042			251						1047
SODIUM	1421	40		24	3056	1043			251			852			1047

HAZARDOUS MATERIALS REFERENCE BOOKS INDEX

CHEMICAL OR MATERIAL NAME	UN/NA Number	Guide Number (DOT)	Firefighter's Hazardous Materials Reference Book	First Aid Manual for Chemical Accidents	Sax's Dangerous Properties of Industrial Materials	Hazardous Chemicals Desk Reference	Rapid Guide to Hazardous Chemicals in the Workplace	Fire Protection Guide on Hazardous Materials (NFPA)	Firefighter's Handbook of Hazardous Materials	Pocket Guide to Chemical Hazards (NIOSH)	Chemical Hazard Response Information System (CHRIS)	Emergency Handling of Hazardous Materials in Surface Transportation (AAR)	Emergency Action Guides (EAG)	Chemical Data Notebook: A User's Manual	Condensed Chemical Dictionary
SODIUM 2-DIAZO-1-NAPHTHOL-4-SULFONATE	3040	72							251						
SODIUM 2-DIAZO-1-NAPHTHOL-5-SULFONATE	3041	72							251						
SODIUM ACETATE					3057	1043									1047
SODIUM ACID PYROPHOSPHATE Synonym: SODIUM PHOSPHATE	9147/3077				3106	1063			255		SPP1054	849			1047
SODIUM ACID SULFATE	1821	60			3058	1043			252			840			1047
SODIUM ACID SULFITE Synonym: SODIUM BISULFITE	1821/3082	60			3065	1046					SBS1017	833			1047
SODIUM ALKYL SULFATES			754								SAS1012				
SODIUM ALKYLBENZENESULFONATES			753								SAB1007				
SODIUM ALUMINATE	1819	60							252			829			1047
SODIUM ALUMINATE	2812	60							252						1047
SODIUM ALUMINUM HYDRIDE	2835	40							252			830			1048
SODIUM ALUMINUM SULPHATE	9188/3077											830			
SODIUM AMALGAM	1424	40							252						1048
SODIUM AMIDE	1425	76	755		3059	1044			252		SAM1010	830			1048
SODIUM AMMONIUM VANADATE	2863	53							252			831			
SODIUM ANILINE ARSONATE	2473	53							252						1048
SODIUM ARSANILATE	2473	53							252			831			1049
SODIUM ARSENATE	1685	53	756						252		SDA1027	831			1049
SODIUM ARSENATE DIBASIC Synonym: SODIUM ARSENATE	1685	53	756						252		SDA1027	831			1049
SODIUM ARSENITE	2027	53	758			1044			252		SAR1011				1049
SODIUM ARSENITE	1686	54	757		3063	1044			252		SAR1011	832			1049

CHEMICAL OR MATERIAL NAME	UN/NA Number	Guide Number (DOT)	Firefighter's Hazardous Materials Reference Book	First Aid Manual for Chemical Accidents	Sax's Dangerous Properties of Industrial Materials	Hazardous Chemicals Desk Reference	Rapid Guide to Hazardous Chemicals In the Workplace	Fire Protection Guide on Hazardous Materials (NFPA)	Firefighter's Handbook of Hazardous Materials	Pocket Guide to Chemical Hazards (NIOSH)	Chemical Hazard Response Information System (CHRIS)	Emergency Handling of Hazardous Materials In Surface Transportation (AAR)	Emergency Action Guides (EAG)	Chemical Data Notebook: A User's Manual	Condensed Chemical Dictionary
SODIUM ATOM Synonym: SODIUM	1428	40	752	24	3056	1042		155–	251		SDU1036	844	S2.0	S 1	1047
SODIUM AZIDE	1687	56	759		3063	1045	198		252		SAZ1014	832			1049
SODIUM BIBORATE Synonym: SODIUM BORATE			763	24	3066	1046	198				SDB1028				1050
SODIUM BICHROMATE Synonym: SODIUM DICHROMATE	1479	35	771	24	3078	1052			253		SCR1025	838			1050
SODIUM BIFLUORIDE	2439	60	760						252		SBF1015				1050
SODIUM BIFLUORIDE SOLUTION	2439	60	760						252		SBF1015				
SODIUM BISULFATE	1821	60							252			840			1050
SODIUM BISULFATE	2837	60							252						1050
SODIUM BISULFIDE	2693	60													1050
SODIUM BISULFIDE Synonym: SODIUM HYDROSULFIDE SOLUTION	2922/1760	59			3086	1055			254		SHR1046	841	S2.3		1050
SODIUM BISULFITE	1821	60									SBS1017				1050
SODIUM BISULFITE	2693/3082	60	760		3065	1046	198		252			833			1050
SODIUM BISULFITE SOLUTION	2693/3082	60	761		3065	1046	198		252			833			
SODIUM BORATE			763	24	3066	1046	198				SDB1028				1050
SODIUM BORATE DECAHYDRATE					3066	1046	199								
SODIUM BOROHYDRIDE	1426	32			3066	1047			252		SBH1016	833			1051
SODIUM BROMATE	1494	42			3067	1047			252			833			1051
SODIUM CACODYLATE	1688/2811	53	765						252		SCD1020	834			1051
SODIUM CETYL SULFATE SOLUTION Synonym: HEXADECYL SULFATE SODIUM SALT											HSS 645				
SODIUM CHLORATE	2428/1479	31		24	3070	1048					SDD1030	835	S2.1	S 2	1052

HAZARDOUS MATERIALS REFERENCE BOOKS INDEX

CHEMICAL OR MATERIAL NAME	UN/NA Number	Guide Number (DOT)	Firefighter's Hazardous Materials Reference Book	First Aid Manual for Chemical Accidents	Sax's Dangerous Properties of Industrial Materials	Hazardous Chemicals Desk Reference	Rapid Guide to Hazardous Chemicals in the Workplace	Fire Protection Guide on Hazardous Materials (NFPA)	Firefighter's Handbook of Hazardous Materials	Pocket Guide to Chemical Hazards (NIOSH)	Chemical Hazard Response Information System (CHRIS)	Emergency Handling of Hazardous Materials in Surface Transportation (AAR)	Emergency Action Guides (EAG)	Chemical Data Notebook: A User's Manual	Condensed Chemical Dictionary
SODIUM CHLORATE	1495	35	766	24	3070	1048		156–	253		SDC1029	834	S2.1	S 2	1052
SODIUM CHLORIDE				24	3070	1048									1052
SODIUM CHLORITE	1908	60		24	3071	1049			253			835			1052
SODIUM CHLORITE	1496	43		24	3071	1049		156–	253			835			1052
SODIUM CHLOROACETATE	2659	53			3071	1049			253			835			1052
SODIUM CHROMATE	9143/3082			24							SCH1021	836			1053
SODIUM CHROMATE	9145/3077		767	24					253		SCH1021	836			1053
SODIUM CUPROCYANIDE	2317	54			3074				253						
SODIUM CUPROCYANIDE	2316/2811	53			3074				253			836			
SODIUM CYANIDE	1689	55	768	24	3074	1050	199	157–	253		SCN1024	837	S2.2	S 3	1053
SODIUM CYANIDE CYANOBRIK Synonym: SODIUM CYANIDE	1689	55	768	24	3074	1050	199	157–	253		SCN1024	837	S2.2	S 3	1053
SODIUM DICHLORO-S-TRIAZINETRIONE	2465	45	770					157–			SDT1035				
SODIUM DICHLOROISOCYANATE	2465	45							253						
SODIUM DICHLOROISOCYANURATE Synonym: SODIUM DICHLORO-s-TRIAZINETRIONE	2465	45	770					157–			SDT1035				1054
SODIUM DICHROMATE	1479	35	771	24	3078	1052			253		SCR1025	838			1054
SODIUM DIFLUORIDE Synonym: SODIUM BIFLUORIDE	2439	60	760						253		SBF1015				1050
SODIUM DIMETHYL ARSENATE	1688	53							253						1055
SODIUM DIMETHYLARSENATE Synonym: SODIUM CACODYLATE	1688/2811	53	765						253		SCD1020	834			1055
SODIUM DINITRO-ortho-CRESOLATE	1348	36							253						
SODIUM DITHIONITE Synonym: SODIUM HYDROSULFITE	1384	37	776		3086	1056			253			838	S2.4		1055

CHEMICAL OR MATERIAL NAME	UN/NA Number	Guide Number (DOT)	Firefighter's Hazardous Materials Reference Book	First Aid Manual for Chemical Accidents	Sax's Dangerous Properties of Industrial Materials	Hazardous Chemicals Desk Reference	Rapid Guide to Hazardous Chemicals in the Workplace	Fire Protection Guide on Hazardous Materials (NFPA)	Firefighter's Handbook of Hazardous Materials	Pocket Guide to Chemical Hazards (NIOSH)	Chemical Hazard Response Information System (CHRIS)	Emergency Handling of Hazardous Materials in Surface Transportation (AAR)	Emergency Action Guides (EAG)	Chemical Data Notebook: A User's Manual	Condensed Chemical Dictionary
SODIUM DODECYL SULFATE Synonym: DODECYL SULFATE SODIUM SALT											DDS 378				
SODIUM DODECYLBENZENESULFONATE	9188/3082	31										838			1055
SODIUM ETHYL XANTHATE					3081										1056
SODIUM FERROCYANIDE			772								SFC1038				1056
SODIUM FLUORIDE	1690	54	773	24	3081	1053	199	158–	253		SDF1031	838			1056
SODIUM FLUORIDE SOLUTION	1690	54	773	24	3082	1053	199	158–	253		SDF1031	838			
SODIUM FLUOROACETATE	2629/2811	53			3082	1054	200		253	196	SAT1013	839			1056
SODIUM FLUOROSILICATE	2674	53							253			839			1057
SODIUM FLUOSILICATE Synonym: SODIUM SILICOFLUORIDE	2674	53	786	24					255		SFR1042				1069
SODIUM HEXAFLUOROSILICATE Synonym: SODIUM SILICOFLUORIDE	2674	53	786						255		SFR1042				1058
SODIUM HYDRATE	1824	60													1058
SODIUM HYDRATE Synonym: SODIUM HYDROXIDE	1823	60	777	24	3086	1056	200	159–	253	198	SHD1044	842	S2.5	S 4	1058
SODIUM HYDRIDE	1427	40	774		3085	1055		158–	254		SDH1032	839			1058
SODIUM HYDROGEN ALKYL SULFATE Synonym: SODIUM ALKYL SULFATES			754								SAS1012				
SODIUM HYDROGEN DIFLUORIDE Synonym: SODIUM BIFLUORIDE	2439	60	760						252		SBF1015				1050
SODIUM HYDROGEN FLUORIDE Synonym: SODIUM BIFLUORIDE	2439	60	760		3086	1055			254		SBF1015	839			1050
SODIUM HYDROGEN SULFATE	1821	60							254			840			
SODIUM HYDROGEN SULFATE SOLUTION	2837	60							254			840			

HAZARDOUS MATERIALS REFERENCE BOOKS INDEX

CHEMICAL OR MATERIAL NAME	UN/NA Number	Guide Number (DOT)	Firefighter's Hazardous Materials Reference Book	First Aid Manual for Chemical Accidents	Sax's Dangerous Properties of Industrial Materials	Hazardous Chemicals Desk Reference	Rapid Guide to Hazardous Chemicals In the Workplace	Fire Protection Guide on Hazardous Materials (NFPA)	Firefighter's Handbook of Hazardous Materials	Pocket Guide to Chemical Hazards (NIOSH)	Chemical Hazard Response Information System (CHRIS)	Emergency Handling of Hazardous Materials In Surface Transportation (AAR)	Emergency Action Guides (EAG)	Chemical Data Notebook: A User's Manual	Condensed Chemical Dictionary
SODIUM HYDROGEN SULFIDE Synonym: SODIUM HYDROSULFIDE SOLUTION	2922/1760	59				1055			254		SHR1046	841	S2.3		1058
SODIUM HYDROSULFIDE	2318	34			3086	1055			254			840	S2.3		1058
SODIUM HYDROSULFIDE	2949	59	775		3086				254			841	S2.3		1058
SODIUM HYDROSULFIDE	2923	59							254				S2.3		1058
SODIUM HYDROSULFIDE SOLUTION	2922/1760	59				1055			254		SHR1046	841	S2.3		
SODIUM HYDROSULFIDE SOLUTION	2949	59	775		3086				254			841	S2.3		
SODIUM HYDROSULFITE	1384	37	776		3086	1056			254			842	S2.4		1058
SODIUM HYDROSULPHITE Synonym: SODIUM HYDROSULFITE	1384	37	776		3086	1056			254			842	S2.4		1058
SODIUM HYDROXIDE	1823	60	777	24	3086	1056	200	159–	254	198	SHD1044	842	S2.5	S 4	1058
SODIUM HYDROXIDE	1824	60		24	3086	1056	200	159–		198		842	S2.5	S 4	1058
SODIUM HYDROXIDE SOLUTION Synonym: CAUSTIC SODA	1824	60	194	24	3086	1057			67		CSS 314	842		S 4	233
SODIUM HYPOCHLORITE	1791	60	778	24	3088	1057			254		SHC1043	843			1059
SODIUM HYPOPHOSPHITE					3089	1057			254						1059
SODIUM IODIDE					3090	1057									1059
SODIUM LAURYL SULFATE Synonym: DODECYL SULFATE SODIUM SALT					3091	1058					DDS 378				1060
SODIUM MERCAPTAN Synonym: SODIUM HYDROSULFIDE	2922/1760	59				1055			254		SHR1046	841	S2.3		1058
SODIUM MERCAPTIDE Synonym: SODIUM HYDROSULFIDE	2922/1760	59				1055			254		SHR1046	841	S2.3		1058
SODIUM META ARSENITE Synonym: SODIUM ARSENITE	1686	54	757		3063	1044			252		SAR1011	832			1049

CHEMICAL OR MATERIAL NAME	UN/NA Number	Guide Number (DOT)	Firefighter's Hazardous Materials Reference Book	First Aid Manual for Chemical Accidents	Sax's Dangerous Properties of Industrial Materials	Hazardous Chemicals Desk Reference	Rapid Guide to Hazardous Chemicals in the Workplace	Fire Protection Guide on Hazardous Materials (NFPA)	Firefighter's Handbook of Hazardous Materials	Pocket Guide to Chemical Hazards (NIOSH)	Chemical Hazard Response Information System (CHRIS)	Emergency Handling of Hazardous Materials in Surface Transportation (AAR)	Emergency Action Guides (EAG)	Chemical Data Notebook: A User's Manual	Condensed Chemical Dictionary
SODIUM METABISULFITE Synonym: SODIUM BISULFITE	1821/3082	60			3094						SBS1017	833			1060
SODIUM METABISULFITE	2693	60			3065	1059	200					843			1060
SODIUM METAL Synonym: SODIUM	1428	40	752	24	3056	1042		155–	254		SDU1036	844	S2.0	S 1	1047
SODIUM METAL DISPERSION IN ORGANIC SOLVENT	1429	40			3057										
SODIUM METAL LIQUID ALLOY	1421	40			3057							852			
SODIUM METHOXIDE Synonym: SODIUM METHYLATE	1431	40	779		3095	1059			254		SML1050	844			1061
SODIUM METHYLATE	1289	26			3095				254			845			1061
SODIUM METHYLATE	1431	40	779		3095	1059			254		SML1050	844			1061
SODIUM MONOXIDE	1825	60			3096	1060			254			845			1062
SODIUM NITRATE	1498	35	780		3097	1060			254		SDN1033	845	S2.6	S 5	1063
SODIUM NITRATE AND POTASH MIXTURE	1478	35							254						
SODIUM NITRATE AND POTASSIUM NITRATE MIXTURE	1499	35							254						
SODIUM NITRITE	1500	35	781		3098	1060			254		SNT1051	846			1063
SODIUM NITROFERRICYANIDE					3101	1061			254						1063
SODIUM OLEATE Synonym: OLEIC ACID SODIUM SALT			656		2630	896					OAC 866				1063
SODIUM ORTHO ARSENITE Synonym: SODIUM ARSENITE	1686	54	757		3063	1044			252		SAR1011	832			1049
SODIUM OXALATE			782		3103						SOX1052				1063
SODIUM OXIDE	1825	60		24					255						1063
SODIUM PENTACHLOROPHENATE	2567	53			3103	1061			255			846			1064
SODIUM PERBORATE	1479	35										847			1064

HAZARDOUS MATERIALS REFERENCE BOOKS INDEX

CHEMICAL OR MATERIAL NAME	UN/NA Number	Guide Number (DOT)	Firefighter's Hazardous Materials Reference Book	First Aid Manual for Chemical Accidents	Sax's Dangerous Properties of Industrial Materials	Hazardous Chemicals Desk Reference	Rapid Guide to Hazardous Chemicals in the Workplace	Fire Protection Guide on Hazardous Materials (NFPA)	Firefighter's Handbook of Hazardous Materials	Pocket Guide to Chemical Hazards (NIOSH)	Chemical Hazard Response Information System (CHRIS)	Emergency Handling of Hazardous Materials in Surface Transportation (AAR)	Emergency Action Guides (EAG)	Chemical Data Notebook: A User's Manual	Condensed Chemical Dictionary
SODIUM PERCARBONATE	2467	35							255			847			1064
SODIUM PERCHLORATE	1502	35		24				160–	255			847			1064
SODIUM PERMANGANATE	1503	35			3104	1062			255			847			1064
SODIUM PEROXIDE	1504	47		24	3104	1062		160–	255			848			1065
SODIUM PERSULFATE	1505	35			3105	1062			255			848			1065
SODIUM PHENATE	2497	60							255						1065
SODIUM PHENOLATE	2497	60							255			848			1065
SODIUM PHENOXIDE	2497	60			3105	1062			255						
SODIUM PHOSPHATE	9147/3077				3106	1063			255		SPP1054	849			1065
SODIUM PHOSPHATE DIBASIC Synonym: SODIUM PHOSPHATE	9147/3077				3106	1063			255		SPP1054	849			1065
SODIUM PHOSPHATE MONOBASIC Synonym: SODIUM PHOSPHATE	9147/3077				3106	1063			255		SPP1054	849			1066
SODIUM PHOSPHATE TRIBASIC Synonym: SODIUM PHOSPHATE	9147/3077				3106	1063			255		SPP1054	849			1066
SODIUM PHOSPHATE TRIBASIC	9148/3077	31	783		3106	1063					SPH1053	849			1066
SODIUM PHOSPHIDE	1432	41			3107	1063			255			849			1066
SODIUM PICRAMATE	1349	33							255			849			1066
SODIUM PICRATE					3107				255						
SODIUM POTASSIUM ALLOY	1422	40						161–				850			1067
SODIUM PYROBORATE Synonym: SODIUM BORATE			763	24	3066	1046	198				SDB1028				1067
SODIUM PYROSULFITE Synonym: SODIUM BISULFITE	1821/3082	60									SBS1017	833			1068
SODIUM RESINATE	1325	32										850			1068

CHEMICAL OR MATERIAL NAME	UN/NA Number	Guide Number (DOT)	Firefighter's Hazardous Materials Reference Book	First Aid Manual for Chemical Accidents	Sax's Dangerous Properties of Industrial Materials	Hazardous Chemicals Desk Reference	Rapid Guide to Hazardous Chemicals in the Workplace	Fire Protection Guide on Hazardous Materials (NFPA)	Firefighter's Handbook of Hazardous Materials	Pocket Guide to Chemical Hazards (NIOSH)	Chemical Hazard Response Information System (CHRIS)	Emergency Handling of Hazardous Materials in Surface Transportation (AAR)	Emergency Action Guides (EAG)	Chemical Data Notebook: A User's Manual	Condensed Chemical Dictionary
SODIUM RHODANIDE Synonym: SODIUM THIOCYANATE			788	24							SCY1026				1068
SODIUM SELENATE	2630	53							255						1068
SODIUM SELENITE	2630	53	784		3112	1064			255		SSE1058	850			1068
SODIUM SILICATE	1759	60	785	24	3113	1065					SSC1057	850			1068
SODIUM SILICOFLUORIDE	2674	53	786						255		SFR1042				1069
SODIUM SOLUTION WASTE	1760	60													
SODIUM SULFATE				24	3114	1065			255						1069
SODIUM SULFHYDRATE Synonym: SODIUM HYDROSULFIDE SOLUTION	2922	59				1055			254		SHR1046	841	S2.3		1070
SODIUM SULFIDE	1849	60						161–	255						1070
SODIUM SULFIDE	1385	34	787		3114	1065		161–	255		SDS1034	851			1070
SODIUM SULFITE					3114	1066					SSF1059				1070
SODIUM SULFOCYANATE Synonym: SODIUM THIOCYANATE			788	24							SCY1026				1070
SODIUM SULFOXYLATE Synonym: SODIUM HYDROSULFITE	1384	37	776		3086	1056			254			842	S2.4		1070
SODIUM SULFURET	1385	34							255						
SODIUM SULHYDRATE Synonym: SODIUM HYDROSULFIDE	2922	59				1055			254		SHR1046	841	S2.3		1058
SODIUM SUPEROXIDE	2547	47							255			851			
SODIUM TETRABORATE ANHYDROUS Synonym: SODIUM BORATE			763	24	3066	1046	198				SDB1028				1070
SODIUM THIOCYANATE			788	24							SCY1026				1071
SODIUM THIOSULFATE				24	3117	1066									1071

HAZARDOUS MATERIALS REFERENCE BOOKS INDEX

CHEMICAL OR MATERIAL NAME	UN/NA Number	Guide Number (DOT)	Firefighter's Hazardous Materials Reference Book	First Aid Manual for Chemical Accidents	Sax's Dangerous Properties of Industrial Materials	Hazardous Chemicals Desk Reference	Rapid Guide to Hazardous Chemicals in the Workplace	Fire Protection Guide on Hazardous Materials (NFPA)	Firefighter's Handbook of Hazardous Materials	Pocket Guide to Chemical Hazards (NIOSH)	Chemical Hazard Response Information System (CHRIS)	Emergency Handling of Hazardous Materials in Surface Transportation (AAR)	Emergency Action Guides (EAG)	Chemical Data Notebook: A User's Manual	Condensed Chemical Dictionary
SODIUM VANADATE	2811	53			3120	1067			256						1072
SODIUM-m-ALUMINATE	1819	60							256						
SODIUM-o-PHOSPHATE	9148								256						
SOILBROM Synonym: ETHYLENE DIBROMIDE	1605	55	371					85 –	143	110	EDB 518	430	E2.4.1		487
SOILFUME Synonym: ETHYLENE DIBROMIDE	1605	55	371					85 –	143	110	EDB 518	430	E2.4.1		1073
SOLAR NITROGEN SOLUTIONS Synonym: AMMONIUM NITRATE UREA SOLUTION											ANU 68				
SOLUBLE GLASS Synonym: SODIUM SILICATE	1759	60	785	24	3113	1065					SSC1057	850			1068
SOLVENT ETHER Synonym: ETHYL ETHER	1155	26	354	16	1614	546	96	87 –52	137	112	EET 524	346	E2.8	E 3	492
SOMAN					3122	1067			256						1076
SORBIT Synonym: SORBITOL			789		3124	1068					SBT1018				1077
SORBITOL			789		3124						SBT1018				1077
SORBOL Synonym: SORBITOL			789		3124	1068					SBT1018				1077
SOYBEAN OIL Synonym: OILS EDIBLE SOYA BEAN											OSB 903				1078
SPECTRACIDE Synonym: DIAZINON	2783/2810	55	261	15	1083	363	64		96		DZN 415	319			366
SPENT MIXED ACID	1826	60							256						1080
SPENT NEUTRALIZING AGENT Synonym: CAUSTIC SODA SOLUTION SPENT	1824	60	194						67		CSS 314		C3.0		

CHEMICAL OR MATERIAL NAME	UN/NA Number	Guide Number (DOT)	Firefighter's Hazardous Materials Reference Book	First Aid Manual for Chemical Accidents	Sax's Dangerous Properties of Industrial Materials	Hazardous Chemicals Desk Reference	Rapid Guide to Hazardous Chemicals in the Workplace	Fire Protection Guide on Hazardous Materials (NFPA)	Firefighter's Handbook of Hazardous Materials	Pocket Guide to Chemical Hazards (NIOSH)	Chemical Hazard Response Information System (CHRIS)	Emergency Handling of Hazardous Materials in Surface Transportation (AAR)	Emergency Action Guides (EAG)	Chemical Data Notebook: A User's Manual	Condensed Chemical Dictionary
SPENT SODIUM HYDROXIDE Synonym: CAUSTIC SODA SOLUTION SPENT	1824	60	194						67		CSS 314		C3.0		
SPIRIT OF ETHER NITRITE Synonym: ETHYL NITRITE	1194	30	360		1643	554		89 −53	138		ETN 570	420			497
SPIRIT OF GLYCERYL TRINITRATE	1204	26			3127	1069			256						
SPIRIT OF HARTSHORN Synonym: ANHYDROUS AMMONIA	1005	15	48	11	220	57	11	25 −	32		AMA 47	78	A9.0	A 9	62
SPIRIT OF TURPENTINE Synonym: TURPENTINE	1299/1993	27	856	26	3458	1162	225	−92	256	222	TPT1144	856			1194
SPIRITS OF WINE Synonym: ETHYL ALCOHOL	1170	26	344	16	1572	525	90	−48	136		EAL 504	408	E2.2	E 2	478
SPOTTING NAPHTHA Synonym: NAPHTHA STODDARD SOLVENT	1268	27	604								NSS 849				
SPRENGEL EXPLOSIVES					3129	1069			256						
STAIN	1263	26							256						1083
STANNIC CHLORIDE	1827	39							256			857			1083
STANNIC CHLORIDE	2440/1759	60							256			857			1083
STANNIC PHOSPHIDE	1433	41							256			857			1084
STANNOUS CHLORIDE	1759	60							256			858			1084
STANNOUS FLOURIDE			790								STF1061				1085
STARCH DUST					3130	1070									
STEAM TURBINE LUBE OIL Synonym: OILS MISCELLANEOUS TURBINE											OTB 910				
STEAM TURBINE OIL Synonym: OILS MISCELLANEOUS TURBINE											OTB 910				
STEARIC ACID			791		3131	1070		−84			SRA1055				1088

HAZARDOUS MATERIALS REFERENCE BOOKS INDEX

CHEMICAL OR MATERIAL NAME	UN/NA Number	Guide Number (DOT)	Firefighter's Hazardous Materials Reference Book	First Aid Manual for Chemical Accidents	Sax's Dangerous Properties of Industrial Materials	Hazardous Chemicals Desk Reference	Rapid Guide to Hazardous Chemicals in the Workplace	Fire Protection Guide on Hazardous Materials (NFPA)	Firefighter's Handbook of Hazardous Materials	Pocket Guide to Chemical Hazards (NIOSH)	Chemical Hazard Response Information System (CHRIS)	Emergency Handling of Hazardous Materials in Surface Transportation (AAR)	Emergency Action Guides (EAG)	Chemical Data Notebook: A User's Manual	Condensed Chemical Dictionary
STEARIC ACID AMMONIUM SALT Synonym: AMMONIUM STEARATE											AMR 58				69
STEARIC ACID LEAD SALT Synonym: LEAD STEARATE	2811/3077	53	508	19			128		179		LSA 708	575			691
STEAROPHANIC ACID Synonym: STEARIC ACID			791		3131	1070		–84			SRA1055				1088
STEARYL ALCOHOL CRUDE Synonym: TALLOW FATTY ALCOHOL									256		TFA1107				
STEARYLDIMETHYLBENZYLAMMONIUM CHLORIDE Synonym: BENZYLDIMETHYLOCTADECYLAMMONIUM CHLO-RIDE											BZO 199				
STEEL SWARF	2793	32							256						
STEINBUHL YELLOW Synonym: CALCIUM CHROMATE	9096/3077		169		664	230			63		CCR 232	180			202
STIBINE	2676/1953	18	792	24	3134	1071	201	162–	256	198		858			1090
STINK DAMP Synonym: HYDROGEN SULFIDE	1053	13	452	18	1903	646	116	100–59	165	128	HDS 632	532	H4.7	H 6	617
STODDARD SOLVENT	1268/1993	27		24	3135	1071	201	–28		198		859			1081
STOLZITE Synonym: LEAD TUNGSTATE			513				128				LTU 717				693
STRAIGHT RUN GASOLINE Synonym: DISTILLATES STRAIGHT RUN	1268	27									DSR 480				
STRAW	1327	32							257						1093
STRONTIUM ALLOY	1434	40							257			859			
STRONTIUM ARSENITE	1691	53			3140	1072			257			859			
STRONTIUM CHLORATE	1506	35			3140				257			860			1094

CHEMICAL OR MATERIAL NAME	UN/NA Number	Guide Number (DOT)	Firefighter's Hazardous Materials Reference Book	First Aid Manual for Chemical Accidents	Sax's Dangerous Properties of Industrial Materials	Hazardous Chemicals Desk Reference	Rapid Guide to Hazardous Chemicals in the Workplace	Fire Protection Guide on Hazardous Materials (NFPA)	Firefighter's Handbook of Hazardous Materials	Pocket Guide to Chemical Hazards (NIOSH)	Chemical Hazard Response Information System (CHRIS)	Emergency Handling of Hazardous Materials in Surface Transportation (AAR)	Emergency Action Guides (EAG)	Chemical Data Notebook: A User's Manual	Condensed Chemical Dictionary
STRONTIUM CHROMATE	9149/3077	31	793		3140	1072			257		SCM1023	860			1095
STRONTIUM CHROMATE (1:1)	9194	45			3140	1072	201				SCM1023				
STRONTIUM DIOXIDE	1509	47							257						1095
STRONTIUM NITRATE	1507	35			3142	1073			257			861			1095
STRONTIUM PERCHLORATE	1508/1479	35							257			861			1095
STRONTIUM PEROXIDE	1509	47			3142	1073			257			861			1096
STRONTIUM PHOSPHIDE	2013	41			3142				257			861			
STRONTIUM YELLOW Synonym: STRONTIUM CHROMATE	9149/3077	31	793		3140	1072			257		SCM1023	860			1095
STRYCHNINE	1692	53	794		3143	1074	202		257	198	STR1063	862			1096
STYPHNIC ACID	0219				3145	1074									1097
STYRENE Synonym: STYRENE MONOMER	2055	27	795	24	3145	1074	202	162–84	257	200	STY1064	862	S3.0	S 6	1097
STYRENE ETHYLBENZENE MIXTURE	1993	27										863			
STYRENE MONOMER	2055	27			3145	1074	202		257		STY1064	862	S3.0	S 6	1097
STYRENE OXIDE								–84	257						1098
STYROL Synonym: STYRENE MONOMER	2055	27	795		3145	1074	202		257		STY1064	862	S3.0	S 6	1097
STYROL Synonym: STYRENE	2055	27	795	24	3145	1074	202	162–84	257	200	STY1064	862	S3.0	S 6	1097
STYROLE Synonym: STYRENE MONOMER	2055	27	795		3145	1074	202		257		STY1064	862	S3.0	S 6	1097
STYROLENE Synonym: STYRENE	2055	27	795	24	3145	1074	202	162–84	257	200	STY1064	862	S3.0	S 6	1097

CHEMICAL OR MATERIAL NAME	UN/NA Number	Guide Number (DOT)	Firefighter's Hazardous Materials Reference Book	First Aid Manual for Chemical Accidents	Sax's Dangerous Properties of Industrial Materials	Hazardous Chemicals Desk Reference	Rapid Guide to Hazardous Chemicals in the Workplace	Fire Protection Guide on Hazardous Materials (NFPA)	Firefighter's Handbook of Hazardous Materials	Pocket Guide to Chemical Hazards (NIOSH)	Chemical Hazard Response Information System (CHRIS)	Emergency Handling of Hazardous Materials in Surface Transportation (AAR)	Emergency Action Guides (EAG)	Chemical Data Notebook: A User's Manual	Condensed Chemical Dictionary
STYROLENE Synonym: STYRENE MONOMER	2055	27	795		3145	1074	202				STY1064	862	S3.0	S 6	1097
STYRON Synonym: STYRENE MONOMER	2055	27	795		3145	1074	202		257		STY1064	862	S3.0	S 6	1098
STYROPOL Synonym: STYRENE MONOMER	2055	27	795		3145	1074	202		257		STY1064	862	S3.0	S 6	1097
SUBERANE	2241	27							257						1098
SUBSTANCES	2813	40							257						1098
SUBSTITUTED NITROPHENOL PESTICIDE	3014/2779	55							257			867			
SUBSTITUTED NITROPHENOL PESTICIDE	2780	28							257			867			
SUBSTITUTED NITROPHENOL PESTICIDE	3013	28							257			867			
SUBSTITUTED NITROPHENOL PESTICIDE	2779	53							257			868			
SUCCINIC ACID					3148	1076			257						1099
SUCCINIC ACID PEROXIDE	2135/3102	49				1076			257			868			1099
SUCCINIC ANHYDRIDE					3149	1076			257						1099
SUCCINONITRILE					3150	1077	202	-84	258						1099
SUCCINYL PEROXIDE	2135	48							258						
SUCROSE			796		3151	1077					SRS1056				1099
SUGAR Synonym: SUCROSE			796		3151	1077					SRS1056				1100
SUGAR OF LEAD Synonym: LEAD ACETATE	1616	53	501	19	2094	708	128		178		LAC 690	570			1100
SULFAMIC ACID	2967	60			3153	1078			258			869			1101
SULFAMIC ACID MONOAMMONIUM SALT Synonym: AMMONIUM SULFAMATE	9089/3077	31	69		234		12		30	38	ASM 92	68			69

CHEMICAL OR MATERIAL NAME	UN/NA Number	Guide Number (DOT)	Firefighter's Hazardous Materials Reference Book	First Aid Manual for Chemical Accidents	Sax's Dangerous Properties of Industrial Materials	Hazardous Chemicals Desk Reference	Rapid Guide to Hazardous Chemicals in the Workplace	Fire Protection Guide on Hazardous Materials (NFPA)	Firefighter's Handbook of Hazardous Materials	Pocket Guide to Chemical Hazards (NIOSH)	Chemical Hazard Response Information System (CHRIS)	Emergency Handling of Hazardous Materials in Surface Transportation (AAR)	Emergency Action Guides (EAG)	Chemical Data Notebook: A User's Manual	Condensed Chemical Dictionary
SULFATE OF COPPER Synonym: COPPER SULFATE	9109		229	14	950	322	55				CSF 311				315
SULFATE TURPENTINE Synonym: TURPENTINE	1299	27	856	26	3458	1162	225	-92	282	222	TPT1144	939			1101
SULFATED NEATSFOOT OIL Synonym: OILS MISCELLANEOUS TANNER'S											OTN 916				
SULFATES MIXED ETHYL	1760	60													
SULFIDE WASTE	1760	60													
SULFOCARBOLIC ACID	1803	60							258						1102
SULFOLANE			797		3158			-84			SFL1040				1102
SULFOLANE W Synonym: SULFOLANE			797		3158			-84			SFL1040				1102
SULFONATED ALKYLBENZENE SODIUM SALT Synonym: SODIUM ALKYLBENZENESULFONATES			753								SAB1007				
SULFONIC ACID Synonym: CHLOROSULFONIC ACID	1754	39	212					55 -			CSA 309	233	C4.5	C 8	273
SULFONYL CHLORIDE	1834	39							258						
SULFOTEP	1704	55			3161	1080	203								
SULFOTEPP	1704	55		24					258						1102
SULFUR	2448	32			3161	1080		-84	258		SXX1072	871	S3.5		1103
SULFUR	1350	32	798		3161	1080		163–84	258			869	S3.5		1103
SULFUR BROMIDE					3162				258						1103
SULFUR CHLORIDE	1828	39	799					-84	258		SFM1041	869	S4.0		1103
SULFUR CHLORIDE PENTAFLUORIDE	9272	15			3162				258						
SULFUR DECAFLUORIDE									258						

HAZARDOUS MATERIALS REFERENCE BOOKS INDEX

CHEMICAL OR MATERIAL NAME	UN/NA Number	Guide Number (DOT)	Firefighter's Hazardous Materials Reference Book	First Aid Manual for Chemical Accidents	Sax's Dangerous Properties of Industrial Materials	Hazardous Chemicals Desk Reference	Rapid Guide to Hazardous Chemicals in the Workplace	Fire Protection Guide on Hazardous Materials (NFPA)	Firefighter's Handbook of Hazardous Materials	Pocket Guide to Chemical Hazards (NIOSH)	Chemical Hazard Response Information System (CHRIS)	Emergency Handling of Hazardous Materials in Surface Transportation (AAR)	Emergency Action Guides (EAG)	Chemical Data Notebook: A User's Manual	Condensed Chemical Dictionary
SULFUR DICHLORIDE	1828	39			3162	1081			258						1104
SULFUR DIOXIDE	1079	16	800	24	3162	1081	203	163–	258	200	SFD1039	870	S4.1	S 7	1104
SULFUR HEXAFLUORIDE	1080	12			3163	1082	203		258			870			1104
SULFUR HYDRIDE Synonym: HYDROGEN SULFIDE	1053	13	452	18	1903	646	116	100–59	165	128	HDS 632	532	H4.7	H 6	617
SULFUR MONOCHLORIDE Synonym: SULFUR CHLORIDE	1828	39	801		3166		204	164–84	258	200	SFM1041	869	S4.0		1105
SULFUR OXIDE Synonym: SULFUR DIOXIDE	1079	16	800	24	3162	1081	203	163–	258	200	SFD1039	870	S4.1	S 7	1104
SULFUR PENTAFLUORIDE					3167	1085	204			202					
SULFUR SUBCHLORIDE Synonym: SULFUR CHLORIDE	1828	39	799					–84	258		SFM1041	869	S4.0		1105
SULFUR TETRACHLORIDE	2418	15			3167				258						
SULFUR TETRAFLUORIDE	1955	15			3168							876			1105
SULFUR TETRAFLUORIDE	2418	15				1085	205		258						1105
SULFUR TRIOXIDE	1829	39		24	3168	1085			258			870			1105
SULFURETTED HYDROGEN Synonym: HYDROGEN SULFIDE	1053	13	452	18	1903	646	116	100–59	258	128	HDS 632	532	H4.7	H 6	1104
SULFURIC ACID	1830	39	802	24		1082	204	164–	259	200	SFA1037	871	S4.2	S 8	1104
SULFURIC ACID	1831	39		24	3163	1083			259	200		872		S 8	1104
SULFURIC ACID CHROMIUM (3) SALT (3-2) Synonym: CHROMIC SULFATE	9100/3082	31	217						80		CHS 249	239			280
SULFURIC ACID SPENT	1832	39	803	24	3163	1082	204			200	SAC1008	872	S4.3	S 8	
SULFURIC ACID THALLIUM SALT Synonym: THALLIUM SULFATE	1707	53	820		3163	1110	212		266		TSU1150				1138

CHEMICAL OR MATERIAL NAME	UN/NA Number	Guide Number (DOT)	Firefighter's Hazardous Materials Reference Book	First Aid Manual for Chemical Accidents	Sax's Dangerous Properties of Industrial Materials	Hazardous Chemicals Desk Reference	Rapid Guide to Hazardous Chemicals in the Workplace	Fire Protection Guide on Hazardous Materials (NFPA)	Firefighter's Handbook of Hazardous Materials	Pocket Guide to Chemical Hazards (NIOSH)	Chemical Hazard Response Information System (CHRIS)	Emergency Handling of Hazardous Materials in Surface Transportation (AAR)	Emergency Action Guides (EAG)	Chemical Data Notebook: A User's Manual	Condensed Chemical Dictionary
SULFURIC AND HYDROFLUORIC ACID MIXTURE	1786	59							259						
SULFURIC ANHYDRIDE	1829	39							259						1105
SULFURIC CHLORHYDRIN Synonym: CHLOROSULFONIC ACID	1754	39	212					55 –			CSA 309	233	C4.5	C 8	273
SULFURIC CHLOROHYDRIN Synonym: CHLOROSULFONIC ACID	1754	39	212					55 –			CSA 309	233	C4.5	C 8	273
SULFURIC ETHER Synonym: ETHYL ETHER	1155	26	354	16	1614	546	96	87 –52	137	112	EET 524	346	E2.8	E 3	492
SULFURIC OXYFLUORIDE	2191	15							259						1105
SULFUROUS ACID	1833	60		24	3166	1084			259			875			1105
SULFUROUS ACID ANHYDRIDE Synonym: SULFUR DIOXIDE	1079	16	800	24	3162	1081	203	163–	259	200	SFD1039	870	S4.1	S 7	1104
SULFUROUS ANHYDRIDE Synonym: SULFUR DIOXIDE	1079	16	800	24	3162	1081	203	163–	258	200	SFD1039	870	S4.1	S 7	1104
SULFUROUS OXIDE Synonym: SULFUR DIOXIDE	1079	16	800	24	3162	1081	203	163–	258	200	SFD1039	870	S4.1	S 7	1104
SULFUROUS OXYCHLORIDE	1836	39							259						
SULFURYL CHLORIDE	1834	39	804		3168	1085		165–	259		SCL1022	875			1105
SULFURYL FLUORIDE	2191	15			3169	1086	205		259	202		876			1105
SULPHOCARBONIC ANHYDRIDE Synonym: CARBON BISULFIDE	1131	28	188				42		65			189	C2.1		218
SULPHUR TETRAFLUORIDE	1955	15										876			
SULPHUR ZINC SULPHATE AND LEAD ARSENATE	1617	53										877			
SULPHURETTED HYDROGEN Synonym: HYDROGEN SULFIDE	1053	13	452	18	1903	646	116	100–59	165	128	HDS 632	532	H4.7	H 6	617

HAZARDOUS MATERIALS REFERENCE BOOKS INDEX

CHEMICAL OR MATERIAL NAME	UN/NA Number	Guide Number (DOT)	Firefighter's Hazardous Materials Reference Book	First Aid Manual for Chemical Accidents	Sax's Dangerous Properties of Industrial Materials	Hazardous Chemicals Desk Reference	Rapid Guide to Hazardous Chemicals in the Workplace	Fire Protection Guide on Hazardous Materials (NFPA)	Firefighter's Handbook of Hazardous Materials	Pocket Guide to Chemical Hazards (NIOSH)	Chemical Hazard Response Information System (CHRIS)	Emergency Handling of Hazardous Materials in Surface Transportation (AAR)	Emergency Action Guides (EAG)	Chemical Data Notebook: A User's Manual	Condensed Chemical Dictionary
SUPERLYSOFORM Synonym: FORMALDEHYDE SOLUTION	1198/2209	29	401	17	1748	591	102	91 −54	150	116	FMS 590	462	F2.0	F 2	
SUPEROXOL Synonym: HYDROGEN PEROXIDE	2015	47	450	18	1901	644	115	100−	165	126	HPO 643	532	H4.5	H 5	616
SUPEROXOL Synonym: HYDROGEN PEROXIDE	2014	45		18	1901	644	115	99 −	165	126		531	H4.5	H 5	616
SWEDISH GREEN Synonym: COPPER ARSENITE	1586	53	223				55		83		CPA 283	257			311
SWEET SPIRIT OF NITRE Synonym: ETHYL NITRITE	1194	30	360		1643	554		89 −53	138		ETN 570	420			497
SYLVAN	2301	26						−84	259						1110
SYNTHETIC PINE OIL Synonym: PINE OIL	1272	26	695		2810	959		−80	235		OPI 895	764	P5.0		918
SYNTHETIC RUBBER LATEX Synonym: LATEX LIQUID SYNTHETIC	3082/9188		497								LLS 701	569			
SYSTOX AND ISOSYSTOX MIXTURE Synonym: DEMETON				15	1039				91	80	DTN 490				353
T E P Synonym: TETRAETHYL PYROPHOSPHATE	1705	15			3220				262		TEP1104	890			1130
T E P P Synonym: TETRAETHYL PYROPHOSPHATE	1705	15		24	3220				262		TEP1104	890			1122
T GAS Synonym: ETHYLENE OXIDE	1040	69	377	17	1611	544	96	87 −52	144	112	EOX 550	434	E2.7	E 6	491
T M L COMPOUND	1649	56							259						1155
T STUFF Synonym: HYDROGEN PEROXIDE	2014	45		18	1901	644	115	99 −		126		531	H4.5	H 5	616

CHEMICAL OR MATERIAL NAME	UN/NA Number	Guide Number (DOT)	Firefighter's Hazardous Materials Reference Book	First Aid Manual for Chemical Accidents	Sax's Dangerous Properties of Industrial Materials	Hazardous Chemicals Desk Reference	Rapid Guide to Hazardous Chemicals in the Workplace	Fire Protection Guide on Hazardous Materials (NFPA)	Firefighter's Handbook of Hazardous Materials	Pocket Guide to Chemical Hazards (NIOSH)	Chemical Hazard Response Information System (CHRIS)	Emergency Handling of Hazardous Materials in Surface Transportation (AAR)	Emergency Action Guides (EAG)	Chemical Data Notebook: A User's Manual	Condensed Chemical Dictionary
T STUFF Synonym: HYDROGEN PEROXIDE	2015	47	450	18	1901	644	115	100–	259	126	HPO 643	532	H4.5	H 5	616
TALC				24	3177	1088	206			202					1113
TALLOW								−84			TLO1125				1114
TALLOW ALCOHOL	1987	26													
TALLOW BENZYL DIMETHYL AMMONIUM CHLORIDE Synonym: BENZYLDIMETHYLOCTADECYLAMMONIUM CHLO-RIDE											BZO 199				
TALLOW FATTY ALCOHOL											TFA1107				
TALLOW OIL Synonym: TALLOW								−84			TLO1125				1114
TANNIC ACID			805		3180	1089		−84			TNA1130				1114
TANNIC CHLORIDE Synonym: TITANIUM TETRACHLORIDE	1838	39	825		3302	1124		169–	269		TTT1158	908	T3.0	T 3	1154
TANNIN Synonym: TANNIC ACID			805		3180	1089		−84			TNA1130				1115
TANTALUM					3180	1089	200			204					1115
TAR	1993	27							260			880			1116
TAR Synonym: ASPHALT	1999	27	105	12	307	97	17	−17	260		ASP 93	880	A12.0		1116
TAR ACIDS Synonym: CRESOL	2076	55	232	14	959	324	55	57 –	85	74	CRS 307	270	C7.0		1116
TAR CAMPHOR	1334	32							260			879			1116
TAR CAMPHOR Synonym: NAPHTHALENE	2304	32		21	2464	835	152	−73	207		NTM 857	675	NO.5.5		1116

HAZARDOUS MATERIALS REFERENCE BOOKS INDEX

CHEMICAL OR MATERIAL NAME	UN/NA Number	Guide Number (DOT)	Firefighter's Hazardous Materials Reference Book	First Aid Manual for Chemical Accidents	Sax's Dangerous Properties of Industrial Materials	Hazardous Chemicals Desk Reference	Rapid Guide to Hazardous Chemicals in the Workplace	Fire Protection Guide on Hazardous Materials (NFPA)	Firefighter's Handbook of Hazardous Materials	Pocket Guide to Chemical Hazards (NIOSH)	Chemical Hazard Response Information System (CHRIS)	Emergency Handling of Hazardous Materials in Surface Transportation (AAR)	Emergency Action Guides (EAG)	Chemical Data Notebook: A User's Manual	Condensed Chemical Dictionary
TAR OIL **Synonym: CREOSOTE COAL TAR**	1993	27	231	14					84		CCT 233		C6.0		1117
TARTAR EMETIC **Synonym: ANTIMONY POTASSIUM TARTRATE**	1551	53	90		274	80			33		APT 81	83			1117
TARTARIC ACID (d,l)					3181	1090		–84							1117
TARTARIC ACID COPPER SALT **Synonym: COPPER TARTRATE**	9111						55				CTT 323				
TARTARIZED ANTIMONY **Synonym: ANTIMONY POTASSIUM TARTRATE**	1551	53	90		274	80			33		APT 81	83			89
TARTRATED ANTIMONY **Synonym: ANTIMONY POTASSIUM TARTRATE**	1551	53	90		274	80			33		APT 81	83			89
TCDD					3183	1090	206								1118
TCE **Synonym: TRICHLOROETHYLENE**	1710	74	837	25	3352	1137	218	174–88	273	216	TCL1085	919	T5.0	T 6	1170
TCP **Synonym: TRICRESYLPHOSPHATE**	2574/2810	55	841						274		TCP1089	922			1118
TDE **Synonym: DDD**	2761	55	254								DDD 373	880			1118
TDI **Synonym: TOLUENE-2,4-DIISOCYANATE**	2078	57	828	25	3313		215	170–87	269	214	TDI1095	910			1119
TDI **Synonym: TOLUENE DIISOCYANATE**	2078	57	828		3312	1128	215		269		TDI1095	910	T4.1	T 5	1119
TEA **Synonym: TRIETHYLALUMINUM**	1102/3051			25	3368	1141		175–88			TAL1074	923			1119
TEAR GAS **Synonym: CHLOROACETOPHENONE (-alpha)**	1697	55	200			271	47	–26	73	64	CRA 297	217			261

CHEMICAL OR MATERIAL NAME	UN/NA Number	Guide Number (DOT)	Firefighter's Hazardous Materials Reference Book	First Aid Manual for Chemical Accidents	Sax's Dangerous Properties of Industrial Materials	Hazardous Chemicals Desk Reference	Rapid Guide to Hazardous Chemicals in the Workplace	Fire Protection Guide on Hazardous Materials (NFPA)	Firefighter's Handbook of Hazardous Materials	Pocket Guide to Chemical Hazards (NIOSH)	Chemical Hazard Response Information System (CHRIS)	Emergency Handling of Hazardous Materials in Surface Transportation (AAR)	Emergency Action Guides (EAG)	Chemical Data Notebook: A User's Manual	Condensed Chemical Dictionary
TEAR GAS	1693	58			762				260			881			
TEAR GAS CANDLE	1700	58							260			881			
TEAR GAS DEVICES	1693	58							260			881			
TEDP	1704	55							260	204					1119
TEFLON					3184	1091			260						1119
TEFLON MONOMER Synonym: TETRAFLUOROETHYLENE	1081	17	817	25	3222	1102		166–85	263		TFE1109	893			1130
TEG Synonym: TRIETHYLENE GLYCOL			848	25	3370	1142		–89			TEG1101				1119
TEKRESOL Synonym: CRESOL	2076	55	232	14	959	324	55	57 –	85	74	CRS 307	270	C7.0		322
TEL Synonym: TETRAETHYL LEAD	1649	56	813	25	3219	1100	209	–85	260	208	TEL1102	889			1119
TELLURIUM					3186	1091	207		260	204					1120
TELLURIUM HEXAFLUORIDE	2195/1955	15		24	3187	1092	207		260	204		882			
TELLURIUM NITRIDE					3187				260						
TELONE Synonym: 1,3-DICHLOROPROPENE	2047	29			1164	393	70	–35	260		DPU 472				1121
TEMEPHOS					3188	1092	207		260						1121
TEN Synonym: TRIETHYLAMINE	1296	68	846	25	3368	1141	220	176–88	275	218	TEN1103	923			1174
TENOX P GRAIN PRESERVATIVE Synonym: PROPIONIC ACID	1848	60	721		2909	997	188	148–81	243		PNA 968	791	P7.1		968
TEPP	2783	55		24			201		260	206					1122
TEREPHTHALIC ACID					3191			–84					T0.0.5		1122

HAZARDOUS MATERIALS REFERENCE BOOKS INDEX

CHEMICAL OR MATERIAL NAME	UN/NA Number	Guide Number (DOT)	Firefighter's Hazardous Materials Reference Book	First Aid Manual for Chemical Accidents	Sax's Dangerous Properties of Industrial Materials	Hazardous Chemicals Desk Reference	Rapid Guide to Hazardous Chemicals in the Workplace	Fire Protection Guide on Hazardous Materials (NFPA)	Firefighter's Handbook of Hazardous Materials	Pocket Guide to Chemical Hazards (NIOSH)	Chemical Hazard Response Information System (CHRIS)	Emergency Handling of Hazardous Materials in Surface Transportation (AAR)	Emergency Action Guides (EAG)	Chemical Data Notebook: A User's Manual	Condensed Chemical Dictionary
TEREPHTHALIC ACID DIMETHYL ESTER Synonym: DIMETHYL TEREPHTHALATE					1418	473		-44			DMT 439				421
TEREPHTHALOYL CHLORIDE								-84							1123
TERGITOL NONIONIC 3-A-6 Synonym: ETHOXYLATED TRIDECANOL											ETD 563				
TERGITOL NONIONIC 45-S-10 Synonym: ETHOXYLATED PENTADECANOL											EOP 548				
TERGITOL NONIONIC 45-S-10 Synonym: ETHOXYLATED TETRADECANOL											EOT 549				
TERGITOL NONIONIC TMN Synonym: ETHOXYLATED DODECANOL											EOD 547				
TERPENE HYDROCARBONS	2319	27							260			883			
TERPENES	2319	27							260						1123
TERPHENYL							208			206					
TERPHENYL-meta					3195	1093	208	-84							
TERPHENYL-ortho					3195	1093	208	-84							
TERPHENYL-para					3195	1093	208								1123
TERPINENE Synonym: DIPENTENE	2052/1993	27	318					-45	126		DPN 468	381			1123
TERPINEOL	2319/1993				3196	1094		-85	260			883			1124
TERPINOLENE	2541	27			3196	1094			260			883			1124
TERPINYL ACETATE					3196	1094		-85	260						1124
TERR O GAS 100 Synonym: METHYL BROMIDE	1062	55	559	20		776	137	110-67	192	146	MTB 805	619	M3.0	M 3	759
TERTIARY ALCOHOL	1987	26										885			1124

CHEMICAL OR MATERIAL NAME	UN/NA Number	Guide Number (DOT)	Firefighter's Hazardous Materials Reference Book	First Aid Manual for Chemical Accidents	Sax's Dangerous Properties of Industrial Materials	Hazardous Chemicals Desk Reference	Rapid Guide to Hazardous Chemicals in the Workplace	Fire Protection Guide on Hazardous Materials (NFPA)	Firefighter's Handbook of Hazardous Materials	Pocket Guide to Chemical Hazards (NIOSH)	Chemical Hazard Response Information System (CHRIS)	Emergency Handling of Hazardous Materials in Surface Transportation (AAR)	Emergency Action Guides (EAG)	Chemical Data Notebook: A User's Manual	Condensed Chemical Dictionary
TESCOL Synonym: ETHYL ALCOHOL	1170	26	344	16	1572	525	90	-48	136		EAL 504	408	E2.2	E 2	478
TESTOSTERONE					3197	1095									1125
TETA Synonym: TRIETHYLENE TETRAMINE	2259/1760	60			3372	1143			275		TET1106	924			1176
TETRA BENZENESULFONIC ACID Synonym: ALKYLBENZENESULFONIC ACID											ABS 11				
TETRAAMMINE COPPER SULFATE Synonym: COPPER SULFATE AMMONIATED	9110										CSN 312				315
TETRABORANE					3201				261						
TETRABROMOETHANE	2504	58							261						
TETRABROMOMETHANE	2516	53							261						1125
TETRABUTYL TITANATE	2920/1760	29	806						261		TBT1078			886	1126
TETRACAP Synonym: TETRACHLOROETHYLENE	1897	74	808	25				166–	262	208	TTE1153	888	T0.5	T 1	1127
TETRACHLOROCARBON Synonym: CARBON TETRACHLORIDE	1846	55	192	13	701	246	43	48 –	66	60	CBT 223	193	C2.3	C 5	221
TETRACHLORODIPHENYL OXIDE					3207	1096	208								
TETRACHLOROETHANE	1702	55	807	25	3207	1096	208		261		TEC1099	887			
TETRACHLOROETHYLENE	1897	74	808	25				166–	262	208	TTE1153	888	T0.5	T 1	1127
TETRACHLOROMETHANE Synonym: CARBON TETRACHLORIDE	1846	55	192	13	701	246	43	48 –	262	60	CBT 223	193	C2.3	C 5	1127
TETRACHLORONAPHTHALENE					3210	1098	209		262	208					1127
TETRACHLOROSILANE Synonym: SILICON CHLORIDE	1818	39	744		3242	1037			262		STC1060	826	S1.0		1127

CHEMICAL OR MATERIAL NAME	UN/NA Number	Guide Number (DOT)	Firefighter's Hazardous Materials Reference Book	First Aid Manual for Chemical Accidents	Sax's Dangerous Properties of Industrial Materials	Hazardous Chemicals Desk Reference	Rapid Guide to Hazardous Chemicals in the Workplace	Fire Protection Guide on Hazardous Materials (NFPA)	Firefighter's Handbook of Hazardous Materials	Pocket Guide to Chemical Hazards (NIOSH)	Chemical Hazard Response Information System (CHRIS)	Emergency Handling of Hazardous Materials in Surface Transportation (AAR)	Emergency Action Guides (EAG)	Chemical Data Notebook: A User's Manual	Condensed Chemical Dictionary
TETRACHLOROZIRCONIUM Synonym: ZIRCONIUM TETRACHLORIDE	2503	39	913				235	185–	290		ZCT1203	973			1253
TETRADECANE					3215	1099		-85							1128
TETRADECANOL Synonym: LINEAR ALCOHOL			809		3215	1100					LAL 692				1128
TETRADECANOL					3215	1100		-85			TTN1156				1128
TETRADECYL ALCOHOL-n Synonym: TETRADECANOL			809		3215	1100		-85			TTN1156				1128
TETRADECYLBENZENE											TDB1093				
TETRAETHOXYPROPANE								-85	262						
TETRAETHYL DITHIONOPYROPHOSPHATE Synonym: TETRAETHYL DITHIOPYROPHOSPHATE	1704	55							262		TED1100	888			1129
TETRAETHYL DITHIOPYROPHOSPHATE	1704	55	811						262		TED1100	888			1129
TETRAETHYL DITHIOPYROPHOSPHATE & COMPRESSED GAS MIXTURE	1703	15							262			888			
TETRAETHYL LEAD	1649	56	813	25	3219	1100	209	-85	262	208	TEL1102	889			1130
TETRAETHYL ORTHOSILICATE Synonym: ETHYL SILICATE	1292	29	365			560	98	-54	139	114	ESC 558	891			1130
TETRAETHYL PYROPHOSPHATE	1705	15			3220				262		TEP1104				1130
TETRAETHYL PYROPHOSPHATE	2784	28				1101			262						1130
TETRAETHYL PYROPHOSPHATE MIXTURE	2783/3018	55	814		3220	1101	209		262			891			
TETRAETHYL SILICATE Synonym: ETHYL SILICATE	1292	29	365		1659	560	98	-54	262	114	ESC 558	891			500
TETRAETHYL-o-SILICATE	1292	29							262						
TETRAETHYL-s,s-METHYLENE DIPHOSPHORODITHIOATE-o,o,o',o'	2783	55							262						

CHEMICAL OR MATERIAL NAME	UN/NA Number	Guide Number (DOT)	Firefighter's Hazardous Materials Reference Book	First Aid Manual for Chemical Accidents	Sax's Dangerous Properties of Industrial Materials	Hazardous Chemicals Desk Reference	Rapid Guide to Hazardous Chemicals in the Workplace	Fire Protection Guide on Hazardous Materials (NFPA)	Firefighter's Handbook of Hazardous Materials	Pocket Guide to Chemical Hazards (NIOSH)	Chemical Hazard Response Information System (CHRIS)	Emergency Handling of Hazardous Materials in Surface Transportation (AAR)	Emergency Action Guides (EAG)	Chemical Data Notebook: A User's Manual	Condensed Chemical Dictionary
TETRAETHYLAMMONIUM PERCHLORATE	1325	32			3218										
TETRAETHYLENE GLYCOL	1993	27	815		3218	1100		-85			TTG1155	892			1129
TETRAETHYLENEPENTAMINE	2320/1719	60	816		3218	1100		-85	262		TTP1157	892			1129
TETRAETHYLTIN				25											1130
TETRAFINOL Synonym: CARBON TETRACHLORIDE	1846	55	192	13	701	246	43	48 –	66	60	CBT 223	193	C2.3	C 5	221
TETRAFLUOROETHYLENE	1081	17	817	25	3222	1102		166–85	263		TFE1109	893			1130
TETRAFLUOROHYDRAZINE	1955	15			3222	1102			263			893			1130
TETRAFLUOROMETHANE	1982/1956	12		25					263			894			1130
TETRAFORM Synonym: CARBON TETRACHLORIDE	1846	55	192	13	701	246	43	48 –	66	60	CBT 223	193	C2.3	C 5	221
TETRAHYDRO-2H-1,4-OXAZINE Synonym: MORPHOLINE	2054	29	597		2447	833	151	120–73	206	156	MPL 788	659			798
TETRAHYDRO-p-OXAZINE Synonym: MORPHOLINE	2054	29	597		2447	833	151	120–73	263	156	MPL 788	659			798
TETRAHYDROBENZALDEHYDE	2498	29			3224	1102			263						
TETRAHYDRODIMETHYLFURAN					3226				263						
TETRAHYDROFURAN	2056	26	818		3227	1103	210	167–86	263	210	THF1116	894	T1.0	T 2	1131
TETRAHYDROFURFURYL ALCOHOL	1993	27						-86	263			895			1131
TETRAHYDROFURFURYLAMINE	2943/1993	26			3228	1103			263			895			1131
TETRAHYDRONAPHTHALENE	1993	27	819		3230			-86	263		THN1118	895			1132
TETRAHYDROPHTHALIC ANHYDRIDE	2698/1759	60			3232				263			896			1132
TETRAHYDROPYRROLE	1922	29							263						
TETRAHYDROTHIOPHENE	2412/1993	26			3233	1104			263			896			1132

CHEMICAL OR MATERIAL NAME	UN/NA Number	Guide Number (DOT)	Firefighter's Hazardous Materials Reference Book	First Aid Manual for Chemical Accidents	Sax's Dangerous Properties of Industrial Materials	Hazardous Chemicals Desk Reference	Rapid Guide to Hazardous Chemicals in the Workplace	Fire Protection Guide on Hazardous Materials (NFPA)	Firefighter's Handbook of Hazardous Materials	Pocket Guide to Chemical Hazards (NIOSH)	Chemical Hazard Response Information System (CHRIS)	Emergency Handling of Hazardous Materials in Surface Transportation (AAR)	Emergency Action Guides (EAG)	Chemical Data Notebook: A User's Manual	Condensed Chemical Dictionary
TETRAHYDROTHIOPHENE-1,1-DIOXIDE Synonym: SULFOLANE			797		3158			-84			SFL1040				1132
TETRAHYDROXYMETHYLMETHANE Synonym: PENTAERYTHRITOL			667		2682	912	172				PET 949				880
TETRALIN Synonym: TETRAHYDRONAPHTHALENE	1993	27	819		3230			-86	264		THN1118	895			1133
TETRALIN HYDROPEROXIDE	2136	48							264						
TETRAMETHOXYSILANE Synonym: METHYL ORTHOSILICATE	2606/1993	57						168–	264			634	M3.6.1		1133
TETRAMETHYL AMMONIUM HYDROXIDE	1835	60			3239	1104			264			897			1133
TETRAMETHYL DIARSINE					3242				264						
TETRAMETHYL ESTER Synonym: METHYL ORTHOSILICATE	2606/1993	57							196			634	M3.6.1		
TETRAMETHYL LEAD	1649	56		25	3245	1105	210	-86	264	210	TML1129				1134
TETRAMETHYL ORTHOSILICATE Synonym: METHYL ORTHOSILICATE	2606/1993	57							196			634	M3.6.1		
TETRAMETHYL SILANE	2749/2924	29			3247				264			898			1134
TETRAMETHYL SILICANE	2749	29							264						
TETRAMETHYL SILICATE Synonym: METHYL ORTHOSILICATE	2606/1993	57							196			634	M3.6.1		
TETRAMETHYL THIURAM DISULFIDE Synonym: THIRAM	2771	55	822		3285	1118	213		268	212	THR1119	904			1135
TETRAMETHYL TIN								-86	265						
TETRAMETHYLENE	2601	22							265						1134
TETRAMETHYLENE CYANIDE	2205	55							265						

CHEMICAL OR MATERIAL NAME	UN/NA Number	Guide Number (DOT)	Firefighter's Hazardous Materials Reference Book	First Aid Manual for Chemical Accidents	Sax's Dangerous Properties of Industrial Materials	Hazardous Chemicals Desk Reference	Rapid Guide to Hazardous Chemicals in the Workplace	Fire Protection Guide on Hazardous Materials (NFPA)	Firefighter's Handbook of Hazardous Materials	Pocket Guide to Chemical Hazards (NIOSH)	Chemical Hazard Response Information System (CHRIS)	Emergency Handling of Hazardous Materials in Surface Transportation (AAR)	Emergency Action Guides (EAG)	Chemical Data Notebook: A User's Manual	Condensed Chemical Dictionary
TETRAMETHYLENE GLYCOL Synonym: 1,4-BUTANEDIOL	1987	26	5		575	189		-20	54		BDO 136				1134
TETRAMETHYLENE OXIDE Synonym: TETRAHYDROFURAN	2056	26	818		3227	1103	210	167-86	265	210	THF1116	894	T1.0	T 2	1131
TETRAMETHYLENE SULFONE Synonym: SULFOLANE			797		3158			-84			SFL1040				1134
TETRAMETHYLMETHYLENEDIAMINE	9069/3077	29							265			898			
TETRAMETHYLMETHYLENEDIAMINE-n,n,n,n	9069	29							265						
TETRAMETHYLOXYSILANE Synonym: METHYL ORTHOSILICATE	2606/1993	57							196			634	M3.6.1		
TETRAMETHYLSTANNANE					3248				265						
TETRAMETHYLSUCCINONITRILE					3248	1105	210		265	210					1134
TETRAMETYLENE OXIDE	2056	26							265						
TETRAMP Synonym: TETRAHYDRONAPHTHALENE	1993	27	819		3230			-86	263		THN1118	895			1132
TETRANAP Synonym: TETRAHYDRONAPHTHALENE	1993	27	819		3230			-86	263		THN1118	895			1132
TETRANITROANILINE	0207				3250	1106			265			899			1135
TETRANITROMETHANE	1510	47		25	3251	1106	211		265	210	TNM1132	899			1135
TETRANITRONAPHTHALENE									265						
TETRAPHENYL TIN								-86	265						1135
TETRAPROPYL-ortho-TITANATE	2413	27							265						
TETRAPROPYLENE Synonym: PROPYLENE TETRAMER	2350/1993	27	730						244		PTT1004	799			1135
TETRAPROPYLENE Synonym: DODECENE	1993	27									DOD 454	390			1135

HAZARDOUS MATERIALS REFERENCE BOOKS INDEX

CHEMICAL OR MATERIAL NAME	UN/NA Number	Guide Number (DOT)	Firefighter's Hazardous Materials Reference Book	First Aid Manual for Chemical Accidents	Sax's Dangerous Properties of Industrial Materials	Hazardous Chemicals Desk Reference	Rapid Guide to Hazardous Chemicals in the Workplace	Fire Protection Guide on Hazardous Materials (NFPA)	Firefighter's Handbook of Hazardous Materials	Pocket Guide to Chemical Hazards (NIOSH)	Chemical Hazard Response Information System (CHRIS)	Emergency Handling of Hazardous Materials in Surface Transportation (AAR)	Emergency Action Guides (EAG)	Chemical Data Notebook: A User's Manual	Condensed Chemical Dictionary
TETRASODIUM PYROPHOSPHATE					3255	1107	211								1136
TETRASOL Synonym: CARBON TETRACHLORIDE	1846	55	192	13	701	246	43	48 –	66	60	CBT 223	193	C2.3	C 5	221
TETRAZENE	0114				3255	1107			266						1136
TETRINE ACID Synonym: ETHYLENEDIAMINE TETRAACETIC ACID	9117/3077		379		1601	539			144		EDT 521	405			486
TETROLE Synonym: FURAN	2389	26			1763	596		-55	266		FUR 600	467	F3.3.1		545
TETRON Synonym: TETRAETHYL PYROPHOSPHATE	1705	15			3220				262		TEP1104	890			1130
TETRYL	0208/0401			25	3256	1108	211		266	212		899			1136
TEXTILE TREATING COMPOUND	1760	60							266			899			
TEXTILE WASTE	1857	32							266						
THALLIC OXIDE									266						
THALLIUM					3258	1108	212								1137
THALLIUM CHLORATE	2573	42			3260		212		266			900			
THALLIUM COMPOUND	1707	53			3260	1109	212		266	212		900			1137
THALLIUM NITRATE	2727/2811	42			3261	1110	212		266		TNI1131	900			1137
THALLIUM SALT	1707	53			3262		212		266						
THALLIUM SULFATE	1707	53	820		3263	1110	212		266		TSU1150	901			1138
THALLOUS SULFATE Synonym: THALLIUM SULFATE	1707	53	820		3263	1110	212		266		TSU1150	901			1138
THANOL PPG Synonym: POLYPROPYLENE GLYCOL								-80			PGC 952				940
THEBAINE					3264										1138

CHEMICAL OR MATERIAL NAME	UN/NA Number	Guide Number (DOT)	Firefighter's Hazardous Materials Reference Book	First Aid Manual for Chemical Accidents	Sax's Dangerous Properties of Industrial Materials	Hazardous Chemicals Desk Reference	Rapid Guide to Hazardous Chemicals in the Workplace	Fire Protection Guide on Hazardous Materials (NFPA)	Firefighter's Handbook of Hazardous Materials	Pocket Guide to Chemical Hazards (NIOSH)	Chemical Hazard Response Information System (CHRIS)	Emergency Handling of Hazardous Materials in Surface Transportation (AAR)	Emergency Action Guides (EAG)	Chemical Data Notebook: A User's Manual	Condensed Chemical Dictionary
THEOBROMINE					3265	1111									1138
THEOBROMINE SODIUM SALICYLATE					3265										
THEOPHYLLINE					3266	1111									1138
THERMIT					3268	1112			266						
THF Synonym: TETRAHYDROFURAN	2056	26	818		3227	1103	210	167–86	263	210	THF1116	894	T1.0	T 2	1141
THIACETIC ACID	2436	26							266						
THIALDINE								–86	266						
THIAPENTANAL	2785	55							266						
THINNER	1263	26							267			901			1143
THIOACETAMIDE					3274	1113			267						1143
THIOACETIC ACID	2436/1993	26			3274	1113			267			901			1143
THIOBUTYL ALCOHOL Synonym: BUTYL MERCAPTAN (-n)	2347	27			623	208	33		57	54	BTM 178	155			187
THIOCARBAMIZINE					3276	1114	212								
THIOCARBONYL CHLORIDE Synonym: THIOPHOSGENE	2474	55	821		3282	1116			267		TPG1138	904			1143
THIOCARBONYL TETRACHLORIDE Synonym: PERCHLOROMETHYL MERCAPTAN	1670	55	674		2702	919	174		267	176	PCM 938	734			884
THIOCYANATES					3277	1114									
THIOCYANIC ACID AMMONIUM SALT Synonym: AMMONIUM THIOCYANATE	9092/3077		73		235	67			30		AMT 60	69			70
THIOCYANOGEN					3278				267						
THIODAN Synonym: ENDOSULFAN	2761	55	332		1519	505	86		267		ESF 559	398			463

CHEMICAL OR MATERIAL NAME	UN/NA Number	Guide Number (DOT)	Firefighter's Hazardous Materials Reference Book	First Aid Manual for Chemical Accidents	Sax's Dangerous Properties of Industrial Materials	Hazardous Chemicals Desk Reference	Rapid Guide to Hazardous Chemicals in the Workplace	Fire Protection Guide on Hazardous Materials (NFPA)	Firefighter's Handbook of Hazardous Materials	Pocket Guide to Chemical Hazards (NIOSH)	Chemical Hazard Response Information System (CHRIS)	Emergency Handling of Hazardous Materials in Surface Transportation (AAR)	Emergency Action Guides (EAG)	Chemical Data Notebook: A User's Manual	Condensed Chemical Dictionary
THIODEMETON Synonym: DISULFOTON	2783	55	325		1484	494	85		267		DIS 421	387			436
THIOETHANOL Synonym: ETHYL MERCAPTAN	2363	27	358		1631	550	97	−53	138	112		418	E2.9		496
THIOETHYL ALCOHOL Synonym: ETHYL MERCAPTAN	2363	27	358		1631	550	97	−53	138	112		418	E2.9		496
THIOETHYL ALCOHOL Synonym: ETHYL MERCAPTAN	1228	28			1631						EMC 543				496
THIOFAC M 50 Synonym: MONOETHANOLAMINE	2491	60	594					118–	206		MEA 749		M5.0		795
THIOGLYCOL	2966/2810	53							267			902			
THIOGLYCOLIC ACID	1940	60			3279	1115	213		267			902			1144
THIOLACTIC ACID	2936/2810	59							267			903			1144
THIOMETHYL ALCOHOL Synonym: METHYL MERCAPTAN	1064	13	574		2338	801	147	−70	195	154	MMC 774	633	M3.5		773
THIONYL CHLORIDE	1836	39			3281	1115	213	168–	267			903			1145
THIONYL FLUORIDE					3281	1116			267						
THIOPHENE	2414/1993	27			3281	1116		−87	267			903			1145
THIOPHENOL	2337	57							267						1145
THIOPHOSGENE	2474	55	821		3282	1116			267		TPG1138	904			1145
THIOPHOSPHORIC ANHYDRIDE Synonym: PHOSPHORUS PENTASULFIDE	1340	41	688		2794	951	181	139–	233	182	PPP 919	757	P4.2		910
THIOPHOSPHORYL CHLORIDE	1837	60			3282	1116			267			904			1145
THIOPHOSPHORYL FLUORIDE					3283				267						
THIOSULFURIC ACID LEAD SALT Synonym: LEAD THIOSULFATE			512				128				LTS 715				693

CHEMICAL OR MATERIAL NAME	UN/NA Number	Guide Number (DOT)	Firefighter's Hazardous Materials Reference Book	First Aid Manual for Chemical Accidents	Sax's Dangerous Properties of Industrial Materials	Hazardous Chemicals Desk Reference	Rapid Guide to Hazardous Chemicals in the Workplace	Fire Protection Guide on Hazardous Materials (NFPA)	Firefighter's Handbook of Hazardous Materials	Pocket Guide to Chemical Hazards (NIOSH)	Chemical Hazard Response Information System (CHRIS)	Emergency Handling of Hazardous Materials in Surface Transportation (AAR)	Emergency Action Guides (EAG)	Chemical Data Notebook: A User's Manual	Condensed Chemical Dictionary
THIOUREA	2877	53							268						1146
THIRAM	2771	55	822		3285	1118	213		268	212	THR1119	904			1146
THIURAM Synonym: THIRAM	2771	55	822		3285	1118	213		268	212	THR1119	904			1146
THORIUM CHLORIDE					3287	1119			268						1147
THORIUM METAL	9170	65							268						1147
THORIUM METAL	2975	65			3286	1119			268			905			1147
THORIUM METAL	2912	62													1147
THORIUM NITRATE	2976	64	823		3287				268			905			1148
THORIUM NITRATE	9171	64				1119			268		TRN1148				1148
THORIUM NITRATE TETRAHYDRATE Synonym: THORIUM NITRATE	9171	64			3287				268		TRN1148				1148
THORIUM ORE	2912	62	824									906	T2.0		
THORIUM OXIDE	2912	62			3287	1119						906			1148
THULIUM						1120			268						
THYMOL					3290	1121			268						1150
TIBA Synonym: TRIISOBUTYL ALUMINUM	1930/3051	40		27	3391	1145		176–89	277		TIA1120	927			1179
TIBAL Synonym: TRIISOBUTYL ALUMINUM	1930/3051	40		27	3391	1145		176–89	268		TIA1120	927			1151
TIN AND COMPOUNDS					3294	1121	214			212					1151
TIN CHLORIDE	1827	39			3295	1122	214		268						
TIN DIFLUORIDE Synonym: STANNOUS FLUORIDE			790				214				STF1061				1085
TIN MONOPHOSPHIDE	1433	41					214		268						

HAZARDOUS MATERIALS REFERENCE BOOKS INDEX

CHEMICAL OR MATERIAL NAME	UN/NA Number	Guide Number (DOT)	Firefighter's Hazardous Materials Reference Book	First Aid Manual for Chemical Accidents	Sax's Dangerous Properties of Industrial Materials	Hazardous Chemicals Desk Reference	Rapid Guide to Hazardous Chemicals in the Workplace	Fire Protection Guide on Hazardous Materials (NFPA)	Firefighter's Handbook of Hazardous Materials	Pocket Guide to Chemical Hazards (NIOSH)	Chemical Hazard Response Information System (CHRIS)	Emergency Handling of Hazardous Materials in Surface Transportation (AAR)	Emergency Action Guides (EAG)	Chemical Data Notebook: A User's Manual	Condensed Chemical Dictionary
TIN PHOSPHIDE	1433	41			3299	1123	214		268						
TIN TETRACHLORIDE	1827	39					214	169–	268						1152
TINCTURE	1293	26							268						1151
TITANIC CHLORIDE Synonym: TITANIUM TETRACHLORIDE	1838	39	825		3302	1124		169–	269		TTT1158	908	T3.0	T 3	1154
TITANIUM	2546	37		25	3300	1123			268						1152
TITANIUM	1352	32		25	3300	1123			268						1152
TITANIUM (IV) CHLORIDE Synonym: TITANIUM TETRACHLORIDE	1838	39	825		3302	1124		169–	269		TTT1158	908	T3.0	T 3	1154
TITANIUM BUTOXIDE Synonym: TETRABUTYL TITANATE			806						261		TBT1078				1126
TITANIUM CHLORIDE Synonym: TITANIUM TETRACHLORIDE	1838	39	825	25	3302	1124		169–	268		TTT1158	908	T3.0	T 3	1154
TITANIUM DICHLORIDE					3301				269						1153
TITANIUM DIOXIDE				25	3301	1124	214			214			T2.5		1153
TITANIUM HYDRIDE	1871	32							269						1153
TITANIUM SPONGE	2878	32							269						1154
TITANIUM SULFATE	1760	60			3302				269			907			1154
TITANIUM TETRABUTOXIDE Synonym: TETRABUTYL TITANATE			806						261		TBT1078				1126
TITANIUM TETRACHLORIDE	1838	39	825		3302	1124		169–	269		TTT1158	908	T3.0	T 3	1154
TITANIUM TRICHLORIDE	2441	37							269			909			1154
TITANIUM TRICHLORIDE MIXTURE	2441	37							269			909			
TITANIUM TRICHLORIDE MIXTURE	2869	60							269			908			

CHEMICAL OR MATERIAL NAME	UN/NA Number	Guide Number (DOT)	Firefighter's Hazardous Materials Reference Book	First Aid Manual for Chemical Accidents	Sax's Dangerous Properties of Industrial Materials	Hazardous Chemicals Desk Reference	Rapid Guide to Hazardous Chemicals in the Workplace	Fire Protection Guide on Hazardous Materials (NFPA)	Firefighter's Handbook of Hazardous Materials	Pocket Guide to Chemical Hazards (NIOSH)	Chemical Hazard Response Information System (CHRIS)	Emergency Handling of Hazardous Materials in Surface Transportation (AAR)	Emergency Action Guides (EAG)	Chemical Data Notebook: A User's Manual	Condensed Chemical Dictionary
TMA Synonym: TRIMETHYLAMINE	1083	19	853	26	3399	1146	221	178–	278		TMA1126	930	T6.0	T 7	1155
TOE PUFFS	1353	32							269						
TOLUENE	1294	27	826	25	3309	1125	215	170–87	269	214	TOL1135	909	T4.0	T 4	1157
TOLUENE DIISOCYANATE	2078	57			3312	1127	215		269		TDI1095	910	T4.1	T 5	1157
TOLUENE o-NITRO Synonym: NITROTOLUENE-ortho	1664	55			2591	882	166	130–75	216		NIE 832	699			832
TOLUENE p-NITRO Synonym: NITROTOLUENE-para	1664	55			2591	882	166	130–75	216		NTT 861				833
TOLUENE SULFONIC ACID	2583	60			3315						TAP1075				
TOLUENE SULFONIC ACID	2586	60			3315				269						
TOLUENE SULFONIC ACID	2585	60			3315				269						
TOLUENE SULFONIC ACID	2584	60			3315							910			
TOLUENE SULFONIC ACID-para	2584	60	827		3314										1158
TOLUENE THIOL-alpha					3315	1129			269						1159
TOLUENE-2,4-DIAMINE	1709	53			3310	1126	215								1157
TOLUENE-2,4-DIISOCYANATE Synonym: TOLUENE DIISOCYANATE	2078	57	828	25	3313	1128	215	170–87	269	214	TDI1095	910	T4.1	T 5	1157
TOLUENE-2,6-DINITRO Synonym: 2,6-DINITROTOLUENE	2038	56			1441				125		DNL 447				
TOLUENE-3,4-DINITRO Synonym: 3,4-DINITROTOLUENE	2038	56			1441				125		DNU 451				
TOLUENEDIAMINE	1709	53			3310	1126			269		TDA1092	910			
TOLUENESULFONIC ACID	2583	60			3314				269		TAP1075				
TOLUENESULFONIC ACID-para	2583	60			3314			–87			TAP1075				1158

HAZARDOUS MATERIALS REFERENCE BOOKS INDEX

CHEMICAL OR MATERIAL NAME	UN/NA Number	Guide Number (DOT)	Firefighter's Hazardous Materials Reference Book	First Aid Manual for Chemical Accidents	Sax's Dangerous Properties of Industrial Materials	Hazardous Chemicals Desk Reference	Rapid Guide to Hazardous Chemicals in the Workplace	Fire Protection Guide on Hazardous Materials (NFPA)	Firefighter's Handbook of Hazardous Materials	Pocket Guide to Chemical Hazards (NIOSH)	Chemical Hazard Response Information System (CHRIS)	Emergency Handling of Hazardous Materials in Surface Transportation (AAR)	Emergency Action Guides (EAG)	Chemical Data Notebook: A User's Manual	Condensed Chemical Dictionary
TOLUIDINE HYDROCHLORIDE-para					3319	1131									
TOLUIDINE-meta	1708	55	829	27	3317	1129	216		269		TOI1134				1159
TOLUIDINE-ortho	1708	55	830	27	3317	1129	216	171–87	269	216	TLI1124				1159
TOLUIDINE-para	1708	55	831	27	3318	1130	217	171–87	269		TOD1133				1159
TOLUOL Synonym: TOLUENE	1294	27	826	25	3309	1125	215	170–87	269	214	TOL1135	909	T4.0	T 4	1160
TOLUOL-ortho Synonym: CRESOL (-ortho)	2076	55	232	14	960	325	55	57 –29	84	74	CRO 305	270	C7.0		322
TOLUOL-para Synonym: CRESOL (-para)	2076	55	232	14	960	325	55	57 –29	85	74	CSO 305	270	C7.0		322
TOLUYLENEDIAMINE	1709	53							270						
TOLYL CHLORIDE-para Synonym: CHLOROTOLUENE-para	2238	27	213		3322	1131			270		CRN 304				274
TOLYL EPOXYPROPYL ETHER Synonym: CRESYL GLYCIDYL ETHER											CGE 242				
TOLYL GLYCIDYL ETHER Synonym: CRESYL GLYCIDYL ETHER					3323				270		CGE 242				
TOLYLENE DIISOCYANATE Synonym: TOLUENE DIISOCYANATE	2078	57	828		3312		215		270		TDI1095	910	T4.1	T 5	1157
TOLYLENE ISOCYANATE Synonym: TOLUENE DIISOCYANATE	2078	57	828		3312		215		269		TDI1095	910	T4.1	T 5	1157
TOSIC ACID Synonym: TOLUENESULFONIC ACID-para	2583	60			3314			–87			TAP1075				1158
TOXADUST	2761	55													
TOXAPHENE	2761	55	832						270		TXP1159	911			1161

CHEMICAL OR MATERIAL NAME	UN/NA Number	Guide Number (DOT)	Firefighter's Hazardous Materials Reference Book	First Aid Manual for Chemical Accidents	Sax's Dangerous Properties of Industrial Materials	Hazardous Chemicals Desk Reference	Rapid Guide to Hazardous Chemicals in the Workplace	Fire Protection Guide on Hazardous Materials (NFPA)	Firefighter's Handbook of Hazardous Materials	Pocket Guide to Chemical Hazards (NIOSH)	Chemical Hazard Response Information System (CHRIS)	Emergency Handling of Hazardous Materials in Surface Transportation (AAR)	Emergency Action Guides (EAG)	Chemical Data Notebook: A User's Manual	Condensed Chemical Dictionary
TOXICHLOR Synonym: CHLORDANE	2762	28	196			263	44		69	60	CDN 237	211			258
TOXILIC ACID Synonym: MALEIC ACID	2215	60	527		2154	736			270		MLI 771	591			724
TOXILIC ANHYDRIDE Synonym: MALEIC ANHYDRIDE	2215	60	528	19	2155	736	130	106–64	270	138	MLA 769	592	M1.0		724
TRANSFORMER OIL	9188							–87	270						1164
TRANSMISSION OIL Synonym: OILS MISCELLANEOUS MOTOR	1270	27									OMT 890				
TRANSMISSION OIL Synonym: OILS MISCELLANEOUS LUBRICATING	1270	27									OLB 884				
TRANSOTE	1993	27							270						
TREFLAN Synonym: TRIFLURALIN	1609								270		TFR1110				1178
TRETHYLENE Synonym: TRICHLOROETHYLENE	1710	74	837	25	3352	1137	218	174–88	273	216	TCL1085	919	T5.0	T 6	1170
TRI (1-AZIRIDINYL)PHOSPHINE OXIDE	2501	55							270						
TRI 6 Synonym: BENZENE HEXACHLORIDE	2761	55	116		360	117			40		BHC 144				129
TRI BENZENESULFONIC ACID Synonym: ALKYLBENZENESULFONIC ACID											ABS 11				
TRI CLOR Synonym: CHLOROPICRIN	1580	56	210	14	865	296	50	54 –	77	70	CPL 290	229	C4.3		271
TRI-n-BUTYL BORANE					3338	1134			270						1167
TRI-n-BUTYL BORATE					3338	1134		–87	270						
TRI-n-BUTYLAMINE	1993	27			3338	1133						913			1167

CHEMICAL OR MATERIAL NAME	UN/NA Number	Guide Number (DOT)	Firefighter's Hazardous Materials Reference Book	First Aid Manual for Chemical Accidents	Sax's Dangerous Properties of Industrial Materials	Hazardous Chemicals Desk Reference	Rapid Guide to Hazardous Chemicals in the Workplace	Fire Protection Guide on Hazardous Materials (NFPA)	Firefighter's Handbook of Hazardous Materials	Pocket Guide to Chemical Hazards (NIOSH)	Chemical Hazard Response Information System (CHRIS)	Emergency Handling of Hazardous Materials in Surface Transportation (AAR)	Emergency Action Guides (EAG)	Chemical Data Notebook: A User's Manual	Condensed Chemical Dictionary
TRI-o-CRESYL PHOSPHATE	2574	55							270						
TRI-p-CRESYL PHOSPHATE Synonym: TRICRESYLPHOSPHATE	2574/2810	55	841						274		TCP1089	922			1173
TRI-p-TOLYL PHOSPHATE Synonym: TRICRESYLPHOSPHATE	2574/2810	55	841		3450				274		TCP1089	922			1173
TRI-sec-BUTYL BORATE					3339				271						
TRIALLYL BORATE	2609	55			3331				271			914			
TRIALLYL CYANURATE					3331				271						1165
TRIALLYLAMINE	2610/1993	29			3331	1132			271			914			1165
TRIAMYL BORATE								-87	271						1165
TRIAMYLAMINE	1993	27						-87	271						1165
TRIAZINE PESTICIDE	2763	55							271			916			
TRIAZINE PESTICIDE	2997	28							271						
TRIAZINE PESTICIDE	2998/2763	55							271			915			
TRIAZINE PESTICIDE	2764	28							271			915			
TRIBROMOSILANE					3338				271						
TRIBUTYL ALUMINUM	1930	40							271						
TRIBUTYL PHOSPHATE			834	25	3340	1134	217	-88	272	216	TBP1077				1167
TRIBUTYL PHOSPHITE					3340			-88	272						1168
TRIBUTYLAMINE	2542/1760	68			3338	1133		172-87	272			916			
TRIBUTYLPHOSPHINE								-88	272						
TRICALCIUM ARSENATE Synonym: CALCIUM ARSENATE	1573	53	165						62	56	CCA 225	178			200
TRICALCIUM ORTHO ARSENATE Synonym: CALCIUM ARSENATE	1573	53	165						272	56	CCA 225	178			1168

CHEMICAL OR MATERIAL NAME	UN/NA Number	Guide Number (DOT)	Firefighter's Hazardous Materials Reference Book	First Aid Manual for Chemical Accidents	Sax's Dangerous Properties of Industrial Materials	Hazardous Chemicals Desk Reference	Rapid Guide to Hazardous Chemicals in the Workplace	Fire Protection Guide on Hazardous Materials (NFPA)	Firefighter's Handbook of Hazardous Materials	Pocket Guide to Chemical Hazards (NIOSH)	Chemical Hazard Response Information System (CHRIS)	Emergency Handling of Hazardous Materials in Surface Transportation (AAR)	Emergency Action Guides (EAG)	Chemical Data Notebook: A User's Manual	Condensed Chemical Dictionary
TRICHLORAN Synonym: TRICHLOROETHYLENE	1710	74	837	25	3352	1137	218	174–88	273	216	TCL1085	919	T5.0	T 6	1170
TRICHLORFON	2783/3018	55	835						272		TRC1146	917			1169
TRICHLORMETHYL SULFUR CHLORIDE Synonym: PERCHLOROMETHYL MERCAPTAN	1670	55	674		2702	919	174		225	176	PCM 938	734			884
TRICHLORO-s-TRIAZINE-2,4,6-(1H,3H,5H)-TRIONE Synonym: TRICHLORO-s-TRIAZINETRIONE	2468	45	836						272		TCT1091	917			
TRICHLORO-S-TRIAZINETRIONE	2468	45	836						272		TCT1091	917			
TRICHLOROACETALDEHYDE	2075	55							272		TCH1084				1169
TRICHLOROACETIC ACID	1839	59		25	3346	1135	217		272			918			1169
TRICHLOROACETIC ACID	2564	59		25	3346	1135	217		272			917			1169
TRICHLOROACETONITRILE					3347				272						1169
TRICHLOROACETYL CHLORIDE	2442	59			3348	1136			272						1169
TRICHLOROAMYLSILANE Synonym: AMYLTRICHLOROSILANE (-n)	1728	29	84						32		ATS 107	78			76
TRICHLOROBENZENE	2321	54			3349						TCB1081	918			
TRICHLOROBORANE Synonym: BORON TRICHLORIDE	1741	15	133		532	175			49		BRT 168	136	B2.9		164
TRICHLOROBUTENE	2322	54							273						
TRICHLOROCYANIDINE	2760	60							273						
TRICHLOROETHANE Synonym: 1,1,1-TRICHLOROETHANE	2831	74		25	3350			173–88	273		TCE1082	919	T4.2		1170
TRICHLOROETHANE-beta	2831	74			3350				273						
TRICHLOROETHANOL					3351				273						1170

HAZARDOUS MATERIALS REFERENCE BOOKS INDEX

CHEMICAL OR MATERIAL NAME	UN/NA Number	Guide Number (DOT)	Firefighter's Hazardous Materials Reference Book	First Aid Manual for Chemical Accidents	Sax's Dangerous Properties of Industrial Materials	Hazardous Chemicals Desk Reference	Rapid Guide to Hazardous Chemicals in the Workplace	Fire Protection Guide on Hazardous Materials (NFPA)	Firefighter's Handbook of Hazardous Materials	Pocket Guide to Chemical Hazards (NIOSH)	Chemical Hazard Response Information System (CHRIS)	Emergency Handling of Hazardous Materials in Surface Transportation (AAR)	Emergency Action Guides (EAG)	Chemical Data Notebook: A User's Manual	Condensed Chemical Dictionary
TRICHLOROETHENE Synonym: TRICHLOROETHYLENE	1710	74	837	25	3352	1137	218	174–88	273	216	TCL1085	919	T5.0	T 6	1170
TRICHLOROETHYL SILANE Synonym: ETHYLTRICHLOROSILANE	1196	29	381		1665	562		90 –54	273		ETS 572	440			501
TRICHLOROETHYL SILICONE Synonym: ETHYLTRICHLOROSILANE	1196	29	381		1665	562		90 –54	140		ETS 572	440			501
TRICHLOROETHYLENE	1710	74	837	25	3352	1137	218	174–88	273	216	TCL1085	919	T5.0	T 6	1170
TRICHLOROFLUOROMETHANE	3082/9188		838	25	3353	1138	219		273		TCF1083	919			1170
TRICHLOROFORM Synonym: CHLOROFORM	1888	55	205	14	815	282	49	52 –	76	68	CRF 301	223	C4.2		266
TRICHLOROIMINOISOCYANURIC ACID Synonym: TRICHLORO-S-TRIAZINETRIONE	2468	45	836						272		TCT1091	917			
TRICHLOROISOCYANURIC ACID Synonym: TRICHLORO-S-TRIAZINETRIONE	2468	45	836		3355	1139		172–	273		TCT1091	920			1170
TRICHLOROMETHANE Synonym: CHLOROFORM	1888	55	205	14	815	282	49	52 –	273	68	CRF 301	223	C4.2	C 7	1171
TRICHLOROMETHANE SULFENYL CHLORIDE	1670	55							273						
TRICHLOROMETHANE SULFURYL CHLORIDE Synonym: PERCHLOROMETHYL MERCAPTAN	1670	55	674		2702	919	174		225	176	PCM 938	734			884
TRICHLOROMETHANESULFENYL CHLORIDE Synonym: PERCHLOROMETHYL MERCAPTAN	1670	55	674		2702	919	174		225	176	PCM 938	734			884
TRICHLOROMETHYL SULFOCHLORIDE Synonym: PERCHLOROMETHYL MERCAPTAN	1670	55	674		2702	919	174		225	176	PCM 938	734			884
TRICHLOROMETHYLSILANE Synonym: METHYL TRICHLOROSILANE	1250	29	581		2408	819		117–72	273		MTS 812	651	M3.6.2		782

CHEMICAL OR MATERIAL NAME	UN/NA Number	Guide Number (DOT)	Firefighter's Hazardous Materials Reference Book	First Aid Manual for Chemical Accidents	Sax's Dangerous Properties of Industrial Materials	Hazardous Chemicals Desk Reference	Rapid Guide to Hazardous Chemicals in the Workplace	Fire Protection Guide on Hazardous Materials (NFPA)	Firefighter's Handbook of Hazardous Materials	Pocket Guide to Chemical Hazards (NIOSH)	Chemical Hazard Response Information System (CHRIS)	Emergency Handling of Hazardous Materials in Surface Transportation (AAR)	Emergency Action Guides (EAG)	Chemical Data Notebook: A User's Manual	Condensed Chemical Dictionary
TRICHLOROMONOSILANE Synonym: TRICHLOROSILANE	1295	38	840		3361	1141		174–88	273		TCS1090	922			1172
TRICHLORONAPHTHALENE					3356	1139	219		273	218					1171
TRICHLORONITROMETHANE Synonym: CHLOROPICRIN	1580	56	210	14	865	296	50	54 –	273	70	CPL 290	229	C4.3		1171
TRICHLOROOXOVANADIUM Synonym: VANADIUM OXYTRICHLORIDE	2443	39	868		3478	1169	226		284		VOT1180	946			1209
TRICHLOROPENTYLSILANE Synonym: AMYLTRICHLOROSILANE (-n)	1728	29	84						32		ATS 107	78			76
TRICHLOROPHENOL	2020	53	839								TPH1139				
TRICHLOROPHENOXYACETIC ACID AMINE	2765	55							274						
TRICHLOROPHENOXYPROPIONIC ACID ESTER	2765	55							274						
TRICHLOROPHOPHORUS OXIDE Synonym: PHOSPHORUS OXYCHLORIDE	1810	39	686		2793	950	181	138–	233		PPO 978	755	P4.1.1		909
TRICHLOROPHOSPHINE OXIDE Synonym: PHOSPHORUS OXYCHLORIDE	1810	39	686		2793	950	181	138–	233		PPO 978	755	P4.1.1		909
TRICHLOROSILANE	1295	38	840		3361	1141		174–88	274		TCS1090	922			1172
TRICHLOROTRIAZINETRIONE Synonym: TRICHLORO-S-TRIAZINETRIONE	2468	45	836						274		TCT1091	917			
TRICHLOROVINYL SILICANE Synonym: VINYL TRICHLOROSILANE	1305	29	877					–94	286		VTS1183	951			1218
TRICHLOROVINYLSILANE Synonym: VINYL TRICHLOROSILANE	1305	29	877					–94	274		VTS1183	951			1218
TRICLENE Synonym: TRICHLOROETHYLENE	1710	74	837	25	3352	1137	218	174–88	273	216	TCL1085	919	T5.0	T 6	1170

HAZARDOUS MATERIALS REFERENCE BOOKS INDEX

CHEMICAL OR MATERIAL NAME	UN/NA Number	Guide Number (DOT)	Firefighter's Hazardous Materials Reference Book	First Aid Manual for Chemical Accidents	Sax's Dangerous Properties of Industrial Materials	Hazardous Chemicals Desk Reference	Rapid Guide to Hazardous Chemicals in the Workplace	Fire Protection Guide on Hazardous Materials (NFPA)	Firefighter's Handbook of Hazardous Materials	Pocket Guide to Chemical Hazards (NIOSH)	Chemical Hazard Response Information System (CHRIS)	Emergency Handling of Hazardous Materials in Surface Transportation (AAR)	Emergency Action Guides (EAG)	Chemical Data Notebook: A User's Manual	Condensed Chemical Dictionary
TRICRESOL **Synonym: CRESOL**	2076	55	232	14	959	324	55	57 –	85	74	CRS 307	270	C7.0		322
TRICRESYLPHOSPHATE	2574/2810	55	841						274		TCP1089	922			1173
TRIDECANOL			842		3365			–88	274		TDN1096				1173
TRIDECANOL **Synonym: LINEAR ALCOHOL**			842		3365				274		LAL 692				1173
TRIDECYL ALCOHOL								–88	274						1173
TRIEN **Synonym: TRIETHYLENE TETRAMINE**	2259/1760	60			3372	1143			275		TET1106	924			
TRIETHANOLAMINE	9151	31	844					–88			TEA1097				1174
TRIETHANOLAMINE DODECEYLBENZENESULFONATE **Synonym: DODECYLBENZENESULFONIC ACID TRIETHANOLAMINE SALT**	9151	31							274		DBS 353				
TRIETHANOLAMINE DODECYLBENZENESULFONATE	9188	31													
TRIETHANOLAMINE LAURYL SULFATE **Synonym: DODECYL SULFATE TRIETHANOLAMINE SALT**											DST 482				1174
TRIETHYL BORATE					3370				274						1175
TRIETHYL CITRATE					3370	1142		–89							1175
TRIETHYL PHOSPHATE	2783	55			3374			–89	275		TPS1143				1177
TRIETHYL PHOSPHINE					3374				275						
TRIETHYL PHOSPHITE	2323	26			3374	1143			275		TPI1140	923			1177
TRIETHYL PHOSPHOROTHIOATE					3374				275						
TRIETHYLALUMINUM	1102/3051	40		25	3368	1141		175–88			TAL1074	923			1174
TRIETHYLAMINE	1296	68	846	25	3368	1141	220	176–88	275	218	TEN1103	923			1174
TRIETHYLBENZENE	1993	27	847		3369				275		TEB1098				

CHEMICAL OR MATERIAL NAME	UN/NA Number	Guide Number (DOT)	Firefighter's Hazardous Materials Reference Book	First Aid Manual for Chemical Accidents	Sax's Dangerous Properties of Industrial Materials	Hazardous Chemicals Desk Reference	Rapid Guide to Hazardous Chemicals in the Workplace	Fire Protection Guide on Hazardous Materials (NFPA)	Firefighter's Handbook of Hazardous Materials	Pocket Guide to Chemical Hazards (NIOSH)	Chemical Hazard Response Information System (CHRIS)	Emergency Handling of Hazardous Materials in Surface Transportation (AAR)	Emergency Action Guides (EAG)	Chemical Data Notebook: A User's Manual	Condensed Chemical Dictionary
TRIETHYLBENZENE-sym Synonym: TRIETHYLBENZENE	1993	27	847		3369				275		TEB1098				
TRIETHYLBORANE					3369	1142		-89	275						1175
TRIETHYLENE GLYCOL			848	25	3370	1142		-89			TEG1101				1175
TRIETHYLENE GLYCOL MONOETHYL ETHER Synonym: ETHOXY TRIGLYCOL					1568	523		-47	135		ETG 565				477
TRIETHYLENE PHOSPHORAMIDE	2501	55							275						1176
TRIETHYLENE TETRAMINE	2259/1760	60			3372	1143			275		TET1106	924			1176
TRIETHYLENETETRAMINE	2259/1760	60			3372	1143		-89	275		TET1106	924			1176
TRIETHYLOLAMINE Synonym: TRIETHANOLAMINE	9151	31	844					-88			TEA1097				1174
TRIFLUOROACETIC ACID	2699/1760	60							275			924			1177
TRIFLUOROACETYL CHLORIDE	3057	16			3378				275			925			
TRIFLUOROBROMOMETHANE	1009	12			3379	1143	220		276	220					1177
TRIFLUOROCHLOROETHANE	1983	12							276						
TRIFLUOROCHLOROETHENE	1082	17							276						
TRIFLUOROCHLOROETHYLENE	1082	17	850					-89	276		TFC1108	925			1177
TRIFLUOROCHLOROMETHANE	1022	12							276						1177
TRIFLUOROETHANE	2035/1954	22		25					276			925			
TRIFLUOROMETHANE	984/1956	12		26					276			926			1177
TRIFLUOROMETHANE	3136	21		26											1177
TRIFLUOROMETHANE AND CHLOROTRIFLUOROMETHANE MIXTURE	2599	12							276						
TRIFLUOROMETHANE AND CHLOROTRIFLUOROMETHANE MIXTURE	1078	12							276			926			

HAZARDOUS MATERIALS REFERENCE BOOKS INDEX

CHEMICAL OR MATERIAL NAME	UN/NA Number	Guide Number (DOT)	Firefighter's Hazardous Materials Reference Book	First Aid Manual for Chemical Accidents	Sax's Dangerous Properties of Industrial Materials	Hazardous Chemicals Desk Reference	Rapid Guide to Hazardous Chemicals in the Workplace	Fire Protection Guide on Hazardous Materials (NFPA)	Firefighter's Handbook of Hazardous Materials	Pocket Guide to Chemical Hazards (NIOSH)	Chemical Hazard Response Information System (CHRIS)	Emergency Handling of Hazardous Materials in Surface Transportation (AAR)	Emergency Action Guides (EAG)	Chemical Data Notebook: A User's Manual	Condensed Chemical Dictionary
TRIFLUOROMETHYLHYPOFLUORITE					3382				276						
TRIFLUOROMONOCHLOROETHYLENE	1082	17	850					-89	276		TFC1108	925			1177
Synonym: TRIFLUOROCHLOROETHYLENE															
TRIFLUOROVINYL CHLORIDE	1082	17	850					-89	276		TFC1108	925			1178
Synonym: TRIFLUOROCHLOROETHYLENE															
TRIFLURALIN	1609								277		TFR1110				1178
TRIGLYCINE			629								NAA 815				1178
Synonym: NITRILOTRIACETIC ACID AND SALTS															
TRIGLYCOL			848	25	3370	1142		-89			TEG1101				1175
Synonym: TRIETHYLENE GLYCOL															
TRIGLYCOL DICHLORIDE					3388			-89	277						1178
TRIGLYCOL MONOETHYL ETHER					1568	523		-47	135		ETG 565				477
Synonym: ETHOXY TRIGLYCOL															
TRIHYDROXYTRIETHYLAMINE	9151	31	844		3390	1144		-88			TEA1097				1174
Synonym: TRIETHANOLAMINE															
TRIISOBUTYL ALUMINUM	1930/3051	40		26	3391	1145		176-89	277		TIA1120	927			1179
TRIISOBUTYL BORATE					3392			-89	277						
TRIISOBUTYLENE	2324	27							277		TIB1121	927			1179
TRIISOCYANATOISOCYANURATE OF ISOPHORONEDIISO-CYANATE	2906/1993	26							277			927			
TRIISOPROPANOLAMINE								-89	277		TIP1122				1179
TRIISOPROPYL BORATE	2616/1993	26							277			928			1180
TRILENE	1710	74	837	25	3352	1137	218	174-88	273	216	TCL1085	919	T5.0	T 6	1170
Synonym: TRICHLOROETHYLENE															
TRIMETHOXYSILANE	9269	57			3397			177-	277						
TRIMETHYL ACETYL CHLORIDE	2438	29							277						

CHEMICAL OR MATERIAL NAME	UN/NA Number	Guide Number (DOT)	Firefighter's Hazardous Materials Reference Book	First Aid Manual for Chemical Accidents	Sax's Dangerous Properties of Industrial Materials	Hazardous Chemicals Desk Reference	Rapid Guide to Hazardous Chemicals in the Workplace	Fire Protection Guide on Hazardous Materials (NFPA)	Firefighter's Handbook of Hazardous Materials	Pocket Guide to Chemical Hazards (NIOSH)	Chemical Hazard Response Information System (CHRIS)	Emergency Handling of Hazardous Materials in Surface Transportation (AAR)	Emergency Action Guides (EAG)	Chemical Data Notebook: A User's Manual	Condensed Chemical Dictionary
TRIMETHYL ALUMINUM	1103/3051	40		26	3398	1146		-89	277			928			1181
TRIMETHYL BISMUTH					3405				277						
TRIMETHYL NORPINANYL HYDROPEROXIDE	2162	51							278						
TRIMETHYL PHOSPHATE						1150	222		278						1184
TRIMETHYL PHOSPHINE									278						
TRIMETHYL PHOSPHITE	2329/1993	26				1150	222	-91	278		TPP1142	929			1184
TRIMETHYLACETYL CHLORIDE	2438	29							277			930			
TRIMETHYLAMINE	1297	29		26	3399	1146	221	-90	278			931	T6.0	T 7	1181
TRIMETHYLAMINE	1083	19	853	26	3399	1146	221	178-	278		TMA1126	930	T6.O	T 7	1181
TRIMETHYLBENZENE	2325	26			3404	1147	221		278						1181
TRIMETHYLBENZYLAMMONIUM CHLORIDE Synonym: BENZYLTRIMETHYLAMMONIUM CHLORIDE					413						BMA 146				138
TRIMETHYLBORATE	2416/1993	26			3406	1148			277			929			1181
TRIMETHYLCARBINOL Synonym: BUTYL ALCOHOL (-tert)	1120	26			594	197	31	-21	277	52	BAT 121	150	B7.1		1182
TRIMETHYLCHLOROSILANE	1298	29	854		3406	1148		-90	278		TMC1127	931			1182
TRIMETHYLCYCLOHEXANOL								-90	278						
TRIMETHYLCYCLOHEXYLAMINE	2326	29							278			931			
TRIMETHYLENE Synonym: CYCLOPROPANE	1027	22	251			339		-31	278		CPR 294	287			1182
TRIMETHYLENE CHLORIDE Synonym: 1,3-DICHLOROPROPANE	1279	27			1163	392			104		DPC 460				
TRIMETHYLENE DICHLORIDE Synonym: 1,3-DICHLOROPROPANE	1279	27			1163	392			278		DPC 460				
TRIMETHYLENE TRINITRAMINE	0072								279						1182

HAZARDOUS MATERIALS REFERENCE BOOKS INDEX

CHEMICAL OR MATERIAL NAME	UN/NA Number	Guide Number (DOT)	Firefighter's Hazardous Materials Reference Book	First Aid Manual for Chemical Accidents	Sax's Dangerous Properties of Industrial Materials	Hazardous Chemicals Desk Reference	Rapid Guide to Hazardous Chemicals in the Workplace	Fire Protection Guide on Hazardous Materials (NFPA)	Firefighter's Handbook of Hazardous Materials	Pocket Guide to Chemical Hazards (NIOSH)	Chemical Hazard Response Information System (CHRIS)	Emergency Handling of Hazardous Materials in Surface Transportation (AAR)	Emergency Action Guides (EAG)	Chemical Data Notebook: A User's Manual	Condensed Chemical Dictionary
TRIMETHYLHEPTANALS	1993	27	470					-61	173		IDA 666				655
Synonym: ISODECALDEHYDE															
TRIMETHYLHEXAMETHYLENE DIISOCYANATE	2328	55							279		THI1117	932			
TRIMETHYLHEXAMETHYLENEDIAMINE	2327	60							279		THA1113	932			
TRIMETHYLOLPROPANE TRIACRYLATE					3413			-91							
TRIMETHYLSILYL CHLORIDE	1298	29	854		3406	1148		-90	278		TMC1127	931			1182
Synonym: TRIMETHYLCHLOROSILANE															
TRINITROACETONITRILE					3422	1151			279						
TRINITROBENZENE	1354	33		26	3423				279			933			
TRINITROBENZOIC ACID	1355	33			3423	1152			280			934			1185
TRINITROMETHANE					3424	1153			280						1185
TRINITROPHENOL	1344	33							280						1185
TRINITROTOLUENE	1356	33		26	3425				280			936			
TRIORTHOCRESYL PHOSPHATE	2574	55			3427	1154	223			220					
TRIOSULFUROUS DICHLORIDE	1828	39	799					-84	258			869	S4.0		1103
Synonym: SULFUR CHLORIDE															
TRIOXANE					3428			-91	280						1186
TRIPHENYL AMINE					3429	1155	223								
TRIPHENYL BORATE					3430				280						
TRIPHENYL PHOSPHATE					3431	1156	223	-91	280	220					1187
TRIPHENYL PHOSPHITE					3432	1156		-91	280						1187
TRIPHENYLPHOSPHINE					3431	1156			280						1187
TRIPOLI					3434	1156	224								1188
TRIPROPYLALUMINUM	2718	40			3435			-92	281						1188

CHEMICAL OR MATERIAL NAME	UN/NA Number	Guide Number (DOT)	Firefighter's Hazardous Materials Reference Book	First Aid Manual for Chemical Accidents	Sax's Dangerous Properties of Industrial Materials	Hazardous Chemicals Desk Reference	Rapid Guide to Hazardous Chemicals in the Workplace	Fire Protection Guide on Hazardous Materials (NFPA)	Firefighter's Handbook of Hazardous Materials	Pocket Guide to Chemical Hazards (NIOSH)	Chemical Hazard Response Information System (CHRIS)	Emergency Handling of Hazardous Materials in Surface Transportation (AAR)	Emergency Action Guides (EAG)	Chemical Data Notebook: A User's Manual	Condensed Chemical Dictionary
TRIPROPYLAMINE	2260/1993	68			3435	1156		−92	281			937			1188
TRIPROPYLENE	2057/1993	27						−92	281			938			1188
TRIPROPYLENE Synonym: NONENE	2057/1993	27						−75	217		NON 844	938			1188
TRIPROPYLENE GLYCOL			855		3436			−92			TGC1111				1188
TRIPROPYLENE GLYCOL METHYL ETHER					3436			−92			TGM1112				
TRIS(1-AZIRIDINYL)PHOSPHINE OXIDE	2501	55			3439	1157			281		TPO1141	938			1188
TRIS(2-CHLOROETHYL)AMINE					3441				281						
TRIS(HYDROXYETHYL)AMINE Synonym: TRIETHANOLAMINE	9151	31	844					−88			TEA1097				1174
TRISILYL ARSINE					3445				281						
TRISODIUM NITRILOTRIACETATE *Synonym: NITRILOTRIACETIC ACID AND SALTS*			629								NAA 815				1190
TRISODIUM ORTHOPHOSPHATE Synonym: SODIUM PHOSPHATE TRIBASIC	9148/3077	31	783		3106	1063					SPH1053	849			1190
TRISODIUM PHOSPHATE Synonym: SODIUM PHOSPHATE TRIBASIC	9148/3077	31	783	26	3106	1063			282		SPH1053	849			1190
TRITHION				26	3449	1159			282						1190
TSA-para Synonym: TOLUENESULFONIC ACID-para	2583	60			3314			−87			TAP1075				1158
TSUMACIDE									282						
TUBERCUPROSE Synonym: COPPER FORMATE											CUF 324				
TUCUM OIL *Synonym: OILS EDIBLE TUCUM*											OTC 911				

HAZARDOUS MATERIALS REFERENCE BOOKS INDEX

CHEMICAL OR MATERIAL NAME	UN/NA Number	Guide Number (DOT)	Firefighter's Hazardous Materials Reference Book	First Aid Manual for Chemical Accidents	Sax's Dangerous Properties of Industrial Materials	Hazardous Chemicals Desk Reference	Rapid Guide to Hazardous Chemicals in the Workplace	Fire Protection Guide on Hazardous Materials (NFPA)	Firefighter's Handbook of Hazardous Materials	Pocket Guide to Chemical Hazards (NIOSH)	Chemical Hazard Response Information System (CHRIS)	Emergency Handling of Hazardous Materials in Surface Transportation (AAR)	Emergency Action Guides (EAG)	Chemical Data Notebook: A User's Manual	Condensed Chemical Dictionary
TUMBLEAF **Synonym: SODIUM CHLORATE**	1495	35	766	24	3070	1048		156–	253		SDC1029	834	S2.1	S 2	1052
TUNGSTEN AND COMPOUNDS					3455	1161	224		282						1192
TUNGSTEN HEXAFLUORIDE	2196	15			3457				282			939			1193
TURBINE OIL **Synonym: OILS MISCELLANEOUS TURBINE**											OTB 910				
TURKEY RED OIL					3457			−92							1194
TURPENTINE	1299	27	856	26	3458	1162	225	−92	282	222	TPT1144	939			1194
TURPENTINE OIL	1299	27			3458	1162			282						1194
TURPENTINE SUBSTITUTE	1300	27							282						
TURPS **Synonym: TURPENTINE**	1299	27	856	26	3458	1162	225	−92	282	222	TPT1144	939			1194
TYRANTON **Synonym: DIACETONE ALCOHOL**	1148	26	260		1059	354	62	−32	94	82	DAA 332	316			359
UCANE ALKYLATE 12 **Synonym: DODECYLBENZENE**					3461						DDB 371				441
UCAR BISPHENOL HP **Synonym: BISPHENOL A**					511	169			48		BPA 154				154
UCON 11 **Synonym: TRICHLOROFLUOROMETHANE**	3082/9188	31	838	25	3353	1138	219		273		TCF1083	919			1170
UCON 12 **Synonym: DICHLORODIFLUOROMETHANE**	1028	12	275	15	1137	380	68		102	86	DCF 360	331	D1.1		379
UCON 22 **Synonym: MONOCHLORODIFLUOROMETHANE**	1018	12	593						205				M4.0		795
UCONN 22 **Synonym: CHLORODIFLUOROMETHANE**	1018	12		14	791	278	48		74		MCF 737				265

CHEMICAL OR MATERIAL NAME	UN/NA Number	Guide Number (DOT)	Firefighter's Hazardous Materials Reference Book	First Aid Manual for Chemical Accidents	Sax's Dangerous Properties of Industrial Materials	Hazardous Chemicals Desk Reference	Rapid Guide to Hazardous Chemicals in the Workplace	Fire Protection Guide on Hazardous Materials (NFPA)	Firefighter's Handbook of Hazardous Materials	Pocket Guide to Chemical Hazards (NIOSH)	Chemical Hazard Response Information System (CHRIS)	Emergency Handling of Hazardous Materials in Surface Transportation (AAR)	Emergency Action Guides (EAG)	Chemical Data Notebook: A User's Manual	Condensed Chemical Dictionary
UDMH **Synonym: 1,1-DIMETHYLHYDRAZINE**	1163	28			1381	459	79	−43	282	96	DMH 432				417
ULTRASENE	1223	27						−92	282						
UN BENZENESULFONIC ACID **Synonym: ALKYLBENESULFONIC ACID**											ABS 11				
UNDECANE	2330/1993	27			3464	1163			282			940			1199
UNDECANOL			857								UND1165				
UNDECYL ALCOHOL **Synonym: UNDECANOL**			857		3466	1164					UND1165				1200
UNDECYLBENZENE (-n)			859								UDB1163				
UNDECYLETHYLENE **Synonym: 1-TRIDECENE**											TDC1094				
UNIFUME **Synonym: ETHYLENE DIBROMIDE**	1605	55	371					85 –	143	110	EDB 518	430	E2.4.1		487
UNIVERM **Synonym: CARBON TETRACHLORIDE**	1846	55	192	13	701	246	43	48 –	66	60	CBT 223	193	C2.3	C 5	221
UNSLAKED LIME **Synonym: CALCIUM OXIDE**	1910	60	174	13	669	233	38	47 –	282	56	CAO 209	183			205
URANIUM	2979	65		26	3466	1164	225		282	222		942			1201
URANIUM (VI) FLUORIDE **Synonym: URANIUM HEXAFLUORIDE**	2977	66	860		3467				283			940	U1.0		1202
URANIUM ACETATE **Synonym: URANYL ACETATE**	3180/2982	62	862						283		URA1167	942			1203
URANIUM ACETATE DIHYDRATE **Synonym: URANYL ACETATE**	3180/2982	62	862						283		URA1167	942			1203
URANIUM BEARING ORE	2912	62										940			

HAZARDOUS MATERIALS REFERENCE BOOKS INDEX

CHEMICAL OR MATERIAL NAME	UN/NA Number	Guide Number (DOT)	Firefighter's Hazardous Materials Reference Book	First Aid Manual for Chemical Accidents	Sax's Dangerous Properties of Industrial Materials	Hazardous Chemicals Desk Reference	Rapid Guide to Hazardous Chemicals in the Workplace	Fire Protection Guide on Hazardous Materials (NFPA)	Firefighter's Handbook of Hazardous Materials	Pocket Guide to Chemical Hazards (NIOSH)	Chemical Hazard Response Information System (CHRIS)	Emergency Handling of Hazardous Materials in Surface Transportation (AAR)	Emergency Action Guides (EAG)	Chemical Data Notebook: A User's Manual	Condensed Chemical Dictionary
URANIUM FLUORIDE Synonym: URANIUM HEXAFLUORIDE	2977	66	860		3467				283			940	U1.0		1202
URANIUM HEXAFLUORIDE	2978	66							283			941	U1.0		1202
URANIUM HEXAFLUORIDE	2977	66	860						283			940	U1.0		1202
URANIUM METAL	2979	65		26	3466	1164	225		282	222		942			1201
URANIUM NITRATE Synonym: URANYL NITRATE	2981	64	863		3469	1165			283		UAN1160	943			1203
URANYL NITRATE HEXAHYDRATE SOLUTION	2980	64							283			943		943	
URANIUM OXIDE (UO4) Synonym: URANIUM PEROXIDE	2918	63	861		3468						URP1169	942			
URANIUM OXIDE PEROXIDE (UO2[O2]) Synonym: URANIUM PEROXIDE			861								URP1169				
URANIUM OXYACETATE DIHYDRATE Synonym: URANYL ACETATE	9180/2982	62	862		3469	1165			283		URA1167	942			1203
URANIUM PEROXIDE			861								URP1169				
URANIUM SULFATE Synonym: URANYL SULFATE			864								URS1170				
URANIUM SULFATE TRIHYDRATE Synonym: URANYL SULFATE			864								URS1170				
URANYL ACETATE	9180/2982	62	862						283		URA1167	942			1203
URANYL ACETATE DIHYDRATE Synonym: URANYL ACETATE	9180/2982	62	862		3469				283		URA1167	942			1203
URANYL NITRATE	2981	64	863		3469	1165					UAN1160	943			1203
URANYL NITRATE HEXAHYDRATE SOLUTION	9178					1165			283						
URANYL SULFATE			864								URS1170				

CHEMICAL OR MATERIAL NAME	UN/NA Number	Guide Number (DOT)	Firefighter's Hazardous Materials Reference Book	First Aid Manual for Chemical Accidents	Sax's Dangerous Properties of Industrial Materials	Hazardous Chemicals Desk Reference	Rapid Guide to Hazardous Chemicals in the Workplace	Fire Protection Guide on Hazardous Materials (NFPA)	Firefighter's Handbook of Hazardous Materials	Pocket Guide to Chemical Hazards (NIOSH)	Chemical Hazard Response Information System (CHRIS)	Emergency Handling of Hazardous Materials in Surface Transportation (AAR)	Emergency Action Guides (EAG)	Chemical Data Notebook: A User's Manual	Condensed Chemical Dictionary
URANYL SULFATE TRIHYDRATE Synonym: URANYL SULFATE			864								URS1170				
UREA			865		3470	1166			283		URE1168		U3.0		1203
UREA HYDROGEN PEROXIDE Synonym: UREA PEROXIDE	1511/1479	35	866						283		UPO1166	944			1204
UREA HYDROGEN PEROXIDE SALT Synonym: UREA PEROXIDE	1511	35	866						283		UPO1166	945			1204
UREA NITRATE	1357	33			3471				283			944			1204
UREA PEROXIDE	1511	35	866						283		UPO1166	945			1204
URETHANE	2979	65			3471	1166	225								1204
UROTROPIN Synonym: HEXAMETHYLENETETRAMINE	1328	32	434		1856	625			160		HMT 639				598
USAF EK 1,597 Synonym: MONOETHANOLAMINE	2491	60	594					118–	206		MEA 749		M5.0		795
VAC Synonym: VINYL ACETATE	1301	26	871	26	3492	1172	227	179–92	285		VAM1172	948	V1.0		1215
VAL DROP Synonym: SODIUM CHLORATE	1495	35	766	24	3037	1048		156–	253		SDC1029	834	S2.1	S 2	1052
VALENTINITE Synonym: ANTIMONY TRIOXIDE	549/3077	60	94								ATX 109	87			1207
VALERAL Synonym: VALERALDEHYDE	2058/1993	26	867		3475	1167	226	-92	284		VAL1171	945			1207
VALERALDEHYDE Synonym: VALERALDEHYDE-n	2058/1993	26	867		3475	1167	226	-92	284		VAL1171	945			1207
VALERALDEHYDE-n Synonym: VALERALDEHYDE	2058/1993	26	867		3475	1167	226		284		VAL1171	945			1207

HAZARDOUS MATERIALS REFERENCE BOOKS INDEX

CHEMICAL OR MATERIAL NAME	UN/NA Number	Guide Number (DOT)	Firefighter's Hazardous Materials Reference Book	First Aid Manual for Chemical Accidents	Sax's Dangerous Properties of Industrial Materials	Hazardous Chemicals Desk Reference	Rapid Guide to Hazardous Chemicals in the Workplace	Fire Protection Guide on Hazardous Materials (NFPA)	Firefighter's Handbook of Hazardous Materials	Pocket Guide to Chemical Hazards (NIOSH)	Chemical Hazard Response Information System (CHRIS)	Emergency Handling of Hazardous Materials in Surface Transportation (AAR)	Emergency Action Guides (EAG)	Chemical Data Notebook: A User's Manual	Condensed Chemical Dictionary
VALERIC ACID	1760	60			3475	1167			284						1208
VALERIC ALDEHYDE Synonym: VALERALDEHYDE (-n)	2058/1993	26	867		3475	1167	226	−92	284		VAL1171	945			1208
VALERYL CHLORIDE	2502	60			3476	1167			284			946			
VALINE ALDEHYDE Synonym: ISOBUTYRALDEHYDE	2045/1993	26	467	18	2022	679		−61			BAD 115	548	11.1		653
VAM Synonym: VINYL ACETATE	1301	26	871	26	3492	1172	227	179–92	285		VAM1172	948	V1.0		1215
VANADIC ANHYDRIDE Synonym: VANADIUM PENTOXIDE	2862	53	869		3479	1169	226		284	222	VOX1181				1209
VANADIUM				26	3477	1168	226								1208
VANADIUM DICHLORIDE					3478		226		284						1209
VANADIUM OXYSULFATE Synonym: VANADYL SULFATE	2931/3077	55	870			1170	226				VSF1182	948			1210
VANADIUM OXYTRICHLORIDE	2443	39	868		3478	1169	226		284		VOT1180	946			1209
VANADIUM OXYTRICHLORIDE & TITANIUM TETRACHLORIDE MIX	2443/1760	39					226		284			946			
VANADIUM PENTAOXIDE Synonym: VANADIUM PENTOXIDE	2862/3077	53	869		3479	1169	226		284	222	VOX1181	947			1209
VANADIUM PENTOXIDE	2862/3077	53	869		3479	1169	226		284	222	VOX1181	947			1209
VANADIUM PENTOXIDE (fume)	2862	53			3479	1169	226			224	VOX1181				
VANADIUM SESQUIOXIDE	2860	53			3479	1170	226								1209
VANADIUM SULFATE	2931	55			3480		226		284						1209
VANADIUM TETRACHLORIDE	2444	39			3480	1170	226	179–	284			947			1210
VANADIUM TRICHLORIDE	2475	60			3480	1170	226					947			1210

CHEMICAL OR MATERIAL NAME	UN/NA Number	Guide Number (DOT)	Firefighter's Hazardous Materials Reference Book	First Aid Manual for Chemical Accidents	Sax's Dangerous Properties of Industrial Materials	Hazardous Chemicals Desk Reference	Rapid Guide to Hazardous Chemicals in the Workplace	Fire Protection Guide on Hazardous Materials (NFPA)	Firefighter's Handbook of Hazardous Materials	Pocket Guide to Chemical Hazards (NIOSH)	Chemical Hazard Response Information System (CHRIS)	Emergency Handling of Hazardous Materials in Surface Transportation (AAR)	Emergency Action Guides (EAG)	Chemical Data Notebook: A User's Manual	Condensed Chemical Dictionary
VANADIUM TRIOXIDE	2860	53					226		284						1210
VANADYL CHLORIDE Synonym: **VANADIUM OXYTRICHLORIDE**	2443	39	868		3478	1169	226		284		VOT1180	946			1210
VANADYL SULFATE	2931/3077	55	870		3481	1170					VSF1182	948			1210
VANADYL SULFATE DIHYDRATE Synonym: **VANADYL SULFATE**	2931/3077	55	870			1170					VSF1182	948			1210
VANADYL TRICHLORIDE Synonym: **VANADIUM OXYTRICHLORIDE**	2443	39	868		3478	1169	226		284		VOT1180	946			1209
VANICIDE Synonym: **CAPTAN**	9099/3082		180		682	238	40		64		CPT 296	186			214
VAPAM					3483				284						1211
VAPOTONE Synonym: **TETRAETHYL PYROPHOSPHATE**	1705	15			3220	1101			262		TEP1104				1130
VARNISH	1263	26							284						1211
VARNISH SHELLAC	1263	26							284						
VASELINE Synonym: **PETROLATUM**			675								PTL 997				889
VATROLITE DITHIONOUS ACID DISODIUM SALT Synonym: **SODIUM HYDROSULFITE**	1384	37	776		3086	1056			254			842	S2.4		1058
VC Synonym: **VINYL CHLORIDE**	1086	17	872	26	3495	1174	227	180-93	285	224	VCM1174	949	V1.1	V 1	1212
VCL Synonym: **VINYL CHLORIDE**	1086	17	872	26	3495	1174	227	180-93	285	224	VCM1174	949	V1.1	V 1	1215
VCM Synonym: **VINYL CHLORIDE**	1086	17	872	26	3495	1174	227	180-93	285	224	VCM1174	949	V1.1	V 1	1215

HAZARDOUS MATERIALS REFERENCE BOOKS INDEX

CHEMICAL OR MATERIAL NAME	UN/NA Number	Guide Number (DOT)	Firefighter's Hazardous Materials Reference Book	First Aid Manual for Chemical Accidents	Sax's Dangerous Properties of Industrial Materials	Hazardous Chemicals Desk Reference	Rapid Guide to Hazardous Chemicals in the Workplace	Fire Protection Guide on Hazardous Materials (NFPA)	Firefighter's Handbook of Hazardous Materials	Pocket Guide to Chemical Hazards (NIOSH)	Chemical Hazard Response Information System (CHRIS)	Emergency Handling of Hazardous Materials in Surface Transportation (AAR)	Emergency Action Guides (EAG)	Chemical Data Notebook: A User's Manual	Condensed Chemical Dictionary
VDC Synonym: VINYLIDENE CHLORIDE	1303	26	878	26	3498	1175	229	181–93	286		VCI1173	952	V1.2		1217
VEGETABLE CARBON Synonym: CHARCOAL	1361	32	195		739	258			68		CHC 244	204			248
VELSICOL Synonym: HEPTACHLOR	2761/3077	55	424		1826	615	106		156	120	HTC 646	505			591
VELSICOL 1068 Synonym: CHLORDANE	2762	28	196		749	263	44		69	60	CDN 237	211			258
VENTOX Synonym: ACRYLONITRILE	1093	30	34	11	68	27	6	19 –12	20	34	ACN 23	13	A5.1	A 7	19
VERMILION Synonym: MERCURIC SULFIDE	2025	53	539						186		MSF 800				1213
VERMOESTRICID Synonym: CARBON TETRACHLORIDE	1846	55	192	13	701	246	43	48 –	66	60	CBT 223	193	C2.3	C 5	221
VERSENE ACID Synonym: ETHYLENEDIAMINE TETRACETIC ACID	9117/3077	31	379		1601	539			144		EDT 521	405			486
VIENNA GREEN Synonym: COPPER ACETOARSENITE	1585	53	222				55		83		CAA 203	257			310
VILRATHANE 4300 Synonym: DIPHENYLMETHANE DIISOCYANATE	2489	55	321						127		DPM 467				
VINAMAR Synonym: VINYL ETHYL ETHER	1302	26	873					–93	285		VEE1175	950			1216
VINEGAR ACID Synonym: ACETIC ACID GLACIAL	2789	29	18	11	15	6	1	14 –11	284	30	AAC 1	3	A2.0	A 2	7
VINEGAR NAPHTHA Synonym: ETHYL ACETATE	1173	26	340	16	1570	523	89	–47	135	104	ETA 560	406	E2.0		477
VINYL 2-CHLOROETHYL ETHER								–93	285						

CHEMICAL OR MATERIAL NAME	UN/NA Number	Guide Number (DOT)	Firefighter's Hazardous Materials Reference Book	First Aid Manual for Chemical Accidents	Sax's Dangerous Properties of Industrial Materials	Hazardous Chemicals Desk Reference	Rapid Guide to Hazardous Chemicals in the Workplace	Fire Protection Guide on Hazardous Materials (NFPA)	Firefighter's Handbook of Hazardous Materials	Pocket Guide to Chemical Hazards (NIOSH)	Chemical Hazard Response Information System (CHRIS)	Emergency Handling of Hazardous Materials in Surface Transportation (AAR)	Emergency Action Guides (EAG)	Chemical Data Notebook: A User's Manual	Condensed Chemical Dictionary
VINYL 2-ETHYLHEXOATE					3497			-93							
VINYL 2-ETHYLHEXYL ETHER								-93	285						
VINYL 2-METHOXYETHYL ETHER					3499			-93	285						
VINYL A MONOMER Synonym: VINYL ACETATE	1301	26	871	26	3492	1172	227	179-92	285		VAM1172	948	V1.0		1215
VINYL ACETATE	1301	26	871	26	3492	1172	227	179-92	285		VAM1172	948	V1.0		1215
VINYL ACETATE MONOMER Synonym: VINYL ACETATE	1301	26	871	26	3492	1172	227	179-92	285		VAM1172	948	V1.0		1215
VINYL ACETYLENE								180-92	285						1215
VINYL ALLYL ETHER								-92	285						
VINYL BROMIDE	085/1954	60			3494	1173	227	-92	285			949			1215
VINYL BUTYL ETHER	1304	26						-92							1215
VINYL BUTYRATE	2838	26			3495	1173		-92	285						1215
VINYL C MONOMER Synonym: VINYL CHLORIDE	1086	17	872	26	3495	1174	227	180-93	285	224	VCM1174	949	V1.1	V 1	1215
VINYL CARBINOL Synonym: ALLYL ALCOHOL	1098	57	38	11	102	33	7	21 -13	285	34	ALA 39	35	A7.0		38
VINYL CHLORIDE	1086	17	872	26	3495	1174	227	180-93	285	224	VCM1174	949	V1.1	V 1	1215
VINYL CHLORIDE MONOMER Synonym: VINYL CHLORIDE	1086	17	872	26	3495	1174	227	180-93	285	224	VCM1174	949	V1.1	V 1	1215
VINYL CHLOROACETATE	2589	57							285						
VINYL CROTONATE					3496			-93	285						
VINYL CYANIDE Synonym: ACRYLONITRILE	1093	30	34	11	68	27	6	19 -12	285	34	ACN 23	13	A5.1	A 7	1216
VINYL CYCLOHEXENE DIOXIDE					3496	1174	228		285						1216

HAZARDOUS MATERIALS REFERENCE BOOKS INDEX

CHEMICAL OR MATERIAL NAME	UN/NA Number	Guide Number (DOT)	Firefighter's Hazardous Materials Reference Book	First Aid Manual for Chemical Accidents	Sax's Dangerous Properties of Industrial Materials	Hazardous Chemicals Desk Reference	Rapid Guide to Hazardous Chemicals in the Workplace	Fire Protection Guide on Hazardous Materials (NFPA)	Firefighter's Handbook of Hazardous Materials	Pocket Guide to Chemical Hazards (NIOSH)	Chemical Hazard Response Information System (CHRIS)	Emergency Handling of Hazardous Materials in Surface Transportation (AAR)	Emergency Action Guides (EAG)	Chemical Data Notebook: A User's Manual	Condensed Chemical Dictionary
VINYL ETHER	1167	30			3497	1175		181–	285						1216
VINYL ETHYL ALCOHOL								–93	285						
VINYL ETHYL ETHER	1302	26	873					–93	285		VEE1175	950			1216
VINYL FLUORIDE	1860	17	874	26	3498	1175	228	–93	285		VFI1176	950			1217
VINYL FORMATE									285						
VINYL FORMIC ACID Synonym: ACRYLIC ACID	2218	29	33		65	26	6	18 –12	20		ACR 25	13	A5.0	A 6	18
VINYL ISOBUTYL ETHER	1304	26						–93	285			950			1217
VINYL ISOOCTYL ETHER								–93	285						
VINYL ISOPROPYL ETHER								–93	285						
VINYL METHYL ETHER	1087	17	875					–93	285		VME1177	951			1217
VINYL PROPIONATE					3500	1176		–93	285						1218
VINYL PYRIDINES	3073	57			3500				285						1218
VINYL TOLUENE	2618	27	876		3500	1176	229	182–94		224	VNT1179	952			1218
VINYL TRICHLOROSILANE	1305	29	877					–94	286		VTS1183	951			1218
VINYLAMINE Synonym: ETHYLENEIMINE	1185	30	380	17	3493	544	95	86 –51	286	110	ETI 567	437	E2.6.1		490
VINYLBENZENE Synonym: STYRENE MONOMER	2055	27	795		3145	1074	202	162–84	257		STY1064	862	S3.0	S 6	1097
VINYLBENZENE Synonym: STYRENE	2055	27	795	24	3145	1174	202	162–84	285	200	STY1064	862	S3.0	S 6	1097
VINYLBENZOL Synonym: STYRENE MONOMER	2055	27	795		3145	1074	202		257		STY1064	862	S3.0	S 6	1097
VINYLETHYLENE Synonym: BUTADIENE	1010	17	139	12	572			40 –	286	48	BDI 134	148	B4.0	B 4	1216

CHEMICAL OR MATERIAL NAME	UN/NA Number	Guide Number (DOT)	Firefighter's Hazardous Materials Reference Book	First Aid Manual for Chemical Accidents	Sax's Dangerous Properties of Industrial Materials	Hazardous Chemicals Desk Reference	Rapid Guide to Hazardous Chemicals in the Workplace	Fire Protection Guide on Hazardous Materials (NFPA)	Firefighter's Handbook of Hazardous Materials	Pocket Guide to Chemical Hazards (NIOSH)	Chemical Hazard Response Information System (CHRIS)	Emergency Handling of Hazardous Materials in Surface Transportation (AAR)	Emergency Action Guides (EAG)	Chemical Data Notebook: A User's Manual	Condensed Chemical Dictionary
VINYLIDENE CHLORIDE	1303	26	878	26	3498	1175	229	181–93	286		VCI1173	952	V1.2		1217
VINYLIDENE CHLORIDE MONOMER Synonym: VINYLIDENE CHLORIDE	1303	26	878	26	3498	1175	229	181–93	286		VCI1173	952	V1.2		1217
VINYLIDENE FLUORIDE	1959	22		26	3499	1176	229	–93	286						1217
VINYLSILICON TRICHLORIDE Synonym: VINYL TRICHLOROSILANE	1305	29	877					–94	286		VTS1183	951			1218
VY AC Synonym: VINYL ACETATE	1301	26	871	26	3492	1172	227	179–92	285		VAM1172	948	V1.0		1215
W 10 Synonym: ETHYLENE DIBROMIDE	1605	55	371					85 –	143	110	EDB 518	430	E2.4.1		487
W 15 Synonym: ETHYLENE DIBROMIDE	1605	55	371					85 –	143	110	EDB 518	430	E2.4.1		487
W 40 Synonym: ETHYLENE DIBROMIDE	1605	55	371					85 –	143	110	EDB 518	430	E2.4.1		487
WARFARIN	3027	55			3510	1180	229		286	224					1223
WATER DISPLACING OIL Synonym: OILS MISCELLANEOUS PENETRATING	1257	27									OPT 749				
WATER GLASS Synonym: SODIUM SILICATE	1759	60	785	24	3113	1065					SSC1057	850			1224
WATER OF AMMONIA Synonym: AMMONIUM HYDROXIDE	2672	60	62	12	228	62	11		29		AMH 52	60	A8.0	A10	66
WATER REACTIVE SOLID	2813	40							286						
WAX	1993	27	879						286						1225
WAXES CARNAUBA											WCA1184				
WAXES PARAFFIN								–94			WPF1185				

HAZARDOUS MATERIALS REFERENCE BOOKS INDEX

CHEMICAL OR MATERIAL NAME	UN/NA Number	Guide Number (DOT)	Firefighter's Hazardous Materials Reference Book	First Aid Manual for Chemical Accidents	Sax's Dangerous Properties of Industrial Materials	Hazardous Chemicals Desk Reference	Rapid Guide to Hazardous Chemicals in the Workplace	Fire Protection Guide on Hazardous Materials (NFPA)	Firefighter's Handbook of Hazardous Materials	Pocket Guide to Chemical Hazards (NIOSH)	Chemical Hazard Response Information System (CHRIS)	Emergency Handling of Hazardous Materials in Surface Transportation (AAR)	Emergency Action Guides (EAG)	Chemical Data Notebook: A User's Manual	Condensed Chemical Dictionary
WEISSPIESSGLANZ Synonym: ANTIMONY TRIOXIDE	1549/3077	60	94								ATX 109	87			90
WELDING FUMES						1180	230								
WHISKEY					3513	1181			286						1227
WHITE ARSENIC Synonym: ARSENIC TRIOXIDE	1561	53	100		299	92	16	33 –	35		ATO 105	93	A11.1	A14	1227
WHITE ASBESTOS Synonym: ASBESTOS, WHITE	2590	31					17		286			99			98
WHITE CAUSTIC Synonym: SODIUM HYDROXIDE	1823	60	777	24	3086	1056	200	159–	286	198	SHD1044	842	S2.5	S 4	1058
WHITE FUMING ACID Synonym: NITRIC ACID FUMING	2032	44		21	2509	856		123–	210	160		684	N2.0	N 1	823
WHITE OIL Synonym: OILS MISCELLANEOUS MINERAL	1270	27									OMN 888				1228
WHITE PHOSPHORIC ACID Synonym: PHOSPHORIC ACID	1805	60	684	22	2786	946	180	137–	232	182	PAC 922	753	P4.0		908
WHITE PHOSPHORUS Synonym: PHOSPHORUS YELLOW & WHITE	1381	38	693		2791	949	180	137–	287	182	PPW 982	760	P4.1		1228
WHITE SPIRIT	1115	26			3514				287						
WHITE VITRIOL Synonym: ZINC SULFATE	9161/3082		907		3546	1196	234		289		ZSF1217	970			1228
WINES HIGH						1181			287						
WINES SHERRY AND PORT						1181									
WITICIZER 300 Synonym: DIBUTYL PHTHALATE	9095		265		1119	375	66	–34		84	DPA 458				374

CHEMICAL OR MATERIAL NAME	UN/NA Number	Guide Number (DOT)	Firefighter's Hazardous Materials Reference Book	First Aid Manual for Chemical Accidents	Sax's Dangerous Properties of Industrial Materials	Hazardous Chemicals Desk Reference	Rapid Guide to Hazardous Chemicals in the Workplace	Fire Protection Guide on Hazardous Materials (NFPA)	Firefighter's Handbook of Hazardous Materials	Pocket Guide to Chemical Hazards (NIOSH)	Chemical Hazard Response Information System (CHRIS)	Emergency Handling of Hazardous Materials in Surface Transportation (AAR)	Emergency Action Guides (EAG)	Chemical Data Notebook: A User's Manual	Condensed Chemical Dictionary
WLNL Q8 Synonym: OCTANOL	1987	26	650		2620						OTA 909	706	O1.0		847
WOLFATOX Synonym: METHYL PARATHION	2783	55	577	20	2369	809	148		196		MPT 790	635			776
WOOD ALCOHOL Synonym: METHYL ALCOHOL	1230	28	554	20	2251	770	136	-66	287	144	MAL 722	613		M 2	1230
WOOD ALCOHOL Synonym: METHANOL	1230	28	549						287		MAL 722	613	M2.0		1230
WOOD CHARCOAL Synonym: CHARCOAL	1361	32	195		739	258			68		CHC 244	204			248
WOOD ETHER Synonym: DIMETHYL ETHER	1033	22	301					74 –	116		DIM 416	361			416
WOOD FILLER	1263	26							287			957			
WOOD NAPHTHA Synonym: METHANOL	1230	28	549						190		MAL 722	613	M2.0		751
WOOD PRESERVATIVE	1306	26							287			957			
WOOD SPIRIT Synonym: METHYL ALCOHOL	1230	28	554	20		770	136	-66	191	144	MAL 722	613		M 2	755
WOOD SPIRIT Synonym: METHANOL	1230	28	549		2251				190		MAL 722	613	M2.0		751
WOOD TURPENTINE Synonym: TURPENTINE	1299	27	856	26	3458	1162	225	-92	282	222	TPT1144	939			1231
WOOL WASTE	1387	32							287						1231
WP Synonym: PHOSPHORUS YELLOW & WHITE	1381	38	693		2791	949	180	137–	232	182	PPW 982	760	P4.1		
XENON	2591	21			3519	1182			287			959			1234

HAZARDOUS MATERIALS REFERENCE BOOKS INDEX

CHEMICAL OR MATERIAL NAME	UN/NA Number	Guide Number (DOT)	Firefighter's Hazardous Materials Reference Book	First Aid Manual for Chemical Accidents	Sax's Dangerous Properties of Industrial Materials	Hazardous Chemicals Desk Reference	Rapid Guide to Hazardous Chemicals in the Workplace	Fire Protection Guide on Hazardous Materials (NFPA)	Firefighter's Handbook of Hazardous Materials	Pocket Guide to Chemical Hazards (NIOSH)	Chemical Hazard Response Information System (CHRIS)	Emergency Handling of Hazardous Materials in Surface Transportation (AAR)	Emergency Action Guides (EAG)	Chemical Data Notebook: A User's Manual	Condensed Chemical Dictionary
XENON	2036	12			3519	1182			287			959			1234
XYLENE	1307	27		26	3520	1182	230	183–	287	226		959	X1.0		1235
XYLENE-alpha,alpha'-DIISOCYANATE-m					3522		232								
XYLENE-meta **Synonym: XYLENE**	1307	27	880	26	3521	1183	231	183–94	287	226	XLM1187	959	X1.0		1235
XYLENE-ortho **Synonym: XYLENE**	1307	27	881	26	3521	1183	231	183–94	287	226	XLO1188	959	X1.0		1235
XYLENE-para **Synonym: XYLENE**	1307	27	882	26	3522	1184	231	183–94	287	226	XLP1189	959	X1.0		1235
XYLENOL	2261	55	883		3523	1184			288		XYL1190	962			
XYLIDINE	1711/2810	55		26	3524	1185	232		288	226		963			1236
XYLIDINE-ortho				26				183–94							
XYLOL **Synonym: XYLENE**	1307	27		26	3522	1182	230	183–	288	226		959	X1.0		1236
XYLOL **Synonym: XYLENE (-para)**	1307	27	882	26	3520	1184	231	183–94	288	226	XLP1189	959	X1.0		1236
XYLOL **Synonym: XYLENE (-meta)**	1307	27	880	26	3521	1183	231	183–94	288	226	XLM1187	959	X1.0		1236
XYLOL **Synonym: XYLENE (-ortho)**	1307	27	881	26	3521	1183	231	183–94	288	226	XLO1188	959	X1.0		1236
XYLOL-meta	1307	27							288						
XYLOL-ortho	1307	27							288						
XYLOL-para	1307	27							288						
XYLYL BROMIDE	1701	55			3528	1188			288			963			1236
XYLYL BROMIDE-meta	1701	55							288						1236

CHEMICAL OR MATERIAL NAME	UN/NA Number	Guide Number (DOT)	Firefighter's Hazardous Materials Reference Book	First Aid Manual for Chemical Accidents	Sax's Dangerous Properties of Industrial Materials	Hazardous Chemicals Desk Reference	Rapid Guide to Hazardous Chemicals in the Workplace	Fire Protection Guide on Hazardous Materials (NFPA)	Firefighter's Handbook of Hazardous Materials	Pocket Guide to Chemical Hazards (NIOSH)	Chemical Hazard Response Information System (CHRIS)	Emergency Handling of Hazardous Materials in Surface Transportation (AAR)	Emergency Action Guides (EAG)	Chemical Data Notebook: A User's Manual	Condensed Chemical Dictionary
XYLYL BROMIDE-ortho	1701	55							288						1236
XYLYL BROMIDE-para	1701	55							288						1236
YARMAR **Synonym: PINE OIL**	1272	26	695		2810	959		-80	235			764	P5.0		918
YELLOW PETROLATUM **Synonym: PETROLATUM**			675								PTL 997				889
YELLOW PHOSPHORUS **Synonym: PHOSPHORUS YELLOW & WHITE**	1381	38	693		2791	949	180	137–	288	182	PPW 982	760	P4.1		1237
YOHIMBINE					3531	1189			288						1238
YOHIMBINE HYDROCHLORIDE					3531				288						
YTTRIUM				26	3533	1189	233		288	226					1238
YTTRIUM NITRATE				26	3533	1190			288						
ZACTRAN **Synonym: ZECTRAN**	1615										ZEC1205				1240
ZECTANE **Synonym: ZECTRAN**	1615										ZEC1205				1240
ZECTRAN	1615										ZEC1205				1240
ZELIO **Synonym: THALLIUM SULFATE**	1707	53	820		3263	1110	212		266		TSU1150	901			1138
ZESET T **Synonym: VINYL ACETATE**	1301	26	871	26	3492	1172	227	179–92	285		VAM1172	948	V1.0		1215
ZEXTRAN **Synonym: ZECTRAN**	1615										ZEC1205				1240
ZINC ACETATE	9153/3077		886		3536	1191			288		ZNA1211	964			1242
ZINC ACETATE DIHYDRATE **Synonym: ZINC ACETATE**	9153/3077		886		3537	1191	234		288		ZNA1211	964			1242

HAZARDOUS MATERIALS REFERENCE BOOKS INDEX

CHEMICAL OR MATERIAL NAME	UN/NA Number	Guide Number (DOT)	Firefighter's Hazardous Materials Reference Book	First Aid Manual for Chemical Accidents	Sax's Dangerous Properties of Industrial Materials	Hazardous Chemicals Desk Reference	Rapid Guide to Hazardous Chemicals in the Workplace	Fire Protection Guide on Hazardous Materials (NFPA)	Firefighter's Handbook of Hazardous Materials	Pocket Guide to Chemical Hazards (NIOSH)	Chemical Hazard Response Information System (CHRIS)	Emergency Handling of Hazardous Materials in Surface Transportation (AAR)	Emergency Action Guides (EAG)	Chemical Data Notebook: A User's Manual	Condensed Chemical Dictionary
ZINC AMMONIUM CHLORIDE	9154/3077		887						288		ZAC1191	964			1242
ZINC AMMONIUM NITRITE	1512	35			3537	1191			288			964			1242
ZINC ARSENATE	1712	53	888		3537	1191			288		ZAR1192	964			1242
ZINC ARSENATE AND ZINC ARSENITE MIXTURE	1712	53							288			964			
ZINC ARSENITE	1712	53	889		3538	1192			288			964			1243
ZINC ASHES	1435	40							288			965			
ZINC BICHROMATE			890								ZBC1193				
ZINC BISULFITE	2693	60							288						
ZINC BORATE	9188/3077		891								ZBO1194	965			1243
ZINC BROMATE	2469	35							288						1243
ZINC BROMIDE	9156/3077		892						288		ZBR1195	965			1243
ZINC CARBONATE	9157/3077		893		3538				288		ZCB1197	965			1243
ZINC CHLORATE	1513	35			3538	1192		184–	288			966			1243
ZINC CHLORIDE	2331	60		26	3538	1192	233		288						1243
ZINC CHLORIDE	1840	60		26	3538	1192	233		288		ZCL1198				1243
ZINC CHROMATE			895		3539	1192	233				ZCR1201				1244
ZINC CHROMATE (VI) HYDROXIDE Synonym: ZINC CHROMATE			895		3539	1192	233				ZCR1201				1244
ZINC CYANIDE	1713	53	896		3540	1194			288		ZCN1199	966			1244
ZINC DIACETATE Synonym: ZINC ACETATE	9153/3077		886		3536	1191	234		288		ZNA1211	964			1242
ZINC DIALKYLDITHIOPHOSPHATE	1893		897								ZDP1204				1244
ZINC DICHROMATE Synonym: ZINC BICHROMATE			890				234				ZBC1193				1244

CHEMICAL OR MATERIAL NAME	UN/NA Number	Guide Number (DOT)	Firefighter's Hazardous Materials Reference Book	First Aid Manual for Chemical Accidents	Sax's Dangerous Properties of Industrial Materials	Hazardous Chemicals Desk Reference	Rapid Guide to Hazardous Chemicals in the Workplace	Fire Protection Guide on Hazardous Materials (NFPA)	Firefighter's Handbook of Hazardous Materials	Pocket Guide to Chemical Hazards (NIOSH)	Chemical Hazard Response Information System (CHRIS)	Emergency Handling of Hazardous Materials in Surface Transportation (AAR)	Emergency Action Guides (EAG)	Chemical Data Notebook: A User's Manual	Condensed Chemical Dictionary
ZINC DICYANIDE Synonym: ZINC CYANIDE	1713	53	896		3540	1194	234		288		ZCN1199	966			1244
ZINC DIETHYL Synonym: DIETHYLZINC	1366	40	294		1243	419		71 -39	288		DEZ 395	353			1244
ZINC DIFLUORIDE Synonym: ZINC FLUORIDE	9158/3077		898			1194	234		289		ZFX1208	967			1244
ZINC DIHEXYLDITHIOPHOSPHATE Synonym: ZINC DIALKYLDITHIOPHOSPHATE	1893		897				234				ZDP1204				1244
ZINC DIHEXYLPHOSPHORODITHIOATE Synonym: ZINC DIALKYLDITHIOPHOSPHATE	1893		897				234				ZDP1204				1244
ZINC DIMETHYL Synonym: DIMETHYLZINC	1370	40	309		1425				289		DMZ 441	373			
ZINC DITHIONITE Synonym: ZINC HYDROSULFITE	1931	32	901				234		289		ZHS1209	968			1244
ZINC ETHYL Synonym: DIETHYLZINC	1366	40	294		1243	419		71 -39	289		DEZ 395	353			1244
ZINC FLUOBORATE SOLUTION Synonym: ZINC FLUOROBORATE			899				234				ZFB1206				1245
ZINC FLUORIDE	9158/3077		898		3541	1194			289		ZFX1208	967			1244
ZINC FLUOROBORATE			899								ZFB1206				1245
ZINC FLUOROSILICATE	2855	53							289						1245
ZINC FLUOSILICATE Synonym: ZINC SILICOFLUORIDE	2855/3077	53	906		3542	1194	234		289		ZSL1218	970			1248
ZINC FORMATE	9159/3077		900						289		ZFM1207	967			1245
ZINC HEXAFLUOROSILICATE Synonym: ZINC SILICOFLUORIDE	2855/3077	53	906				234		289		ZSL1218	970			1248

HAZARDOUS MATERIALS REFERENCE BOOKS INDEX

CHEMICAL OR MATERIAL NAME	UN/NA Number	Guide Number (DOT)	Firefighter's Hazardous Materials Reference Book	First Aid Manual for Chemical Accidents	Sax's Dangerous Properties of Industrial Materials	Hazardous Chemicals Desk Reference	Rapid Guide to Hazardous Chemicals in the Workplace	Fire Protection Guide on Hazardous Materials (NFPA)	Firefighter's Handbook of Hazardous Materials	Pocket Guide to Chemical Hazards (NIOSH)	Chemical Hazard Response Information System (CHRIS)	Emergency Handling of Hazardous Materials in Surface Transportation (AAR)	Emergency Action Guides (EAG)	Chemical Data Notebook: A User's Manual	Condensed Chemical Dictionary
ZINC HYDROSULFITE	1931	32	901						289		ZHS1209	968			1245
ZINC m-ARSENITE	1712	53			3538	1192									
ZINC MERCURY CHROMATE COMPLEX					3542	1194									
ZINC METAL	1436	76			3536		234		289			968			1242
ZINC METHYL	1370	40	309		1425				123		DMZ 441	373			
Synonym: DIMETHYLZINC															
ZINC NITRATE	1514	35	902		3543	1194			289		ZNT1212	968			1246
ZINC NITRATE HEXAHYDRATE	1514	35	902		3543	1194	234		289		ZNT1212	968			1246
Synonym: ZINC NITRATE															
ZINC O,O-DI-n-BUTYL PHOSPHORODITHIOATE	1893		897				234				ZDP1204				1244
Synonym: ZINC DIALKYLDITHIOPHOSPHATE															
ZINC OXIDE					3544	1195	234			228					1246
ZINC p-PHENOLSULFONATE	9160/3077	31	903				234		289		ZPS1216	970			
Synonym: ZINC PHENOLSULFONATE															
ZINC PERMANGANATE	1515	35			3544	1195			289			968			1247
ZINC PEROXIDE	1516	47			3545	1195			289			969			1247
ZINC PHENOLSULFONATE	9160/3077	31	903						289		ZPS1216	969			
ZINC PHENOLSULFONATE OCTAHYDRATE	9160/3077	31	903				234		289		ZPS1216	969			
Synonym: ZINC PHENOLSULFONATE															
ZINC PHOSPHIDE	1714	41	904		3545	1196		184–	289		ZPP1215	969			1247
ZINC POTASSIUM CHROMATE			905								ZPC1213				1247
ZINC POWDER	1383	37			3536		234		289						
ZINC POWDER	1436	76			3536		234		289			968			
ZINC RESINATE	2714	32							289			970			1248
ZINC SELENATE	2630	53							289						

CHEMICAL OR MATERIAL NAME	UN/NA Number	Guide Number (DOT)	Firefighter's Hazardous Materials Reference Book	First Aid Manual for Chemical Accidents	Sax's Dangerous Properties of Industrial Materials	Hazardous Chemicals Desk Reference	Rapid Guide to Hazardous Chemicals in the Workplace	Fire Protection Guide on Hazardous Materials (NFPA)	Firefighter's Handbook of Hazardous Materials	Pocket Guide to Chemical Hazards (NIOSH)	Chemical Hazard Response Information System (CHRIS)	Emergency Handling of Hazardous Materials in Surface Transportation (AAR)	Emergency Action Guides (EAG)	Chemical Data Notebook: A User's Manual	Condensed Chemical Dictionary
ZINC SELENITE	2630	53							289						
ZINC SILICOFLUORIDE	2855/3077	53	906						289		ZSL1218	970			1248
ZINC SILICOFLUORIDE HEXAHYDRATE Synonym: ZINC SILICOFLUORIDE	2855/3077	53	906				234		289		ZSL1218	970			1248
ZINC STEARATE					3546	1196	234	-94							1248
ZINC SULFATE	9161/3082		907		3546	1196			289		ZSF1217	970			1248
ZINC SULFATE HEPTAHYDRATE Synonym: ZINC SULFATE	9161/3082		907		3547	1196	234		289		ZSF1217	970			1248
ZINC SULFOCARBOLATE Synonym: ZINC PHENOLSULFONATE	9160/3077	31	903				234		289		ZPS1216	969			
ZINC SULFOPHENATE Synonym: ZINC PHENOLSULFONATE	9160/3077	31	903				234		289		ZPS1216	969			
ZINC VITRIOL Synonym: ZINC SULFATE	9161/3082		907		3546	1196	234		289		ZSF1217	970			1248
ZINC YELLOW Synonym: ZINC CHROMATE			895		3539	1192	233				ZCR1201				1249
ZINC YELLOW Y-539-D Synonym: ZINC POTASSIUM CHROMATE			905				234				ZPC1213				1247
ZIRCONIUM ACETATE			908								ZCA1196				1250
ZIRCONIUM ACETATE SOLUTION Synonym: ZIRCONIUM ACETATE			908				235				ZCA1196				1250
ZIRCONIUM CHLORIDE Synonym: ZIRCONIUM TETRACHLORIDE	2503	39	913		3548	1197	235	185-	290		ZCT1203	973			1251
ZIRCONIUM DIBROMIDE					3549				290						
ZIRCONIUM HYDRIDE	1437	40			3549	1198			290			971			1251
ZIRCONIUM METAL	1358	32			3547		235		289	228		973			1250

HAZARDOUS MATERIALS REFERENCE BOOKS INDEX

CHEMICAL OR MATERIAL NAME	UN/NA Number	Guide Number (DOT)	Firefighter's Hazardous Materials Reference Book	First Aid Manual for Chemical Accidents	Sax's Dangerous Properties of Industrial Materials	Hazardous Chemicals Desk Reference	Rapid Guide to Hazardous Chemicals in the Workplace	Fire Protection Guide on Hazardous Materials (NFPA)	Firefighter's Handbook of Hazardous Materials	Pocket Guide to Chemical Hazards (NIOSH)	Chemical Hazard Response Information System (CHRIS)	Emergency Handling of Hazardous Materials in Surface Transportation (AAR)	Emergency Action Guides (EAG)	Chemical Data Notebook: A User's Manual	Condensed Chemical Dictionary
ZIRCONIUM METAL	2858	32			3547		235		290						1250
ZIRCONIUM METAL	2009	37			3547		235		289			973			1250
ZIRCONIUM METAL	2008	37			3547		235		289	228		971			1250
ZIRCONIUM METAL	1308	26			3547		235		289			974			1250
ZIRCONIUM NITRATE	2728	35	909		3550	1198			290		ZIR1210	972			1252
ZIRCONIUM NITRATE PENTAHYDRATE	2728	35	909		3550	1198	235		290		ZIR1210	972			1252
Synonym: ZIRCONIUM NITRATE															
ZIRCONIUM OXIDE CHLORIDE			910		3550	1198	235				ZCO1200				1252
Synonym: ZIRCONIUM OXYCHLORIDE															
ZIRCONIUM OXYCHLORIDE			910		3550	1198					ZCO1200				1252
ZIRCONIUM OXYCHLORIDE HYDRATE			910		3550	1198	235				ZCO1200				1252
Synonym: ZIRCONIUM OXYCHLORIDE															
ZIRCONIUM PICRAMATE	1517	33							290			972			
ZIRCONIUM POTASSIUM FLUORIDE	9162/3077	31	911						290		ZPF1214	972			1252
ZIRCONIUM POWDER	1358	32			3547		235		289	228		973			1250
ZIRCONIUM POWDER	2008	37			3547		235		289	228		971			1250
ZIRCONIUM SCRAP	1932	32							290			971			
ZIRCONIUM SULFATE	9163	31	912		3551	1198			290		ZCS1202	973			1253
ZIRCONIUM SULFATE TETRAHYDRATE	9163	31	912		3551	1198	235		290		ZCS1202	973			1253
Synonym: ZIRCONIUM SULFATE															
ZIRCONIUM SUSPENDED IN A LIQUID	1308	26					235								
ZIRCONIUM TETRACHLORIDE	2503	39	913					185–	290		ZCT1203	973			1253
ZIRCONIUM TETRACHLORIDE SOLID	2503	39	913				235	185–	290		ZCT1203	973			1253
Synonym: ZIRCONIUM TETRACHLORIDE															

CHEMICAL OR MATERIAL NAME	UN/NA Number	Guide Number (DOT)	Firefighter's Hazardous Materials Reference Book	First Aid Manual for Chemical Accidents	Sax's Dangerous Properties of Industrial Materials	Hazardous Chemicals Desk Reference	Rapid Guide to Hazardous Chemicals In the Workplace	Fire Protection Guide on Hazardous Materials (NFPA)	Firefighter's Handbook of Hazardous Materials	Pocket Guide to Chemical Hazards (NIOSH)	Chemical Hazard Response Information System (CHRIS)	Emergency Handling of Hazardous Materials In Surface Transportation (AAR)	Emergency Action Guides (EAG)	Chemical Data Notebook: A User's Manual	Condensed Chemical Dictionary
ZIRCONYL CHLORIDE Synonym: ZIRCONIUM OXYCHLORIDE			910		3550	1198	235				ZCO1200				1253

Printed and bound by CPI Group (UK) Ltd, Croydon, CR0 4YY
16/09/2026
14963447-0006